Laboratory Manual to accompany

Biology

Thirteenth Edition

Sylvia S. Mader

Contributor

Jason Carlson

St. Cloud Technical and Community College

LABORATORY MANUAL TO ACCOMPANY BIOLOGY, THIRTEENTH EDITION

Published by McGraw-Hill Education, 2 Penn Plaza, New York, NY 10121.

Some ancillaries, including electronic and print components, may not be available to customers outside the United States.

This book is printed on acid-free paper.

1 2 3 4 5 6 7 8 9 LMN 21 20 19 18

ISBN 978-1-260-17986-6
MHID 1-260-17986-9

Executive Brand Manager: *Michelle Vogler*
Product Developer: *Anne Winch*
Marketing Manager: *Britney Ross*
Content Project Manager: *Mary Jane Lampe*
Buyer: *Sandy Ludovissy*
Cover Design: *David Hash*
Content Licensing Specialist: *Lori Hancock*
Cover Image: *©iStockphoto/Getty Images*
Compositor: *Aptara®, Inc.*

The Internet addresses listed in the text were accurate at the time of publication. The inclusion of a website does not indicate an endorsement by the authors or McGraw-Hill Education, and McGraw-Hill Education does not guarantee the accuracy of the information presented at these sites.

mheducation.com/highered

Each laboratory exercise was carefully revised to ensure that directions to students are clear and follow common laboratory standards and practices.

Lab 15 Bacteria and Protists Evolutionary relationships and supergroup classification among the protists was updated to match the most current hypothesis.

Lab 16 Fungi Evolutionary relationships among the fungi were updated to match the most current hypothesis. The section about lichens was updated with new information to explain that some lichens include two fungal partners along with a photosynthetic partner.

Lab 22 Reproduction in Flowering Plants Organization and clarity were improved by placing examination of flower structure as the first exercise and adding brief reviews of the alternation of generations life cycle and double-fertilization.

Lab 23 Introduction to Invertebrates Evolutionary relationships among invertebrates were updated to match the most current hypothesis.

Customized Editions

With McGraw-Hill Create™ you can easily rearrange the labs in this manual and combine them with content from other sources, including your own syllabus or teaching notes. Go to create.mheducation.com to learn how to create your own version of the *Biology Laboratory Manual.*

Laboratory Resource Guide

The *Laboratory Resource Guide*, an essential aid for instructors and laboratory assistants, free to adopters of the *Biology Laboratory Manual*, is available in the Instructor Resources area of Connect. The answers to the Laboratory Review questions are in the Resource Guide.

To the Student

Special care has been taken in preparing the *Biology Laboratory Manual* to enable you to **enjoy** the laboratory experience as you **learn** from it. The instructions and discussions are written clearly so that you can understand the material while working through it. Student aids are designed to help you focus on important aspects of each exercise.

Student Learning Aids

Student learning aids are carefully integrated throughout this manual. The Learning Outcomes set the goals of each laboratory session and help you review the material for a laboratory practical or any other kind of exam. The major sections of each laboratory are numbered, and the Learning Outcomes are grouped according to these topics. This system allows you to study the chapter in terms of the outcomes presented.

The Introduction reviews much of the necessary background information required for comprehending upcoming experiments. Color bars bring attention to exercises that require your active participation by highlighting Observations and Experimental Procedures, and an icon indicates a timed experiment. Throughout, space is provided for recording answers to questions and the results of investigations and experiments. Each laboratory ends with a set of review questions covering the day's work.

Appendices at the end of the book provide useful information on preparing a laboratory report and the metric system.

Laboratory Preparation

Read each exercise before coming to the laboratory. ***Study*** the introductory material and the procedures. If necessary, to obtain a better understanding, read the corresponding chapter in your text. If your text is *Biology,* by Sylvia S. Mader and Michael Windelspecht, see the "text chapter reference" column in the table of contents at the beginning of the *Laboratory Manual.*

Explanations and Conclusions

Throughout a laboratory, you are often asked to formulate explanations or conclusions. To do so, you will need to synthesize information from a variety of sources, including the following:

1. Your experimental results and/or the results of other groups in the class. If your data are different from those of other groups in your class, do not erase your answer; add the other groups' answers in parentheses.
2. Your knowledge of underlying principles. Obtain this information from the laboratory Introduction or the appropriate section of the laboratory and from the corresponding chapter of your text.
3. Your understanding of how the experiment was conducted and/or the materials used. *Note:* Ingredients can be contaminated or procedures incorrectly followed, resulting in reactions that seem inappropriate. If this occurs, consult with other students and your instructor to see if you should repeat the experiment.

In the end, be sure you are truly writing an explanation or conclusion based on your data and not just restating the observations you have made.

Color Bars, Time Icon, and Safety Boxes

Throughout each laboratory, a color bar designates either an Observation or an Experimental Procedure.

Observation: An activity in which models, slides, and preserved or live organisms are observed to achieve a learning outcome.

Experimental Procedure: An activity in which a series of steps uses laboratory equipment to gather data and come to a conclusion.

 **Time:** An icon is used to designate when time is needed for an Experimental Procedure. Allow the designated amount of time for this activity. Start these activities at the beginning of the laboratory, proceed to other activities, and return to these when the designated time is up.

⚠ A **Safety Icon** throughout the manual alerts you to any specific activity that requires a cautionary approach. Read these boxes, and follow the advice given in the box and/or your instructor when performing the activity.

Laboratory Review

Each laboratory ends with 17 questions that will help you determine whether you have accomplished the learning outcomes for the laboratory. Do all the review questions as an aid to understanding the laboratory. Your instructor may require you to hand in these questions for credit.

Laboratory Safety

The following is a list of practices required for safety purposes in the biology laboratory and in outdoor activities. Following rules of lab safety and using common sense throughout the course will enhance your learning experience by increasing your confidence in your ability to safely use chemicals and equipment. Pay particular attention to oral and written safety instructions given by the instructor. If you do not understand a procedure, ask the instructor, rather than a fellow student, for clarification. Be aware of your school's policy regarding accident liability and any medical care needed as a result of a laboratory or outdoor accident.

The following rules of laboratory safety should become a habit:

1. To prevent possible hazards to eyes or contact lenses, wear safety glasses or goggles during exercises in which glassware and solutions are heated, or when dangerous fumes may be present.
2. Assume that all reagents are poisonous and act accordingly. Read the caution boxes in your laboratory manual and know the nature of the chemical you are using. If chemicals come into contact with skin, wash immediately with water.
3. **DO NOT**
   - **a.** ingest any reagents.
   - **b.** eat, drink, or smoke in the laboratory. Toxic material may be present, and some chemicals are flammable.
   - **c.** carry reagent bottles around the room.
   - **d.** pipet anything by mouth.
   - **e.** put chemicals in the sink or trash unless instructed to do so.
   - **f.** pour chemicals back into containers unless instructed to do so.
   - **g.** operate any equipment until you are instructed in its use.
4. **DO**
   - **a.** note the location of emergency equipment such as a first aid kit, eyewash bottle, fire extinguisher, switch for ceiling showers, fire blanket(s), sand bucket, and telephone (911).
   - **b.** become familiar with the experiments you will be doing before coming to the laboratory. This will increase your understanding, enjoyment, and safety during exercises. Confusion is dangerous. Completely follow the procedure set forth by the instructor.
   - **c.** keep your work area neat, clean, and organized. Before beginning, remove everything from your work area except the lab manual, pen, and equipment used for the experiment. Wash hands and desk area, including desk top and edge, before and after each experiment. Use clean glassware at the beginning of each exercise, and wash glassware at the end of each exercise or before leaving the laboratory.
   - **d.** wear clothing that, if damaged, would not be a serious loss, or use aprons or laboratory coats, since chemicals may damage fabrics.
   - **e.** wear shoes as protection against broken glass or spillage that may not have been adequately cleaned up.
   - **f.** handle hot glassware with a test tube clamp or tongs. Use caution when using heat, especially when heating chemicals. Do not leave a flame unattended; do not light a Bunsen burner near a gas tank or cylinder; do not move a lit Bunsen burner; do keep long hair and loose clothing well away from the flame; do make certain gas jets are off when the Bunsen burner is not in use. Use proper ventilation and hoods when instructed.
   - **g.** read chemical bottle labels; be aware of the hazards of all chemicals used. Know the safety precautions for each.
   - **h.** stopper all reagent bottles when not in use. Immediately wash reagents off yourself and your clothing if they spill on you, and immediately inform the instructor. If you accidentally get any reagent in your mouth, rinse the mouth thoroughly, and immediately inform your instructor.
   - **i.** use extra care and wear disposable gloves when working with glass tubing and when using dissection equipment (scalpels, knives, or razor blades), whether cutting or assisting.
   - **j.** administer first aid immediately to clean, sterilize, and cover any scrapes, cuts, and burns where the skin is broken and/or where there may be bleeding. Wear bandages over open skin wounds.
   - **k.** report all accidents to the instructor immediately, and ask your instructor for assistance in cleaning up broken glassware and spills.
   - **l.** report to the instructor any condition that appears unsafe or hazardous.
   - **m.** use caution during any outdoor activities. Watch for snakes, poisonous insects or spiders, stinging insects, poison oak, poison ivy, and so on. Be careful near water.

I understand the safety rules as presented. I agree to follow them and all other instructions given by the instructor.

Name ______________________________ Date ______________________________

Laboratory Class and Time: ______________________________

LABORATORY

1
Scientific Method

Learning Outcomes

Introduction
- In general, describe pillbug external anatomy and lifestyle.

1.1 Using the Scientific Method
- Outline the steps of the scientific method.
- Distinguish among observations, hypotheses, conclusions, and theories.

1.2 Observing a Pillbug
- Observe and describe the external anatomy of a pillbug, *Armadillidium vulgare.*
- Observe and describe how a pillbug moves.

1.3 Formulating Hypotheses
- Formulate a hypothesis based on appropriate observations.

1.4 Performing the Experiment and Coming to a Conclusion
- Design an experiment that can be repeated by others.
- Reach a conclusion based on observation and experimentation.

Introduction

This laboratory will provide you with an opportunity to use the scientific method in the same manner as scientists. Today your subject will be the pillbug, *Armadillidium vulgare,* a type of crustacean that lives on land.

Pillbugs have an exoskeleton consisting of overlapping "armored" plates that make them look like little armadillos. As pillbugs grow, they molt (shed the exoskeleton) four or five times during a lifetime. A pillbug can roll up into such a tight ball that its legs and head are no longer visible, earning it the nickname "roly-poly." They have three body parts: head, thorax, and abdomen. The head bears compound eyes and two pairs of antennae. The thorax bears pairs of walking legs; gills for gas exchange are located at the top of the first five pairs. The gills must be kept slightly moist, which explains why pillbugs are usually found in damp places. The final pair of appendages, the uropods, which are sensory and defensive in function, project from the abdomen of the animal.

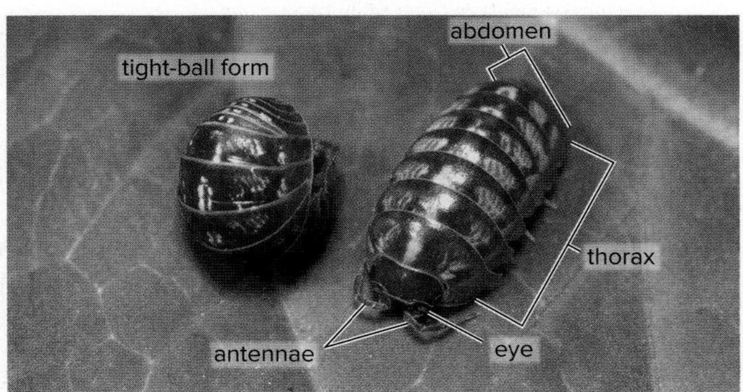

Pillbugs on leaf

Pillbugs are commonly found in damp leaf litter, under rocks, and in basements or crawl spaces under houses. Following an inactive winter, pillbugs mate in the spring. Several weeks later, the eggs hatch and remain for six weeks in a brood pouch on the underside of the female's body. Once they leave the pouch, they eat primarily dead organic matter, including decaying leaves. Therefore, they are easy to find and to maintain in a moist terrarium with leaf litter, rocks, and wood chips. You are encouraged to collect some for your experiment. Since they live in the same locations as snakes, be careful when collecting them.

1.1 Using the Scientific Method

Some scientists work alone, but often scientists belong to a community of scientists who are working together to study some aspect of the natural world. For example, many scientists from different institutions come together to study Arctic ecology (Fig. 1.1).

Figure 1.1 Scientists work together. These scientists are studying the ecology of the Arctic. ©Arnulf Husmo/The Image Bank/Getty Images

You will share your study of pillbugs with the other members of the class.

Even though the methodology can vary, scientists often use the **scientific method** (Fig. 1.2) when doing research. The scientific method involves these steps:

Making observations. Observations help scientists begin their study of a particular topic.

To learn about pillbugs you will visually observe one. You could also do a Google search of the Web or talk to someone who has worked with pillbugs for a long time.

Why does the scientific method begin with observations? ______________________________

____

Formulating a hypothesis. Based on their observations, scientists come to a tentative decision, called a hypothesis, about their topic. Formulating hypotheses helps scientists decide how an experiment will be conducted.

Based on your observations you might hypothesize that a pillbug will be attracted to fruit juice.

Now you know what you will actually do. What is the benefit of formulating a hypothesis? ______________

____

Testing the hypothesis. Scientists make further observations or perform experiments in order to test the hypothesis.

You could decide to expose a pillbug to fruit juice and observe its reaction.

A well-designed experiment must have a **negative control**—that is, a sample or event—that is not exposed to the testing procedure. If the negative control and the test sample produce the same results, either the procedure is flawed or the hypothesis is false.

Water can substitute for fruit juice and be the control in your experiment.

Scientists call the results of their experiments the **data.** It is very important for scientists to keep accurate records of all their data.

You will record your data in a table that can be easily examined by another person.

When another person repeats the same experiment, and the data is the same, both experiments have merit.

Why must a scientist keep a complete record of an experiment? ______________________________

Coming to a conclusion. Scientists come to a conclusion as to whether their data support or do not support the hypothesis.

If a pillbug is attracted to fruit juice, your hypothesis is supported. If the pillbug is not attracted to fruit juice, your hypothesis is not supported.

A scientist never says that a hypothesis has been proven true because, after all, some future knowledge might have a bearing on the experiment. What is the purpose of the conclusion? ______________________________

____

Developing a scientific theory. A *theory* in science is an encompassing conclusion based on many individual conclusions in the same field. For example, the gene theory states that organisms inherit coded information that controls their anatomy, physiology, and behavior. It takes many years for scientists to develop a theory and, therefore, we will not be developing any theories today. How is a scientific theory different from a conclusion? ______________________________

____

Figure 1.2 Flow diagram for the scientific method. Scientists can predict the results by formulating hypotheses that are based on prior observations. Experiments are conducted to test the hypotheses, which are supported or not supported by the analysis of data collected. Modification of the hypotheses can lead to more experiments. The conclusions of many related experiments can lead to the development of a theory.

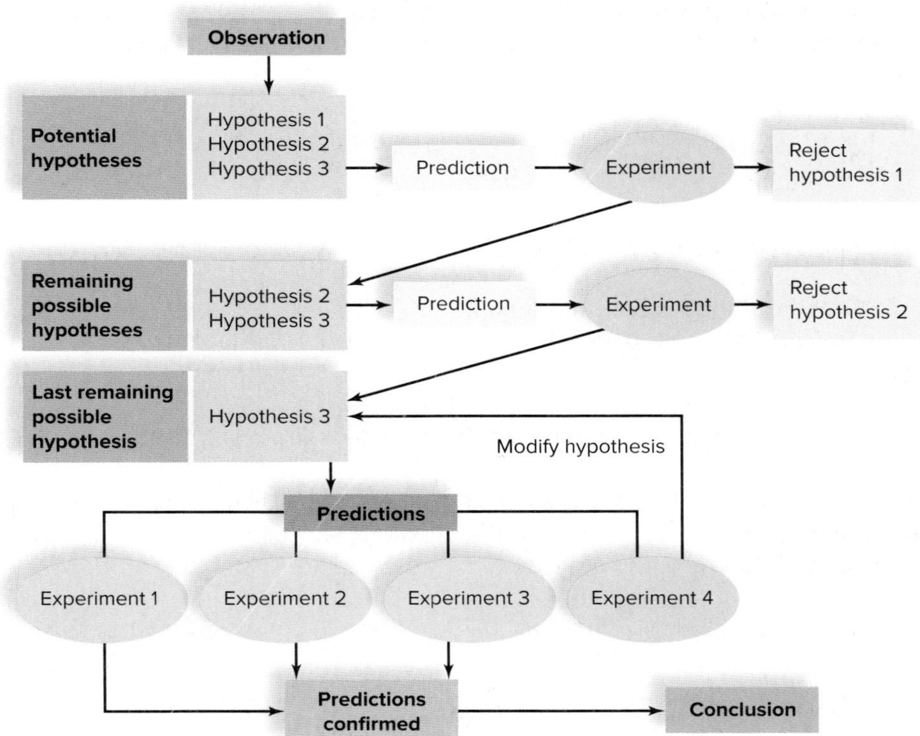

1.2 Observing a Pillbug

Wash your hands before and after handling pillbugs. Please handle them carefully so they are not crushed. When touched, they roll up into a ball or "pill" shape as a defense mechanism. They will soon recover if left alone.

Observation: Pillbug's External Anatomy

Obtain a pillbug that has been numbered with nail polish, white correction fluid, or tape tags. Put the pillbug in a small glass or plastic dish to keep it contained.

1. Examine the exterior of the pillbug with the unaided eye and with a magnifying lens or dissecting microscope.

- How can you recognize the head end of a pillbug? ______________________________
- How many segments and pairs of walking legs are in the thorax? ____________________
- The abdomen ends in uropods, appendages with a sensory and defense function. (Females have leaflike growths at the base of some legs where developing eggs and embryos are held in pouches.)

2. In the following space, draw an outline of your pillbug (at least 7 cm long). Label the head, thorax, abdomen, antennae, eyes, uropods, and one of the seven pairs of legs.

3. Draw a pillbug rolled into a ball.

Observation: Pillbug's Motion

1. Watch a pillbug's underside as the pillbug moves up a transparent surface, such as the side of a graduated cylinder or beaker.
   a. Describe the action of the feet and any other motion you see. ______
   b. Allow a pillbug to crawl on your hand. Describe how it feels. ______
   c. Does a pillbug have the ability to move directly forward? ______
   d. Do you see evidence of mouthparts on the underside of the pillbug? ______
2. As you watch the pillbug, identify
   a. the anatomical parts that allow a pillbug to identify and take in food. ______
   b. behaviors that will help the pillbug acquire food. For example, is the ability of the pillbug to move directly forward a help in acquiring food? Explain. ______
      What other behaviors allow a pillbug to acquire food? ______
   c. a behavior that helps a pillbug avoid dangerous situations. ______
      If the pillbug rolls up into a ball, wait a few minutes and it may uncurl itself.
3. Measure the speed of three pillbugs.
   a. Place each pillbug on a metric ruler, and use a stopwatch to measure the number of seconds (sec) it takes for the pillbug to move several centimeters (cm). Quickly record here the number of cm moved and the time in sec.
      pillbug 1 ______
      pillbug 2 ______
      pillbug 3 ______
   b. Knowing that 10 millimeters (mm) are in a cm, convert the number of cm traveled to mm and record this in the first column of Table 1.1. Record the total time taken in the second column.
   c. Use the space above to calculate the speed of each pillbug in mm/sec and record the speed for each pillbug in the last column of Table 1.1.
   d. Add the three speeds together and divide by 3 to determine the average speed for the three pillbugs. Record the average speed of pillbug motion in Table 1.1.
      When you conduct the experiment in section 1.4 you will have to be patient with your pillbug as it moves toward or away from a substance.

Table 1.1 Pillbug Speed

Pillbug	Millimeters (mm) Traveled	Time (sec)	Speed (mm/sec)
1			
2			
3			
		Average speed:	

1.3 Formulating Hypotheses

You will be testing whether pillbugs are attracted to (move toward and eat), repelled by (move away from), or unresponsive to (don't move away from and do not move toward and eat) the particular substances, which are potential foods. If a pillbug simply rolls into a ball, nothing can be concluded, and you may wish to choose another pillbug or wait a minute or two to check for further response.

1. Choose
   a. two dry substances, such as flour, cornstarch, coffee creamer, or baking soda. Fine sand will serve as a control for dry substances. Record your "dry" choices as 1, 2, and 3 in the first column of Table 1.2.
   b. two liquids, such as milk, orange juice, ketchup, applesauce, or carbonated beverage. Water will serve as a control for liquid substances. Record your "wet" choices as 4, 5, and 6 in the first column of Table 1.2.
2. In the second column of Table 1.2, hypothesize how you expect the pillbug to respond to each substance. Use a plus (+) sign if you hypothesize that the pillbug will move toward and eat the substance, a minus (−) sign if you hypothesize that the pillbug will be repelled by the substance, and a zero (0) if you expect the pillbug to show neither behavior.
3. In the third column of Table 1.2 offer a reason for your hypothesis based on your knowledge of pillbugs from the introduction and your examination of the animal.

Table 1.2 Hypotheses About Pillbug's Response to Potential Foods

Substance		Hypothesis About Pillbug's Response	Reason for Hypothesis
1			
2			
3	(control)		
4			
5			
6	(control)		

1.4 Performing the Experiment and Coming to a Conclusion

A good experimental design would be to keep your pillbug in a petri dish to test its reaction to the chosen substances. During your experiment, no substance must be put directly on the pillbug, nor can the pillbug be placed directly onto the substances.

Experimental Procedure: Pillbug's Response to Potential Foods

1. Before testing the pillbug's reaction, fill in the first column of Table 1.3. It will look exactly like the first column of Table 1.2.
2. Since pillbugs tend to walk around the edge of a petri dish, you could put the wet or dry substance there; or for the wet substance you could put liquid-soaked cotton in the pillbug's path.
3. Rinse your pillbug between procedures by spritzing it with distilled water from a spray bottle. Then put it on a paper towel to dry it off.
4. Watch the pillbug's response to each substance, and record it in Table 1.3, using +, −, or 0 as before.

Table 1.3 Pillbug's Response to Potential Foods		
Substance	**Pillbug's Response**	**Hypothesis Supported?**
1		
2		
3 (control)		
4		
5		
6 (control)		

Conclusion

5. Do your results support your hypotheses? Answer yes or no in the last column of Table 1.3.
6. Are there any hypotheses that were not supported by the experimental results (data)? Does this difference give you more insight into pillbug behavior? Explain. ______________________________

____

____

7. **Class Results.** Compare your results with those of other students who tested the same substance. Calculate the proportional response to each potential food (%+, %−, %0) and record your calculations in Table 1.4. As a group, your class can decide what proportion is needed to designate this response as typical. For example, if the pillbugs as a whole were attracted to a substance 70% or more of the time, you can call that response the "typical response."

Table 1.4 Pillbug's Response to Potential Foods: Class Results				
Substance	**Pillbug's Response**			**Hypothesis Supported?**
1	%+	%−	%0	
2	%+	%−	%0	
3 (control)	%+	%−	%0	
4	%+	%−	%0	
5	%+	%−	%0	
6 (control)	%+	%−	%0	

8. On the basis of the class data do you need to revise your conclusion for any particular pillbug response? ______ Scientists prefer to come to conclusions on the basis of many trials. Why is this the best methodology? ______

____

9. Did the pillbugs respond as expected to the controls, that is, did not eat them? ______________ If they did not respond as expected, what can you conclude about your experimental results? ______________

____

Laboratory Review 1

______________ **1.** What kind of animal is a pillbug?

______________ **2.** What kind of skeleton does a pillbug have?

______________ **3.** What structures do pillbugs use for gas exchange, and what condition must these structures be in to function properly?

______________ **4.** What do scientists do first when they begin to study a specific topic?

______________ **5.** What is a tentative decision about the outcome of an experiment?

______________ **6.** What do we call the sample that lacks the factor being tested and goes through all the experimental steps?

______________ **7.** What do scientists call the information they collect while doing experiments or making observations?

______________ **8.** Which is made by a scientist following experiments and observations, a theory or a conclusion?

______________ **9.** What do scientists develop after many years of experimentation and a lot of similar individual conclusions?

______________ **10.** If your hypothesis was that your pillbug would be attracted to applesauce and your pillbug moved toward the applesauce, what would you say about your hypothesis?

For 11 and 12, indicate whether the statements are hypotheses, conclusions, or scientific theories.

______________ **11.** All organisms are made of cells.

______________ **12.** The data show that trans fat intake raises cholesterol and contributes to heart disease.

______________ **13.** If a pillbug travels 3 mm in 30 seconds, what is its rate of speed?

______________ **14.** If a pillbug moves toward a substance, is it attracted to or repelled by that substance?

Thought Questions

15. Why is a theory more comprehensive than a conclusion?

16. Why is it important to have a control substance for an experiment?

17. Why is it important to test a pillbug's response using one substance at a time?

LABORATORY

2

Metric Measurement and Microscopy

Learning Outcomes

2.1 The Metric System
- Use metric units of measurement for length, weight, volume, and temperature.

2.2 Microscopy
- Describe similarities and differences between the stereomicroscope (dissecting microscope), the compound light microscope, and the electron microscope.

2.3 Stereomicroscope (Dissecting Microscope)
- Identify the parts and tell how to focus the stereomicroscope.

2.4 Use of the Compound Light Microscope
- Identify and give the function of the basic parts of the compound light microscope.
- List, in proper order, the steps for bringing an object into focus with the compound light microscope.
- Describe how the image is inverted by the compound light microscope.
- Calculate the total magnification and the diameter of field for both low- and high-power lens systems.
- Explain how a slide of colored threads provides information on the depth of field.

2.5 Microscopic Observations
- Name and describe the three kinds of cells studied in this exercise.
- State two differences between human epithelial cells and onion epidermal cells.
- Examine a wet mount of *Euglena* and pond water. Contrast the organisms observed in pond water.

Introduction

This laboratory introduces you to the metric system, which biologists use to indicate the sizes of cells and cell structures. This laboratory also examines the features, functions, and use of the compound light microscope and the stereomicroscope (dissecting microscope). Transmission and scanning electron microscopes are explained, and micrographs produced using these microscopes appear throughout this lab manual. The stereomicroscope and the scanning electron microscope view the surface and/or the three-dimensional structure of an object. The compound light microscope and the transmission electron microscope can view only extremely thin sections of a specimen. If a subject was sectioned lengthwise for viewing, the interior of the projections at the top of the cell, called cilia, would appear in the micrograph. A lengthwise cut through any type of specimen is called a **longitudinal section (ls).** On the other hand, if the subject in Figure 2.1 was sectioned crosswise below the area of the cilia, you would see other portions of the interior of the subject. A crosswise cut through any type of specimen is called a **cross section (cs).**

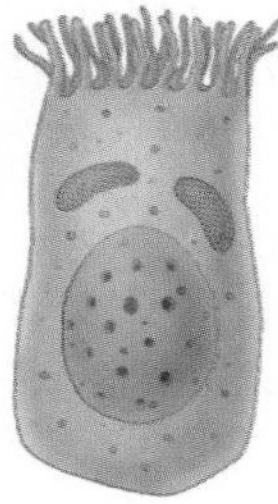
a. The cell

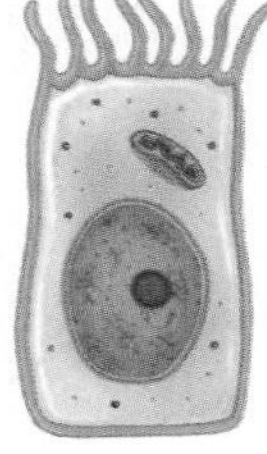
b. Longitudinal section

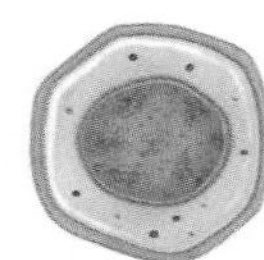
c. Cross section

Figure 2.1 Longitudinal and cross sections. **a.** Transparent view of a cell. **b.** A longitudinal section would show the cilia at the top of the cell. **c.** A cross section shows only the interior where the cut is made.

2.1 The Metric System

The **metric system** is the standard system of measurement in the sciences, including biology, chemistry, and physics. It has tremendous advantages because all conversions, whether for volume, mass (weight), or length, can be in units of ten. Refer to Appendix B, page B–1, for an in-depth look at the units of the metric system.

Length

Metric units of length measurement include the **meter (m), centimeter (cm), millimeter (mm), micrometer (μm),** and **nanometer (nm)** (Table 2.1). The prefixes milli- (10^{-3}), micro- (10^{-6}), and nano (10^{-9}) are used with length, weight, and volume.

Table 2.1 Metric Units of Length Measurement

Unit	Meters	Centimeters	Millimeters	Relative Size
Meter (m)	1 m	100 cm	1,000 mm	Largest
Centimeter (cm)	0.01 (10^{-2}) m	1 cm	10 mm	
Millimeter (mm)	0.001 (10^{-3}) m	0.1 cm	1.0 mm	
Micrometer (μm)	0.000001 (10^{-6}) m	0.0001 (10^{-4}) cm	0.001 (10^{-3}) mm	
Nanometer (nm)	0.000000001 (10^{-9}) m	0.0000001 (10^{-7}) cm	0.000001 (10^{-6}) mm	Smallest

Experimental Procedure: Length

1. Obtain a small ruler marked in centimeters and millimeters. How many centimeters are represented? ________ One centimeter equals how many millimeters? ________ To express the size of small objects, such as cell contents, biologists use even smaller units of the metric system than those on the ruler. These units are the micrometer (μm) and the nanometer (nm).

   According to Table 2.1, 1 μm = ________ mm, and 1 nm = ________ mm.

   Therefore, 1 mm = ________ μm = ________ nm.

2. Measure the diameter (a line passing through the center with endpoints touching the circle) of the circle shown below to the nearest millimeter. This circle's diameter is

   ________ mm = ________ μm = ________ nm.

For example, to convert mm to μm:

$$\_\_\_\_\_\_ \text{ mm} \times \frac{1{,}000 \text{ μm}}{\text{mm}} = \_\_\_\_\_\_ \text{ μm}$$

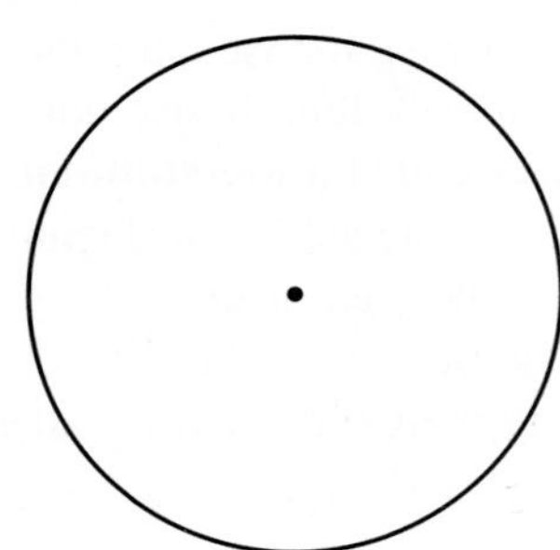

3. Obtain a meter stick. On one side, find the numbers 1 through 39, which denote inches. One meter equals 39.37 inches; therefore, 1 meter is roughly equivalent to 1 yard. Turn the meter stick over, and observe the metric subdivisions. How many centimeters are in a meter? __________ How many millimeters are in a meter? __________ The prefix *milli* means __________.
4. Use the meter stick and the method shown in Figure 2.2 to measure the length of two long bones from a disarticulated human skeleton. Lay the meter stick flat on the lab table. Place a long bone next to the meter stick between two pieces of cardboard (each about 10 cm × 30 cm), held upright at right angles to the stick. The narrow end of each piece of cardboard should touch the meter stick. The length between the cards is the length of the bone in centimeters. For example, if the bone measures from the 22 cm mark to the 50 cm mark, the length of the bone is __________ cm. If the bone measures from the 22 cm mark to midway between the 50 cm and 51 cm marks, its length is __________ mm, or __________ cm.
5. Record the length of two bones. First bone: __________ cm =__________ mm. Second bone: __________ cm = __________ mm.

Figure 2.2 Measurement of a long bone.
How to measure a long bone using a meter stick.

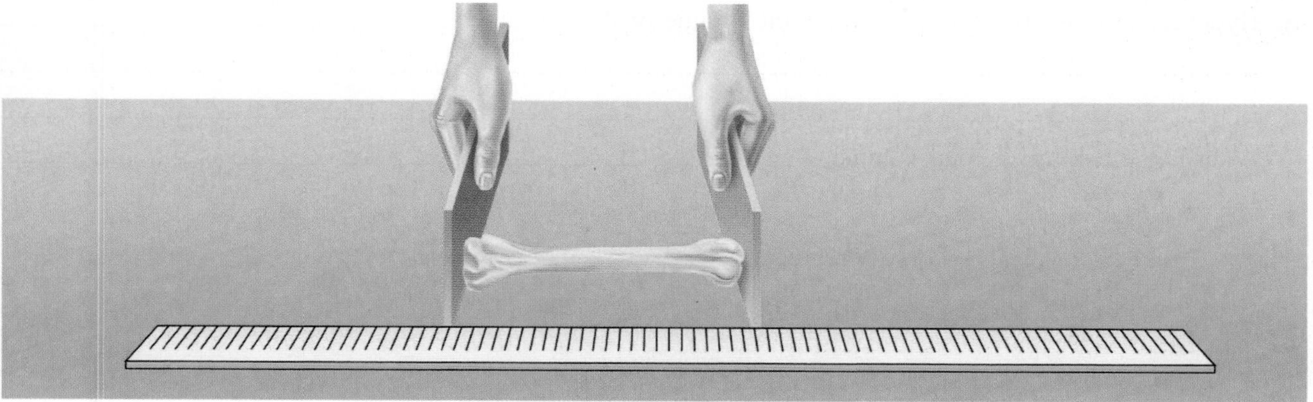

Weight

Two metric units of weight are the **gram (g)** and the **milligram (mg).** A paper clip weighs about 1 g, which equals 1,000 mg. 2 g = __________ mg; 0.2 g = __________ mg; and 2 mg = __________ g.

Experimental Procedure: Weight

1. Use a balance scale to measure the weight of a wooden block small enough to hold in the palm of your hand.
2. Measure the weight of the block to the tenth of a gram. The weight of the wooden block is __________ g = __________ mg.
3. Measure the weight of an item small enough to fit inside the opening of a 50 ml graduated cylinder. The item, a(n) __________, is __________ g = __________ mg.

Volume

Two metric units of volume are the **liter** (**l**) and the **milliliter** (**ml**). One liter = 1,000 ml.

Experimental Procedure: Volume

1. Volume measurements can be related to those of length. For example, use a millimeter ruler to measure the wooden block used in the previous Experimental Procedure to get its length, width, and depth.

   length = __________ cm; width = __________ cm; depth = __________ cm

   The volume, or space, occupied by the wooden block can be expressed in cubic centimeters (cc or cm^3) by multiplying: length × width × depth = __________ cm^3. For purposes of this Experimental Procedure, 1 cubic centimeter equals 1 milliliter; therefore, the wooden block has a volume of __________ ml.
2. In the biology laboratory, liquid volume is usually measured directly in liters or milliliters with appropriate measuring devices. For example, use a 50 ml graduated cylinder to add 20 ml of water to a test tube. First, fill the graduated cylinder to the 20 ml mark. To do this properly, you have to make sure that the lowest margin of the water level, or the **meniscus** (Fig. 2.3), is at the 20 ml mark. Place your eye directly parallel to the level of the meniscus, and add water until the meniscus is at the 20 ml mark. (Having a dropper bottle filled with water on hand can help you do this.) A large, blank, white index card held behind the cylinder can also help you see the scale more clearly. Now pour the 20 ml of water into the test tube.
3. Hypothesize how you could find the total volume of the test tube. ______________________________

   ____

   ____

   What is the test tube's total volume? ______________________________

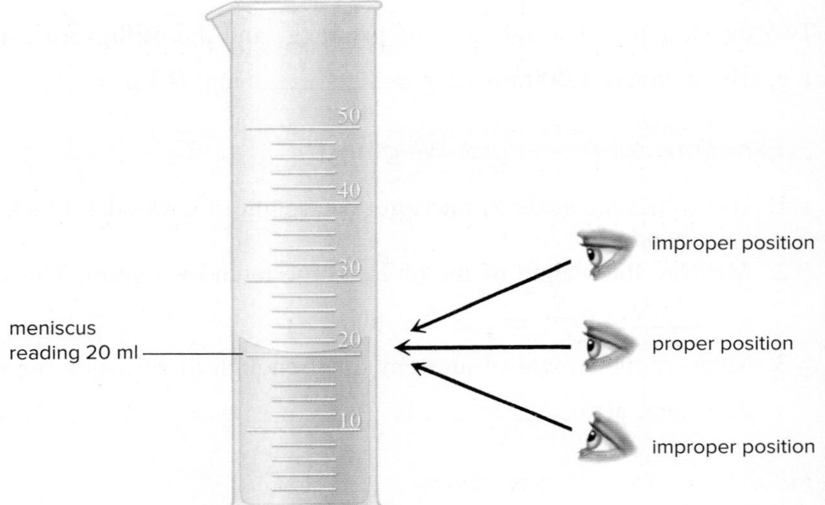

Figure 2.3 Meniscus.
The proper way to view the meniscus.

4. Fill a 50 ml graduated cylinder with water to about the 20 ml mark. Hypothesize how you could use this setup to calculate the volume of an object. ______________________________

____

Now perform the operation you suggested. The object, __________, has a volume of __________ ml.

5. Hypothesize how you could determine how many drops from the pipette of the dropper bottle equal 1 ml. ______________________________

Now perform the operation you suggested. How many drops from the pipette of the dropper bottle equal 1 ml? ______________________________

6. Some pipettes are graduated and can be filled to a certain level as a way to measure volume directly. Your instructor will demonstrate this. Are pipettes customarily used to measure large or small volumes? _____

Temperature

There are two temperature scales: the **Fahrenheit (F)** and **Celsius (centigrade, C)** scales (Fig. 2.4). Scientists use the Celsius scale.

Experimental Procedure: Temperature

1. Study the two scales in Figure 2.4, and complete the following information:

   a. Water freezes at either __________ °F or __________ °C.

   b. Water boils at either __________ °F or __________ °C.

2. To convert from the Fahrenheit to the Celsius scale, use the following equation:

$$°C = (°F - 32°)/1.8$$

or

$$°F = (1.8°C) + 32$$

Human body temperature of 98°F is what temperature on the Celsius scale? __________

3. Record any two of the following temperatures in your lab environment. In each case, allow the end bulb of the Celsius thermometer to remain in or on the sample for one minute.

   Room temperature = __________ °C

   Surface of your skin = __________ °C

   Cold tap water in a 50 ml beaker = __________ °C

   Hot tap water in a 50 ml beaker = __________ °C

   Ice water = __________ °C

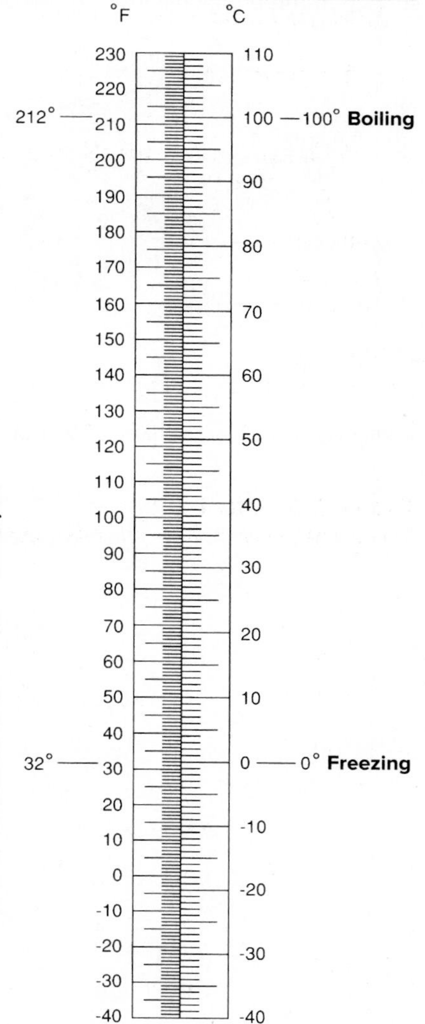

Figure 2.4 Temperature scales.
The Fahrenheit (°F) scale is on the left, and the Celsius (°C) scale is on the right.

2.2 Microscopy

Because biological objects can be very small, we often use a microscope to view them. Many kinds of instruments, ranging from the hand lens to the electron microscope, are effective magnifying devices. A short description of two kinds of light microscopes and two kinds of electron microscopes follows.

Light Microscopes

Light microscopes use light rays passing through lenses to magnify the object. The **stereomicroscope** (**dissecting microscope**) is designed to study entire objects in three dimensions at low magnification. The **compound light microscope** is used for examining small or thinly sliced sections of objects under higher magnification than that of the stereomicroscope. The term **compound** refers to the use of two sets of lenses: the ocular lenses located near the eyes and the objective lenses located near the object. Illumination is from below, and visible light passes through clear portions but does not pass through opaque portions. To improve contrast, the microscopist uses stains or dyes that bind to cellular structures and absorb light. Photomicrographs, also called light micrographs, are images produced by a compound light microscope (Fig. 2.5*a*).

Figure 2.5 Comparative micrographs of a lymphocyte.
A lymphocyte is a type of white blood cell. **a.** A photomicrograph (light micrograph) of a lymphocyte shows less detail than a **(b)** transmission electron micrograph (TEM). **c.** A scanning electron micrograph (SEM) of a lymphocyte shows the cell surface in three dimensions. (a) ©Michael Ross/Science Source; (b) ©CNRI/SPL/Science Source; (c) ©Steve Gschmeissner/Science Source

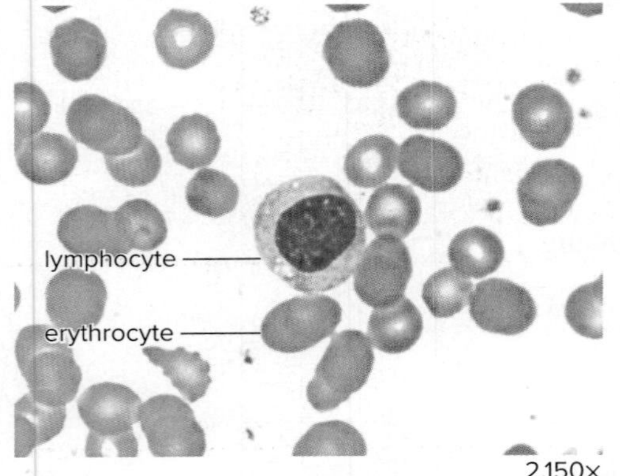

a. Photomicrograph or light micrograph (LM)

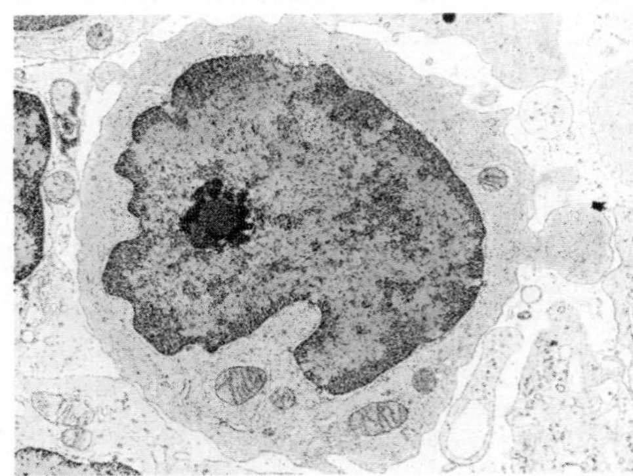

b. Transmission electron micrograph (TEM)

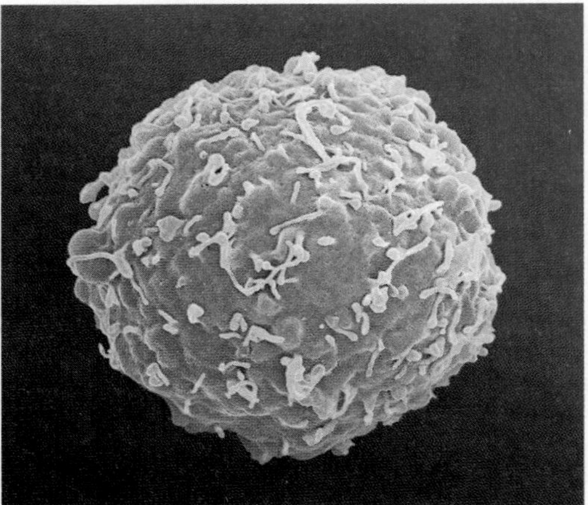

c. Scanning electron micrograph (SEM)

Electron Microscopes

Electron microscopes use beams of electrons to magnify the object. The beams are focused on a photographic plate by means of electromagnets. The **transmission electron microscope** is analogous to the compound light microscope. The object is ultra-thinly sliced and treated with heavy metal salts to improve contrast. Figure 2.5*b* is a micrograph produced by this type of microscope. The **scanning electron microscope** is analogous to the dissecting light microscope. It gives an image of the surface and dimensions of an object, as is apparent from the scanning electron micrograph in Figure 2.5*c*.

The micrographs in Figure 2.5 demonstrate that an object is magnified more with an electron microscope than with a compound light microscope. The difference between these two types of microscopes, however, is not simply a matter of magnification; it is also the electron microscope's ability to show detail. The electron microscope has greater resolving power. **Resolution** is the minimum distance between two objects at which they can still be seen, or resolved, as two separate objects. The use of high-energy electrons rather than light gives electron microscopes a much greater resolving power since two objects that are much closer together can still be distinguished as separate points. Table 2.2 lists several other differences between the compound light microscope and the transmission electron microscope.

Table 2.2 Comparison of the Compound Light Microscope and the Transmission Electron Microscope

Compound Light Microscope	Transmission Electron Microscope
1. Glass lenses	1. Electromagnetic lenses
2. Illumination by visible light	2. Illumination due to beam of electrons
3. Resolution ≅ 200 nm	3. Resolution ≅ 0.1 nm
4. Magnifies to 2,000×	4. Magnifies to 1,000,000×
5. Costs up to tens of thousands of dollars	5. Costs up to hundreds of thousands of dollars

Conclusions: Microscopy

- Which two types of microscopes view the surface of an object? ____________________
- Which two types of microscopes view objects that have been sliced and treated to improve contrast? ____________________
- Of the microscopes just mentioned, which one resolves the greater amount of detail?

____

2.3 Stereomicroscope (Dissecting Microscope)

The **stereomicroscope** (**dissecting microscope,** Fig. 2.6) allows you to view objects in three dimensions at low magnifications. It is used to study entire small organisms, any object requiring lower magnification, and opaque objects that can be viewed only by reflected light. It is called a stereomicroscope because it produces a three-dimensional image.

Identifying the Parts

After your instructor has explained how to carry a microscope, obtain a stereomicroscope and a separate illuminator, if necessary, from the storage area. Place it securely on the table. Plug in the power cord, and turn on the illuminator. There is a wide variety of stereomicroscope styles, and your instructor will discuss the specific style(s) available to you. Regardless of style, the following features should be present:

1. **Binocular head:** Holds two eyepiece lenses that move to accommodate for the various distances between different individuals' eyes.
2. **Eyepiece lenses:** The two lenses located on the binocular head. What is the magnification of your eyepieces? __________ Some models have one **independent focusing eyepiece** with a knurled knob to allow independent adjustment of each eye. The nonadjustable eyepiece is called the **fixed eyepiece.**
3. **Focusing knob:** A large, black or gray knob located on the arm; used for changing the focus of both eyepieces together.

Figure 2.6 Binocular dissecting microscope (stereomicroscope).
Label this stereomicroscope with the help of the text material. ©Leica Microsystems GmbH

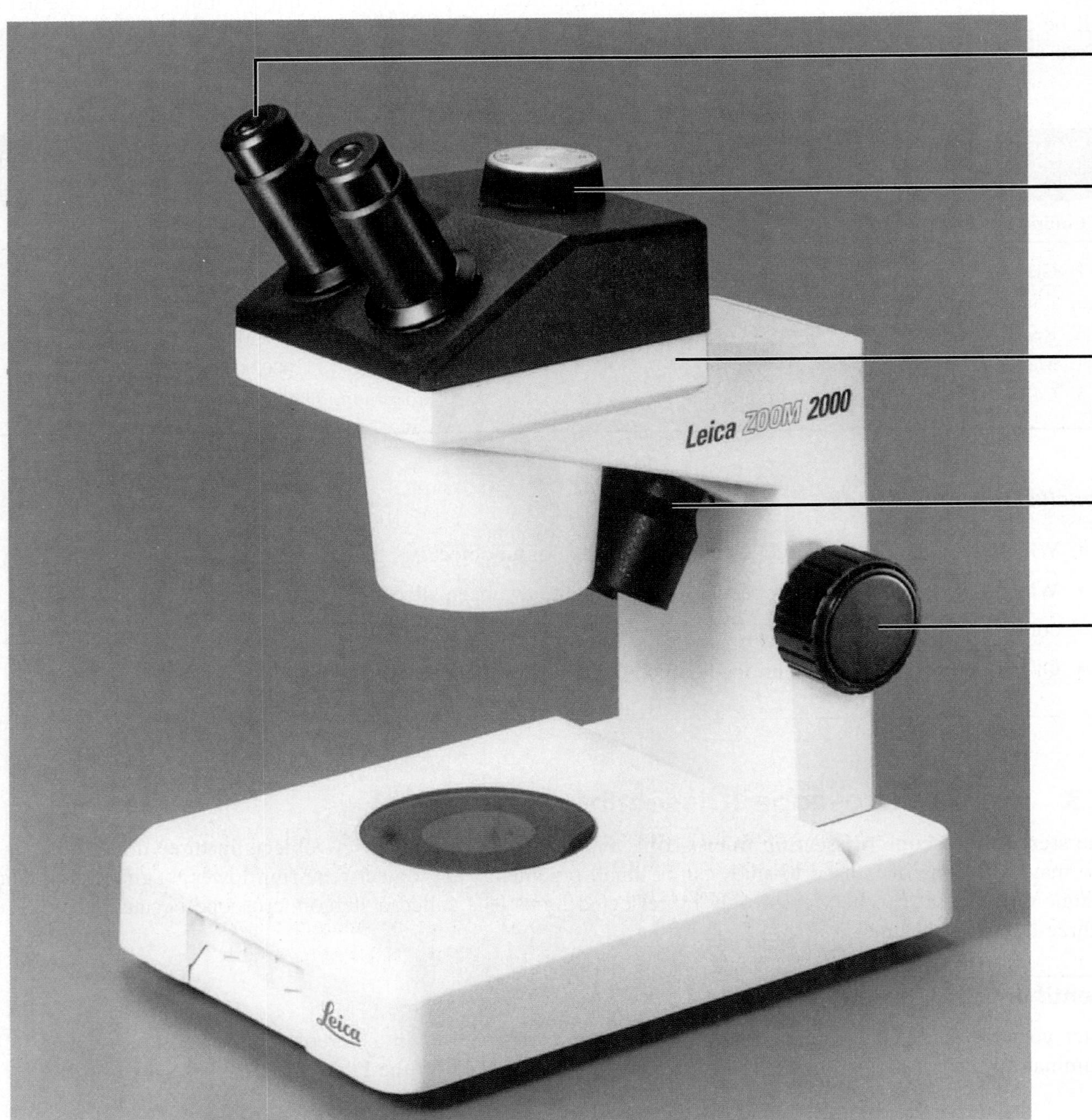

4. **Magnification changing knob:** A knob, often built into the binocular head, used to change magnification in both eyepieces simultaneously. This may be a **zoom** mechanism or a **rotating lens** mechanism of different powers that clicks into place.
5. **Illuminator:** Used to illuminate an object from above; may be built into the microscope or separate.

Locate each of these parts on your stereomicroscope, and label them on Figure 2.6.

Focusing the Stereomicroscope

1. In the center of the stage, place a plastomount that contains small organisms.
2. Adjust the distance between the eyepieces on the binocular head so that they comfortably fit the distance between your eyes. You should be able to see the object with both eyes as one three-dimensional image.
3. Use the focusing knob to bring the object into focus.
4. Does your microscope have an independent focusing eyepiece? __________ If so, use the focusing knob to bring the image in the fixed eyepiece into focus, while keeping the eye at the independent focusing eyepiece closed. Then adjust the independent focusing eyepiece so that the image is clear, while keeping the other eye closed. Is the image inverted? ______________________________
5. Turn the magnification changing knob, and determine the kind of mechanism on your microscope. A zoom mechanism allows continuous viewing while changing the magnification. A rotating lens mechanism blocks the view of the object as the new lenses are rotated. Be sure to click each lens firmly into place. If you do not, the field will be only partially visible. What kind of mechanism is on your microscope? _____
6. Set the magnification changing knob on the lowest magnification. Sketch the object in the following circle as though this represents your entire field of view:

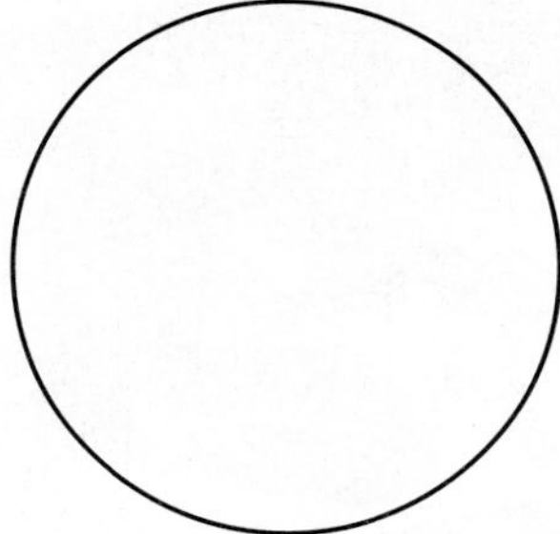

7. Rotate the magnification changing knob to the highest magnification. Draw another circle within the one provided to indicate the reduction of the field of view.
8. Experiment with various objects at various magnifications until you are comfortable with using the stereomicroscope.

2.4 Use of the Compound Light Microscope

As mentioned, the name **compound light microscope** indicates that it uses two sets of lenses and light to view an object. The two sets of lenses are the ocular lenses located near the eyes and the objective lenses located near the object. Illumination is from below, and the light passes through clear portions but does not pass through opaque portions. This microscope is used to examine small or thinly sliced sections of objects under higher magnification than would be possible with the stereomicroscope.

Identifying the Parts

Obtain a compound light microscope from the storage area, and place it securely on the table. *Identify the following parts on your microscope, and label them in Figure 2.7 with the help of the text material.*

Figure 2.7 Compound light microscope.
Compound light microscope with binocular head and mechanical stage. Label this microscope with the help of the text material.

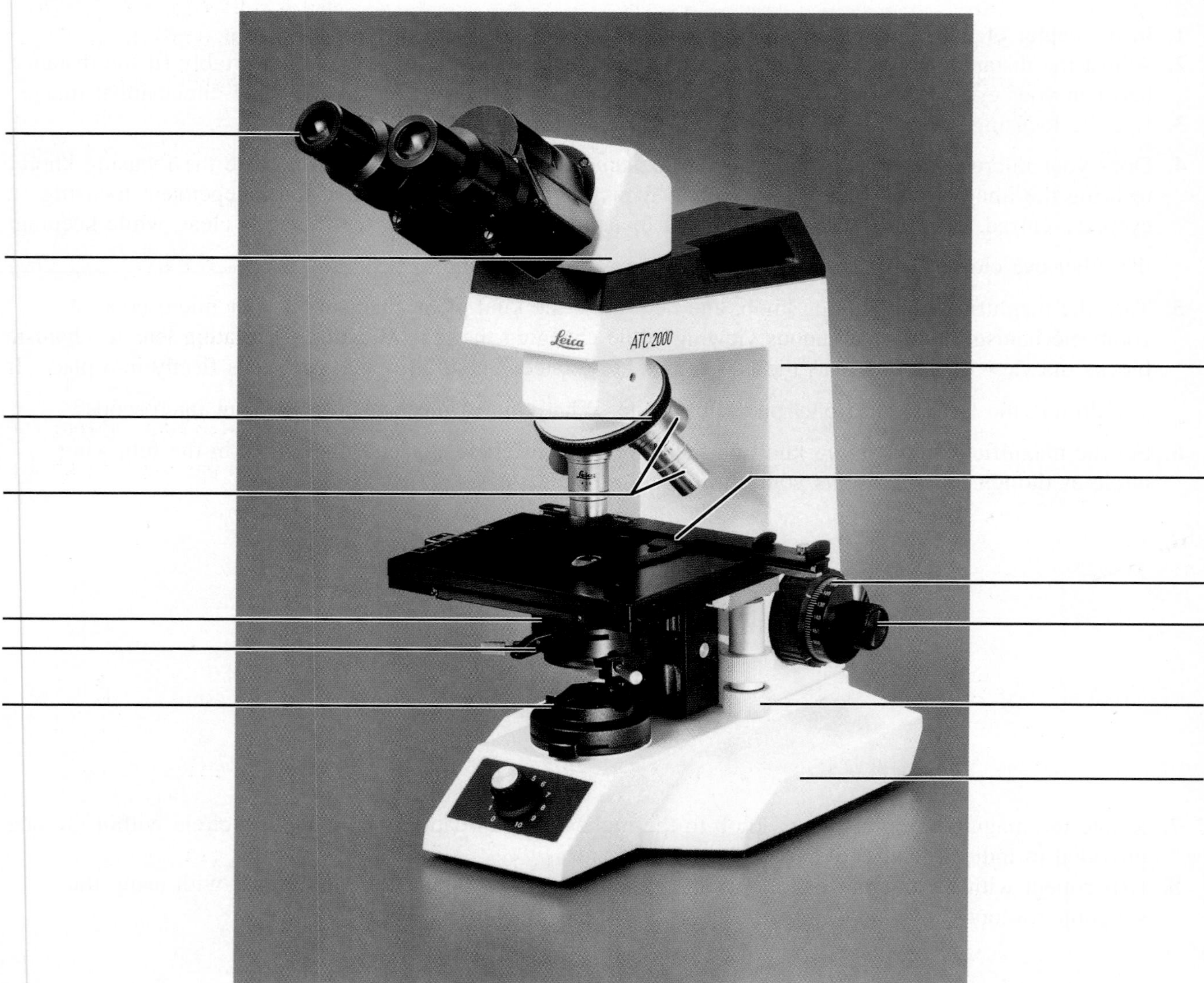

1. **Eyepieces** (ocular lenses): What is the magnifying power of the ocular lenses on your microscope? ________
2. **Viewing head:** Holds the ocular lenses.
3. **Arm:** Supports upper parts and provides carrying handle.
4. **Nosepiece:** Revolving device that holds objectives.
5. **Objectives** (objective lenses):
   a. **Scanning objective:** This is the shortest of the objective lenses and is used to scan the whole slide. The magnification is stamped on the housing of the lens. It is a number followed by an ×. What is the magnifying power of the scanning objective lens on your microscope? _______

b. **Low-power objective:** This lens is longer than the scanning objective lens and is used to view objects in greater detail. What is the magnifying power of the low-power objective lens on your microscope? __________

c. **High-power objective:** If your microscope has three objective lenses, this lens will be the longest. It is used to view an object in even greater detail. What is the magnifying power of the high-power objective lens on your microscope? __________

d. **Oil immersion objective** (on microscopes with four objective lenses): Holds a 95× (to 100×) lens and is used in conjunction with immersion oil to view objects with the greatest magnification. Does your microscope have an oil immersion objective? __________ If this lens is available, your instructor will discuss its use when the lens is needed.

6. **Stage:** Platform that holds and supports microscope slides. A mechanical stage is a movable stage that aids in the accurate positioning of the slide. Does your microscope have a mechanical stage? __________
   a. **Stage clips:** Clips that hold a slide in place on the stage.
   b. **Mechanical stage control knobs:** Two knobs that control forward/reverse movement and right/left movement, respectively.
7. **Coarse-adjustment knob:** Knob used to bring object into approximate focus; used only with low-power objective.
8. **Fine-adjustment knob:** Knob used to bring object into final focus.
9. **Condenser:** Lens system below the stage used to focus the beam of light on the object being viewed.
   a. **Diaphragm** or **diaphragm control lever:** Lever that controls the amount of light passing through the condenser.
10. **Light source:** An attached lamp that directs a beam of light up through the object.
11. **Base:** The flat surface of the microscope that rests on the table.

Rules for Microscope Use

Observe the following rules for using a microscope:

1. The lowest power objective (scanning or low) should be in position at both the beginning and the end of microscope use.
2. Use only lens paper for cleaning lenses.
3. Do not tilt the microscope because the eyepieces could fall out, or wet mounts could be ruined.
4. Keep the stage clean and dry to prevent rust and corrosion.
5. Do not remove parts of the microscope.
6. Keep the microscope dust-free by covering it after use.
7. Report any malfunctions.

Focusing the Compound Light Microscope—Lowest Power

1. Turn the nosepiece so that the *lowest* power objective on your microscope is in straight alignment over the stage.
2. Always begin focusing with the *lowest* power objective on your microscope (4× [scanning] or 10× [low power]).
3. With the coarse-adjustment knob, lower the stage (or raise the objectives) until it stops.

4. Place a slide of the letter *e* on the stage, and stabilize it with the clips. (If your microscope has a mechanical stage, pinch the spring of the slide arms on the stage, and insert the slide.) Center the *e* as best you can on the stage or use the two control knobs located below the stage (if your microscope has a mechanical stage) to center the *e*.
5. Again, be sure that the lowest-power objective is in place. Then, as you look from the side, decrease the distance between the stage and the tip of the objective lens until the lens comes to an automatic stop or is no closer than 3 mm above the slide.
6. While looking into the eyepiece, rotate the diaphragm (or diaphragm control lever) to give the maximum amount of light.
7. Using the coarse-adjustment knob, slowly increase the distance between the stage and the objective lens until the object—in this case, the letter *e*—comes into view, or focus.
8. Once the object is seen, you may need to adjust the amount of light. To increase or decrease the contrast, rotate the diaphragm slightly.
9. Use the fine-adjustment knob to sharpen the focus if necessary.
10. Practice having both eyes open when looking through the eyepiece, as this greatly reduces eyestrain.

Inversion

Inversion refers to the fact that a microscopic image is upside down and reversed.

Observation: Inversion

1. Draw the letter *e* as it appears on the slide (with the unaided eye, not looking through the eyepiece). ________
2. Draw the letter *e* as it appears when you look through the eyepiece. ________
3. What differences do you notice? ________
4. Move the slide to the right. Which way does the image appear to move? ________
5. Move the slide toward you. Which way does the image appear to move? ________

Focusing the Compound Light Microscope—Higher Powers

Compound light microscopes are **parfocal;** that is, once the object is in focus with the lowest power, it should also be almost in focus with the higher power.

1. Bring the object into focus under the lowest power by following the instructions in the previous section.
2. Make sure that the letter *e* is centered in the field of the lowest objective.
3. Move to the next higher objective (low power [10×] or high power [40×]) by turning the nosepiece until you hear it click into place. Do not change the focus; parfocal microscope objectives will not "hit" normal slides when changing the focus if the lowest objective is initially in focus. (If you are on low power [10×], proceed to high power [40×] before going on to step 4.)
4. If any adjustment is needed, use only the *fine*-adjustment knob. (*Note:* Always use only the fine-adjustment knob with high power, and do not use the coarse-adjustment knob.)
5. On a drawing of the letter *e* to the right, *draw a circle around the portion of the letter that you are now seeing with high-power magnification.* The letter *e* will not disappear because your microscope is parcentric (the focus remains near the center). ________
6. When you have finished your observations of this slide (or any slide), rotate the nosepiece until the lowest-power objective clicks into place, and then remove the slide.

Total Magnification

Total magnification is calculated by multiplying the magnification of the ocular lens (eyepiece) by the magnification of the objective lens. The magnification of a lens is imprinted on the lens casing.

Observation: Total Magnification

Calculate total magnification figures for your microscope, and record your findings in Table 2.3.

Table 2.3 Total Magnification

Objective	Ocular Lens	Objective Lens	Total Magnification
Scanning power (if present)			
Low power			
High power			
Oil immersion (if present)			

Field of View

A microscope's **field of view** is the circle visible through the lenses. The **diameter of field** is the length of the field from one edge to the other.

Observation: Field of View

Low-Power (10×) Diameter of Field

1. Place a clear plastic ruler across the stage so that the edge of the ruler is visible as a horizontal line along the diameter of the low-power (not scanning) field. Be sure that you are looking at the millimeter side of the ruler.
2. Estimate the number of millimeters, to tenths, that you see along the field: ________________ mm. (*Hint:* Start by placing any millimeter marker at the edge of the field.) Convert the observed number of millimeters to micrometers: ____________ μm. This is the **low-power diameter of field** (**LPD**) for your microscope in micrometers.

High-Power (40×) Diameter of Field

1. To compute the **high-power diameter of field (HPD),** substitute these data into the formula given:
   a. LPD = low-power diameter of field (in micrometers) = ________________
   b. LPM = low-power total magnification (from Table 2.3) = ________________
   c. HPM = high-power total magnification (from Table 2.3) = ________________

Example: If the diameter of field is about 2 mm, then the LPD is 2,000 μm. Using the LPM and HPM values from Table 2.3, the HPD would be 500 μm.

$$HPD = LPD \times \frac{LPM}{HPM}$$

$$HPD = (\quad\quad) \times \frac{(\quad\quad)}{(\quad\quad)} = \_\_\_\_\_\_$$

Pond Water

Examination of pond water will also test your ability to observe objects with the microscope, to utilize depth of field, and to control illumination to heighten contrast. Use Figure 2.13 to help identify the organisms in pond water.

Figure 2.13 Microorganisms found in pond water (not actual size).

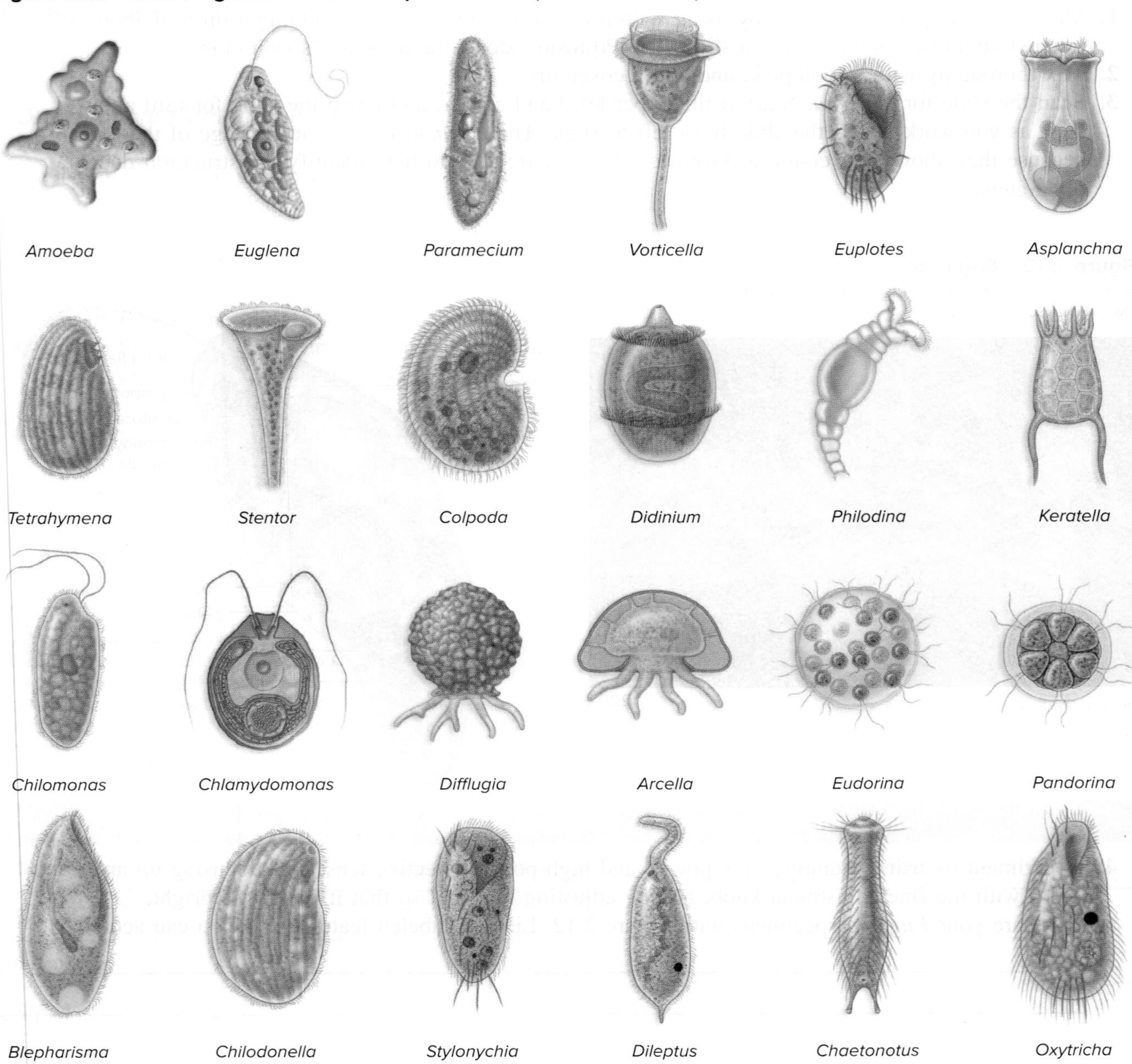

Laboratory Review 2

_______________ 1. 19 mm equals how many cm?

_______________ 2. 880 mm equals how many m?

_______________ 3. 2,700 mg equals how many grams?

_______________ 4. 3.4 l equals how many ml?

_______________ 5. To properly measure 20 ml of water, what must be at the 20 ml mark of the graduated cylinder?

_______________ 6. 90°F equals how many degrees Celsius?

_______________ 7. What kind of microscope would be used to study a whole or opaque object?

_______________ 8. What is the name for the lenses located near the eye?

_______________ 9. If the total magnification of a slide is 400× and the ocular lenses are 10×, what is the magnifying power of the objective being used?

_______________ 10. If the amount of light passing through the condenser needs to be decreased, what microscope part should be adjusted?

_______________ 11. What word describes a microscope that remains in focus when the objective lenses are changed?

_______________ 12. If a slide is being viewed with the high-power objective, which adjustment knob should be used to sharpen the focus?

_______________ 13. Which objective should be in place when the microscope is put away?

_______________ 14. If threads are layered from top to bottom, brown, green, red, which layer will come into focus first if you are using the microscope properly?

Thought Questions

15. Which type of microscope should be used to view a virus that is 50 nm in size? Justify your choice.

16. Justify your choice of an objective to use when starting your observation of the *Euglena* wet mount; if the *Euglena* are swimming to the left, which way should you move your slide to keep the organism in view?

17. If you use the coarse-adjustment knob to focus on an object with the high-power objective, what problems will you encounter? Explain.

LABORATORY

3

Chemical Composition of Cells

Learning Outcomes

Introduction

- Describe how cells build and break down polymers.
- Distinguish between positive and negative controls and between positive and negative results.

3.1 Proteins

- State the function of various types of proteins.
- Identify the subunit of a polypeptide, and explain how this subunit connects to others to form a polypeptide.
- Describe a test for the detection of a protein versus a peptide.

3.2 Carbohydrates

- Identify the subunit of starch, and distinguish sugars from starch in terms of subunit number.
- Describe a test for the detection of starch and another for the detection of sugars.
- Explain the varied results of the Benedict's test.

3.3 Lipids

- Name several types of lipids and state their components.
- Describe a simple test for the detection of fat and tell how an emulsifier works.

3.4 Testing Foods and Unknowns

- Explain a procedure for testing the same food for all three components—carbohydrates, protein, and fat.

Introduction

Planning Ahead To save time, your instructor may have you start the boiling water bath needed during the test for sugars and testing food for unknowns.

All organisms consist of basic units of matter called **atoms.** Molecules form when atoms bond with one another. Inorganic molecules are often associated with nonliving things. Organic biomolecules are associated with organisms. In this laboratory, you will be studying the biomolecules of cells: **proteins, carbohydrates** (monosaccharides, disaccharides, polysaccharides), and **lipids** (i.e., fat).

Large biomolecules, sometimes called macromolecules, form during dehydration reactions when smaller molecules bond as water is given off. During *hydrolysis reactions,* bonds are broken as water is added.

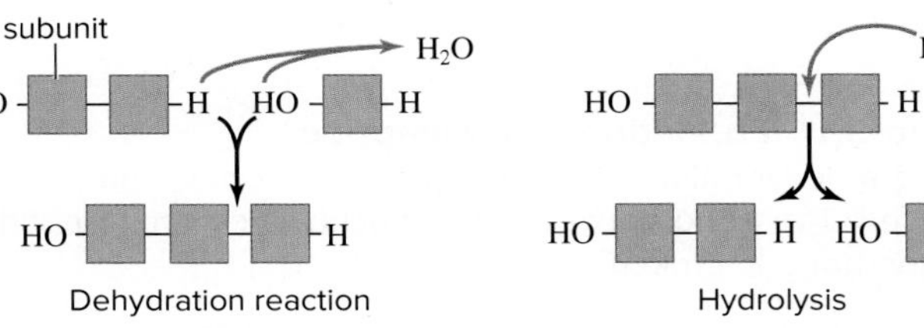

A fat contains one glycerol and three fatty acids. Proteins and some carbohydrates (called polysaccharides) are **polymers** because they are made up of smaller molecules called **monomers.** Proteins contain a large number of amino acids (the monomer) joined together by a peptide bond. A polysaccharide, such as starch, contains a large number of glucose molecules joined together. Various chemicals will be used in this laboratory to test for the presence of cellular biomolecules. If a color change is observed, the test is said to be *positive* because it indicates that the molecule is present. If the color change is not observed, the test is said to be *negative* because it indicates that the molecule is not present.

What Is a Control?

The experiments in today's laboratory have both a positive control and a negative control, *which should be saved for comparison purposes until the experiment is complete*. The **positive control** goes through all the steps of the experiment and does contain the substance being tested. Therefore, positive results are expected. The **negative control** goes through all the steps of the experiment except it does not contain the substance being tested. Therefore, negative results are expected.

For example, if a test tube contains glucose (the substance being tested) and Benedict's reagent (blue) is added, a red color develops upon heating. This test tube is the positive control. The red color indicates a positive test for glucose. If a test tube does not contain glucose and Benedict's reagent is added, Benedict's is expected to remain blue. This test tube is the negative control; it tests negative for glucose.

What benefit is a positive control? Positive controls give you a standard by which to tell if the substance being tested is present (or acting properly) in an unknown sample. Negative controls ensure that the experiment is giving reliable results. If a negative control should happen to give a positive result, then the entire experiment may be faulty and unreliable.

3.1 Proteins

Proteins have numerous functions in cells. Antibodies are proteins that combine with pathogens so that the pathogens are destroyed by the body. Transport proteins combine with and move substances from place to place. Hemoglobin transports oxygen throughout the body. Albumin is another transport protein in our blood. Regulatory proteins control cellular metabolism in some way. For example, the hormone insulin regulates the amount of glucose in blood so that cells have a ready supply. Structural proteins include keratin, found in hair, and myosin, found in muscle. **Enzymes** are proteins that speed chemical reactions. A reaction that could take days or weeks to complete can happen within an instant if the correct enzyme is present. Amylase is an enzyme that speeds the breakdown of starch in the mouth and small intestine.

Proteins are made up of **amino acids** (the subunits) joined together. About 20 different common amino acids are found in cells. All amino acids have an acidic group (—COOH) and an amino group (H_2N—). They differ by the ***R* group** (remainder group) attached to a carbon atom, as shown in Figure 3.1. The *R* groups have varying sizes, shapes, and chemical activities.

A chain of two or more amino acids is called a **peptide,** and the bond between the amino acids is called a **peptide bond.** A **polypeptide** is a very long chain of amino acids. A protein can contain one or more polypeptide chains. A single chain forms insulin, while four chains form hemoglobin. A protein has a particular shape, which is important to its function. The shape comes about because the *R* groups of the polypeptide chain(s) can interact with one another in various ways.

Figure 3.1 Formation of a dipeptide.
During a dehydration reaction, a dipeptide forms when an amino acid joins with an amino acid as a water molecule is removed. The bond between amino acids is called a peptide bond. During a hydrolysis reaction, water is added and the peptide bond is broken.

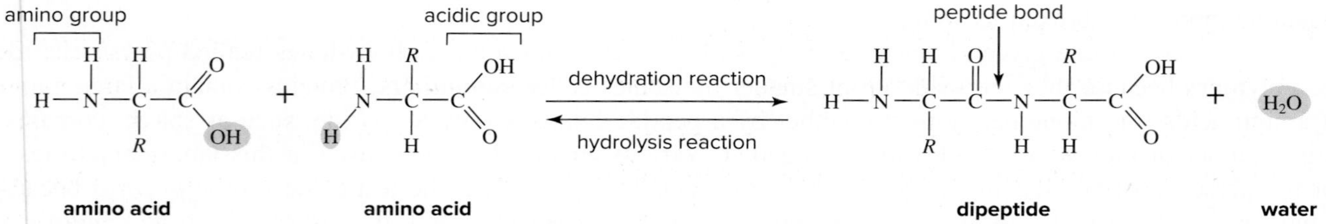

Test for Proteins

Biuret test for protein and peptides
©David S. Moyer RF

Biuret reagent (blue color) contains a strong solution of sodium or potassium hydroxide (NaOH or KOH) and a small amount of dilute copper sulfate ($CuSO_4$) solution. The reagent changes color in the presence of proteins or peptides because the peptide bonds of the protein or peptide chemically combine with the copper ions in biuret reagent (Table 3.1).

Table 3.1 Biuret Test for Protein and Peptides

	Protein	Peptides
Biuret reagent (blue)	Purple	Pinkish-purple

Experimental Procedure: Test for Proteins

1. Label four clean test tubes (1 to 4).
2. Using the designated graduated transfer pipets, add 1 ml of the experimental solutions listed in Table 3.2 to the test tubes according to their numbers.
3. Then add five drops of biuret reagent to the tubes, swirling to mix.
4. The reaction is almost immediate. Record your observations in Table 3.2.

> ⚠ **Biuret reagent** Biuret reagent is highly corrosive. Exercise care in using this chemical. If any should spill on your skin, wash the area with mild soap and water. Follow your instructor's directions for its disposal.

Table 3.2 Biuret Test for Protein

Tube	Contents	Final Color	Conclusions (+ or −)
1	Distilled water	lighter blue	
2	Albumin	Dark purple	
3	Pepsin	light purple	
4	Starch	foggy blue color	

Conclusions: Proteins

- From your test results, conclude if a protein is present (+) or absent (−). Enter your conclusions in Table 3.2.
- Pepsin is an enzyme. Enzymes are composed of what type of organic molecule? ____________
- According to your results, is starch a protein? ____________
- Which of the four tubes is the negative control sample? ________ Why? ____________
____
- Why do experimental procedures include control samples? ____________
____
- If your results are not as expected, inform your instructor, who will advise you how to proceed.

3.2 Carbohydrates

Carbohydrates include sugars and molecules that are chains of sugars. **Glucose,** which has only one sugar unit, is a monosaccharide; **maltose,** which has two sugar units, is a disaccharide (Fig. 3.2). Glycogen, starch, and cellulose are polysaccharides, made up of chains of glucose units (Fig. 3.3).

Glucose is used by all organisms as an energy source. Energy is released when glucose is broken down to carbon dioxide and water. This energy is used by the organism to do work. Animals store glucose as glycogen and plants store glucose as starch. Plant cell walls are composed of cellulose.

Figure 3.2 Formation of a disaccharide.
During a dehydration reaction, a disaccharide, such as maltose, forms when a glucose joins with a glucose as a water molecule is removed. During a hydrolysis reaction, the components of water are added, and the bond is broken.

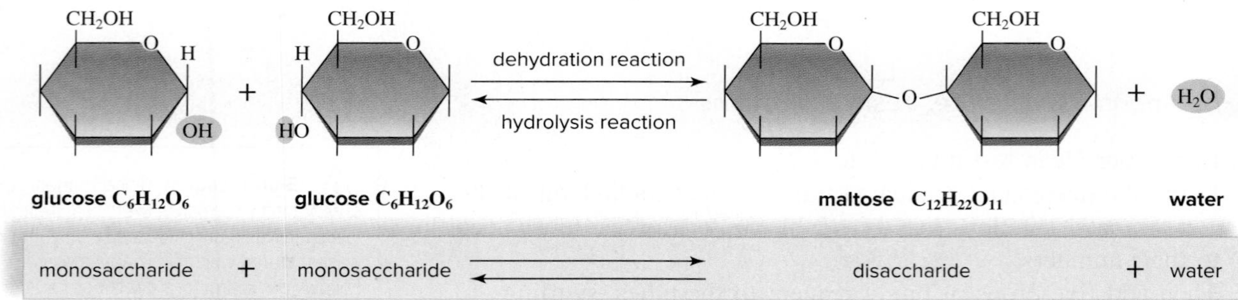

Figure 3.3 Starch.
Starch is a polysaccharide composed of many glucose units. **a.** Photomicrograph of starch granules in cells of a potato. **b.** Structure of starch. Starch consists of amylose that is nonbranched and amylopectin that is branched.
(a) ©Jeremy Burgess/SPL/Science Source

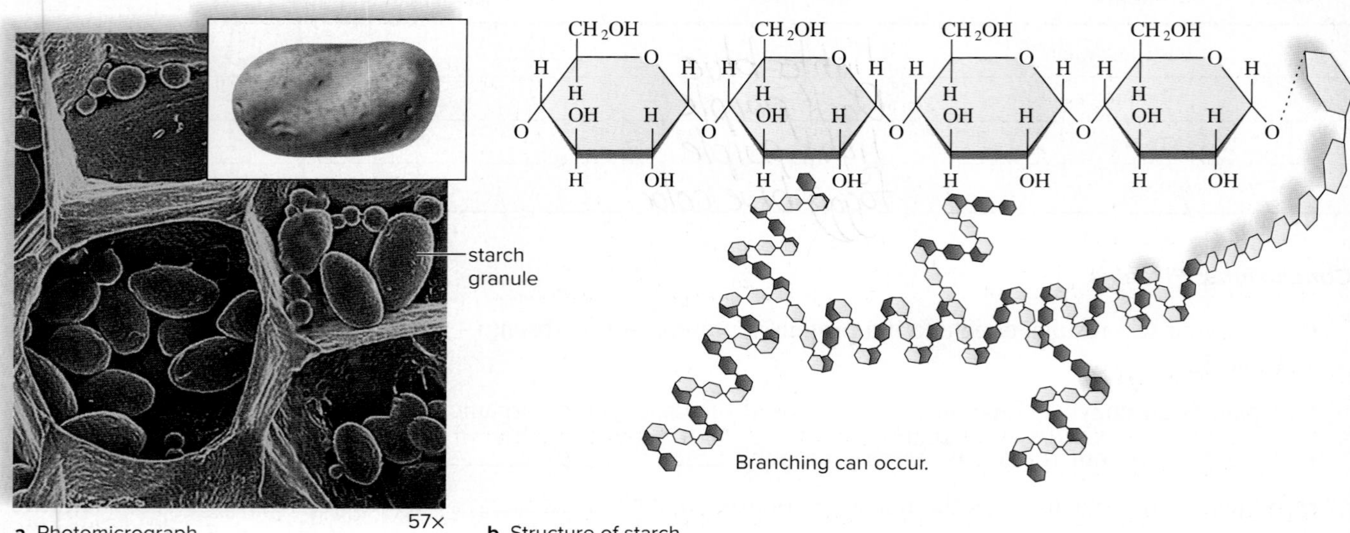

a. Photomicrograph

b. Structure of starch

Test for Starch

In the presence of starch, iodine solution (yellowish-brown) reacts chemically with starch to form a blue-black color (Table 3.3).

Table 3.3 Iodine Test for Starch

	Starch
Iodine solution	Blue-black

Experimental Procedure: Test for Starch

1. Label five clean test tubes (1 to 5).
2. Using the designated graduated transfer pipets, add 1 ml of the experimental solutions listed in Table 3.4 to the test tubes according to their numbers.
3. Then add five drops of iodine solution to the tubes at the same time.
4. Note the final color changes and record your observations in Table 3.4.

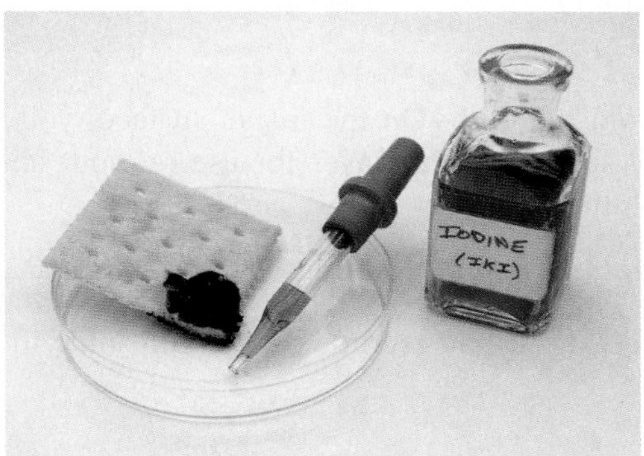

Iodine test for starch
©Martin Shields/Science Source

Table 3.4 Iodine (IKI) Test for Starch

Tube	Contents	Color	Conclusions
1	Water	Dark red	
2	Starch suspension	Black	
3	Onion juice	Pale brown	
4	Potato juice	light brown	
5	Glucose solution	light brown	

Conclusions: Starch

- From your test results, draw conclusions about what organic compound is present in each tube. Write these conclusions in Table 3.4.
- Does the potato or the onion store glucose as starch? __________ How do you know? ________________

  ____
- If your results are not as expected, offer an explanation. Then inform your instructor, who will advise you how to proceed.

Experimental Procedure: Microscopic Study

Potato

1. With a scalpel, slice a very thin piece of potato. Place it on a microscope slide, add a drop of water and a coverslip, and observe under low power with your compound light microscope. Compare your slide with the photomicrograph of starch granules (see Fig. 3.3*a*). Find the cell wall (large, geometric compartments) and the starch grains (numerous clear, oval-shaped objects).
2. Without removing the coverslip, place two drops of iodine solution onto the microscope slide so that the iodine touches the coverslip. Draw the iodine under the coverslip by placing a small piece of paper towel in contact with the water on the **opposite** side of the coverslip.
3. Microscopically examine the potato again on the side closest to where the iodine solution was applied.

   What is the color of the small, oval bodies? ______________________________

   What is the chemical composition of these oval bodies? ______________________________

Onion

1. Peel a single layer of onion from the bulb. On the inside surface, you will find a thin, transparent layer of onion skin. Peel off a small section of this layer for use on your slide.
2. Add a large drop of iodine solution.
3. Does onion contain starch? ______________________________
4. Are these results consistent with those you recorded for onion juice in Table 3.4? ______________________________

Test for Sugars

> ⚠ **Benedict's reagent** Benedict's reagent is highly corrosive. Exercise care in using this chemical. If any should spill on your skin, wash the area with mild soap and water. Follow your instructor's directions for disposal of this chemical.

Monosaccharides and some disaccharides will react with **Benedict's reagent** after being heated in a boiling water bath. In this reaction, copper ion (Cu^{2+}) in the Benedict's reagent reacts with part of the sugar molecule, causing a distinctive color change. The color change can range from green to red, and increasing concentrations of sugar will give a continuum of colored products (Table 3.5).

Table 3.5 Benedict's Test for Sugars (Some Typical Reactions)

Chemical	Chemical Category	Benedict's Reagent (After Heating)
Water	Inorganic	Blue (no change)
Glucose	Monosaccharide (carbohydrate)	Varies with concentration: very low—green low—yellow moderate—yellow-orange high—orange very high—orange-red
Maltose	Disaccharide (carbohydrate)	Varies with concentration—see "Glucose"
Starch	Polysaccharide (carbohydrate)	Blue (no change)

Benedict's test for sugar

Experimental Procedure: Test for Sugars

1. Prepare a boiling water bath and label five clean test tubes (1 to 5).
2. Using the designated graduated transfer pipets, add 1 ml of the experimental solutions listed in Table 3.6 to the test tubes according to their numbers.
3. Then add five drops of Benedict's reagent to all the tubes at this time.
4. Place the tubes into the boiling water bath at the same time.
5. When, after a few minutes, you see a change of colors, remove all of the tubes from the water bath and record your observations in Table 3.6.
6. Save your tubes for comparison purposes when you do section 3.4.

Table 3.6 Benedict's Test for Sugars

Tube	Contents	Color (After Heating)	Conclusions
1	Water	Clear light blue	
2	Glucose solution	light glucose	
3	Starch suspension	light blue	
4	Onion juice	light brown	
5	Potato juice	Sickly Green	low Glucose

Conclusions: Sugars

- From your test results, conclude what kind of chemical is present. Enter your conclusions in Table 3.6.
- Which tube served as a negative and which as a positive control? ______________________

- Compare Table 3.4 with Table 3.6. Sugars are an immediate energy source in cells. In plant cells, glucose (a primary energy molecule) is often stored in the form of starch. Is glucose stored as starch in the potato? __________ Is glucose stored as starch in the onion? __________ Does this explain your results in Table 3.6? __________ Why? ____
____

3.3 Lipids

Lipids are compounds that are insoluble in water and soluble in solvents, such as alcohol and ether. Lipids include fats, oils, phospholipids, steroids, and cholesterol. Typically, **fat,** such as in the adipose tissue of animals, and **oils,** such as the vegetable oils from plants, are composed of three molecules of fatty acids bonded to one molecule of glycerol (Fig. 3.4). **Phospholipids** have the same structure as fats, except that in place of the third fatty acid there is a phosphate group (a grouping that contains phosphate). **Steroids** are derived from **cholesterol.** These molecules have skeletons of four fused rings of carbon atoms, but they differ by functional groups (attached side chains). Fat, as we know, is long-term stored energy in the human body. Phospholipids are found in the plasma membrane of a cell. Cholesterol, a molecule transported in the blood, has been implicated in causing cardiovascular disease. Regardless, steroids are very important compounds in the body. For example, the sex hormones, like testosterone and estrogen, are steroids.

Figure 3.4 Formation of a fat.
During a dehydration reaction, a fat molecule forms when glycerol joins with three fatty acids as three water molecules are removed. During a hydrolysis reaction, water is added, and the bonds are broken between glycerol and the three fatty acids.

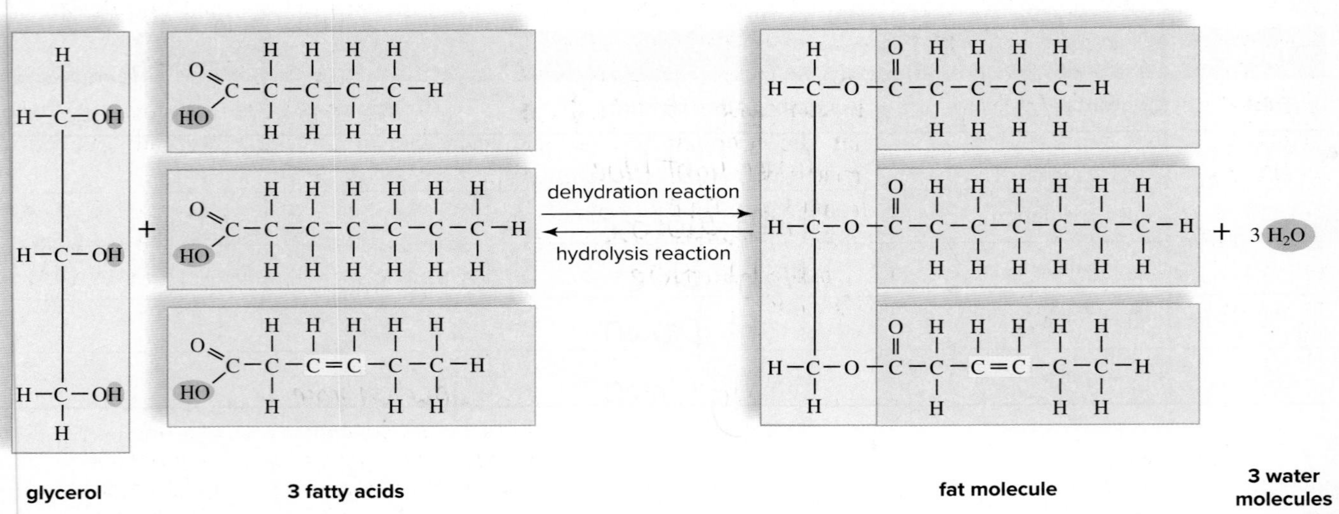

Test for Fat

Fats and oils do not evaporate from brown paper or loose-leaf paper; instead, they leave an oily spot.

Experimental Procedure: Paper Test for Fat

1. Place a small drop of water on a square of brown paper or loose-leaf paper. Describe the immediate effect. ______________________________
2. Place a small drop of vegetable oil on a square of the paper. Describe the immediate effect. ______________________________
3. Wait at least 15 minutes for the paper to dry. Evaluate which substance penetrates the paper and which is subject to evaporation. Record your observations and conclusions in Table 3.7. Save the paper for comparison use with section 3.4.

Table 3.7 Paper Test for Fat

Sample	Observations	Conclusions
Water spot		
Oil spot		

Emulsification of Oil

Some molecules are **polar,** meaning that they have charged groups or atoms, and some are **nonpolar,** meaning that they have no charged groups or atoms. A water molecule is polar, and therefore, water is a good solvent for other polar molecules. When the charged ends of water molecules interact with the charged groups of polar molecules, these polar molecules disperse in water.

Water is not a good solvent for nonpolar molecules, such as fats. A fat has no polar groups to interact with water molecules. An **emulsifier,** however, can cause a fat to disperse in water. An emulsifier contains molecules with both polar and nonpolar ends. When the nonpolar ends interact with the fat and the polar ends interact with the water molecules, the fat disperses in water, and an **emulsion** results (Fig. 3.5).

Bile salts (emulsifiers found in bile produced by the liver) are used in the digestive tract. Today milk, such as 1% milk, has been homogenized so that fat droplets do not congregate and rise to the top of the container. Homogenization requires the addition of natural emulsifiers such as phospholipids—the phosphate part of the molecule is polar and the lipid portion is nonpolar.

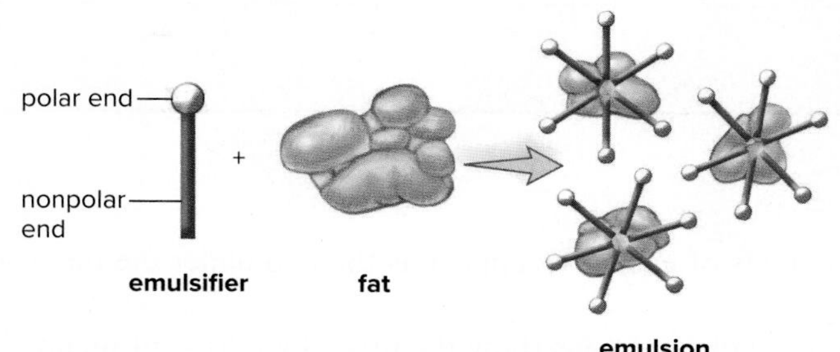

Figure 3.5 Emulsification. An emulsifier contains molecules with both a polar and a nonpolar end. The nonpolar ends are attracted to the nonpolar fat, and the polar ends are attracted to the water. This causes droplets of fat molecules to disperse.

Experimental Procedure: Emulsification of Lipids

Label three clean test tubes (1 to 3), and use the appropriate graduated pipet to add solutions to the test tubes as follows:

Tube 1

1. Add 3 ml of water and 1 ml of vegetable oil. Shake.
2. Observe for the initial dispersal of oil, followed by rapid separation into two layers. Is vegetable oil soluble in water? ______
3. Let the tube settle for 5 minutes. Label a microscope slide as 1.
4. Use a dropper to remove a sample of the solution that is just below the layer of oil. Place the drop on the slide, add a coverslip, and examine with the low power of your compound light microscope.
5. Record your observations in Table 3.8.

Tube 2

1. Add 2 ml of water, 3 ml of vegetable oil, and 1 ml of the available emulsifier (Tween or bile salts). Shake.
2. Describe how the distribution of oil in tube 2 compares with the distribution in tube 1. ______

______

______
3. Let the tube settle for 5 minutes. Label a microscope slide as 2.
4. Use a different dropper to remove a sample of the solution that is just below the layer of oil. Place the drop on the slide, add a coverslip, and examine with the low power of your compound light microscope.
5. Record your observations in Table 3.8.

Tube 3

1. Add 1 ml of milk and 2 ml of water. Shake well.
2. Use a different dropper to remove a sample of the solution that is just below the layer of oil. Place the drop on the slide, add a coverslip, and examine with the low power of your compound light microscope.
3. Record your observations in Table 3.8.

Table 3.8 Emulsification

Tube	Contents	Observations	Conclusions
1	Oil Water		
2	Oil Water Emulsifier		
3	Milk Water		

Conclusions: Emulsification

- From your observations, conclude why the contents of each tube appear as they do under the microscope. Record your conclusions in Table 3.8.
- Explain the correlation between your macroscopic observations (how the tubes look to your unaided eye) and your microscopic observations. ______

______

3.4 Testing Foods and Unknowns

It is common for us to associate the term *organic* with the foods we eat, including carbohydrate foods (Fig. 3.6), protein foods (Fig. 3.7), and lipid foods (Fig. 3.8). Though we may recognize foods as being organic, often we are not aware of what specific types of compounds are found in what we eat. In the following Experimental Procedure, you will use the same tests you used previously to determine the composition of everyday foods and unknowns.

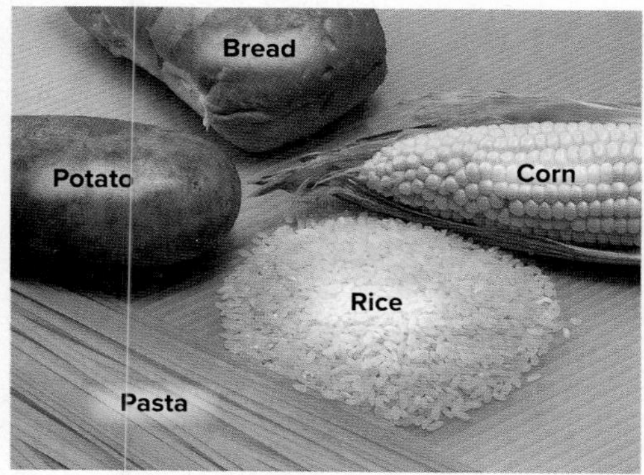

Figure 3.6 Carbohydrate foods.

3.6–3.8 ©McGraw-Hill Education/John Thoeming, photographer

Figure 3.7 Protein foods.

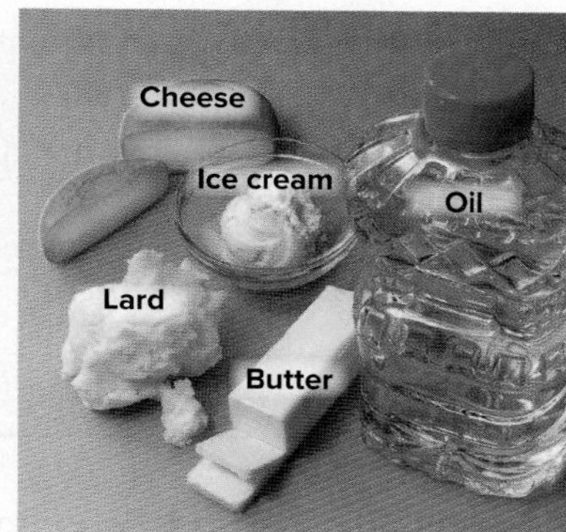

Figure 3.8 Lipid foods.

Experimental Procedure: Testing Foods and Unknowns

Your instructor will provide you with several everyday foods including unknowns, and your task is to

1. State how you will test substances for protein, carbohydrates, and fat.

 ____

 ____

 ____

 ____

2. Have your instructor okay your procedures, and then conduct the necessary tests.
3. Record your results as positive (+) or negative (–) in Table 3.9.

Table 3.9 Testing Foods and Unknowns

Sample Name	Protein (Biuret)	Starch (Iodine)	Sugar (Benedict's)	Fat (Brown or loose-leaf paper)
Unknown A				
Unknown B				

Conclusions: Testing Foods and Unknowns

- What foods tested positive for only one of the organic compounds? ____________________
- What does more than one positive test tell you about these foods? ____________________

Laboratory Review 3

Hydrolysis **1.** What kind of reaction adds water to break large biomolecules into subunits?

Enzymes **2.** What kind of protein is lactase, the biomolecule that speeds up the breakdown of lactose?

dipeptide **3.** What kind of bond forms during a dehydration synthesis reaction involving two amino acids?

nucleotides **4.** If maltose undergoes hydrolysis, what subunits result?

Starch **5.** As what molecule do plants store glucose?

Black-blue **6.** If starch is present, what color will the iodine turn?

Blue **7.** If a solution contains a large amount of glucose, what color will the Benedict's reagent become?

lipids **8.** What kinds of biomolecules are insoluble in water but soluble in alcohol or ether?

________ **9.** If two fatty acids, glycerol, and a phosphate group undergo a dehydration reaction, what biomolecule forms?

________ **10.** What is the name used for plant fats?

~~Blue~~ Bile salts **11.** What must be present to successfully break down fats during digestion?

Protein **12.** If you test an unknown substance with the biuret reagent and it turns purple, what is present?

it will become dark blue **13.** If you test a sample of potato with the biuret reagent, what do you expect the results to be?

Benedicts solution **14.** If you want to test a sample to see if glucose is present, what reagent should you use?

Thought Questions

15. If amylase is added to a sample of potato and you test it with Benedict's reagent 10 minutes later, what do you expect the results of the test to be? Justify your reply.

16. After enjoying a meal of fish and chips, you notice your shirt has an oily stain on it. Why won't the oily stain come out when you dab at it with just water?

17. If the sample of water you test with biuret reagent turns purple, what should you conclude about the results of your experiment? Explain.

LABORATORY

4

Cell Structure and Function

Learning Outcomes

4.1 Prokaryotic Versus Eukaryotic Cells

- Distinguish between prokaryotic and eukaryotic cells by description and examples.

4.2 Animal Cell and Plant Cell Structure

- Label an animal cell diagram, and state a function for the structures labeled.
- Label a plant cell diagram, and state a function for the unique structures labeled.
- Use microscopic techniques to observe plant cell structure.

4.3 Diffusion

- Define diffusion, and describe the process of diffusion as affected by the medium.
- Predict and observe which substances will or will not diffuse across a plasma membrane.

4.4 Osmosis: Diffusion of Water Across Plasma Membrane

- Define osmosis, and explain the movement of water across a membrane.
- Define isotonic, hypertonic, and hypotonic solutions, and give examples in terms of NaCl concentrations.
- Predict the effect of different tonicities on animal (e.g., red blood) cells and on plant (e.g., *Elodea*) cells.

4.5 pH and Cells

- Predict the change in pH before and after the addition of an acid to nonbuffered and buffered solutions.
- Explain the pH scale and predict a method by which it is possible to test the effectiveness of antacid medications.

Introduction

The basic units of life are cells. The **cell theory** states that all organisms are composed of cells and that cells come only from other cells. While we are accustomed to considering the heart, the liver, or the intestines as enabling the human body to function, it is actually cells that do the work of these organs.

Figure 3.8 shows human cheek epithelial cells as viewed by an ordinary compound light microscope available in general biology laboratories. It shows that the content of a cell, called the **cytoplasm,** is enclosed by a **plasma membrane.** The plasma membrane regulates the movement of molecules into and out of the cytoplasm. In this lab, we will study how the passage of water into a cell depends on the difference in concentration of solutes (particles) between the cytoplasm and the surrounding medium or solution. The well-being of cells also depends upon the pH of the solution surrounding them. We will see how a buffer can maintain the pH within a narrow range and how buffers within cells can protect them against damaging pH changes.

> **Planning Ahead** To save time, your instructor may have you start a boiling water bath (page 51) and the potato strip experiment (page 55) at the beginning of the laboratory.

Because a photomicrograph shows only a minimal amount of detail, it is necessary to turn to the electron microscope to study the contents of a cell in greater depth. The models of plant and animal cells available in the laboratory today are based on electron micrographs.

4.1 Prokaryotic Versus Eukaryotic Cells

All living cells are classified as either prokaryotic or eukaryotic. One of the basic differences between the two types is that **prokaryotic cells** do not contain nuclei (*pro* means "before"; *karyote* means "nucleus"), while eukaryotic cells do contain nuclei (*eu* means "true"; *karyote* means "nucleus"). Only bacteria (including cyanobacteria) and archaea are prokaryotes; all other organisms are eukaryotes.

Prokaryotes also don't have the organelles found in **eukaryotic cells** (Fig. 4.1). **Organelles** are small, membranous bodies, each with a specific structure and function. Prokaryotes do have **cytoplasm,** the material enclosed by a plasma membrane and cell wall. The cytoplasm contains ribosomes, small granules that coordinate the synthesis of proteins; thylakoids (only in cyanobacteria) that participate in photosynthesis; and innumerable enzymes. Prokaryotes also have a nucleoid, a region in the bacterial cell interior in which the DNA is physically organized but not enclosed by a membrane.

Figure 4.1 Prokaryotic cell.
Prokaryotic cells lack membrane-bound organelles, as well as a nucleus. Their DNA is in a nucleoid region.
©Sercomi/Science Source

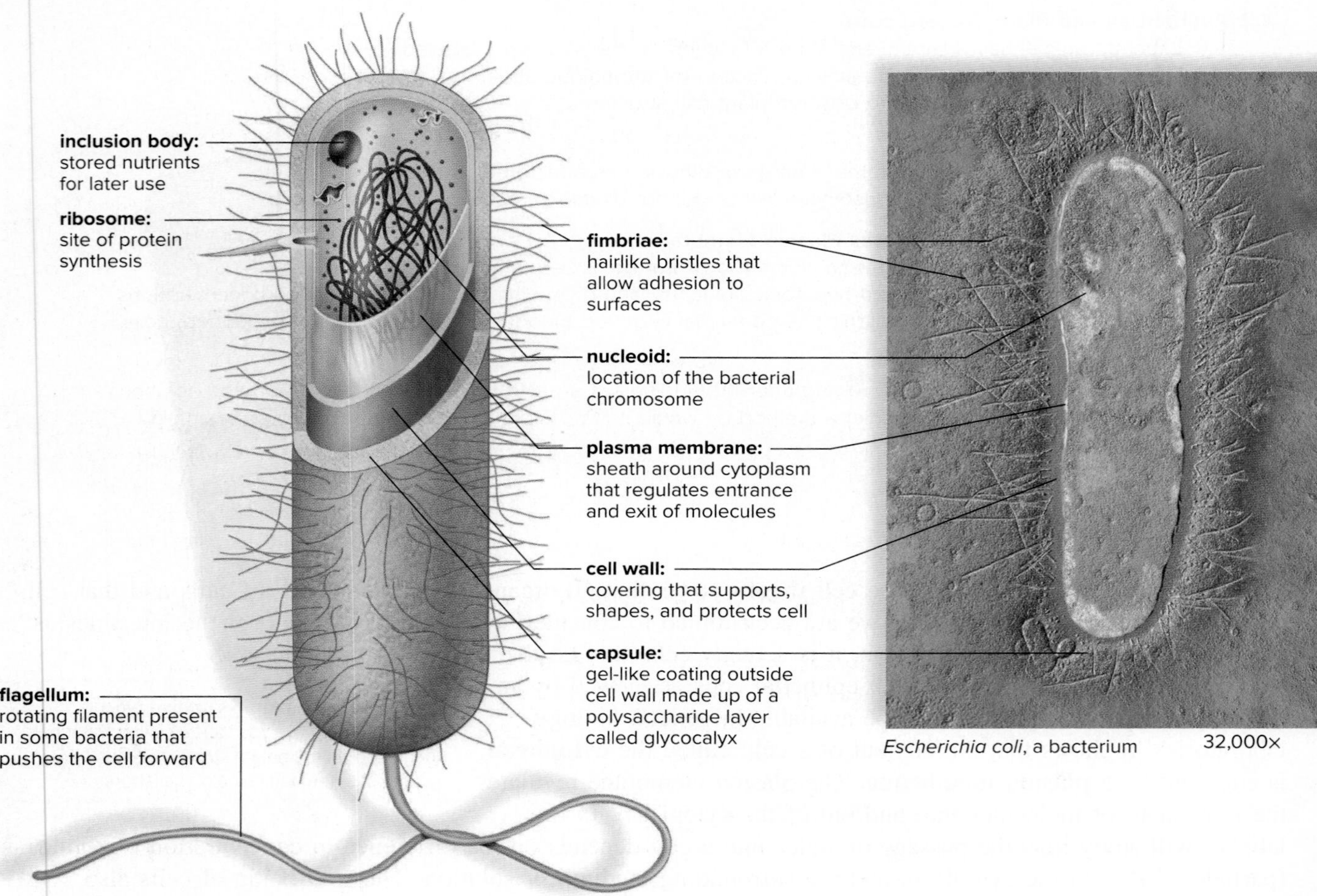

inclusion body: stored nutrients for later use
ribosome: site of protein synthesis
fimbriae: hairlike bristles that allow adhesion to surfaces
nucleoid: location of the bacterial chromosome
plasma membrane: sheath around cytoplasm that regulates entrance and exit of molecules
cell wall: covering that supports, shapes, and protects cell
capsule: gel-like coating outside cell wall made up of a polysaccharide layer called glycocalyx
flagellum: rotating filament present in some bacteria that pushes the cell forward
Escherichia coli, a bacterium
32,000×

Observation: Prokaryotic/Eukaryotic Cells

Two microscope slides on display will show you the main difference between prokaryotic and eukaryotic cells.

1. Examine a prepared slide of a bacterium. Is there a nucleus visible? ____________________
2. Examine a prepared slide of cuboidal cells from a human kidney. Is there a nucleus visible?
____

4.2 Animal Cell and Plant Cell Structure

Table 4.1 lists the structures found in animal and plant cells. The **nucleus** in a eukaryotic cell is enclosed by a **nuclear envelope** and contains **nucleoplasm.** The *cytoplasm,* found between the plasma membrane and the nucleus, consists of a background fluid and the organelles, such as the nucleolus, endoplasmic reticulum, Golgi apparatus, vacuoles and vesicles, lysosomes, peroxisome, mitochondrion, and chloroplast.

Table 4.1 Eukaryotic Structures in Animal Cells and Plant Cells

Name	Composition	Function
Cell wall*	Contains cellulose fibrils	Provides support and protection
Plasma membrane	Phospholipid bilayer with embedded proteins	Outer cell surface that regulates entrance and exit of molecules
Nucleus	Enclosed by nuclear envelope; contains chromatin (threads of DNA and protein)	Storage of genetic information; synthesis of DNA and RNA
Nucleolus	Concentrated area of chromatin	Produces subunits of ribosomes
Ribosome	Protein and RNA in two subunits	Coordinates protein synthesis
Endoplasmic reticulum (ER)	Membranous, flattened channels and tubular canals; rough ER and smooth ER	Synthesizes and/or modifies proteins and other substances; transport by vesicle formation
Rough ER	Studded with ribosomes	Protein synthesis
Smooth ER	Lacks ribosomes	Synthesis of lipid molecules
Golgi apparatus	Stack of membranous saccules	Processes, packages, and distributes proteins and lipids
Vesicle/vacuole	Membrane-bound sac	Stores and transports substances
Lysosome	Vesicle containing hydrolytic enzymes	Digests macromolecules and cell parts
Peroxisome	Vesicle containing specific enzymes	Breaks down fatty acids and converts resulting hydrogen peroxide to water; various other functions
Mitochondrion	Membranous cristae enclosed by double membrane	Carries out cellular respiration Carries out producing energy (ATP molecules)
Chloroplast*	Membranous thylakoids enclosed by double membrane	Carries out photosynthesis, producing sugars
Cytoskeleton	Microtubules, intermediate filaments, actin filaments	Maintains cell shape and assists movement of cell parts
Cilia and flagella	Attachments supported by microtubules	Movement of cell, substances
Centrioles in centrosome**	Microtubule-containing, cylindrically shaped organelle in a structure of complex composition	Centrioles organize microtubules in cilia and flagella; centrosome organizes microtubules in cell

*Plant cells only

**Animal cells only

Study Table 4.1 to determine structures that are unique to plant cells and unique to animal cells, and write them below the examples given.

	Plant Cells	**Animal Cells**
Unique structures:	1. Large central vacuole	1. Small vacuoles
	2. ____________	2. ____________
	3. ____________	

Animal Cell Structure

Label Figure 4.2. With the help of Table 4.1, give a function for each labeled structure.

Structure **Function**

Plasma membrane ______________________________

Nucleus ______________________________

Nucleolus ______________________________

Ribosome ______________________________

Endoplasmic reticulum ______________________________

Rough ER ______________________________

Smooth ER ______________________________

Golgi apparatus ______________________________

Vesicles ______________________________

Lysosome ______________________________

Mitochondrion ______________________________

Centrioles in centrosome ______________________________

Cytoskeleton ______________________________

Figure 4.2 Animal cell structure.

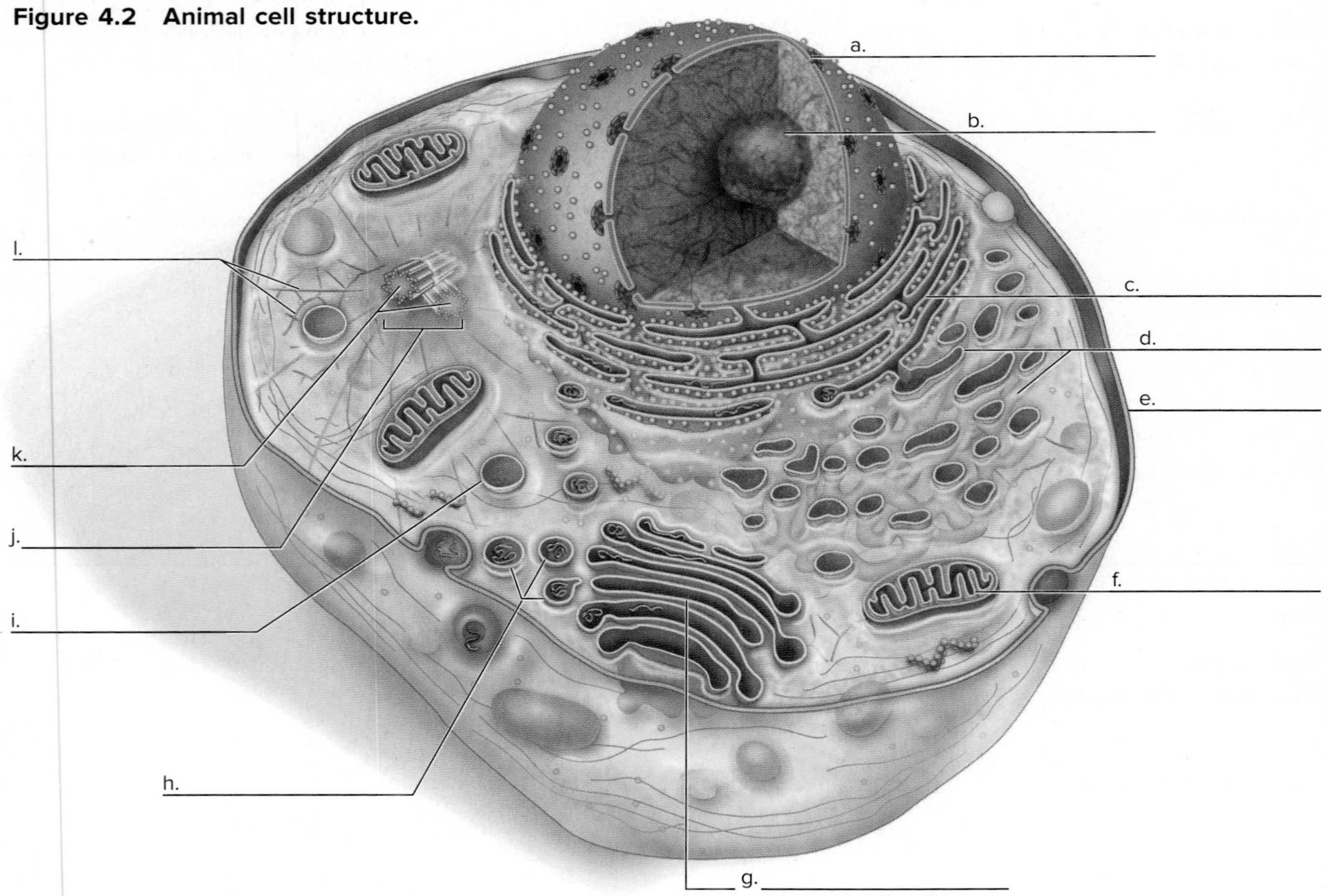

Plant Cell Structure

Label Figure 4.3. With the help of Table 4.1, give a function for each labeled structure unique to plant cells.

Structure	**Function**
Cell wall	______
Central vacuole, large	______
Chloroplast	______

Figure 4.3 Plant cell structure.

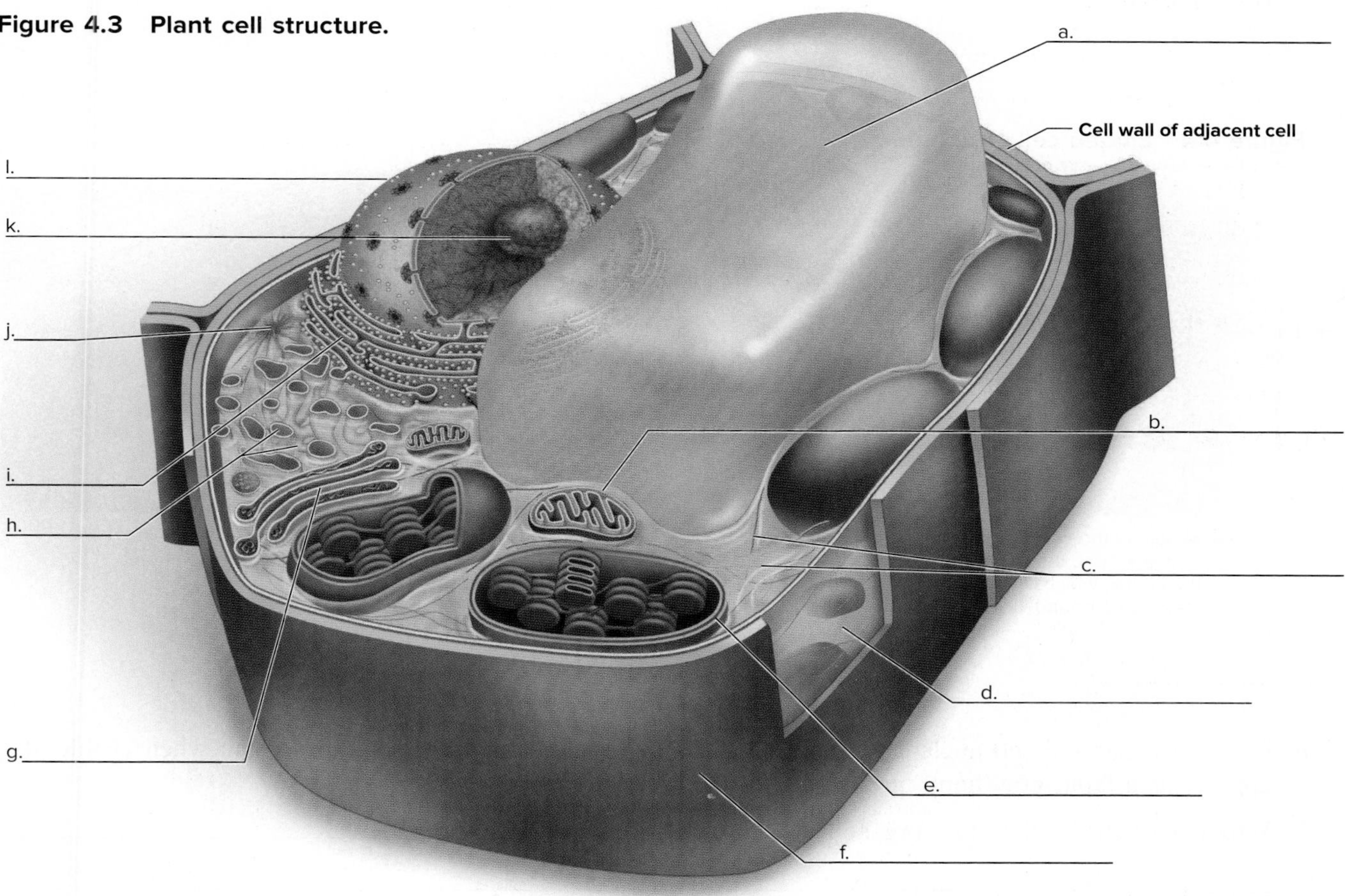

Observation: Microscopic Study of Plant Cells

1. Prepare a wet mount of a small piece of young *Elodea* leaf in fresh water. *Elodea* is a multicellular, eukaryotic plant found in freshwater ponds and lakes.
2. Have the drop of *water* ready on your slide so that the leaf does not dry out, even for a few seconds. Take care that the leaf is mounted with its top side up.
3. Examine the slide using low power, focusing sharply on the leaf surface.
4. Select a cell with numerous chloroplasts for further study, and switch to high power.
5. Carefully focus on the side and end walls of the cell. The chloroplasts appear to be only along the sides of the cell because the large, fluid-filled, membrane-bound central vacuole pushes the cytoplasm against the cell walls (Fig. 4.4*a*). Then focus on the surface and notice an even distribution of chloroplasts (Fig. 4.4*b*).

Figure 4.4 ***Elodea*** **cell structure.**
(a–b) ©Ray F. Evert/University of Wisconsin

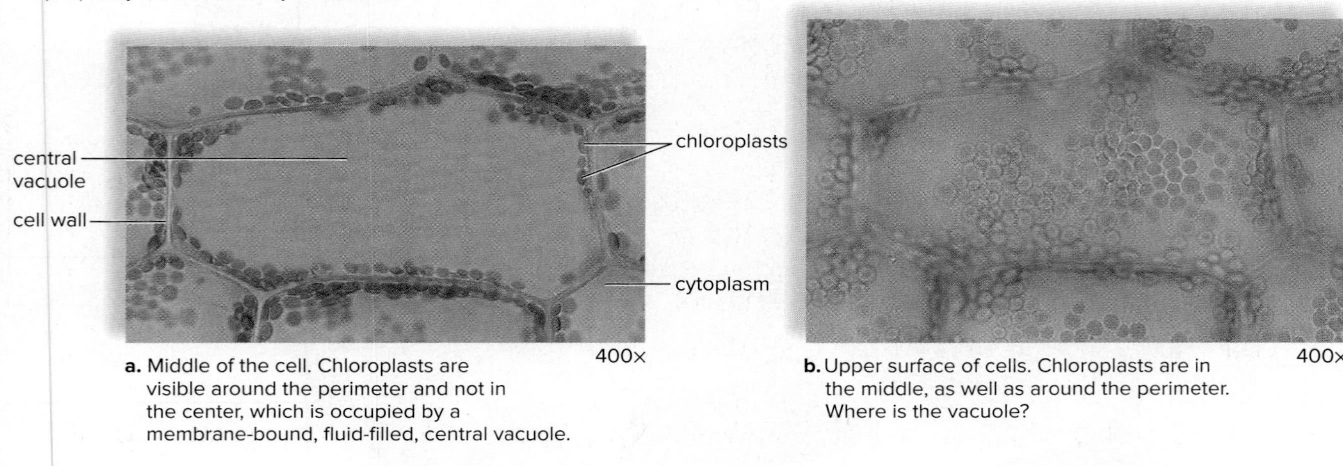

a. Middle of the cell. Chloroplasts are visible around the perimeter and not in the center, which is occupied by a membrane-bound, fluid-filled, central vacuole.

b. Upper surface of cells. Chloroplasts are in the middle, as well as around the perimeter. Where is the vacuole?

6. Can you locate the cell nucleus? __________ It may be hidden by the chloroplasts, but when visible, it appears as a faint, gray lump on one side of the cell.
7. Why can't you see the other organelles featured in Figure 4.3?______________________________________

____
8. Can you detect movement of chloroplasts in this cell or any other cell? __________ The chloroplasts are not moving under their own power but are being carried by a streaming of the nearly invisible cytoplasm.
9. Save your slide for use later in this laboratory.

4.3 Diffusion

Diffusion is the movement of molecules from a higher to a lower concentration until equilibrium is achieved and the molecules are distributed equally (Fig. 4.5). At equilibrium, molecules may still be moving back and forth, but there is no net movement in any one direction.

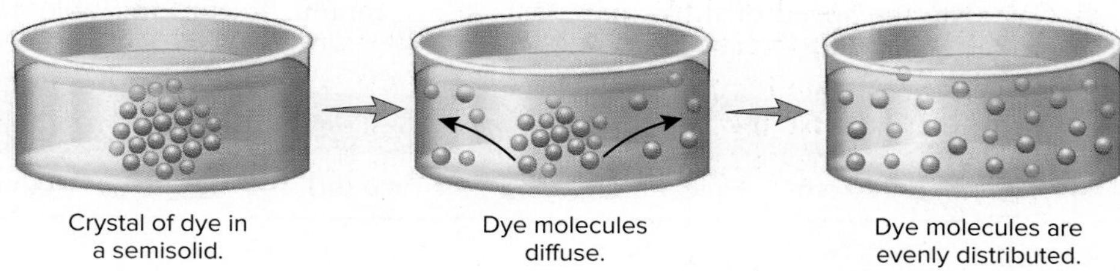
Crystal of dye in a semisolid.
Dye molecules diffuse.
Dye molecules are evenly distributed.

Figure 4.5 Process of diffusion. Diffusion is apparent when dye molecules have equally dispersed.

Diffusion is a general phenomenon in the environment. The speed of diffusion is dependent on such factors as the temperature, the size of the molecule, and the type of medium through which the molecules move.

Experimental Procedure: Speed of Diffusion

Solute Diffusion Through a Semisolid

> ⚠ **Potassium permanganate ($KMnO_4$)** $KMnO_4$ is highly poisonous and is a strong oxidizer. Avoid contact with skin and eyes and with combustible materials. If spillage occurs, wash all surfaces thoroughly. $KMnO_4$ will also stain clothing.

1. Observe a petri dish containing 1.5% gelatin (or agar) to which potassium permanganate ($KMnO_4$) was added in the center depression at the beginning of the lab.
2. Obtain time zero from your instructor, and record time zero and the final time (now) in Table 4.2. What is the length of time in hours and minutes? 15 min What is the time in hours only? ___________hrs.
3. Using a ruler placed over the petri dish, measure (in mm) the movement of color from the center of the depression outward in one direction: __________ mm.
4. Calculate the speed of diffusion: __________ mm/60 min = mm/hr.
5. Record all data in Table 4.2.

Solute Diffusion Through a Liquid

1. Add enough water to cover the bottom of a glass petri dish.
2. Place the petri dish over a thin, flat ruler. Position the petri dish directly over a mm measurement line.
3. With tweezers, add a crystal of potassium permanganate ($KMnO_4$) directly over the mm measurement line. Note time zero in Table 4.2.
4. After 10 minutes, note the distance the color has moved. Record the final time, length of time in hours, and the distance moved in Table 4.2.
5. Multiply the length of time and the distance moved by 6 to calculate the speed of diffusion: __________ mm/hr. Record in Table 4.2.

Solute Diffusion Through Air

1. Measure the distance from a spot designated by your instructor to your laboratory work area today. Record this distance under Distance Moved in Table 4.2.
2. Record time zero in Table 4.2 when a perfume or similar substance is released into the air.
3. Note the time when you can smell the perfume. Record this as the final time in Table 4.2. Calculate the length of time since the perfume was released, and record it in Table 4.2.
4. Calculate the speed of diffusion: __________ mm/hr. Record in Table 4.2.

Table 4.2 Speed of Diffusion

Medium	Time Zero	Final Time	Length of Time (hr)	Distance Moved (mm)	Speed of Diffusion (mm/hr)
Semisolid					
Liquid					
Air					

Conclusions: Solute Diffusion

- In which experiment was diffusion the fastest? ____________________
- What accounts for the difference in speed? ____________________

Solute Diffusion Across the Plasma Membrane

Some molecules can diffuse across a plasma membrane, and some cannot. In general, small, noncharged molecules can cross a membrane by simple diffusion, but large molecules cannot diffuse across a membrane. The dialysis tube membrane in the Experimental Procedure simulates a plasma membrane.

Experimental Procedure: Solute Diffusion Across Plasma Membrane

At the start of the experiment,

1. Cut a piece of dialysis tubing approximately 40 cm (approx. 16 in.) long. Soak the tubing in water until it is soft and pliable.
2. Close one end of the dialysis tubing with two knots.
3. Fill the bag halfway with glucose solution.
4. Add four full droppers of starch solution to the bag.
5. Hold the open end while you mix the contents of the dialysis bag. Rinse off the outside of the bag with distilled water.
6. Fill a beaker two-thirds full with distilled water.
7. Add droppers of iodine solution (IKI) to the water in the beaker until an amber (tealike) color is apparent.
8. Record the color of the solution in the beaker in Table 4.3.
9. Place the bag in the beaker with the open end hanging over the edge. Secure the open end of the bag to the beaker with a rubber band as shown (Fig. 4.6). Make sure the contents do not spill into the beaker.

Figure 4.6 Placement of dialysis bag in water containing iodine.

After about 5 minutes, at the end of the experiment:

> **Benedict's reagent** Exercise care in using this chemical. It is highly corrosive. If any should spill on your skin, wash the area with mild soap and water. Follow your instructor's directions for its disposal.

10. You will note a color change. Record the color of the bag contents in Table 4.3.
11. Obtain a small test tube. Using a graduated transfer pipet, draw 1 ml from the bottom of the beaker (near the bag) and place it in the test tube. Using a designated transfer pipet, add 3 ml of Benedict's solution. Heat in a boiling water bath for 5 to 10 minutes, observe any color change, and record your results as positive or negative in Table 4.3. (Optional use of glucose test strip: Dip glucose test strip into beaker. Compare stick with chart provided by instructor.)
12. Remove the dialysis bag from the beaker. Dispose of it and the used Benedict's reagent solution in the manner directed by your instructor.

Table 4.3 Solute Diffusion Across Plasma Membrane

At Start of Experiment			At End of Experiment		
	Contents	**Color**	**Color**	**Benedict's Test**	**Conclusion**
Bag	Glucose Starch			______________	
Beaker	Water Iodine				

Conclusions: Solute Diffusion Across the Plasma Membrane

- Based on the color change noted in the bag, conclude what solute diffused across the dialysis membrane from the beaker to the bag, and record your conclusion in Table 4.3.
- From the results of the Benedict's test on the beaker contents, conclude what solute diffused across the dialysis membrane from the bag to the beaker, and record your conclusion in Table 4.3.
- Which solute did not diffuse across the dialysis membrane from the bag to the beaker? ______________

  How do you know? ____

  ____

  ____

4.4 Osmosis: Diffusion of Water Across Plasma Membrane

Osmosis is the diffusion of water across the plasma membrane of a cell. Just like any other molecule, water follows its concentration gradient and moves from the area of higher concentration to the area of lower concentration.

Experimental Procedure: Osmosis

To demonstrate osmosis, a thistle tube is covered with a differentially permeable membrane at its lower opening and partially filled with 50% starch solution. The whole apparatus is placed in a beaker containing distilled water, as described in the legend for Figure 4.7. Therefore, the water concentration in the beaker is 100%. Water molecules can move freely between the thistle tube and the beaker.

Figure 4.7 Osmosis demonstration.
a. A thistle tube, covered at the broad end by a differentially permeable membrane, contains a starch solution. The beaker contains distilled water. **b.** The solute (starch) is unable to pass through the membrane (red), but the water (arrows) passes through in both directions. There is a net movement of water toward the inside of the thistle tube, where there is a higher solute (and lower water) concentration. **c.** Due to the incoming water molecules, the level of the solution rises in the thistle tube.

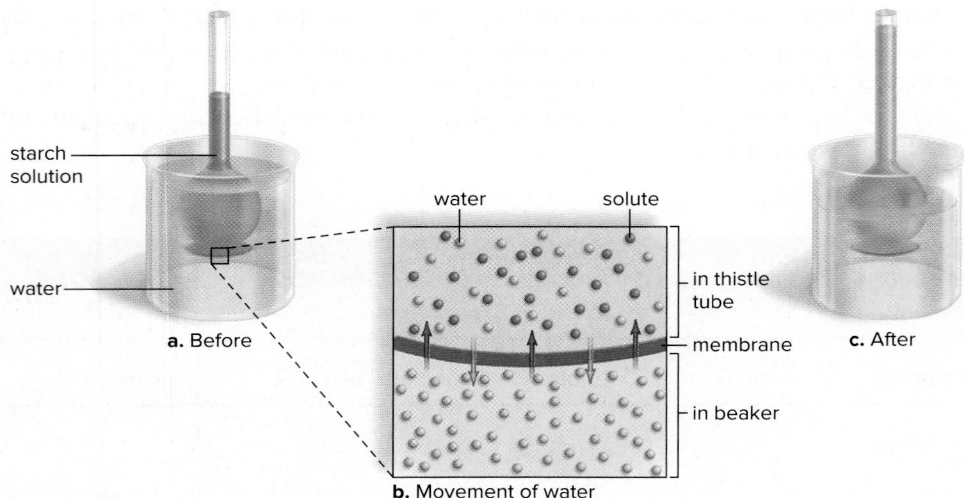

1. Note the level of liquid in the thistle tube, and measure how far it travels up the thistle tube in 10 minutes: __________ mm
2. Calculate the speed of osmosis under these conditions: __________ mm/hr

Conclusions: Osmosis

- In which direction was there a net movement of water? ______________________________
  Explain what is meant by "net movement" after examining the arrows in Figure 4.7*b*. ______________
  ____
- If the starch molecules moved from the thistle tube to the beaker, would there have been a net movement of water into the thistle tube? __________ Why wouldn't large starch molecules be able to move across the membrane from the thistle tube to the beaker? ______________________________
  ____
- Explain why the water level in the thistle tube rose: In terms of solvent concentration, water moved from the area of __________ water concentration to the area of __________ water concentration across a differentially permeable membrane.

Tonicity in Cells

Tonicity is the relative concentration of solute (particles), and therefore also of solvent (water), outside the cell compared with inside the cell.

- An **isotonic solution** has the same concentration of solute (and therefore of water) as the cell. When cells are placed in an isotonic solution, there is no net movement of water.

- A **hypertonic solution** has a higher solute (therefore, lower water) concentration than the cell. When cells are placed in a hypertonic solution, water moves out of the cell into the solution.
- A **hypotonic solution** has a lower solute (therefore, higher water) concentration than the cell. When cells are placed in a hypotonic solution, water moves from the solution into the cell.

Animal Cells (Red Blood Cells)

A solution of 0.9% NaCl is isotonic to red blood cells. In such a solution, red blood cells maintain their normal appearance (Fig. 4.8*a*). A solution greater than 0.9% NaCl is hypertonic to red blood cells. In such a solution, the cells shrivel up, a process called **crenation** (Fig. 4.8*b*). A solution of less than 0.9% NaCl is hypotonic to red blood cells. In such a solution, the cells swell to bursting, a process called **hemolysis** (Fig. 4.8*c*).

Figure 4.8 Tonicity and red blood cells.
(a–c): ©David M. Phillips/Science Source

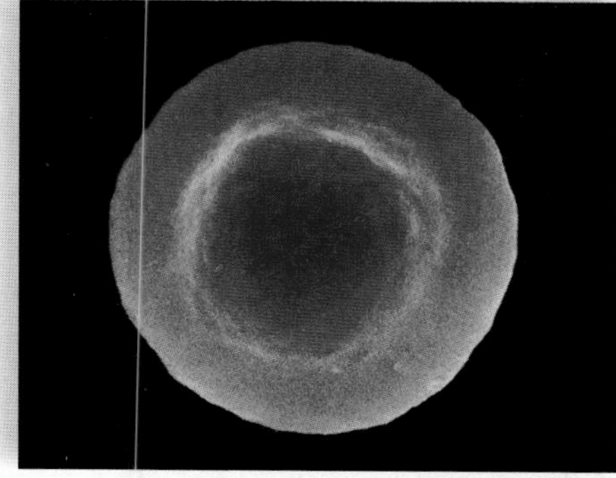
15,000×

a. Isotonic solution. Red blood cell has normal appearance due to no net gain or loss of water.

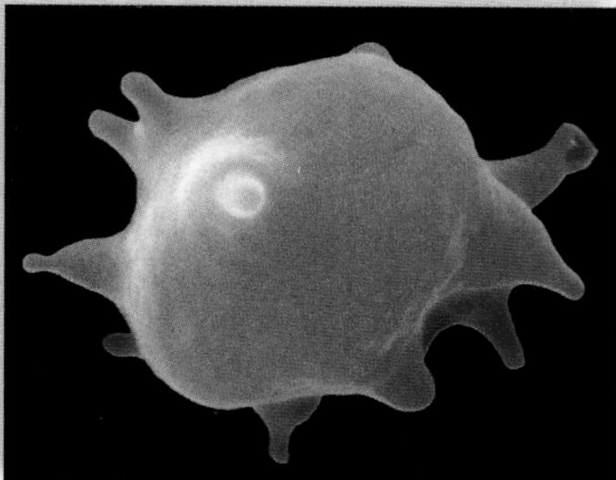
15,000×

b. Hypertonic solution. Red blood cell shrivels due to loss of water.

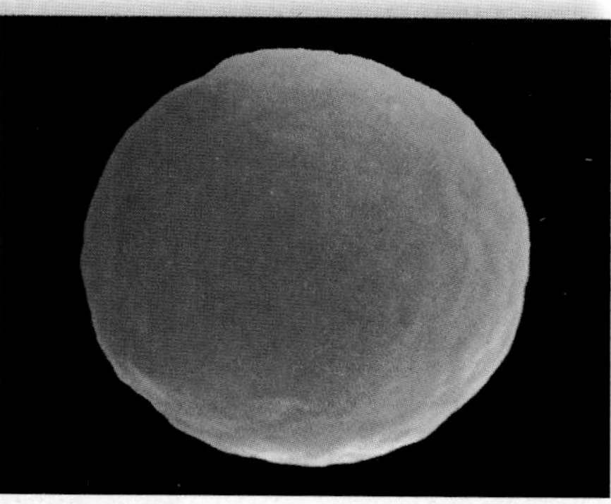
15,000×

c. Hypotonic solution. Red blood cell fills to bursting due to gain of water.

Experimental Procedure: Demonstration of Tonicity in Red Blood Cells

> ⚠ Do not remove the stoppers of test tubes during this procedure.

Three stoppered test tubes on display have the following contents:

Tube 1: 0.9% NaCl plus a few drops of whole sheep blood
Tube 2: 10% NaCl plus a few drops of whole sheep blood
Tube 3: 0.9% NaCl plus distilled water and a few drops of whole sheep blood

1. In the second column of Table 4.4, record the tonicity of each tube in relation to red blood cells.
2. Hold each tube in front of one of the pages of your lab manual. Determine whether you can see the print on the page through the tube. Record your findings in the third column of Table 4.4.
3. In the fourth column of Table 4.4, relate print visibility to the effect of tonicity on cells.

Table 4.4 Effect of Tonicity on Red Blood Cells

Tube	Tonicity	Print Visibility	Explanation
1			
2			
3			

Plant Cells

When plant cells are in a hypotonic solution, such as fresh water, the large central vacuole gains water and exerts pressure, called **turgor pressure.** The cytoplasm, including the chloroplasts, is pushed up against the cell wall. You observed turgor pressure in Figure 4.4.

When plant cells are in a hypertonic solution, such as 10% NaCl, the central vacuole loses water, and the cytoplasm, including the chloroplasts, pulls away from the cell wall. This is called **plasmolysis.** You will observe plasmolysis in the following Experimental Procedure.

Experimental Procedure: Tonicity in Elodea *Cells*

1. If possible, use the *Elodea* slide you prepared earlier in this laboratory. If not, prepare a new wet mount of a small *Elodea* leaf using fresh water. Your slide should look like Figure 4.4.
2. Complete the portion of Table 4.5 that pertains to a hypotonic solution.
3. Prepare a new wet mount of a small *Elodea* leaf using a 10% NaCl solution.
4. After several minutes, focus on the surface of the cells. Note that plasmolysis has occurred, and the cell contents are now in the center in most cells because the cytoplasm has pulled away from the cell wall due to loss of water from the large central vacuole.
5. Complete the portion of Table 4.5 that pertains to a hypertonic solution.

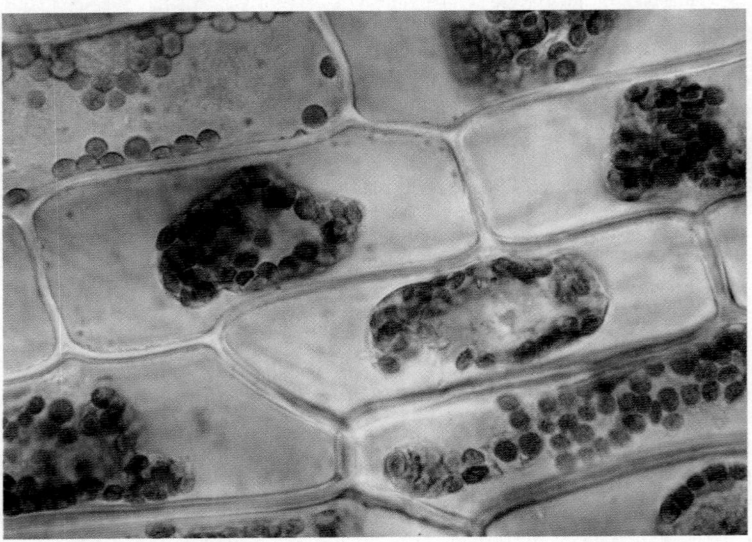

Plasmolysis is visible in most cells.

Table 4.5 Effect of Tonicity on *Elodea* Cells

Tonicity	Appearance of Cells	Due to (Scientific Term)
Hypotonic		
Hypertonic		

Experimental Procedure: Tonicity in Potato Strips

(This Experimental Procedure runs for one hour. Prior setup can maximize time efficiency.)

1. Cut two strips of potato, each about 7 cm long and 1.5 cm wide.
2. *Label two test tubes 1 and 2.* Place one potato strip in each tube.
3. Fill tube 1 with water to cover the potato strip.
4. Fill tube 2 with 10% sodium chloride (NaCl) to cover the potato strip.
5. After 1 hour, remove the potato strips from the test tubes and place them on a paper towel. Observe each strip for limpness (water loss) or stiffness (water gain). Which tube has the limp potato strip? __________ Use tonicity to explain why water diffused out of the potato strip in this tube. ____

   ____

   Which tube has the stiff potato strip? __________ Use tonicity to explain why water diffused into the potato strip in this tube. ____

   ____

6. Use this space to create a table to display your results. Give your table a title and columns for tube number and contents, tonicity, results, and explanation.

Sticker = Water

No Sticker = NaCl

Conclusions: Tonicity

- In a hypotonic solution, animal cells ________________. In red blood cells, this is called ________________. In a hypertonic solution, animal cells ________________. In red blood cells this is called ________________.
- In a hypotonic solution, the central vacuole of *Elodea* cells exerts ________________ pressure, and chloroplasts are seen ________________. In a hypertonic solution, the central vacuole loses water and ________________ occurs. The cytoplasm plus the chloroplasts are seen ________________.
- In a hypotonic solution, potato strips ________________ water and become ________________; in a hypertonic solution, potato strips ________________ water and become ________________.

4.5 pH and Cells

The pH of a solution indicates its hydrogen ion concentration $[H^+]$. The **pH scale** ranges from 0 to 14. A pH of 7 is neutral (Fig. 4.9). A pH lower than 7 indicates that the solution is acidic (has more hydrogen ions than hydroxide ions), whereas a pH greater than 7 indicates that the solution is basic (has more hydroxide ions than hydrogen ions).

The concept of pH is important in biology because organisms are very sensitive to hydrogen ion concentrations. For example, in humans the pH of the blood must be maintained at about 7.4 or we become ill. All organisms need to maintain the hydrogen ion concentration, or pH, at a constant level. A **buffer** is a system of chemicals that takes up excess hydrogen ions or hydroxide ions, as appropriate.

Why are cells and organisms buffered? ______________________________

____

Figure 4.9 The pH scale.
The proportion of hydrogen ions (H^+) to hydroxide ions (OH^-) is indicated by the diagonal line.

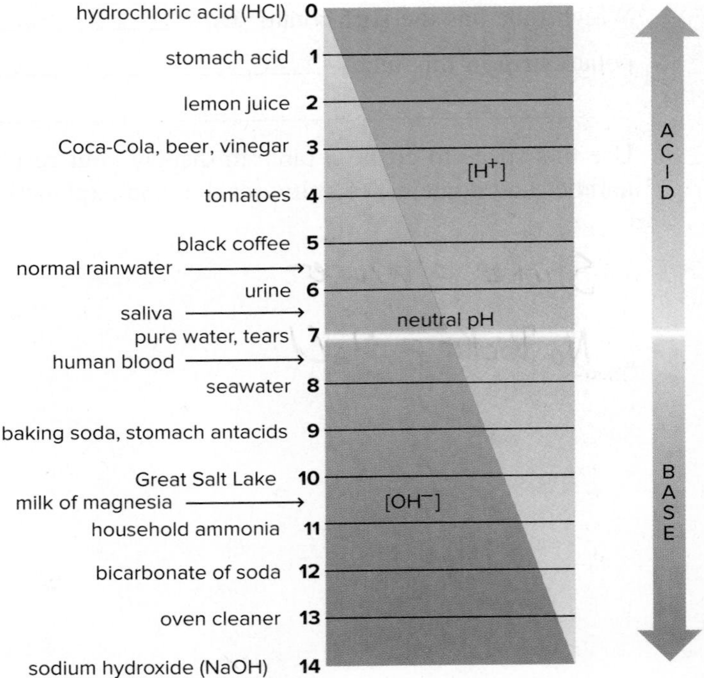

Experimental Procedure: pH and Cells

> **Hydrochloric acid (HCl)** used to produce an acid pH is a strong, caustic acid. Exercise care in using this chemical. If any HCl spills on your skin, rinse immediately with clear water. Follow your instructor's directions for disposal of tubes that contain HCl.

1. Label three test tubes, and fill them to the halfway mark as follows: tube 1: water; tube 2: buffer (a buffered inorganic solution); and tube 3: simulated cytoplasm (a buffered protein solution).
2. Use pH paper to determine the pH of each tube. Dip the end of a stirring rod into the solution, and then touch the stirring rod to a 5 cm strip of pH paper. Read the current pH by matching the color observed with the color code on the pH paper package. Record your results in the "pH Before Acid" column in Table 4.6.
3. Add 0.1 N hydrochloric acid (HCl) dropwise to each tube until you have added 5 drops—shake or swirl after each drop. Use pH paper as in step 2 to determine the new pH of each solution. Record your results in the "pH After Acid" column in Table 4.6.

Table 4.6 pH and Cells

Tube	Contents	pH Before Acid	pH After Acid	Explanation
1	Water			
2	Buffer			
3	Cytoplasm			

Conclusions: pH and Cells

- Enter your explanations in the last column of Table 4.6.
- Why would you expect cytoplasm to be as effective as the buffer in maintaining pH? ______________________________

Experimental Procedure: Effectiveness of Antacids

Perform this procedure to test the ability of commercial products such as Alka-Seltzer®, Rolaids, Tums®, or antacid tablets to absorb excess H^+.

1. Use a mortar and pestle to grind up the amount of antacid that is listed as one dose.
2. For each antacid tested, use 100 ml of phenol red solution diluted to a faint pink to wash the antacid into a 250 ml beaker. Phenol red solution is a pH indicator that turns yellow in an acid and red in a base. Use a stirring rod to get the powder to dissolve.
3. Add and count the number of 0.1 N HCl drops it takes for the solution to turn light yellow.
4. Record your results in Table 4.7.

Table 4.7 Effectiveness of Antacids

Antacid	Drops of Acid Needed to Reach End Point	Evaluation
1		
2		
3		

Conclusions: Effectiveness of Antacids

- Participate with others in concluding which of the antacids tested neutralizes the most acid.
- Did a difference in dosage (convert to mg) have any effect on the results? ______________________________
- Which of the substances on the label could be a buffer? ______________________________

Laboratory Review 4

_______________ **1.** What regulates the movement of molecules into and out of the cytoplasm?

_______________ **2.** What kinds of cells lack a nucleus?

_______________ **3.** What structures associated with prokaryotic and eukaryotic cells carry out protein synthesis?

_______________ **4.** What two things are found in the nucleus of a eukaryotic cell?

_______________ **5.** What plant cell organelle carries out photosynthesis?

_______________ **6.** What molecule is associated with plant cell walls?

_______________ **7.** What energy (ATP)–producing organelle is found in animal and plant cells?

_______________ **8.** What is the movement of molecules from an area of higher to lower concentration called?

_______________ **9.** What is the movement of water across the plasma membrane called?

_______________ **10.** Does water move into or out of cells that are placed in a hypotonic solution?

_______________ **11.** What kind of solution causes crenation to happen to red blood cells: isotonic, hypertonic, or hypotonic?

_______________ **12.** What kind of solution causes chloroplasts to be pushed outward against the cell wall: isotonic, hypertonic, or hypotonic?

_______________ **13.** If a solution has more hydrogen ions than hydroxide ions, is the solution acidic or basic?

_______________ **14.** What is present in human blood that ensures blood pH is maintained at about 7.4?

Thought Questions

15. What cellular components are common to prokaryotic and eukaryotic cells?

16. Ovarian follicle cells produce estrogen, which is a steroid. What organelle will be present in abundance so that the follicle cells might perform this function? Justify your reply.

17. Contact lens solution is described as a sterile, buffered, isotonic aqueous solution. Explain the importance of the adjectives *buffered* and *isotonic* to the person buying the solution.

LABORATORY

5

How Enzymes Function

Learning Outcomes

Introduction

- Describe how an enzyme functions.

5.1 Catalase Activity

- Identify the substrate, enzyme, and product in the reaction studied today.

5.2 Effect of Temperature on Enzyme Activity

- Predict the effect of temperature on an enzymatic reaction.

5.3 Effect of Concentration on Enzyme Activity

- Predict the effect of enzyme and substrate concentration on an enzymatic reaction.

5.4 Effect of pH on Enzyme Activity

- Predict the effect of pH on an enzymatic reaction.

5.5 Factors That Affect Enzyme Activity

- Tell how temperature, concentration of substrate or enzyme, and pH can promote or inhibit enzyme activity.

Introduction

The cell carries out many chemical reactions. A possible chemical reaction can be indicated like this:

$$\underset{\text{reactants}}{A + B} \longrightarrow \underset{\text{products}}{C + D}$$

In all chemical reactions, the **reactants** are molecules that undergo a change, which results in the **products.** The arrow stands for the change that produced the product(s). For example, A + B change and produce C + D. The numbers of reactants and products can vary; in the one you are studying today, a single reactant breaks down to two products. All the reactions that occur in a cell have an enzyme. **Enzymes** are organic catalysts that speed metabolic reactions. Because enzymes are specific and speed only one type of reaction, they are given names. In today's laboratory, you will be studying the action of the enzyme **catalase.** The reactants in an enzymatic chemical reaction are called **substrate(s)** (Fig. 5.1).

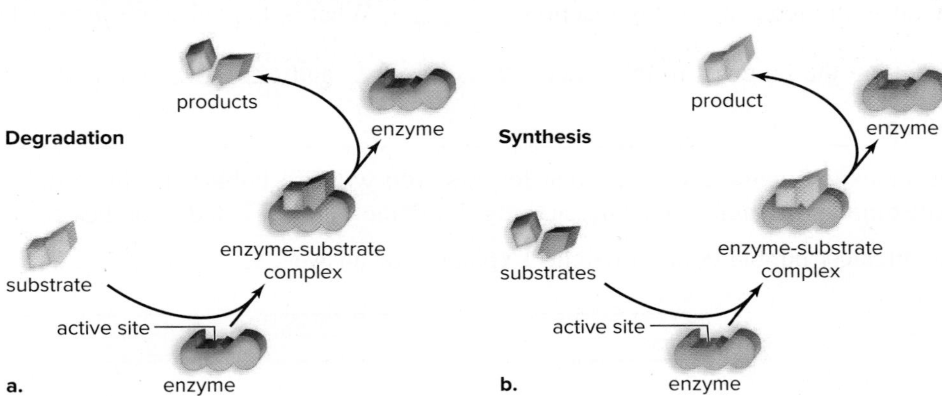

Figure 5.1 Enzymatic action. The reaction occurs on the surface of the enzyme at the active site. The enzyme is reusable. **a.** Degradation: substrate is broken down. **b.** Synthesis: substrates are combined.

Enzymes are specific because they have a shape that accommodates the shape of their substrates. Enzymatic reactions can be indicated like this:

$$E + S \longrightarrow ES \longrightarrow E + P$$

In this reaction, E = enzyme, ES = enzyme-substrate complex, and P = product.

Two types of enzymatic reactions in cells are shown in Figure 5.1. During a **degradation reaction,** the substrate is broken down to the product(s), and during a **synthesis reaction,** substrates are joined to form a product. A number of other types of reactions also occur in cells. The location where the enzyme and substrate form an enzyme-substrate complex is called the **active site** because the reaction occurs here. At the end of the reaction, the product is released, and the enzyme can then combine with its substrate again. A cell needs only a small amount of an enzyme because enzymes are used over and over. Some enzymes have turnover rates well in excess of a million product molecules per minute.

Planning Ahead To save time, your instructor may have you start a boiling water bath at the beginning of the laboratory.

5.1 Catalase Activity

Catalase is involved in a degradation reaction: Catalase speeds the breakdown of hydrogen peroxide (H_2O_2) in nearly all organisms including bacteria, plants, and animals. A cellular organelle called a peroxisome, which contains catalase, is present in every plant and animal organ. This means that we could use any plant or animal organ as our source of catalase today. Commonly, school laboratories use the potato as a source of catalase because potatoes are easily obtained and cut up.

Catalase performs a useful function in organisms because hydrogen peroxide is harmful to cells. Hydrogen peroxide is a powerful oxidizer that can attack and denature cellular molecules like DNA! Knowing its harmful nature, humans use hydrogen peroxide as a commercial antiseptic to kill germs (Fig. 5.2). In reduced concentration, hydrogen peroxide is a whitening agent used to bleach hair and teeth. Skillful technicians use it to provide oxygen to aquatic plants and fish, but it is also used industrially to clean most anything from tubs to sewage. It's even put in glow sticks, where it reacts with a dye that then emits light.

When catalase speeds the breakdown of hydrogen peroxide, water and oxygen are released.

$$\underset{\text{hydrogen peroxide}}{2\ H_2O_2} \xrightarrow{\text{catalase}} \underset{\text{water}}{2\ H_2O} + \underset{\text{oxygen}}{O_2}$$

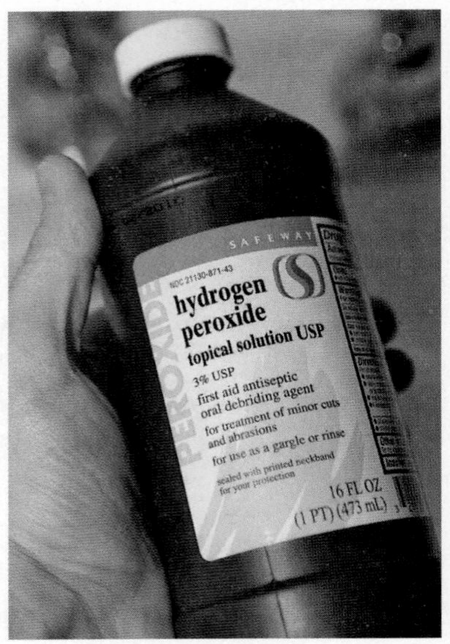

Figure 5.2 Hydrogen peroxide.
Bubbling occurs when you apply hydrogen peroxide to a cut because oxygen is being released when catalase, an enzyme present in the body's cells, degrades hydrogen peroxide.
©McGraw-Hill Education/Jill Braaten, photographer

What is the reactant in this reaction? ________ What is the substrate for catalase? ____________________

What are the products in this reaction? __________ and __________ Bubbling occurs as the reaction proceeds.

Why? ____

In the experimental procedure that follows, you will use bubble height to indicate the amount of product per unit time and therefore enzyme activity. Examine Table 5.1 and hypothesize which tube (1, 2, or 3) will have the greatest bubble column height. Explain your answer. ______________________________

____

____

Experimental Procedure: Catalase Activity

Number three clean test tubes and use the appropriate graduated transfer pipet to add solutions to the test tubes as follows:

Tube 1
1. Add 1 ml of catalase buffered at pH 7.0, the optimum pH for catalase.
2. Add 4 ml of hydrogen peroxide. Swirl well to mix, and wait at least 20 seconds for bubbling to develop.
3. Measure the height of the bubble column (in millimeters), and record your results in Table 5.1.

Tube 2
1. Add 1 ml of water.
2. Add 4 ml of hydrogen peroxide. Swirl well to mix, and wait at least 20 seconds for bubbling to develop.
3. Measure the height of the bubble column (in millimeters), and record your results in Table 5.1.

Tube 3
1. Add 1 ml of catalase buffered at pH 7.0, the optimum pH for catalase.
2. Add 4 ml of sucrose solution. Swirl well to mix, and wait at least 20 seconds for bubbling to develop.
3. Measure the height of the bubble column (in millimeters), and record your results in Table 5.1.

Table 5.1 Catalase Activity

Tube	Contents	Bubble Column Height (mm)	Explanation
1	Catalase Hydrogen peroxide		
2	Water Hydrogen peroxide		
3	Catalase Sucrose solution		

Conclusions: Catalase Activity

- Which tube showed the amount of bubbling you expected? __________ Record your explanation in Table 5.1.
- Which tube is a negative control? __________ If this tube showed bubbling, what could you conclude about your procedure? ____

  Record your explanation in Table 5.1.
- Enzymes are specific; they speed only a reaction that contains their substrate. Which tube exemplifies this characteristic of an enzyme? __________ Record your explanation in Table 5.1.

5.2 Effect of Temperature on Enzyme Activity

The active sites of enzymes increase the likelihood that substrate molecules will find each other and interact. Therefore, enzymes lower the energy of activation (the temperature needed for a reaction to occur). Still, increasing the temperature is expected to increase the likelihood that active sites will be occupied because molecules move about more rapidly as the temperature rises. In this way, a warm temperature increases enzyme activity.

The shape of an enzyme and its active site must be maintained or else they will no longer be functional. A very high temperature, such as the one that causes water to boil, is likely to cause weak bonds of a protein to break. If this occurs, the enzyme **denatures**—it loses its original shape and the active site will no longer function to bring reactants together. Now enzyme activity plummets.

With this information in mind, examine Table 5.2 and hypothesize which tube (1, 2, or 3) will have the most product per unit time as judged by bubble column height. ____ Explain your answer. ________________

____

Experimental Procedure: Effect of Temperature

Number three clean test tubes and use the appropriate graduated pipet to add solutions to the test tubes as follows:

1. To each tube add 1 ml of catalase buffered at pH 7.0, the optimum pH for catalase.
2. Place tube 1 in a refrigerator or cold water bath, tube 2 in an incubator or warm water bath, and tube 3 in a boiling water bath. Complete the second column in Table 5.2. Wait 15 minutes.
3. As soon as you remove the tubes one at a time from the refrigerator, incubator, and boiling water, add 4 ml of hydrogen peroxide.
4. Swirl well to mix, and wait 20 seconds.
5. Measure the height of the bubble column (in millimeters) in each tube, and record your results in Table 5.2. Plot your results in Figure 5.3.

Table 5.2 Effect of Temperature

Tube	Temperature °C	Bubble Column Height (mm)	Explanation
1 Refrigerator			
2 Incubator			
3 Boiling water			

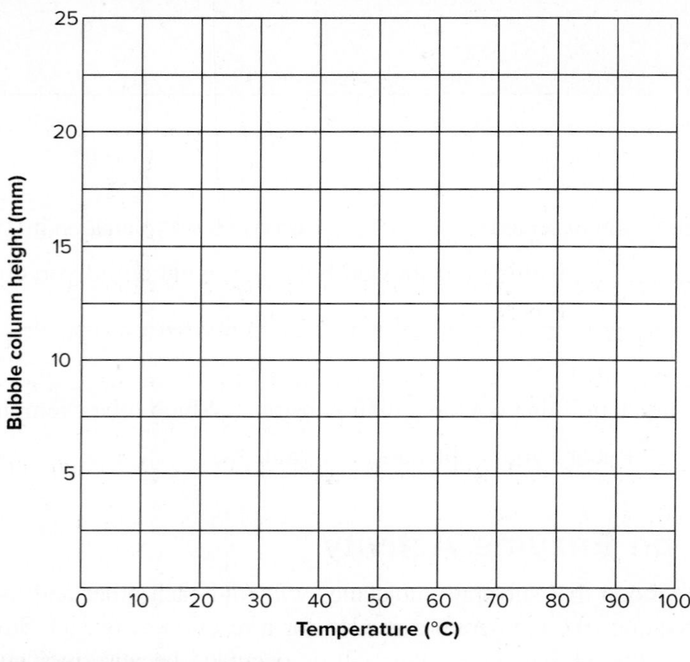

Figure 5.3 Effect of temperature on enzyme activity.

Conclusions: Effect of Temperature

- The bubble column height indicates the degree of enzyme activity. Was your hypothesis supported? ________ Explain in Table 5.2 the degree of enzyme activity per tube.
- What is your conclusion concerning the effect of temperature on enzyme activity? ________________

____

5.3 Effect of Concentration on Enzyme Activity

Consider that if you increase the number of caretakers per number of children, it is more likely that each child will have quality time with a caretaker. So it is if you increase the amount of enzyme per amount of substrate: It is more likely that substrates will find the active site of an enzyme and a reaction will take place. With this in mind, examine Table 5.3 and hypothesize which tube (1, 2, or 3) will have the most product per unit time as judged by bubble column height. ____________ Explain your answer. ____________________________

____

Experimental Procedure: Effect of Enzyme Concentration

Number three clean test tubes and use the appropriate transfer pipet to add solutions to the test tubes as follows:

1. To each tube, add 4 ml of hydrogen peroxide.
2. To tube 1, add 1 ml of water.
3. To tube 2, add 1 ml of buffered catalase.
4. To tube 3, add 3 ml of buffered catalase.
5. Swirl well to mix before measuring the height of the bubble column.
6. Measure the height of the bubble columns, and record your results in Table 5.3.

Table 5.3 Effect of Enzyme Concentration

Tube	Amount of Enzyme	Bubble Column Height (mm)	Explanation
1	none		
2	1 ml		
3	3 ml		

Conclusions: Effect of Concentration

- The bubble column height indicates the degree of enzyme activity. Was your hypothesis supported? ____________ Explain in Table 5.3 the degree of enzyme activity per tube.
- If unlimited time were allotted, would the results be the same in all tubes? ____________________ Explain why or why not. ____
- Would you expect similar results if the substrate concentration were varied in the same manner as the enzyme concentration? __________ Why or why not? ____________________________
- What is your conclusion concerning the effect of concentration on enzyme activity? ______________

  ____

5.4 Effect of pH on Enzyme Activity

> ⚠ **Hydrochloric acid (HCl)** used to produce an acid pH is a strong, caustic acid, and sodium hydroxide (NaOH) used to produce a basic pH is a strong, caustic base. Exercise care in using these chemicals, and follow your instructor's directions for disposal of tubes that contain these chemicals. If any acidic or basic solutions spill on your skin, rinse immediately with clear water.

Each enzyme has a pH at which the speed of the reaction is optimum (occurs best). Any higher or lower pH affects hydrogen bonding and the structure of the enzyme, leading to reduced activity.

Catalase is an enzyme found in cells where the pH is near 7 (called neutral pH). Other enzymes prefer different pHs. The pancreas secretes a slightly basic (pH 8–8.3) juice into the digestive tract, and the stomach wall releases a very acidic digestive juice, which can be as low as pH 2. With this information about catalase in mind, examine Table 5.4 and hypothesize which tube (1, 2, or 3) will have the most product per unit time as judged by the bubble column height. ________

Explain your answer. ____

____

Experimental Procedure: Effect of pH

Number three clean test tubes and use the appropriate transfer pipet to add solutions to the test tubes as follows:

1. To each tube, add 1 ml of catalase.
2. To tube 1, add 2 ml of water adjusted to pH 3.
3. To tube 2, add 2 ml of water adjusted to pH 7.
4. To tube 3, add 2 ml of water adjusted to pH 11.
5. Wait 1 minute. Now add 4 ml hydrogen peroxide to each tube and swirl to mix before noting the height of the bubble column.
6. Record your results in Table 5.4 and plot your results in Figure 5.4.

Table 5.4 Effect of pH

Tube	pH	Bubble Column Height (mm)	Explanation
1	3		
2	7		
3	11		

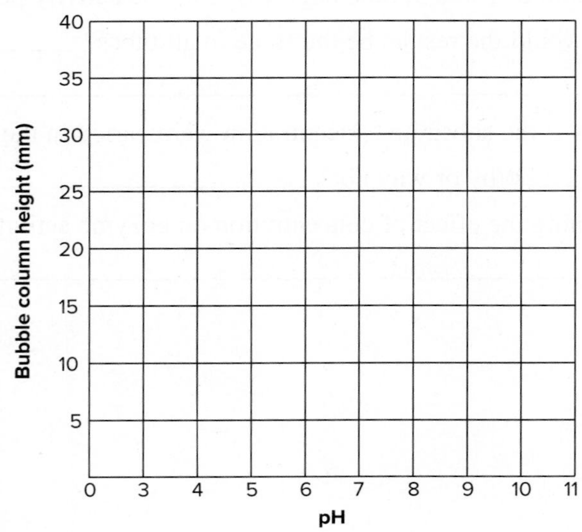

Figure 5.4 Effect of pH on enzyme activity.

Conclusions: Effect of pH

- The amount of bubble column height indicates the degree of enzyme activity. Was your hypothesis supported? ______________________ Explain in Table 5.4 the degree of enzyme activity per tube.
- What is your conclusion concerning the effect of pH on enzyme activity? ______________________

____

5.5 Factors That Affect Enzyme Activity

In Table 5.5, summarize what you have learned about factors that affect the speed of an enzymatic reaction. For example, in general, what type of temperature promotes enzyme activity, and what type inhibits enzyme activity? Answer similarly for enzyme or substrate concentration and pH.

Table 5.5 Factors That Affect Enzyme Activity

Factors	Promote Enzyme Activity	Inhibit Enzyme Activity
Enzyme specificity		
Temperature		
Enzyme or substrate concentration		
pH		

Conclusions: Factors That Affect Enzyme Activity

- Why does enzyme specificity promote enzyme activity? ______________________

____

- Why does a warm temperature promote enzyme activity? ______________________

____

- Why does increasing enzyme concentration promote enzyme activity? ______________________

____

- Why does optimum pH promote enzyme activity? ______________________

____

Laboratory Review 5

_________________ **1.** How is "produces" represented in a chemical reaction?

_________________ **2.** Are the substrates in an enzymatic reaction reactants or products?

_________________ **3.** What is the name of the location on an enzyme where its substrate or substrates bind?

_________________ **4.** What cellular organelle contains catalase and is present in all plant and animal organs?

_________________ **5.** What did measuring the bubble column heights in all the experimental procedures indicate?

_________________ **6.** What purpose did the tube containing water and hydrogen peroxide serve in the catalase activity experimental procedure?

_________________ **7.** What word describes the loss of an enzyme's original shape and ability of its active site to function?

_________________ **8.** What happens to enzyme activity if the temperature increases slightly?

_________________ **9.** What did the negative control tube lack when the effect of enzyme concentration was studied?

_________________ **10.** If cells produce more catalase, will there be more or less water and oxygen produced from the breakdown of hydrogen peroxide by catalase?

_________________ **11.** What word describes the pH at which the speed of a reaction involving an enzyme occurs best?

_________________ **12.** What do changes in pH (too low or too high) do to an enzyme that reduce its activity?

_________________ **13.** In the Experimental Procedure on the effect of pH on enzyme activity, which tube contained water with an acidic pH?

_________________ **14.** Since enzymes are specific for certain substrates and often have similar names, what is the name of the enzyme used in tube 3 of the catalase activity experiment that facilitates the breakdown of sucrose?

Thought Questions

15. People with Crohn's disease may develop lactose intolerance caused by damage to their small intestine. These people may take a commercial product called Lactaid to prevent digestive upset when consuming dairy products. What is in Lactaid that helps these people avoid those symptoms? Explain how this relates to what you learned about the specificity of enzymes for substrates in this lab.

16. High-grade fevers (in humans, 103°F or higher) are of concern, and medical care should be obtained. Explain why, based on your understanding of enzyme function.

17. Pancreatic amylase is a digestive enzyme that digests starch in the small intestine. Pancreatic amylase's optimum pH is slightly basic, which the presence of $NaHCO_3$ provides. Indicate which test tubes (1–3) would show digestion of starch following incubation. Explain why the others would not.

Tube 1 water, starch ____________________

Tube 2 water, starch, pancreatic amylase, HCl ____________________

Tube 3 water, starch, pancreatic amylase, $NaHCO_3$ ____________________

LABORATORY

6

Photosynthesis

Learning Outcomes

6.1 Photosynthetic Pigments
- Describe the separation of plant pigments by paper chromatography.

6.2 Solar Energy
- Describe how white light is composed of many colors of light and the benefit of a plant's utilizing several photosynthetic pigments.
- Explain an experiment that indicates white light promotes photosynthesis.
- Relate the effectiveness of various colors of light for photosynthesis to the various photosynthetic pigments.

6.3 Carbon Dioxide Uptake
- Describe an experiment that indicates carbon dioxide is utilized during photosynthesis.

6.4 The Light Reactions and the Calvin Cycle Reactions
- Use the overall equation for photosynthesis to relate the light reactions to the Calvin cycle reactions.

Introduction

Photosynthesis involves the use of solar energy to produce a carbohydrate:

$$CO_2 + H_2O \xrightarrow{\text{solar energy}} (CH_2O) + O_2$$

In this equation, (CH_2O) represents any general carbohydrate. Sometimes, this equation is multiplied by 6 so that glucose $(C_6H_{12}O_6)$ appears as an end product of photosynthesis.

Photosynthesis takes place in chloroplasts (Fig. 6.1). Here membranous thylakoids are stacked in grana surrounded by the stroma. During the light reactions, pigments within the thylakoid membranes absorb solar energy, water is split, and oxygen is released. The Calvin cycle reactions occur within the stroma. During these reactions, carbon dioxide (CO_2) is reduced and solar energy is now stored in a carbohydrate (CH_2O).

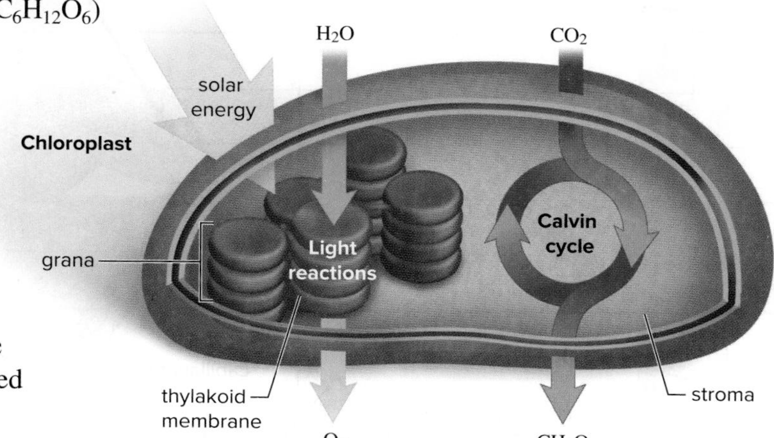

Figure 6.1 Overview of photosynthesis.
Photosynthesis includes the light reactions when energy is collected and O_2 is released and the Calvin cycle reactions when CO_2 is reduced and an energy-rich carbohydrate (CH_2O) is formed.

6.2 Solar Energy

During light reactions of photosynthesis, solar energy is absorbed by the photosynthetic pigments and is transformed into the chemical energy of a carbohydrate (CH_2O). Without solar energy, photosynthesis would be impossible.

Verify that photosynthesis releases oxygen by reviewing the overall equation for photosynthesis on page 69. The release of oxygen from a plant indicates that the light reactions of photosynthesis are occurring. The oxygen released during photosynthesis is taken up by a plant when cellular respiration occurs. This must be taken into account when the rate of photosynthesis is calculated.

Role of White Light

White (sun) light contains different colors of light, as is demonstrated when white light passes through a prism (Fig. 6.3). White light is the best for photosynthesis because it contains all the colors of light.

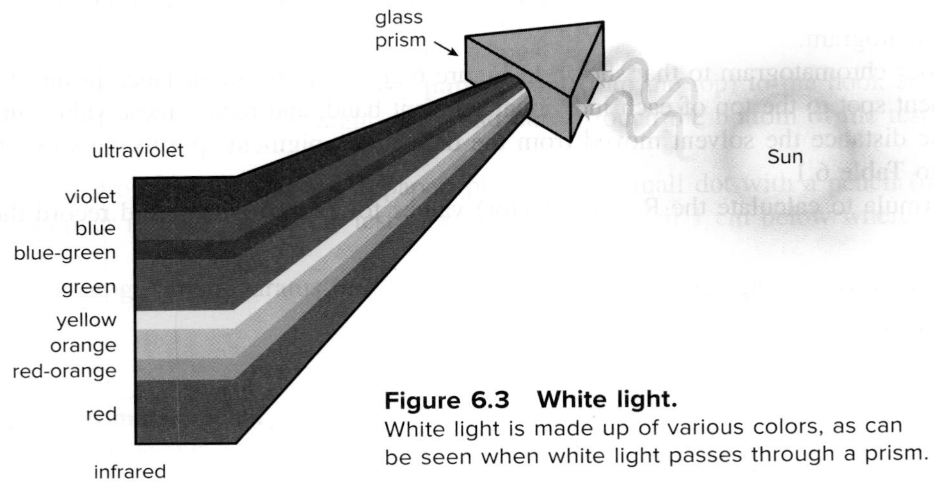

Figure 6.3 White light.
White light is made up of various colors, as can be seen when white light passes through a prism.

Experimental Procedure: White Light

1. Place a generous quantity of *Elodea* with the cut end up (make sure the cuts are fresh) in a test tube with a rubber stopper containing a piece of glass tubing, as illustrated in Figure 6.4. When assembled, this is your volumeter for studying the need for light in photosynthesis. (Do not hold the volumeter in your hand, as body heat will also drive the reaction forward.) Your instructor will show you how to fix the volumeter in an upright position.
2. Before stoppering the test tube, add sufficient 3% sodium bicarbonate ($NaHCO_3$) solution so that, when the rubber stopper is inserted into the tube, the solution comes to rest at about one-fourth the length of the upright glass tubing. Mark this location on the glass tubing with a wax pencil.
3. Place a beaker of plain water next to the *Elodea* tube to serve as a heat absorber. Place a lamp (150 watt) next to the beaker. The tube, beaker, and lamp should be as close to one another as possible.
4. Turn on the lamp. As soon as the edge of the solution in the tubing begins to move, time the reaction for 10 minutes. Be careful not to bump the tubing or to readjust the stopper, or your readings will be altered. After 10 minutes, mark the edge of the solution, and measure in millimeters the distance the edge moved upward: __________ mm/10 min. This is **net photosynthesis,** a measurement that does not take into account the oxygen that was used up for cellular respiration. Record your results in Table 6.2. Why did the edge move upward? ______________________________

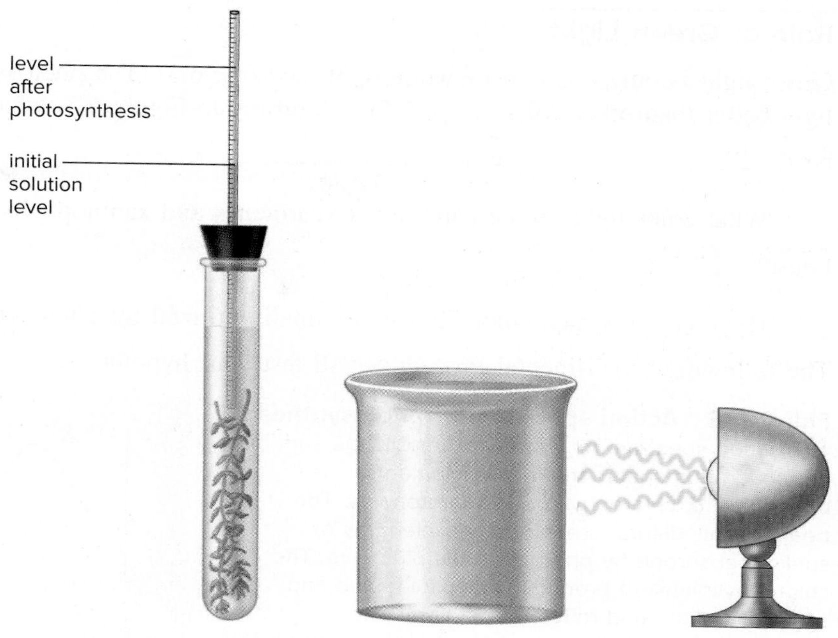

Figure 6.4 Volumeter. A volumeter apparatus is used to study the role of light in photosynthesis.

5. Carefully wrap the tube containing *Elodea* in aluminum foil, and record here the length of time it takes for the edge of the solution in the tubing to move downward 1 mm: __________. Convert your measurement to __________ mm/10 min, and record this value for **cellular respiration** in Table 6.2. (Do not use a minus sign, even though the edge moved downward.) Why does cellular respiration, which occurs in a plant whether it is light or dark, cause the edge to move downward? ______________________________

____

6. If the *Elodea* had *not* been respiring in step 4, how far would the edge have moved upward? __________ mm/10 min. This is **gross photosynthesis** (net photosynthesis + cellular respiration). Record this number in Table 6.2.

7. Calculate the **rate of photosynthesis** (mm/hr) by multiplying gross photosynthesis (mm/10 min) by 6 (that is, 10 min × 6 = 60 min = 1 hr): __________ mm/hr. Record this value in Table 6.2.

Table 6.2 Rate of Photosynthesis (White Light)

	Movement of Edge (mm/10 min)	Rate of Photosynthesis (mm/hr)
Net photosynthesis (white light)		__________
Cellular respiration (no light)		__________
Gross photosynthesis (net + cellular respiration)		

Role of Green Light

Green light is only one part of white light (see Fig. 6.3). The photosynthetic pigments absorb certain colors of light better than other colors (Fig. 6.5). According to Figure 6.5, what color light do the chlorophylls absorb best? _______________ Least? _______________

What color light do the carotenoids (carotenes and xanthophylls) absorb best? _______________ Least? _______________

Hypothesize which color light is minimally utilized for photosynthesis. _______________

The following Experimental Procedure will test your hypothesis.

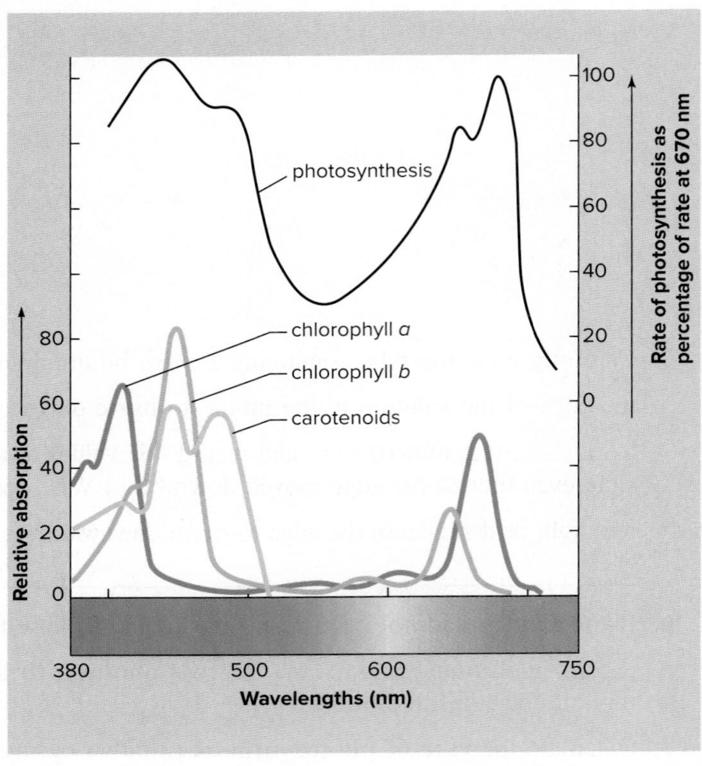

Figure 6.5 Action spectrum for photosynthesis. The action spectrum for photosynthesis is the sum of the absorption spectrums for the pigments chlorophyll *a*, chlorophyll *b*, and carotenoids. The peaks in this diagram represent wavelengths of sunlight absorbed by photosynthetic pigments. The chlorophylls absorb predominantly violet-blue and orange-red light and reflect green light. The carotenoids (carotenes and xanthophylls) absorb mostly blue-green light and reflect yellow-red light.

Experimental Procedure: Green Light

1. Add three drops of green dye (or use a green cellophane wrapper) to the beaker of water used in the previous Experimental Procedure until there is a distinctive green color. Remove all previous wax pencil marks from the glass tubing.
2. Record in Table 6.3 your data for gross photosynthesis (mm/10 min) and for rate of photosynthesis for white light (mm/hr) from Table 6.2.
3. Turn on the lamp. Mark the location of the edge of the solution on the glass tubing. As soon as the edge begins to move, time the reaction for 10 minutes. After 10 minutes, mark the edge of the solution, and measure in millimeters the distance the edge moved. Net photosynthesis for green light = __________ mm/10 min. As before, net photosynthesis does not take into account that oxygen was used for cellular respiration.
4. Carefully wrap the tube containing *Elodea* in aluminum foil, and record here the length of time it takes for the edge of the solution in the tubing to recede 1 mm: __________. Convert your measurement to __________ mm/10 min. As before, this reading shows how much oxygen was used for cellular respiration.

5. Calculate gross photosynthesis for green light (mm/10 min) as you did for white light, and record your data in Table 6.3. (As before, add your data for net photosynthesis to that for cellular respiration.)
6. Calculate rate of photosynthesis for green light (mm/hr) as you did for white light, and record your data in Table 6.3. (As before, convert your data for mm/10 min to mm/hr.)
7. Collect the white and green light rates of photosynthesis (mm/hr) from each group in your lab; then average all the rates including your own. In the last column of Table 6.3 record the average for white and green light rates of photosynthesis (mm/hr).

Table 6.3 Rate of Photosynthesis (White and Green Light)

Gross photosynthesis	**Your Data**	**Class Data**
White (from Table 6.2)	mm/10 min	________
Green	mm/10 min	________
Rate of photosynthesis		
White (from Table 6.2)	mm/hr	mm/hr
Green	mm/hr	mm/hr

8. Calculate the rate of photosynthesis (green light) as a percentage of the rate of photosynthesis (white light) using this equation.

$$\text{Percentage} = \frac{\text{rate of photosynthesis (green light)}}{\text{rate of photosynthesis (white light)}} \times 100$$

Percentage based on your data recorded in Table 6.3 = __________.

Percentage based on class data recorded in Table 6.3 = __________.

Conclusions: Rate of Photosynthesis

- Do your results support the hypothesis that green light is minimally used by a land plant for photosynthesis? __________ Explain, with reference to Figure 6.5. ____
- How does the percentage based on your data differ from that based on class data? ______________________________
  Explain any difference in the percentage. ____

6.3 Carbon Dioxide Uptake

During the Calvin cycle reactions of photosynthesis, the plant takes up carbon dioxide (CO_2) and reduces it to a carbohydrate, such as glucose ($C_6H_{12}O_6$). Therefore, the carbon dioxide in the solution surrounding *Elodea* should disappear as photosynthesis takes place.

Experimental Procedure: Carbon Dioxide Uptake

> **Phenol red** Avoid ingestion, inhalation, and contact with skin, eyes, and mucous membranes. Follow your instructor's directions for disposal of this chemical. Use protective eyewear when performing this experiment.

1. Temporarily remove the *Elodea* from the test tube. Empty the sodium bicarbonate ($NaHCO_3$) solution from the test tube, rinse the test tube thoroughly, and fill with a phenol red solution diluted to a faint pink. (Add more water if the solution is too dark.) Phenol red is a pH indicator that turns yellow in an acid and red in a base.
2. Blow *lightly* on the surface of the solution. Stop blowing as soon as the surface color changes to yellow. Then shake the test tube until the rest of the solution turns yellow.

   Blowing onto the solution adds what gas to the test tube? __________ When carbon dioxide combines with water, it forms carbonic acid; therefore, the solution appears yellow.
3. Thoroughly rinse the *Elodea* with distilled water, return it to the test tube, which now contains a yellow solution, and assemble your volumeter as before.
4. The water in the beaker used to absorb heat should be clear.
5. Turn on the lamp, and wait until the edge of the solution just begins to move upward. Note the time.

   Observe until you note a change in color. How long did it take for the color to change? ______________
6. Why did the solution eventually turn red? ____

   ____
7. The carbon cycle includes all the many ways that organisms exchange carbon dioxide with the atmosphere. Figure 6.6 notes the relationship between cellular respiration and photosynthesis. Animals produce carbon dioxide used by plants to carry out photosynthesis. Plants produce the food (and oxygen) that they and animals require to carry out cellular respiration. Therefore, the same carbon atoms pass between animals and plants and between plants and animals (Fig. 6.6).

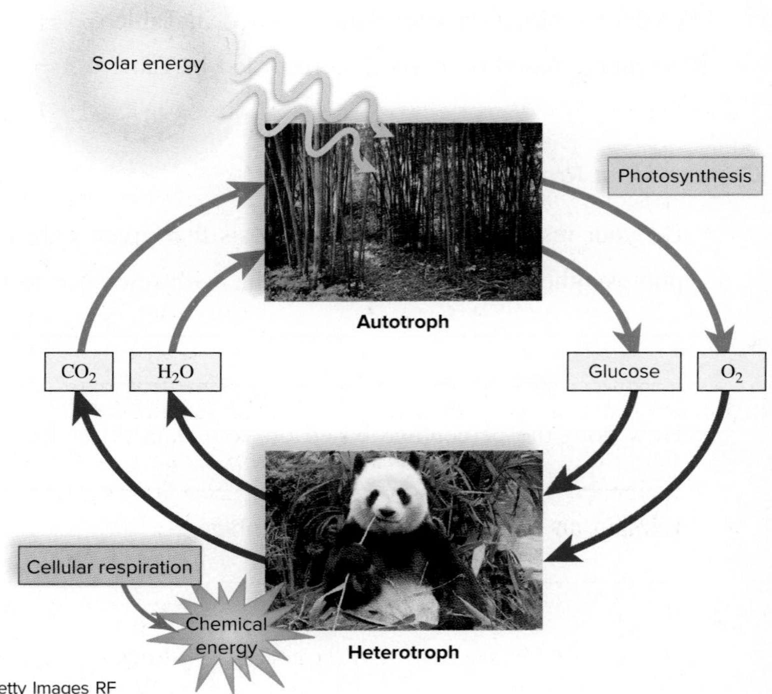

Figure 6.6 Photosynthesis and cellular respiration.
Animals are dependent on plants for a supply of oxygen, and plants are dependent on animals for a supply of carbon dioxide.
(bamboo trees) ©Pan Xunbin/Shutterstock RF; (panda) ©leungchopan/Getty Images RF

6.4 The Light Reactions and the Calvin Cycle Reactions

Review the introduction to this laboratory. Note the overall equation for photosynthesis and that photosynthesis consists of the light reactions and the Calvin cycle reactions. Solar energy is absorbed by photosynthetic pigments during the light reactions and energy is used during the Calvin cycle reactions to reduce carbon dioxide to a carbohydrate.

Light Reactions

1. Photosynthetic pigments
   a. What is the function of the photosynthetic pigments in photosynthesis? ______
   b. How does it benefit a plant to have a variety of photosynthetic pigments? (See Fig. 6.5.) ______
   ______
   c. Green light minimally promotes photosynthesis. Account for this observation with reference to the photosynthetic pigments. (See Fig. 6.5.) ______
   d. Does this explain why leaves are green? ______ How so? ______
2. Water and oxygen
   a. What happens to water during the light reactions? ______
   b. What happens to the released oxygen? ______
3. Location of the light reactions
   Fill in the blank: The light reactions take place in the ______ membranes.
4. In your own words, summarize the light reactions based on this laboratory.

Calvin Cycle Reactions

1. Carbon dioxide and carbohydrate
   What happens to carbon dioxide after it is taken up during the Calvin cycle reactions? ______
2. Location of the Calvin cycle reaction
   Fill in the blank: The Calvin cycle reactions take place in the ______.
3. In your own words, summarize the Calvin cycle reactions based on this laboratory.

Light Reactions and the Calvin Cycle Reactions

1. Examine the overall equation for photosynthesis and show that there is a relationship between the light reactions and the Calvin cycle reactions by drawing an arrow between the hydrogen atoms in water and the hydrogen atoms in the carbohydrate.

$$CO_2 + H_2O \longrightarrow (CH_2O) + O_2$$

2. Only because solar energy splits water can hydrogen atoms be used to reduce carbon dioxide. In this sense, solar energy is now stored in the carbohydrate. This energy sustains all the organisms in the biosphere.

Laboratory Review 6

_______________ **1.** What are the reactants for photosynthesis?

_______________ **2.** What organelle, present in eukaryotic photosynthetic organisms, is the location of photosynthesis?

_______________ **3.** During photosynthesis, where is water split and oxygen released?

_______________ **4.** Where is CO_2 reduced during the Calvin cycle reactions?

_______________ **5.** What technique was used to separate the leaf pigments?

_______________ **6.** Besides chlorophyll, what pigments are found in leaves?

_______________ **7.** Based on distance traveled up the chromatography paper, which pigment is the most nonpolar pigment?

_______________ **8.** What color of light is *not* absorbed by green plants?

_______________ **9.** What color of light is absorbed by carotenes?

_______________ **10.** What gas is released during the light reactions of photosynthesis?

_______________ **11.** What is the gas released during photosynthesis used to do (by animals and plants)?

_______________ **12.** If net photosynthesis is 15 mm/10 min and cellular respiration is 1.5 mm/10 min, how much is gross photosynthesis?

_______________ **13.** When CO_2 combines with H_2O, what is the product that makes phenol red turn yellow?

_______________ **14.** In which molecule do the carbons from CO_2 end up after photosynthesis?

Thought Questions

15. Deforestation decreases the number of plants available to take up CO_2 and use it to make glucose during photosynthesis. What impact could this have on the pH of the oceans? Explain your reply.

16. Cyanobacteria are photosynthetic bacteria and appear green in color. As prokaryotic organisms, they lack chloroplasts but have photosynthetic pigments. Predict where their photosynthetic pigments are located and what photosynthetic pigment gives them their color.

17. What is the importance of the light reactions of photosynthesis since they do not directly produce energy-rich carbohydrates?

LABORATORY

7

Cellular Respiration

Learning Outcomes

Introduction

- Give the overall equation for cellular respiration and fermentation.
- Distinguish between aerobic respiration and anaerobic respiration.
- Explain the relationship between these processes and ATP molecules.

7.1 Cellular Respiration

- Describe and explain the cellular respiration experiment.
- Relate the overall equation for cellular respiration to the cellular respiration experiment.
- State and explain the effects of germination and nongermination of soybeans on the results of the cellular respiration experiment.

7.2 Fermentation

- Describe and explain the fermentation experiment.
- Relate the overall equation for fermentation to the fermentation experiment.
- State and explain the effects of food source on fermentation by yeast.

Introduction

In this laboratory, you will study **cellular respiration** in germinating soybeans. Cellular respiration is an ATP-generating process that involves the complete breakdown most often of glucose to carbon dioxide and water. In eukaryotes, glucose breakdown begins in the cytoplasm, but is completed in mitochondria. Cellular respiration is aerobic and requires oxygen, and it results in a buildup of ATP (Fig. 7.1). This equation represents cellular respiration:

> **Planning Ahead** You may wish to start the fermentation experiment in section 7.2 first, to allow time for incubation.

$$\underset{\text{glucose}}{C_6H_{12}O_6} + \underset{\text{oxygen}}{6\,O_2} \longrightarrow \underset{\text{carbon dioxide}}{6\,CO_2} + \underset{\text{water}}{6\,H_2O} + ATP$$

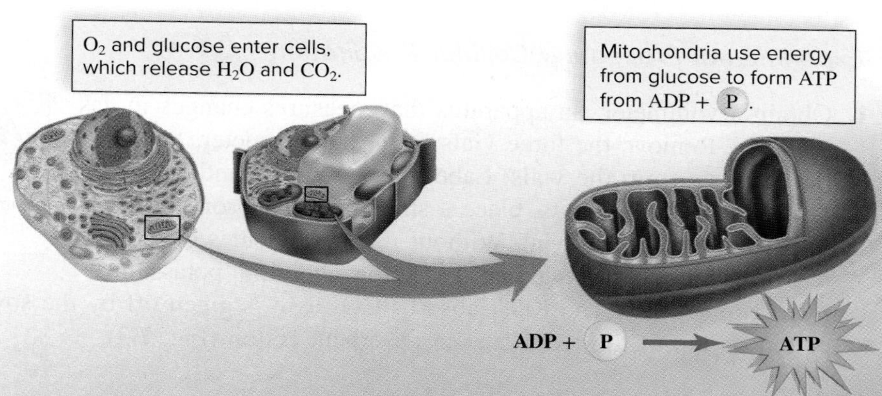

O_2 and glucose enter cells, which release H_2O and CO_2.
Mitochondria use energy from glucose to form ATP from ADP + P.
ADP + P
ATP

Figure 7.1 Cellular respiration. Both plant cells and animal cells have mitochondria and carry on cellular respiration, a process that utilizes the equation above.

In this laboratory, you will also study **ethanol fermentation,** which is an ATP-generating process. When an organism, such as yeast, breaks down glucose to ethanol and carbon dioxide, only 2 ATP result but the process is anaerobic and does not require oxygen. Fermentation occurs in the cytoplasm and mitochondria are not involved. This reaction represents yeast fermentation:

$$\underset{\text{glucose}}{C_6H_{12}O_6} \longrightarrow \underset{\text{carbon dioxide}}{2\,CO_2} + \underset{\text{ethanol}}{2\,C_2H_5OH} + 2\,ATP$$

Fermentation by animal cells produces lactate instead of ethanol and carbon dioxide.

7.1 Cellular Respiration

When germination occurs and plants begin to grow, cellular respiration can provide them with the ATP they need to produce all the molecules that allow them to grow. The soybean seed germinating in Figure 7.2 must produce the molecules needed for mitosis and photosynthesis so that a mature soybean plant can develop.

In the Experimental Procedure that follows, we are going to measure the amount of oxygen uptake as evidence that germinating soybeans are carrying on cellular respiration. The need for oxygen by a germinating soybean will be compared to the need by nongerminating soybeans. State a hypothesis for this experiment here:

Hypothesis: ____

Figure 7.2 Germination of a soybean seed.

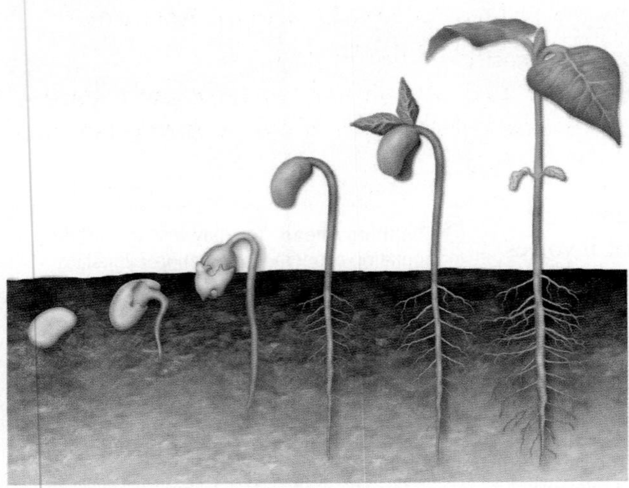

Experimental Procedure: Cellular Respiration

> ⚠ **Potassium hydroxide (KOH)** is a strong, caustic base. Exercise care in using this chemical. If any KOH spills on your skin, rinse immediately with clear water. Follow your instructor's directions for disposal of tubes that contain KOH.

1. Obtain a volumeter, an apparatus that measures changes in gas volumes. Remove the three vials from the volumeter. Remove the stoppers from the vials. Label the vials 1, 2, and 3.
2. Using the same amounts, place a small wad of absorbent cotton in the bottom of each vial. Without getting the sides of the vials wet, use a dropper to saturate the cotton with 15% potassium hydroxide (KOH). The KOH absorbs CO_2 as it is given off by the soybeans. Place a small wad of dry cotton on top of the KOH-soaked absorbent cotton (Fig. 7.3).

Figure 7.3 Vials.
In this experiment, three vials are filled as noted.

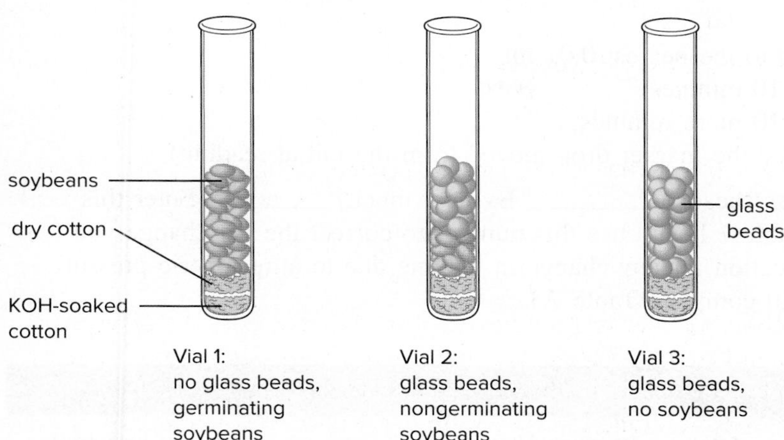

3. Count 25 germinating soybean seeds and add to vial 1. Count 25 dry (nongerminating) soybean seeds and add to vial 2. Add glass beads to vial 2 so that the volume occupied in vial 2 is approximately the same as in vial 1. Add only glass beads to vial 3 to bring to approximately the same volume.
4. Replace the vials in the volumeter and firmly place the stoppers in the vials (Fig. 7.4). Each stopper has a vent (outlet tube) ending in rubber tubing held shut with a clamp. Remove the clamps. Adjust the graduated side arm until only about 5 mm to 1 cm protrudes through the stopper.
5. Use a Pasteur pipet to inject Brodie manometer fluid (or water colored with vegetable dye and a small amount of detergent) into each graduated side arm so that approximately 1 cm of dye is drawn into each side arm.

Figure 7.4 Volumeter containing three respirometers.
In this experiment, the respirometers are vials filled as per Figure 7.3 with graduated side arms attached. Oxygen uptake is measured by movement of a marker drop in each side arm.

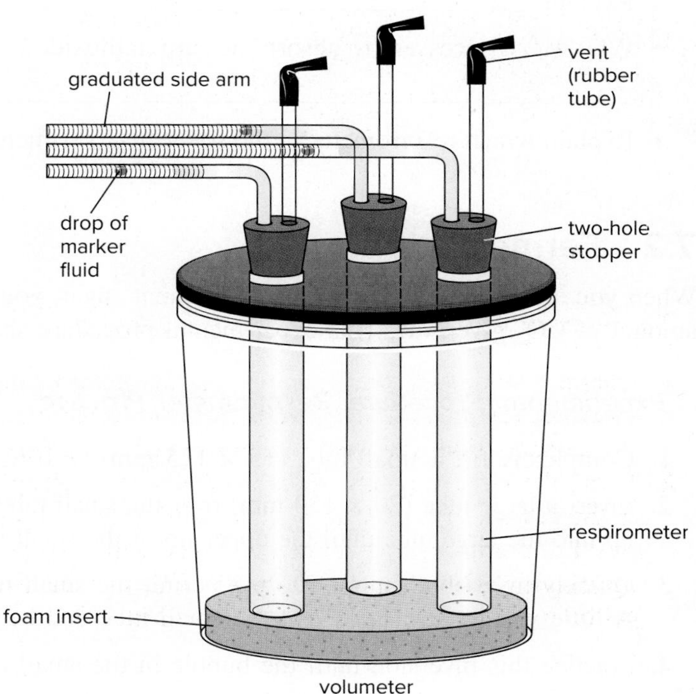

6. After waiting one minute for equilibrium, reattach the pinch clamp to the outlet rubber tube and mark the position of the dye with a wax pencil. The dye should be near the outlet and as O_2 is taken up, the marker drop will move toward the vials.
7. Record in Table 7.1 for each vial:
   - the initial position of the marker drop to the nearest 0.01 ml
   - the position of the marker drop after 10 minutes
   - the position of the marker drop after 10 more minutes
   - the net change of position (the distance the marker drop moved from the initial reading)
8. Did the marker drop change in vial 3 (glass beads)? ________ By how much? ________ Enter this number in the "Correction" column of Table 7.1, and use this number to correct the net change you observed in vials 1 and 2. (This is a correction for any change in volume due to atmospheric pressure changes or temperature changes.) This will complete Table 7.1.

Table 7.1 Cellular Respiration

Vial	Contents	Initial Reading	Reading After 10 Minutes	Reading After 20 Minutes	Net Change	Correction	(Corrected) Net Change
1	Germinating soybeans						
2	Dry (nongerminating) soybeans Glass beads						
3	Glass beads						

Conclusions: Cellular Respiration

- Do your results support or fail to support the hypothesis? ________________________

  Explain. ________________________
- Why was it necessary to absorb the carbon dioxide? ________________________

  ____
- Explain which respirometer in the soybean experiment was a negative control. ________________________

7.2 Fermentation

When you study the ability of yeast to ferment sugar, you will need to use a respirometer to measure the amount of CO_2 given off. This experimental procedure shows you how to prepare your **respirometer.**

Experimental Procedure: Respirometer Practice

1. Completely fill a small tube (15 × 125 mm) with *water only* (Fig. 7.5).
2. Invert a large tube (20 × 150 mm) over the small tube, and with your finger or a pencil, push the small tube up into the large tube until the upper lip of the small tube is in contact with the bottom of the large tube.
3. Quickly invert both tubes. Do not permit the small tube to slip away from the bottom of the large tube. A little water will leak out of the small tube and be replaced by an air bubble.
4. Practice this inversion until the bubble in the small tube is as small as you can make it.

Figure 7.5 Respirometer for fermentation. Place a small tube inside a large tube. Hold the small tube in place as you rotate the entire apparatus, and an air bubble will form in the small tube.

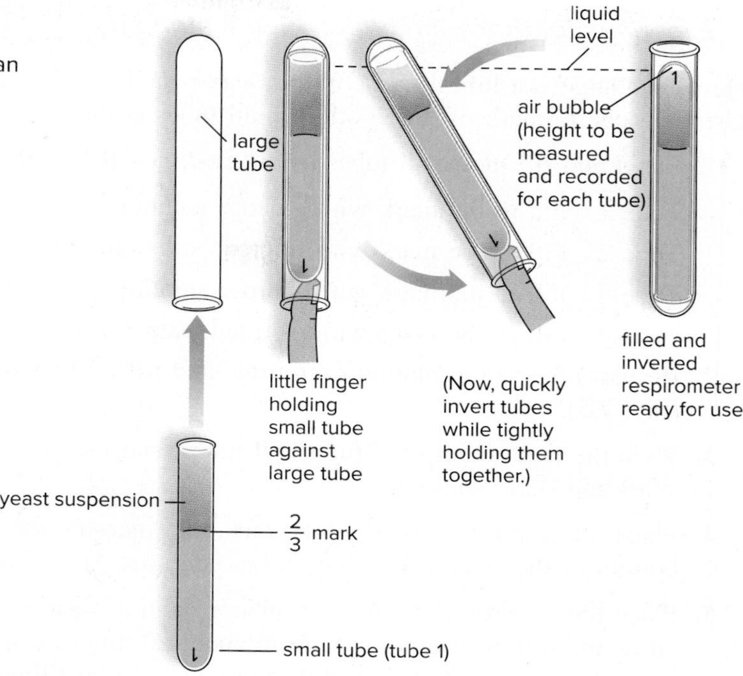

Ethanol Fermentation

Yeast fermentation to produce ethanol (C_2H_5OH) and carbon dioxide (CO_2) has long been utilized by humans to produce wine and bread. During the production of wine, it is the ethanol that is desired, while bread and other baked goods rise when yeast gives off carbon dioxide (Fig. 7.6). Recently, there has been a great deal of interest in using ethanol produced by yeast fermentation of corn as a substitute for gasoline in cars.

Figure 7.6 Products of fermentation.

Testing Sugars

In the Experimental Procedure that follows, we are going to test which of several sugars is the best food source for yeast when they ferment. It will be assumed that the ease of sugar fermentation correlates with the amount of carbon dioxide given off within a certain time limit. As background data, observe the structure of the three sugars involved:

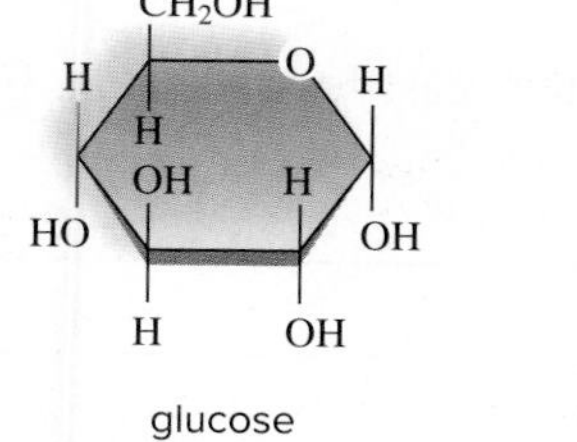

glucose

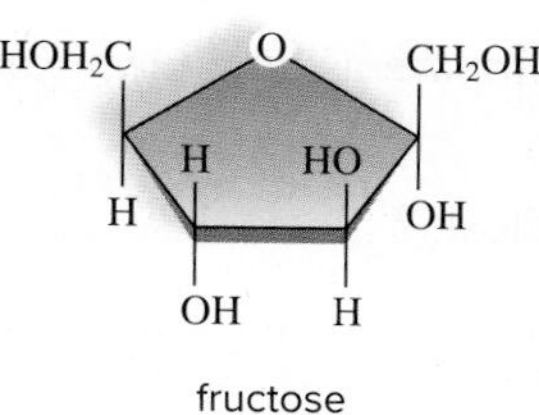

fructose

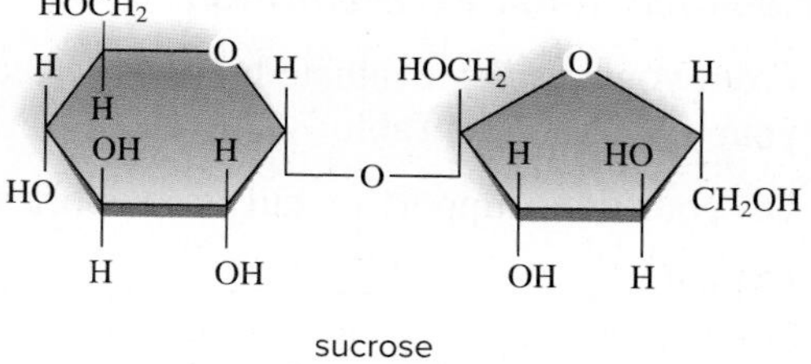

sucrose

State a hypothesis for your experiment in which you sequence the sugars according to how easily you expect yeast to ferment them:

Hypothesis: ______________________________

Experimental Procedure: Yeast Fermentation

Have four large test tubes ready. With a wax pencil, label and mark off a small test tube at the 2/3-full level. Use this tube to mark off three other small tubes at the same level.

1. Label and fill the small tubes as directed, and record the sugar content in Table 7.2.

   Tube 1 Fill to the mark with glucose solution.

   Tube 2 Fill to the mark with fructose solution.

   Tube 3 Fill to the mark with sucrose solution.

   Tube 4 Fill to the mark with distilled water.
2. Resuspend a yeast solution each time, and fill all four tubes to the top with yeast suspension (Fig. 7.5).
3. Slide the large tubes over the small tubes, and invert them in the way you practiced. This will mix the yeast and sugar solutions.
4. Place the respirometers in a tube rack, and measure the initial height of the air space in the rounded bottom of the small tube. Record the height in Table 7.2.
5. Place the respirometers in an incubator or in a warm water bath maintained at 37°C. Note the time, and allow the respirometers to incubate about 20 minutes (incubator) or one hour (water bath). However, watch your respirometers and if they appear to be filling with gas quite rapidly, stop the incubation when appropriate.
6. At the end of the incubation period, measure the final height of the gas bubbles, and record it in Table 7.2. Calculate the net change, and record it in Table 7.2.

Table 7.2 Fermentation by Yeast

Tube	Sugar	Initial Gas Height	Final Gas Height	Net Change	Ease of Fermentation
1					
2					
3					
4					

Conclusions: Yeast Fermentation

- From your results, evaluate how the tested sugars compare as an energy source for yeast fermentation. Enter your evaluation in Table 7.2.
- Do your data support or fail to support your hypothesis? ____________________

  Explain. ____________________

  ____
- Can your results be correlated with the comparative structure of the sugars? (See page 83.) ____________________

  Explain. ____________________
- Explain which respirometer was a negative control. ____________________

Laboratory Review 7

_______________ 1. What molecule is mostly typically broken down during cellular respiration and fermentation?

_______________ 2. What organelle, present in animal and plant cells, completes the breakdown of glucose during cellular respiration?

_______________ 3. What kind of biomolecule is glucose?

_______________ 4. What reactant must be present for cellular respiration to occur but is absent from fermentation?

_______________ 5. What gas is produced by cellular respiration and ethanol fermentation?

_______________ 6. What molecule is formed using energy in glucose and ADP + Ⓟ during cellular respiration?

_______________ 7. Where does fermentation take place in a cell?

_______________ 8. What molecule is made when animal cells do fermentation?

_______________ 9. What were the contents of the vial that served as the negative control for the soybean experiment?

_______________ 10. What were the contents of the vial used to correct for changes in volume caused by atmospheric pressure or temperature changes?

_______________ 11. What does movement of the marker in the side arm of a respirometer toward a tube containing germinating soybeans indicate?

_______________ 12. What biomolecule is the product of fermentation done by yeast?

_______________ 13. What was measured in the yeast fermentation experiment to determine the amount of CO_2 produced?

_______________ 14. What kind of biomolecule is the sucrose used during the fermentation experiment?

Thought Questions

15. Mature plants do photosynthesis to make glucose, which is then used to make the ATP needed to grow and reproduce. Where does the germinating soybean get the glucose it needs for making ATP?

16. Why would "no net change for vial 1" have been the results in the soybean experiment if you forgot to soak the cotton with KOH?

17. Why is it reasonable to hypothesize that sucrose will be more difficult than glucose for yeast to ferment during the fermentation experiment?

8.1 The Cell Cycle

As stated in the Introduction, the period of time between cell divisions is known as interphase. Because early investigators noted little visible activity between cell divisions, they dismissed this period of time as a resting state. But when later investigators discovered that DNA replication and chromosome duplication occur during interphase, the **cell cycle** concept was proposed. The cell cycle is divided into the four stages noted in Figure 8.2.

Figure 8.2 The cell cycle.
Immature cells go through a cycle that consists of four stages: G_1, S (for synthesis), G_2, and M. The cell divides during the M stage, which consists of mitosis and cytokinesis.

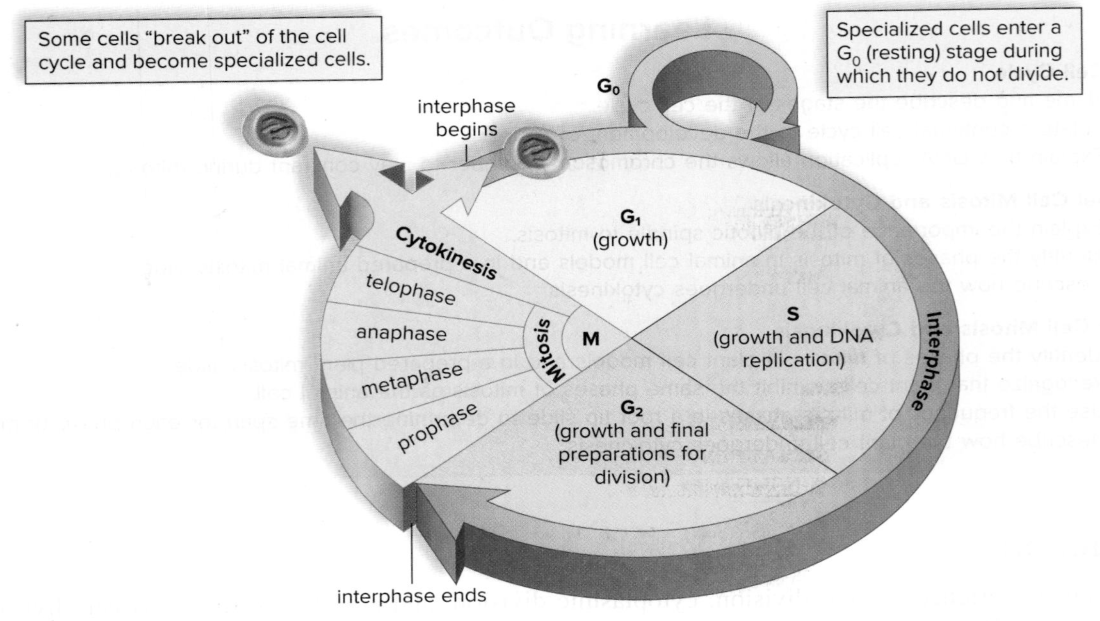

State the events of each stage of the cell cycle:

G_1 ____________________

S ____________________

G_2 ____________________

M ____________________

Explain why the entire process is called the "cell cycle." ____________________

The Cell Cycle and Cancer

Ordinarily, animal cells require about 18 to 24 hours for one cell cycle. Consult Figure 8.2, and notice that interphase consists of G_1, S, and G_2 stages of the cell cycle. When cancer occurs, cell cycle regulation is disturbed; interphase is severely shortened and very abnormal cells repeatedly undergo the cell cycle. A tumor develops and treatment consists of shrinking or removing the tumor.

The S Stage of the Cell Cycle

While each stage of the cell cycle is critical to successful mitosis (division of the nucleus), the S stage is particularly important because it is the stage during which DNA replicates. At the completion of replication, there are two double helices and each chromosome is duplicated.

When mitosis is about to occur, we can see each duplicated chromosome because chromatin has condensed and compacted to form two chromatids held together at a centromere. Consult Table 8.1, and *label the sister chromatids and centromere in Figure 8.3.*

What happens to the sister chromatids during mitosis? During mitosis, the sister chromatids separate, and then they are called daughter chromosomes. This is the manner in which DNA replication causes each body (somatic) cell of an organism to contain the same number of chromosomes. In humans, each cell in the body contains 46 chromosomes.

It is customary to call the cell that divides the **parent cell** and the resulting two cells, the **daughter cells.** If a parent cell undergoing mitosis has 18 chromosomes, each daughter cell will have ____________ chromosomes.

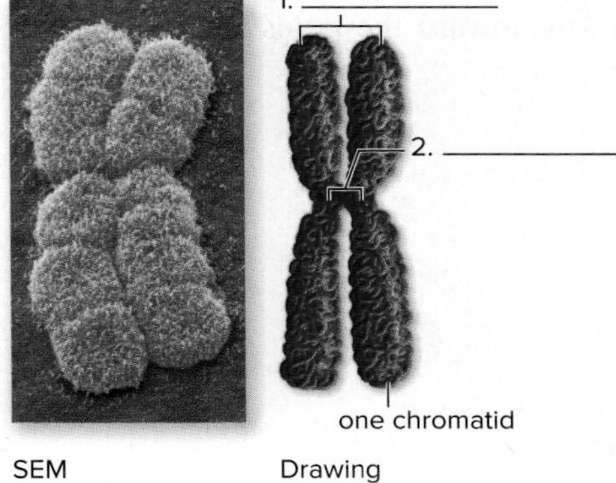

Figure 8.3 Duplicated chromosomes. DNA replication results in duplicated chromosomes that consist of two sister chromatids held together at a centromere.

The M Stage of the Cell Cycle

Study the terms in Table 8.1. These terms all pertain to mitosis. During mitosis sister chromatids (now called daughter chromosomes) move into the daughter nuclei. The process (mitosis) requires several phases. Consult Figure 8.2 and write the phases of mitosis here: ____

____

Table 8.1 Structures Associated with Mitosis

Structure	Description
Nucleus	A large organelle containing the chromosomes and acting as a control center for the cells
Nucleolus	An organelle found inside the nucleus that produces the subunits of ribosomes
Chromosome	Rod-shaped body in the nucleus that is seen during mitosis and meiosis and that contains DNA, and therefore the hereditary units, or genes
Chromatids	The two identical parts of a chromosome following DNA replication
Centromere	A constriction where duplicates (sister chromatids) of a chromosome are held together
Kinetochore	A body that appears during cell division on either side of a centromere; it attaches a chromatid to a spindle fiber.
Spindle	Microtubule structure that brings about chromosome movement during cell division
Metaphase plate	The center of the fully formed spindle
Centrosome	The central microtubule-organizing center of cells; in animal cells, contains two centrioles.
Centrioles*	Short, cylindrical organelles at the spindle poles in animal cells
Aster*	Short, radiating fibers surrounding the centrioles in dividing cells

*Animal cells only

8.2 Animal Cell Mitosis and Cytokinesis

When an animal cell divides, it first undergoes mitosis, and then it undergoes cytokinesis, which is division of the cytoplasm. Mitosis is called duplication division because the daughter cells have the same chromosome makeup as the parent cell. As we now know, a spindle, sometimes called the mitotic spindle, occurs during mitosis. A **spindle** is composed of spindle fibers (microtubules) formed by the centrosomes, which are located at the **poles** of the spindle (Fig. 8.4). In animal cells, the poles contain centrioles surrounded by an **aster,** which is an array of fibers. The kinetochores of the duplicated chromosomes are attached to spindle fibers at the metaphase plate of the spindle. When spindle fibers shorten the chromatids (now daughter chromosomes) move toward the poles.

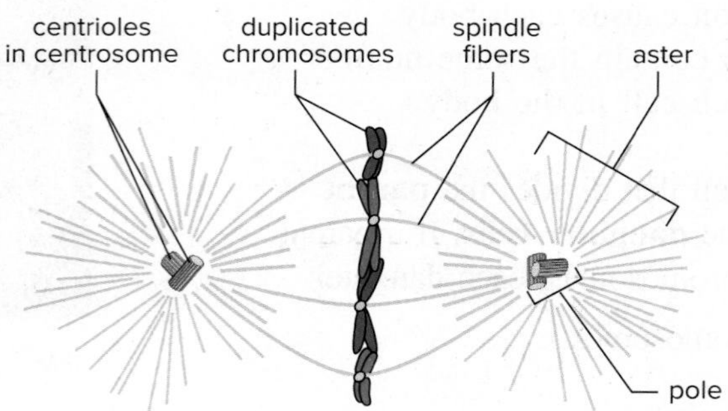

Figure 8.4 The mitotic spindle. The kinetochores of a duplicated chromosome are attached to spindle fibers at the metaphase plate of the spindle. When spindle fibers shorten, the sister chromatids, now called daughter chromosomes, move toward opposite spindle poles.

Animal Cell Mitosis

You will now have the opportunity to observe the stages of mitosis in animal cell models and a prepared slide. Mitosis is the type of nuclear division that occurs when embryos develop, organisms grow larger, and when injuries heal. Without mitosis, none of these important events could occur.

Observation: Animal Cell Mitosis

Animal Mitosis Models

1. Each species has its own chromosome number. Counting the number of centromeres tells you the number of chromosomes in models or slides. What is the number of chromosomes in the parent cell and in the daughter cells in this model series? ______________________
2. Examine the phases of mitosis as depicted by the models.
3. Do these models show the *spindle,* which is illustrated in Figure 8.4? ____________ In animal cells, the centrioles are surrounded by an *aster,* an array of fibers, at the poles of the spindle.
4. Name two ways you would be able to recognize animal cell mitosis: What is the shape of animal cells?

   ____

   What is the appearance of the spindle pole? ______________________
   Plant cells do not have centrioles; therefore, their spindle poles lack centrioles and their poles do not have asters.

Whitefish Blastula Slide

1. Examine a prepared slide of whitefish blastula cells undergoing mitosis. The blastula is an early embryonic stage in the development of animals.
2. Try to find a cell in each phase of mitosis using Figure 8.5 as a guide. Have a partner or your instructor check your ability to identify these cells.
3. Match these statements to the correct phase of animal cell mitosis in Figure 8.5 and *write the correct statements on the lines provided.*

   ***Statements*:**
   - Daughter chromosomes are moving to the poles of the spindle.
   - Duplicated chromosomes have no particular arrangement in the cell.
   - Two daughter cells are now forming.
   - Duplicated chromosomes are aligned at the metaphase plate of the spindle.
4. The prophase cell in Figure 8.5 has the same number of chromosomes as the telophase nuclei. Explain the different appearance of the chromosomes. ______________________________

____

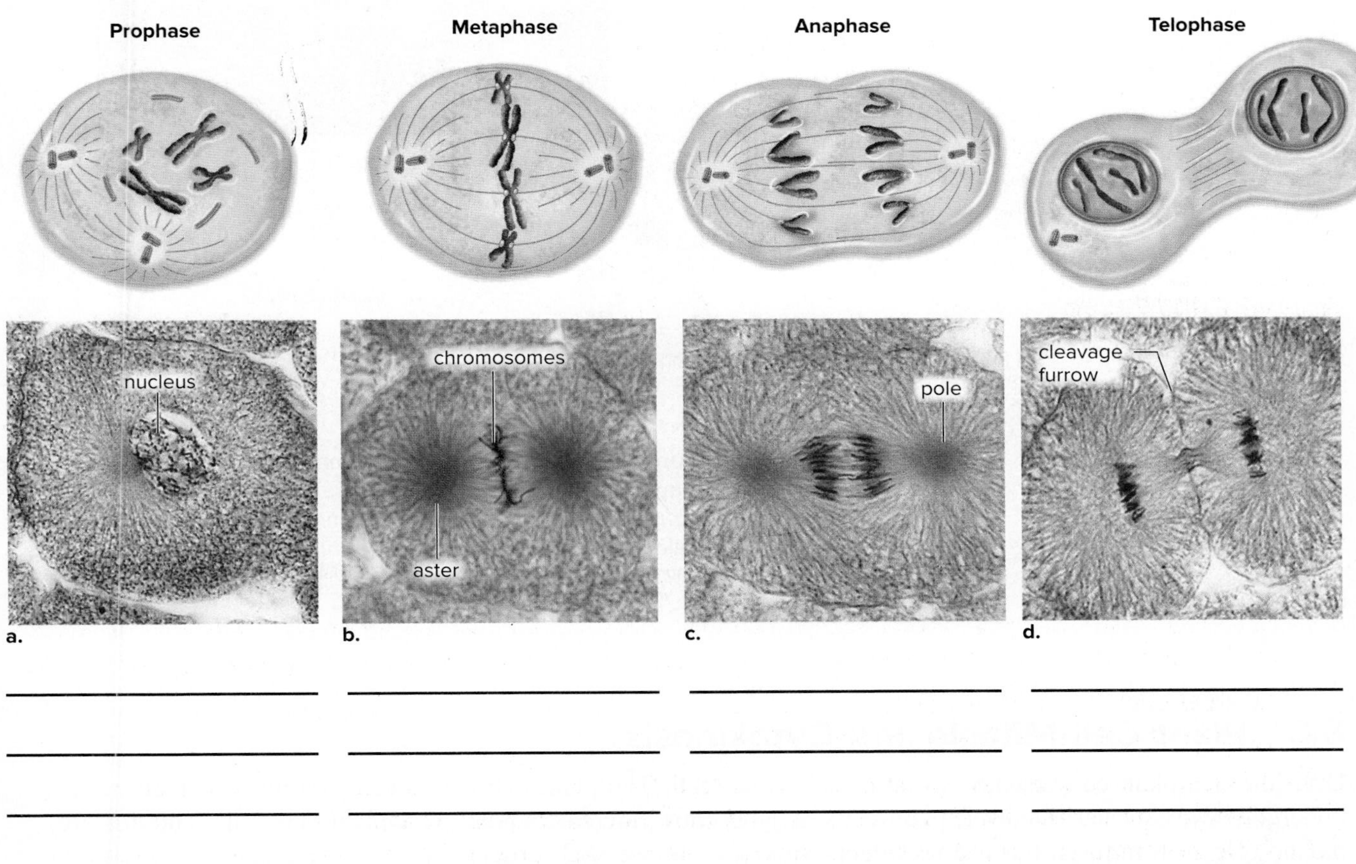

Figure 8.5 The phases of animal cell mitosis.
Mitosis always has these four main phases. Others, designated by the terms “early” or “late,” as in early prophase or late anaphase, can also be cited.
(a, c–d) ©Ed Reschke; (b) ©Ed Reschke/Getty Images

Cytokinesis in Animal Cells

Cytokinesis, division of the cytoplasm, usually accompanies mitosis. During cytokinesis, each daughter cell receives a share of the organelles that duplicated during interphase. Cytokinesis begins in anaphase, continues in telophase, and reaches completion by the start of the next interphase.

In animal cells, a **cleavage furrow,** an indentation of the membrane between the daughter nuclei, begins as anaphase draws to a close (Fig. 8.6). The cleavage furrow deepens as a band of actin filaments called the contractile ring slowly constricts the cell, forming two daughter cells. Note in Figure 8.5 that "cleavage furrow" is labeled in telophase. Are any of the cells in your whitefish blastula slide undergoing cytokinesis? ____________________

Do you see any cleavage furrows? __________

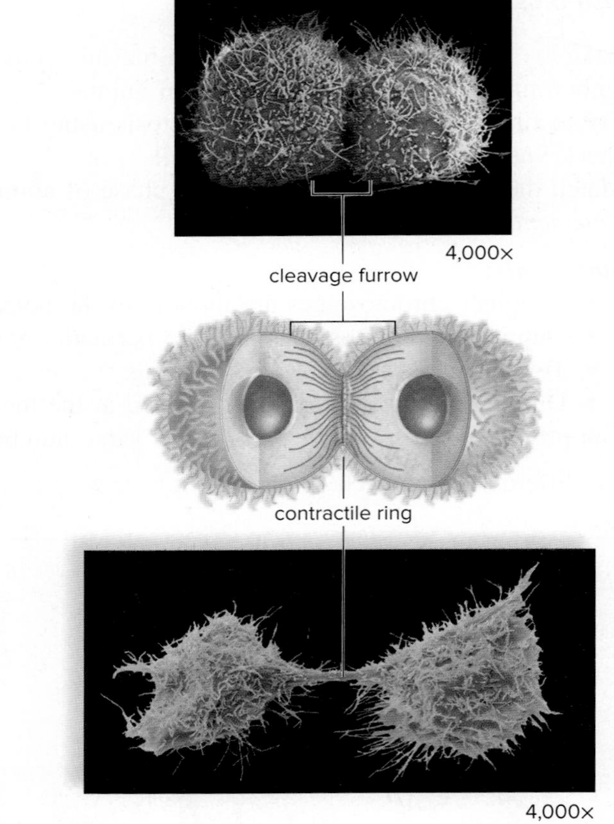

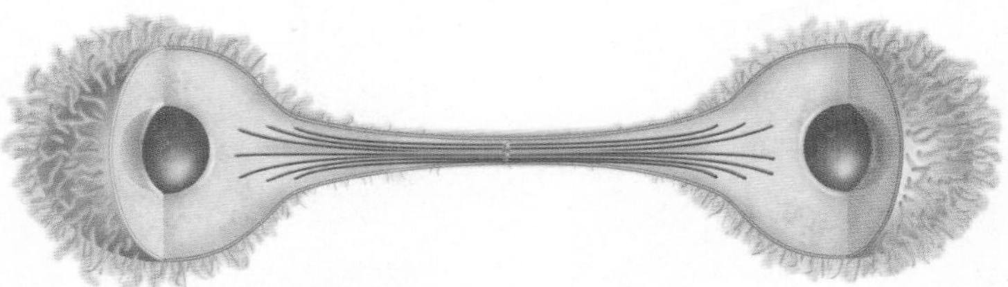

Figure 8.6 Cytokinesis in animal cells.
A single cell becomes two cells by a furrowing process. A contractile ring composed of actin filaments gradually gets smaller, and the cleavage furrow pinches the cell into two cells.
(a) Source: National Institutes of Health (NIH)/USHHS; (b) ©Steve Gschmeissner/SPL/ Getty Images RF

8.3 Plant Cell Mitosis and Cytokinesis

Division of a plant cell resembles that of an animal cell. The plant cell first undergoes mitosis, then it undergoes cytokinesis. However, plant cells, as you know, are surrounded by a plant cell wall. This feature has no effect on mitosis, but it does affect cytokinesis as we shall observe.

Plant Cell Mitosis

Although plant cells do utilize a spindle to divide duplicated chromosomes, they do not have well-defined spindle poles because they lack centrioles and asters.

Observation: Plant Cell Mitosis

Note: Interphase is not a phase *of* mitosis, but a preliminary stage *to* mitosis.

Plant Mitosis Models

1. Identify interphase and the phases of plant cell mitosis using models of plant cell mitosis and Figure 8.7 as a guide.
2. As mentioned previously, plant cells do not have centrioles and asters, which are short radiating fibers produced by centrioles.
3. What is the number of chromosomes in each of the cells in your model series? ________________

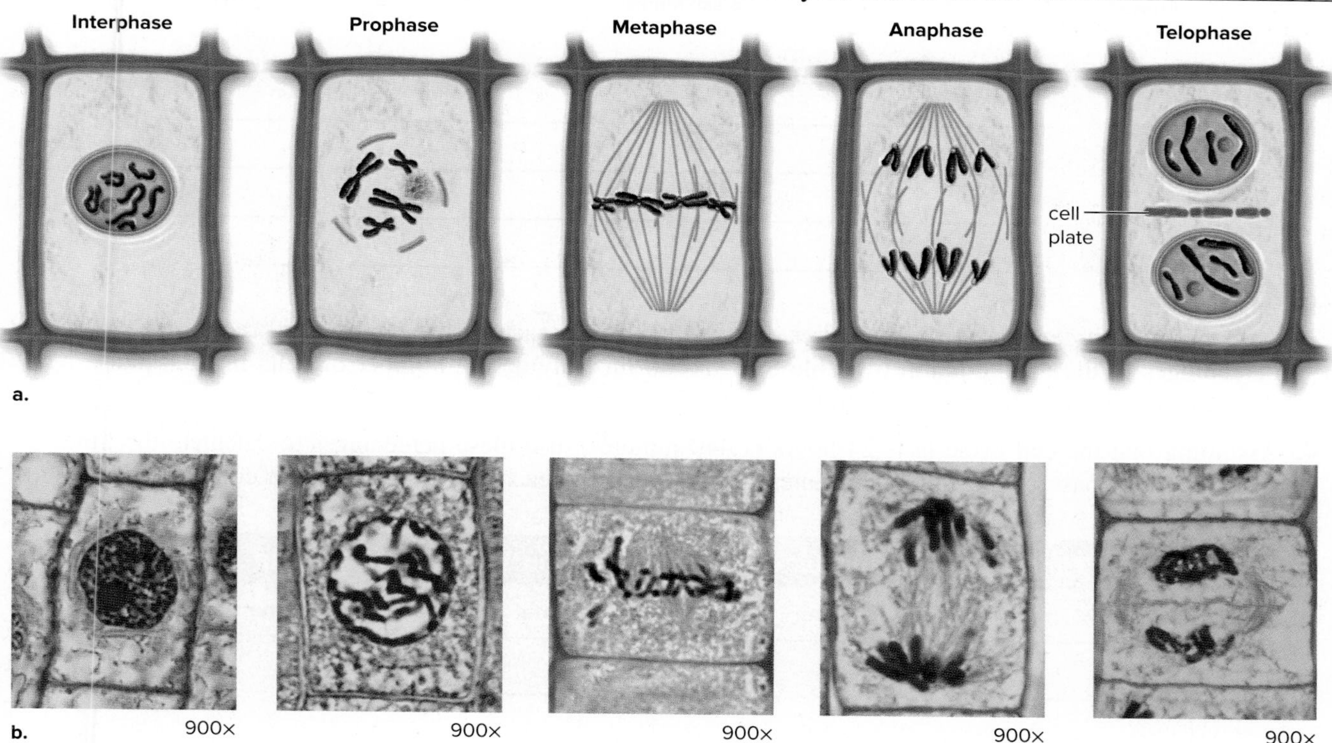

Figure 8.7 Phases of plant cell mitosis.
Mitosis has these phases in all eukaryotic cells. Plant cells are recognizable by the presence of a cell wall and the absence of centrioles and asters. **a.** Drawings of the phases to be identified. **b.** Micrographs of these phases.
b (left) ©Ed Reschke; b (center left–right) ©Kent Wood/Science Source

Onion Root Tip Slide

1. Examine a prepared slide of onion root tip cells (*Allium*) undergoing mitosis. In plants, the root tip contains tissue that is continually dividing and producing new cells.
2. Focus in low power and then switch to high power. Practice identifying phases that correspond to those shown in Figure 8.7.
3. Using high power, focus up and down on a cell in telophase. You may be able to just make out the cell plate, the region where a plasma membrane is forming between the two prospective daughter cells. Later, cell walls will appear in this region.

Time Span for Phases of the Cell Cycle in the Onion Root Tip

Knowing that the cell cycle consists of interphase plus four phases of mitosis and that the cell cycle typically lasts about 24 hours (1,440 minutes), hypothesize how many minutes the cell spends during each of these phases of the cell cycle.

Interphase ______ Prophase ______ Metaphase ______ Anaphase ______ Telophase ______

Experimental Procedure: Time Span for Phases of the Cell Cycle in the Onion Root Tip

1. Select an area of the onion root tip slide that contains cells in all phases of mitosis and also interphase. Concentrate on examining a confined area of about 20 to 30 cells.
2. Don't stray beyond a confined region, and as you identify phases, put a slash mark on the lines provided in this chart. Preferably work with a partner who will enter the slash marks as you call out observed phases in a confined region of the root tip. Convert your 20 to 30 slashes to Arabic numbers on the lines provided under the title Total. Also record these numbers in the second column of Table 8.2.

Phase	Slash Marks	Total
Interphase	______________________________	______
Prophase	______________________________	______
Metaphase	______________________________	______
Anaphase	______________________________	______
Telophase	______________________________	______

3. Calculate the percentage of total number of cells that are in each of the phases, and record the percentages in the third column of Table 8.2. (To do this, divide the number of cells in each phase by the total number of cells observed and multiply by 100.)
4. Assuming that the cell cycle lasts 24 hours (1,440 minutes), use these percentages to calculate the time span for each phase of the cell cycle. Enter the time span for each phase in the fourth column of Table 8.2.

Table 8.2 Time Span for Phases of the Cell Cycle in the Onion Root Tip

Phase	Number Seen	% of Total	Time Span (min)
Interphase			
Prophase			
Metaphase			
Anaphase			
Telophase			
Total			

Conclusions: Time Span for Phases of the Cell Cycle in the Onion Root Tip

- Were your hypotheses supported or not supported by your observation of onion root tip cells undergoing the cell cycle? ______________________ Describe any specific discrepancies.
____
____
- Suggest a possible explanation for the length of time a cell spends on different phases of the cell cycle. ____
____

Cytokinesis in Plant Cells

After mitosis, the cytoplasm divides by cytokinesis. In plant cells, membrane vesicles derived from the Golgi apparatus migrate to the center of the cell and form a **cell plate** (Fig. 8.8), which is the location of a new plasma membrane for each daughter cell. Later, individual cell walls appear in this area. Were any of the cells of the onion root tip slide undergoing cytokinesis, as shown in Figure 8.7, during:

Telophase? ______________________________

How do you know? ______________________________

Offer an explanation for why Figure 8.8 is so detailed. ______________________________

____

Would you predict that the vesicles of the cell plate lay down the new cell wall inside or outside the vesicles? Explain your answer. ______________________________

Figure 8.8 Electron micrograph showing cytokinesis in plant cells. During plant cell cytokinesis, vesicles fuse to form a cell plate that separates the daughter nuclei. Later, the cell plate gives rise to a new cell wall.

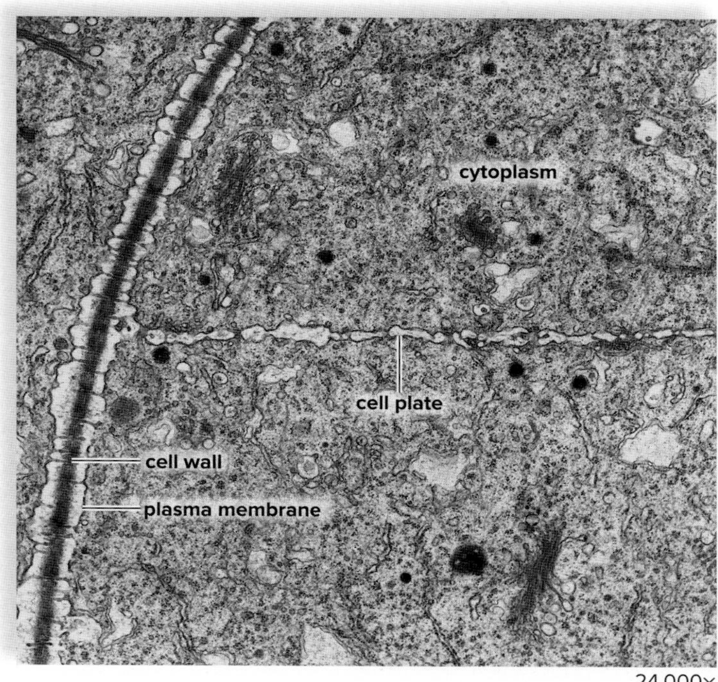

Laboratory Review 8

_______________ 1. What is the name for the period of time when a cell is performing its usual functions and replicating DNA?

_______________ 2. When does nuclear division occur?

_______________ 3. Why do organisms perform mitosis?

_______________ 4. List the four stages of the cell cycle.

_______________ 5. What occurs during the S stage of interphase?

_______________ 6. What is the term used to describe duplicated chromosomes?

_______________ 7. If asters are observed in cells undergoing mitosis, are the cells animal or plant cells?

_______________ 8. Name the structure that helps chromosomes move and is formed during prophase.

_______________ 9. When do chromosomes align at the equator of the spindle?

_______________ 10. What happens to sister chromatids during anaphase of mitosis?

_______________ 11. During which phase of mitosis does the nuclear envelope reform and the nucleolus reappear?

_______________ 12. When are organelles divided up between daughter cells?

_______________ 13. In which phase of the cell cycle do cells spend most of their time?

_______________ 14. What structure gives rise to new plant cell walls?

Thought Questions

15. Why it necessary for chromosomes to duplicate before mitosis?

16. Why is it important for sister chromatids to be attached to each other during the beginning phases of mitosis?

17. List several parts of your body where you would expect to find actively dividing cells and several parts where you would find mostly resting cells.

LABORATORY

9

Meiosis: Sexual Reproduction

Learning Outcomes

9.1 Meiosis: Reduction Division

- Name and describe the phases of meiosis I and meiosis II.
- Describe the mechanism for chromosome reduction during meiosis I, and explain why a reduced chromosome number is beneficial in gametes.
- Recognize the various phases of meiosis when examining slides of meiosis.
- Describe the chromosome composition in the daughter cells following meiosis II.

9.2 Production of Variation During Meiosis

- Using pop bead chromosomes, demonstrate two ways meiosis introduces variations among the daughter cells of meiosis and therefore the gametes.

9.3 Human Life Cycle

- Draw the human life cycle, showing the occurrence of mitosis and meiosis.
- State the function of mitosis and meiosis in the human life cycle.
- Contrast the process and the events of mitosis and meiosis.

Introduction

In sexually reproducing organisms, meiosis is a part of or preparatory to gametogenesis, the production of gametes (sex cells). The gametes are sperm (the smaller gamete) and egg (the larger gamete). Fusion of sperm and egg results in a zygote that develops into a new individual (Fig. 9.1).

Meiosis is nuclear division that reduces the chromosome number so the gametes have half the species number of chromosomes. At the start of meiosis, the parent cells have the full number of chromosomes and each is duplicated. Following two divisions called **meiosis I** and **meiosis II,** the four daughter cells have only one copy of each type of chromosome, and these chromosomes consist of only one chromatid.

Not only does meiosis reduce chromosome number, it also introduces variation, and this laboratory will show you exactly how the chromosomes are shuffled during meiosis and how the genetic material is recombined during the process of meiosis.

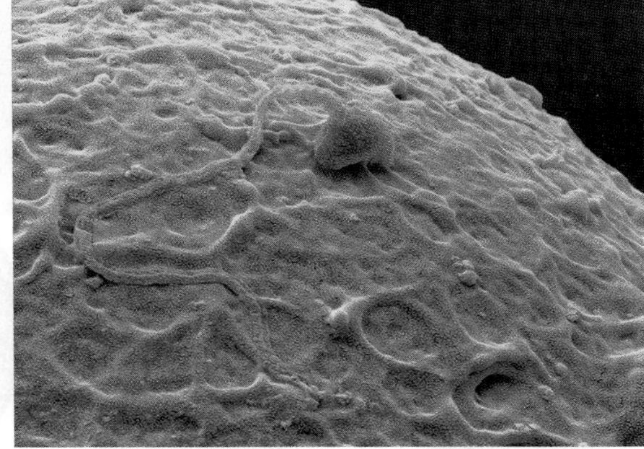

Figure 9.1 Zygote formation.
During sexual reproduction, union of the sperm and egg produces a zygote that becomes a new individual.

9.2 Production of Variation During Meiosis

The following experimental procedure is designed to show that during meiosis, crossing-over and independent separation of homologues lead to diversity of genetic material in the gametes.

Experimental Procedure: Production of Variation During Meiosis

First, you will build four chromosomes: two pairs of homologues, as in Figure 9.5. In other words, the parent cell is 2n = 4.

Building Chromosomes

1. Obtain the following materials: 48 pop beads of one color (e.g., red) and 48 pop beads of another color (e.g., blue) for a total of 96 beads; eight magnetic centromeres.
2. Build a homologue pair of duplicated chromosomes using Figure 9.5*a* as a guide. Each chromatid will have 16 beads. Be sure to bring the centromeres of the same color together so that they form one duplicated chromosome.
3. Build another homologue pair of duplicated chromosomes using Figure 9.5*b* as a guide. Each chromatid will have eight beads.
4. Note that your chromosomes look the same as those in Figure 9.5.

Figure 9.5 Two pairs of homologues.
The red chromosomes were inherited from one parent, and the blue chromosomes were inherited from the other parent. Color does not signify homologues; size and shape signify homologues.

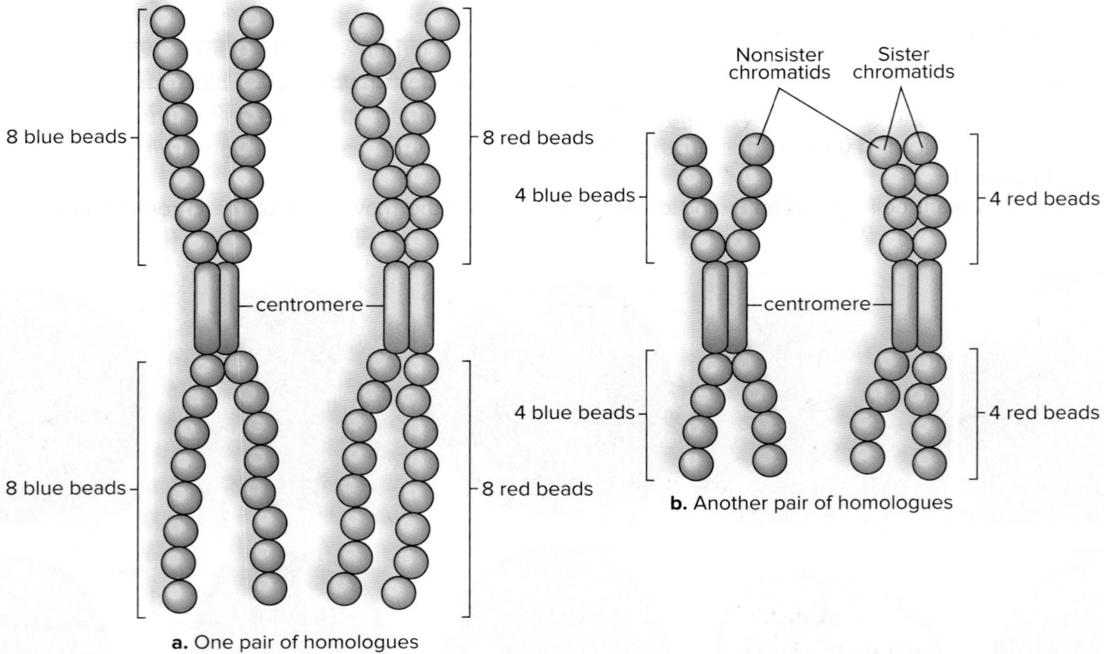

Meiosis I

You will now have your chromosomes undergo meiosis I.

Prophase I

5. Put all four of your chromosomes in the center of your work area: this area represents the nucleus. Synapsis, a very important event, occurs during meiosis. To simulate synapsis, place the long blue chromosome next to the long red chromosome and the short blue chromosome next to the short red chromosome to show that the homologues pair up as in Figure 9.3 (prophase I). Now **crossing-over** occurs. During crossing-over an exchange of genetic material occurs between nonsister chromatids. Perform crossing-over by switching some blue beads for red beads between the inside chromatids of the homologues. The genetic material contains genes and therefore crossing-over recombines genes.

Genetic Variation Due to Meiosis As a result of crossing-over, the genetic material on a chromosome in a gamete can be different from that in the parent cell.

Metaphase I

6. Keep the homologues together and align them at the metaphase plate. Add centrioles to represent the poles of a spindle apparatus. These are animal cells.

Anaphase I and Telophase I

7. Separate the homologues, so that each pole of the spindle receives one chromosome from each pair of homologues. This separation causes each pole to receive the haploid number of chromosomes.

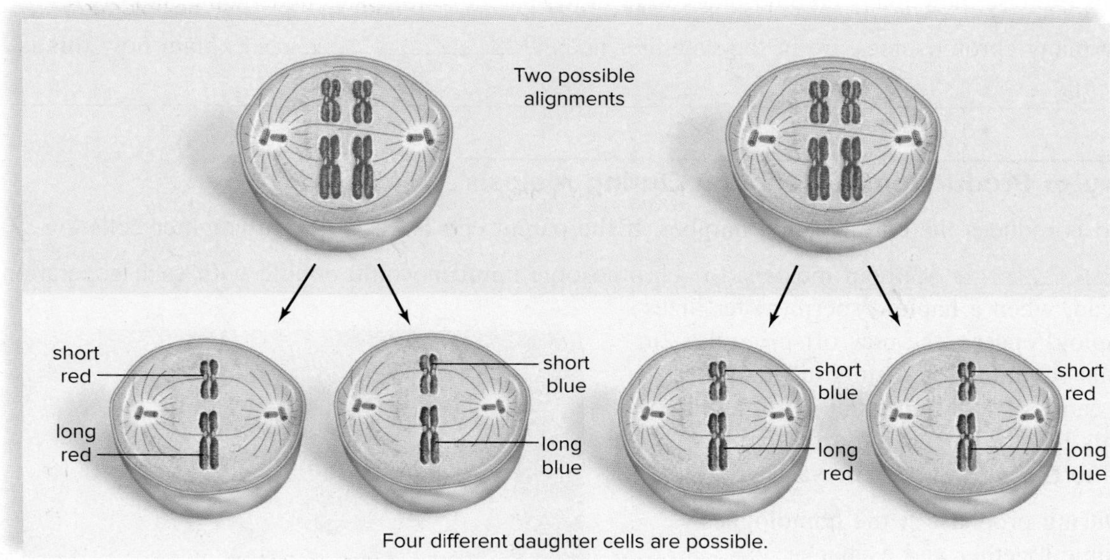

Only two daughter cells result from meiosis I, but two possible alignments of chromosomes can occur at the metaphase plate during metaphase I. Therefore, four possible combinations of haploid chromosomes are possible in the daughter cells following meiosis I in a $2n = 4$ parent cell. The possible combinations increase as the chromosome number increases.

Genetic Variation Due to Meiosis As a result of independent homologue separation, all possible combinations of the haploid number of chromosomes can occur among the gametes.

Meiosis II

Follow these directions to simulate meiosis II.

Prophase II

8. Choose one daughter nucleus (see step 7, page 101) to be the parent nucleus undergoing meiosis II.

Metaphase II

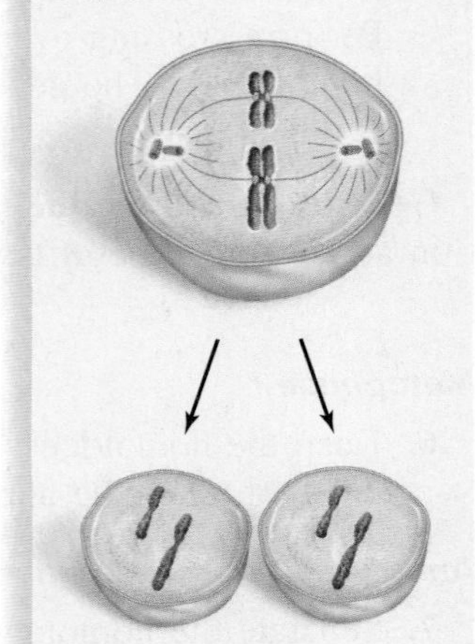

9. Move the duplicated chromosomes to the metaphase II metaphase plate, as shown in the art to the right.

Anaphase II

10. Pull the two magnets of each duplicated chromosome apart. What does this action represent? ______________________________

____

Telophase II

11. Put the chromosomes—each having one chromatid—at the poles near the centrioles. At the end of telophase, the daughter nuclei reform.

- You chose only one daughter nucleus from meiosis I to be the nucleus that divides. In reality, both daughter nuclei go on to divide again. Therefore, how many nuclei are usually present when meiosis II is complete? __________
- In this exercise, how many chromosomes were in the parent cell nucleus undergoing meiosis II? __________
- How many chromosomes are in the daughter nuclei? ______________ Explain how this is possible. ______________________________

Summary of Production of Variation During Meiosis

1. Meiosis reduces the chromosome number. If the parent cell is $2n = 4$, the daughter cells are $n =$ __________. Without meiosis, the chromosome number would double with each generation. Instead, when a haploid sperm(n) fertilizes a haploid egg(n), the new offspring are 2n.
2. Sexual reproduction results in offspring that can look very different as represented in this illustration. Why do the puppies born to these parents show variation?

©American Images, Inc/Getty Images

   a. During prophase I, the homologues come together and exchange genetic material. Now the inherited chromosomes will be different from those in the parent cell. This process is called ______________________.

   b. During metaphase I, the homologues align ____________________ and therefore differently. This is the second process that produces genetic variation through sexual reproduction.

   c. During fertilization variant sperm fertilize variant eggs, further helping to ensure that the new individual inherits different ____________________ of homologues than a parent had.

9.3 Human Life Cycle

The term **life cycle** in sexually reproducing organisms refers to all the reproductive events that occur from one generation to the next. The human life cycle involves both mitosis and meiosis (Fig. 9.6). As you read the following text, *fill in boxes in Figure 9.6 with the terms "mitosis" or "meiosis."*

During development and after birth, mitosis is involved in the continued growth of the child and the repair of tissues at any time. As a result of mitosis, each somatic (body) cell has the diploid number of chromosomes (2n), which is 46 chromosomes.

During gamete formation, meiosis reduces the chromosome number from the diploid to the haploid number (n) in such a way that the gametes (sperm and egg) have one chromosome derived from each pair of homologues. In males, meiosis is a part of **spermatogenesis,** which occurs in the testes and produces sperm. In females, meiosis is a part of **oogenesis,** which occurs in the ovaries and produces eggs. After the sperm and egg join during fertilization, the resulting zygote is 2n. The zygote then undergoes mitosis with differentiation of cells to become a fetus, and eventually a new human being.

Meiosis keeps the number of chromosomes constant between the generations, and it also, as we have seen, causes the gametes to be different from one another. Therefore, due to sexual reproduction, there are more variations among individuals.

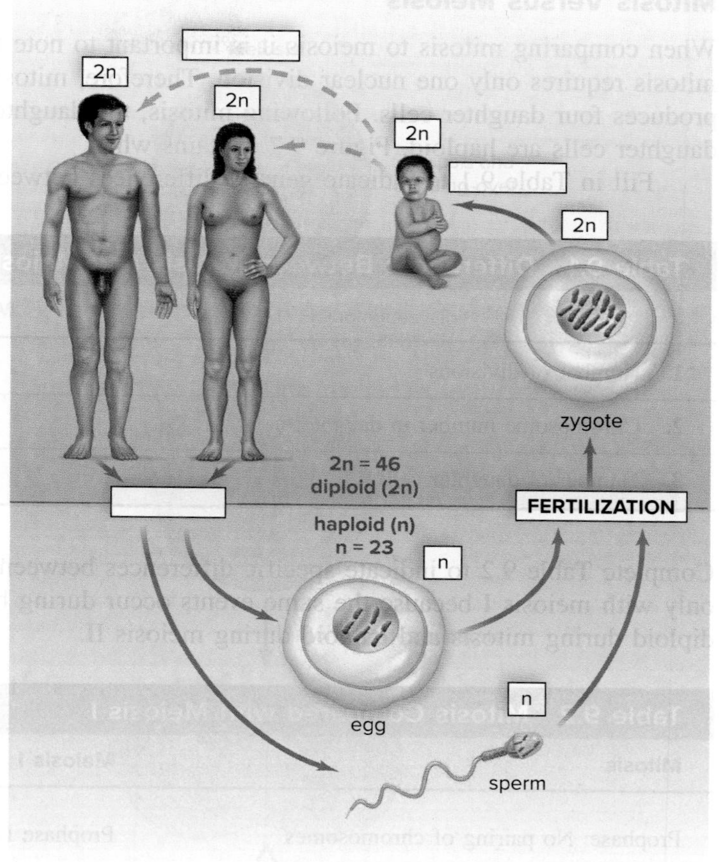

Figure 9.6 Life cycle of humans.

Summary of Human Life Cycle

Fill in the blanks to ensure your understanding of the role of meiosis and mitosis in humans.

1. Name of organ that produces gametes in males ______________ in females ______________
2. Name of process that produces gametes in males ______________ in females ______________
3. Type of cell division involved in process in males ______________ in females ______________
4. Name of gamete in males ______________ in females ______________
5. Number of chromosomes in gamete in males ______________ in females ______________
6. Results of fertilization ______________
7. Number of chromosomes provided ______________

Answer these questions.

One-Trait Genetics Problems

1. In pea plants, purple flowers (*P*) is dominant and white flowers (*p*) is recessive. What is the genotype of pure-breeding white plants? Pure-breeding means that they produce plants with only one phenotype. ____ If pure-breeding purple plants are crossed with these white plants, what phenotype is expected? ________
2. In pea plants, tall (*T*) is dominant and short (*t*) is recessive. A heterozygous tall plant is crossed with a short plant. What is the expected phenotypic ratio? ____________
3. Unexpectedly to the farmer, two tall plants have some short offspring. What is the genotype of the parent plants and the short offspring? parent ____________ offspring ____________
4. In horses, two trotters are mated to each other and produce only trotters; two pacers are mated to each other and produce only pacers. When one of these trotters is mated to one of the pacers, all the horses are trotters. Create a key and show the cross. key ________________________________ cross ______
5. A brown dog is crossed with two different black dogs. The first cross produces only black dogs, and the second cross produces equal numbers of black and brown dogs. What is the genotype of the brown dog? ____________ the first black dog? ____________ the second black dog? ____________
6. In pea plants, green pods (*G*) is dominant and yellow pods (*g*) is recessive. When two pea plants with green pods are crossed, 25% of the offspring have yellow pods. What is the genotype of all plants involved? plants with green pods ____________ plants with yellow pods ____________
7. A breeder wants to know if a dog is homozygous black or heterozygous black. If the dog is heterozygous, which cross is more likely to produce a brown dog, *Bb* × *bb* or *Bb* × *Bb?* Explain your answer. ____

____

____
8. If the cross in problem 6 produces 220 plants, how many offspring have green pods and how many have yellow pods? ____________ If the cross in problem 2 produces 220 plants, how many offspring are tall and how many are short? ____________

10.2 Two-Trait Crosses

Two-trait crosses involve two pairs of alleles. Mendel found that during a **dihybrid cross,** when two dihybrid individuals (*AaBb*) reproduce, the phenotypic ratio among the offspring is 9:3:3:1, representing four possible phenotypes. He realized that these results could be obtained only if the alleles of the parents segregated independently of one another when the gametes were formed. From this, Mendel formulated his second law of inheritance:

Law of Independent Assortment

Members of an allelic pair segregate (assort) independently of members of another allelic pair. Therefore, all possible combinations of alleles can occur in the gametes.

The FOIL method is a way to determine the gametes. FOIL stands for *F*irst two alleles from each trait; *O*uter two alleles from each trait; *I*nner two alleles from each trait; *L*ast two alleles from each trait. Here is how the FOIL method can help you determine the gametes for the genotype *PpSs:*

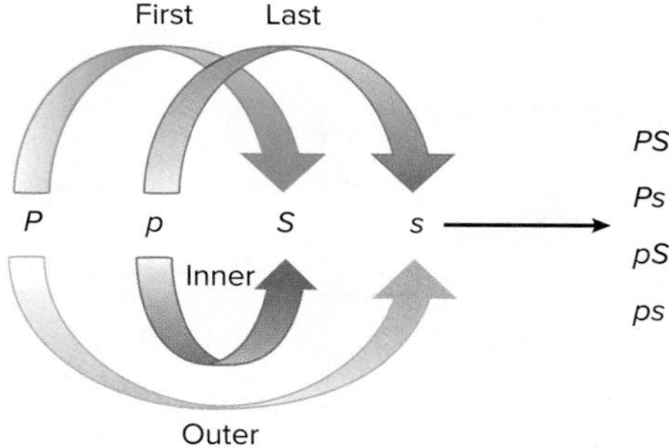

Color and Texture of Corn

In corn plants, the allele for purple kernel (*P*) is dominant over the allele for yellow kernel (*p*), and the allele for smooth kernel (*S*) is dominant over the allele for rough kernel (*s*) (Fig. 10.5).

Figure 10.5 Dihybrid cross.
Four types of kernels are seen on an ear of corn following a dihybrid cross (*PpSs* × *PpSs*): purple smooth, purple rough, yellow smooth, and yellow rough.

Experimental Procedure: Color and Texture of Corn

1. Obtain an ear of corn from the supply table. You will be examining the results of the cross *PpSs* × *PpSs*.
2. Do the Punnett square on page 111, in order to state the expected phenotypic ratio among the offspring. ______________________
3. Count the number of kernels of each possible phenotype listed in Table 10.3. Record the sample number and your results in Table 10.3. Use three samples, and total your results for all samples. Also record the class data (i.e., the number of kernels that are the four phenotypes per class).

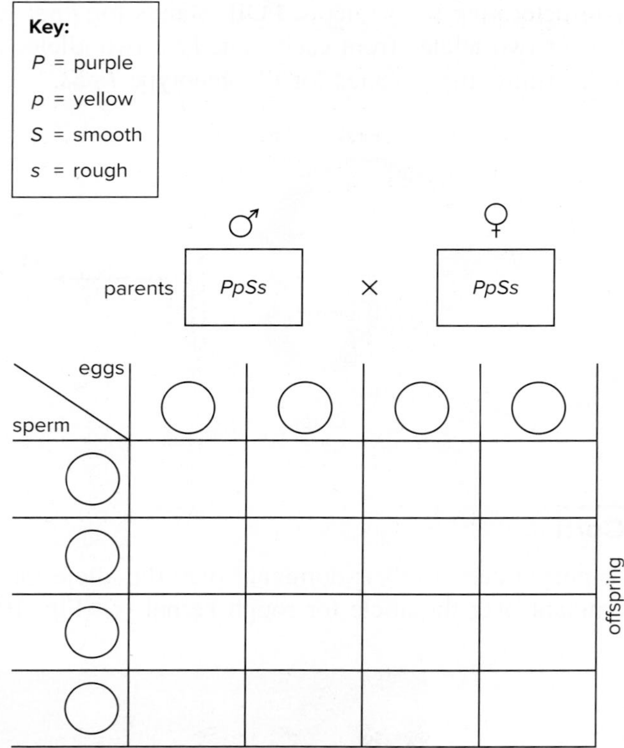

Table 10.3 Color and Texture of Corn					
	Number of Kernels				
	Purple Smooth	**Purple Rough**	**Yellow Smooth**	**Yellow Rough**	**Phenotypic Ratio**
Sample # ________					
Sample # ________					
Sample # ________					
Totals					
Class data					

Conclusions: Color and Texture of Corn

- Calculate the actual phenotypic ratios based on the data and record them in Table 10.3. Do the results differ from the expected ratio per individual data? ____________ Per class data? ____________ If so, explain your answer. ____

Wing Length and Body Color in *Drosophila*

Drosophila are the tiny flies you often see flying around ripe fruit; therefore, they are called fruit flies. If a culture bottle of fruit flies is on display, take a look at it. Because so many flies can be grown in a small culture bottle, fruit flies have contributed substantially to our knowledge of genetics. If you were to examine *Drosophila* flies under the stereomicroscope, they would appear like this:

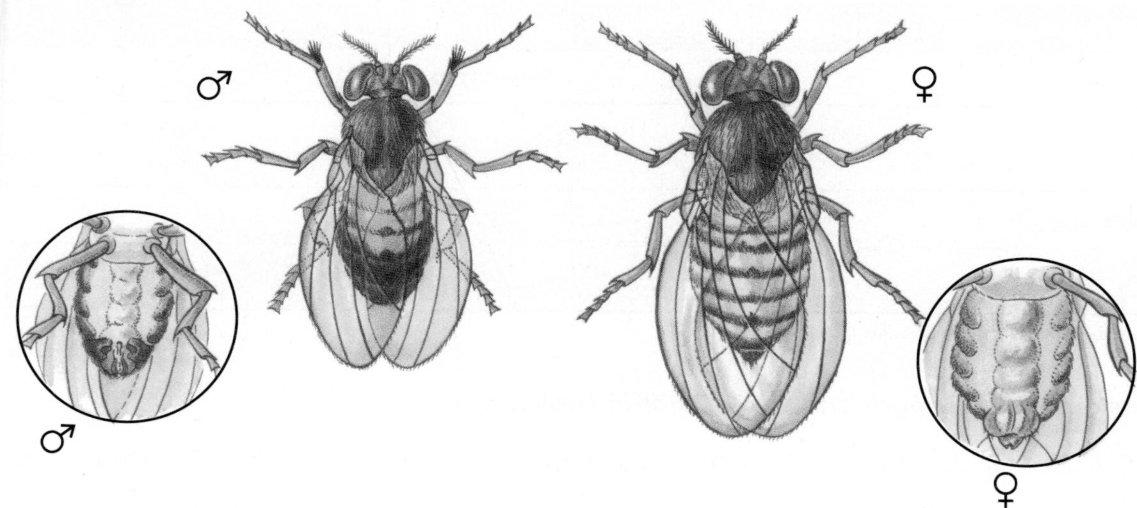

In *Drosophila,* long wings (*L*) are dominant over short (vestigial) wings (*l*), and gray body (*G*) is dominant over black (ebony) body (*g*). Consider the cross *LlGg* × *llgg* and complete this Punnett square:

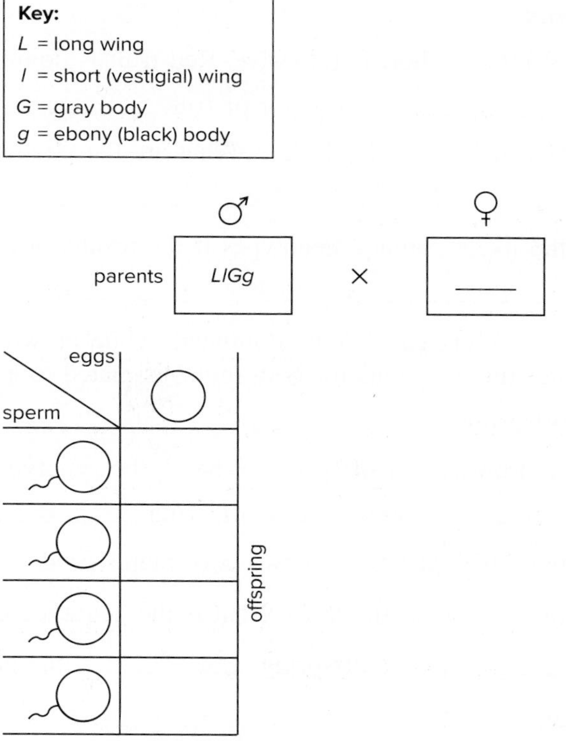

What is the expected phenotypic ratio for this cross? ____________________

Experimental Procedure: Wing Length and Body Color in Drosophila

If your instructor has frozen flies available, cross out the numbers in Table 10.4 and use the stereomicroscope or a hand lens to count the flies of each type given in Table 10.4. Otherwise simply use the data supplied for you in Table 10.4.

Table 10.4 Wing Length and Body Color in *Drosophila**

	Phenotypes				
	Long Gray	**Long Ebony**	**Short Gray**	**Short Ebony**	**Phenotypic Ratio**
Number of offspring	28	32	28	30	
Class data	128	120	120	120	

*Wings and body are understood in this table.

Conclusions: Wing Length and Body Color in Drosophila

- Calculate the actual phenotypic ratio based on the data and record in Table 10.4. Do the results differ from the expected ratio per individual data? ____________ Per class data? ____________ Explain your answer. ____

____

Two-Trait Genetics Problems

1. In tomatoes, tall is dominant and short is recessive. Red fruit is dominant and yellow fruit is recessive. Choose a key for height ____________ for color of fruit ____________ What is the genotype of a plant heterozygous for both traits? ____________ What are the possible gametes for this plant? ________

____

2. Using words, what are the likely parental genotypes if the results of a two-trait problem are 1:1:1:1 among the offspring? ______________________________ × ______________________________
3. In horses, black (*B*) and a trotting gait (*T*) are dominant, while brown (*b*) and a pacing gait (*t*) are recessive. If a black trotter (homozygous for both traits) is mated to a brown pacer, what phenotypic ratio is expected among the offspring? ____________________________
4. Two black trotters have a brown pacer offspring. What is the genotype of all horses involved? black trotter parents ____________ brown pacer offspring ____________
5. The phenotypic ratio among the offspring for two corn plants producing purple and smooth kernels is 9:3:3:1. (See page 114 of this lab for the key.) What is the genotype of these plants? parental plants ____ the 9 offspring ____________ 3 of the offspring ____________ the other 3 ____________ and the 1 offspring? ____________

6. Which matings could produce at least some fruit flies heterozygous in both traits? Write yes or no beside each. (You do not need a key.)

ggLl × *Ggll* ____________ *GGLl* × *ggLl* ____________ *GGLL* × *ggll* ____________

Explain your answer. ____

____

7. State two new crosses that could not produce fruit flies heterozygous in both traits.

____________ × ____________ ____________ × ____________

8. Chimpanzees are not deaf if they inherit both an allele *E* and an allele *G*. A cross between two deaf chimpanzees produces only chimpanzees that can hear. What are the genotypes of all chimpanzees involved? parents ____________ × ____________ offspring ____________

10.3 X-Linked Crosses

In animals such as fruit flies, chromosomes differ between the sexes. All but one pair of chromosomes in males and females are the same; these are called **autosomes** because they do not actively determine sex. The pair that is different is called the **sex chromosomes.** In fruit flies and humans, the sex chromosomes in females are XX and those in males are XY.

Some alleles on the X chromosome have nothing to do with gender, and these genes are said to be X-linked. The Y chromosome does not carry these genes and indeed carries very few genes. Males with a normal chromosome inheritance are never heterozygous for X-linked alleles, and if they inherit a recessive X-linked allele, it will be expressed.

Red/White Eye Color in *Drosophila*

In fruit flies, red eyes (X^R) are dominant over white eyes (X^r). You will be examining the results of the cross $X^RY \times X^RX^r$. Complete this Punnett square and state the expected phenotypic ratio for this cross.

females ____________________ males ____________________

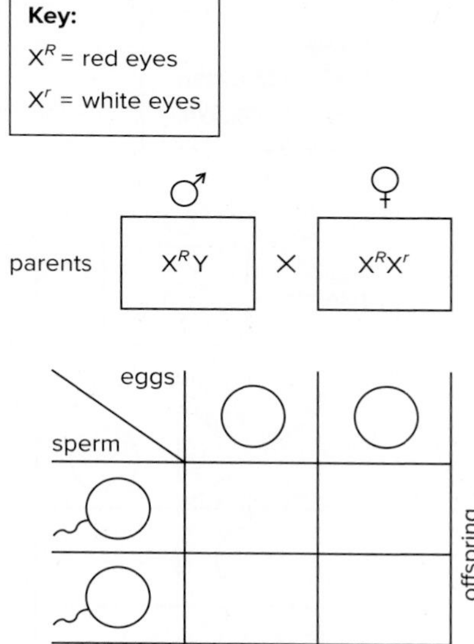

Experimental Procedure: Red/White Eye Color in Drosophila

If your instructor has frozen flies available, cross out the numbers in Table 10.5 and use the stereomicroscope or a hand lens to count the flies of each type given in Table 10.5. Use the art on page 115 to tell males from females, and record male and female data separately. If frozen flies are not available, simply use the data supplied for you in Table 10.5.

Table 10.5 Red/White Eye Color in *Drosophila*

	Number of Offspring		
Your Data:	**Red Eyes**	**White Eyes**	**Phenotypic Ratio**
Males	16	17	
Females	63	0	
Class Data:			
Males	45	48	
Females	215	0	

Conclusions: Red/White Eye Color in Drosophila

- Calculate the phenotypic ratios based on the data for males and females separately and record them in Table 10.5. Do the results differ from the expected ratio per individual data? ____________ Per class data? ____________ If so, explain your answer. ____
____
____
- Using the Punnett square provided, calculate the expected phenotypic results for the cross $X^RY \times X^rX^r$. What is the expected phenotypic ratio among the offspring? males ____________ females ____________

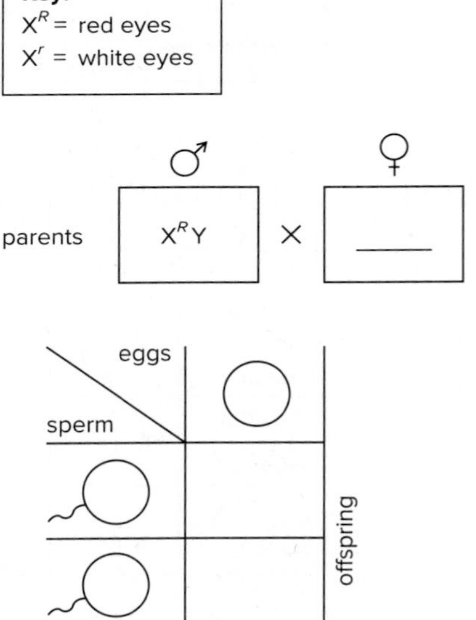

X-Linked Genetics Problems

1. State the genotypes and gametes for each of these fruit flies:

	genotype	gamete(s)
white-eyed male	______	______
white-eyed female	______	______
red-eyed male	______	______
homozygous red-eyed female	______	______
heterozygous red-eyed female	______	______

2. What are the phenotypic ratios if a white-eyed female is crossed with a red-eyed male?
   males ______ females ______
3. Regardless of any type of cross, do white-eyed males inherit the allele for white eyes from their father or their mother? ______ Explain your answer. ______
4. In sheep, horns are sex linked; H = horns and h = no horns. Using symbols, what cross do you recommend if a farmer wants to produce hornless males? ______ × ______
5. In *Drosophila,* bar eye is sex linked; B = bar eye and b = no bar eye. What are the genotypes of the parents and the phenotypic ratios for these crosses?
   bar-eyed male × non–bar-eyed female ______, ______
   bar-eyed male × heterozygous female ______, ______
   non–bar-eyed male × heterozygous female ______, ______
6. A female fruit fly has white eyes. What is the genotype of the father? ______ What could be the genotype of the mother? ______ or ______
7. In a cross between fruit flies, all the males have white eyes and the females are 1:1. What is the genotype of the parents? female parent ______ male parent ______
8. In a cross between fruit flies, a white-eyed male and a red-eyed female produce no offspring that have white eyes. What is the genotype of the parents? male parent ______ female parent ______
9. Make up a sex-linked genetic cross using words and parental genotypes. Create a key and show a Punnett square and the phenotypic ratios.

10.4 Chi-Square Analysis

Your experimental results can be evaluated using the **chi-square** (χ^2) test. This is the statistical test most frequently used to determine whether data obtained experimentally provide a "good fit," or approximation, to the expected data. Basically, the chi-square test can determine whether any deviations from the expected values are due to chance. Chance alone can cause the actual observed ratio to vary somewhat from the calculated ratio for a genetic cross. For example, a ratio of exactly 3:1 for a monohybrid cross is only rarely observed. Actual results will differ, but at some point, the difference is so great as to be unexpected. The chi-square test indicates this point. After this point, the original hypothesis (for example, that a monohybrid cross gives a 3:1 ratio) is not supported by the data.

The formula for the test is $\chi^2 = \Sigma(d^2/e)$

where χ^2 = chi-square
Σ = sum of
d = difference between expected and observed results (most often termed the *deviation*)
e = expected results

For example, in a monohybrid cross involving fruit flies with long wings (dominant) and fruit flies with short wings (recessive), a 3:1 ratio is expected. Therefore, if you count 160 flies, 40 are expected to have short wings, and 120 are expected to have long wings. But if you count 44 short-winged flies and 116 long-winged flies, then the value for the chi-square test would be as calculated in Table 10.6.

Table 10.6 Calculation of Chi-Square (Example)

Phenotype	Observed Number	Expected Results (*e*)	Difference (*d*)	d^2	Partial Chi-Square (d^2/e)
Short wings	44	40	4	16	16/40 = 0.400
Long wings	116	120	4	16	16/120 = 0.133
			Chi-square = $\chi^2 = \Sigma(d^2/e) = 0.400 + 0.133 = 0.533$		

Now look up this chi-square value (χ^2) in a table that indicates whether the probability (p) is that the differences noted are due only to chance in the form of a random sampling error or whether the results should be explained on the basis of a different prediction (hypothesis). In Table 10.7, the notation C refers to the number of "classes," which in this laboratory would be determined by the number of phenotypic traits studied. The example involves two classes: short wings and long wings. However, as indicated in the table by $C - 1$, it is necessary to subtract 1 from the total number of classes. In the example, $C - 1 = 1$. Therefore, the χ^2 value (0.533) would fall in the first line of Table 10.7 between 0.455 and 1.074. These correspond with p values of 0.50 and 0.30, respectively. This means that, by random chance, this difference between the actual count and the expected count would occur between 30% and 50% of the time. In biology, it is generally accepted that a p value greater than 0.10 is acceptable and that a p value lower than 0.10 indicates that the results cannot be due to random sampling and therefore do not support (do not "fit") the original prediction (hypothesis). A chi-square analysis is used to *refute* (falsify) a hypothesis, not to *prove* it.

Table 10.7 Values of Chi-Square

	Hypothesis Is Supported						***Hypothesis Is Not Supported***			
	Differences Are Insignificant						**Differences Are Significant**			
p	0.99	0.95	0.80	0.50	0.30	0.20	0.10	0.05	0.02	0.01
$C - 1$										
1	0.00016	0.0039	0.064	0.455	1.074	1.642	2.706	3.841	5.412	6.635
2	0.0201	0.103	0.446	1.386	2.408	3.219	4.605	5.991	7.824	9.210
3	0.115	0.352	1.005	2.366	3.665	4.642	6.251	7.815	9.837	11.345
4	0.297	0.711	1.649	3.357	4.878	5.989	7.779	9.488	11.668	13.277
5	0.554	1.145	2.343	4.351	6.064	7.289	9.236	11.070	13.388	15.086

Source: Keeton, W. T. et al., *Laboratory Guide for Biological Science,* New York: Norton, 1968, p. 189.

Use Table 10.8 for performing a chi-square analysis of your results from a previous Experimental Procedure in this laboratory. If you performed a one-trait (monohybrid) cross, you will use only the first two lines. If you performed a two-trait (dihybrid) cross, you will use four lines.

χ^2 = ____________________

$C - 1$ = ____________________

p (from Table 10.7) = ____________________

Table 10.8 Calculation of Chi-Square

Phenotype	Observed Number	Expected Results (e)	Difference (*d*)	d^2	Partial Chi-Square (d^2/e)
					=
					=
					=
					=
			Chi-square = $\chi^2 = \Sigma(d^2/e)$ =		

Conclusions: Chi-Square Analysis

- Do your results support your original prediction? ____________________
- If not, how can you account for this? ____________________

____

Laboratory Review 10

________________ 1. What are the alternate forms of a gene called?

________________ 2. What word is used to describe the possession of one dominant allele and one recessive allele?

________________ 3. What word is used to refer to an individual's appearance?

________________ 4. Which of Mendel's laws says alleles separate during gamete formation so that each gamete contains only one allele for each trait?

________________ 5. What is used to determine the results of a cross between individuals with known genotypes?

________________ 6. If the phenotypic ratio among the offspring is 3:1, what is the genotype of the parents?

________________ 7. How would a tobacco seedling with the genotype *cc* for chlorophyll be described?

________________ 8. If the allele for a purple corn kernel (*P*) is dominant over the allele for a yellow corn kernel (*p*), what is the genotype of the yellow kernel?

________________ 9. What is the cross between two individuals with *GgRr* genotypes called?

________________ 10. How many of the offspring of the cross between two *GgRr* individuals will show the recessive phenotype for both traits?

________________ 11. Which of Mendel's laws says the separation of one pair of alleles into gametes does not impact the separation of a second pair of alleles into gametes?

________________ 12. How many different kinds of gametes can be produced by a parent with a *Ggrr* genotype?

________________ 13. What is the name given to the chromosomes that do not actively determine the gender of an organism?

________________ 14. Can a female fruit fly that is homozygous dominant for red eyes have any offspring (male or female) with white eyes?

Thought Questions

15. If an organism displays the dominant trait phenotype (e.g., green color in tobacco plants), it is uncertain whether the organism's genotype is homozygous dominant (*CC*) or heterozygous (*Cc*). What kind of cross (a known genotype) could be done to determine the genotype of the unknown organism? Explain your reply by describing the expected offsprings' phenotypes.

16. If you count 52 purple rough corn kernels and 50 yellow smooth corn kernels, how many purple smooth corn kernels should you expect to count? How many yellow rough corn kernels should you expect to count? Explain your reply.

17. If a trait is an X-linked recessive trait, why is it unlikely that many females with the recessive trait (e.g., white eyes in fruit flies) phenotype will exist?

LABORATORY

11

Human Genetics

Learning Outcomes

11.1 Determining the Genotype

- Determine genotypes by observation of individuals and their relatives.

11.2 Determining Inheritance

- Complete genetic problems involving autosomal dominant, autosomal recessive, and X-linked recessive alleles.
- Complete genetic problems involving multiple allele inheritance and use blood type to help determine paternity.

11.3 Genetic Counseling

- Analyze a karyotype to determine if a person's chromosomal inheritance is as expected or whether a chromosome anomaly has occurred.
- Analyze a pedigree to determine if the pattern of inheritance is autosomal dominant, autosomal recessive, or X-linked recessive.
- Construct a pedigree to determine the chances of inheriting a particular phenotype when provided with generational information.

Introduction

In this laboratory, you will discover that the same principles of genetics apply to humans as they do to plants and fruit flies. A gene has two alternate forms, called **alleles,** for any trait, such as hairline, finger length, and so on. One possible allele, designated by a capital letter, is **dominant** over the **recessive** allele, designated by a lowercase letter. An individual can be **homozygous dominant** (two dominant alleles, *EE*), **homozygous recessive** (two recessive alleles, *ee*), or **heterozygous** (one dominant and one recessive allele, *Ee*). **Genotype** refers to an individual's alleles, and **phenotype** refers to an individual's appearance (Fig. 11.1). Homozygous dominant and also heterozygous individuals show the dominant phenotype; homozygous recessive individuals show the recessive phenotype.

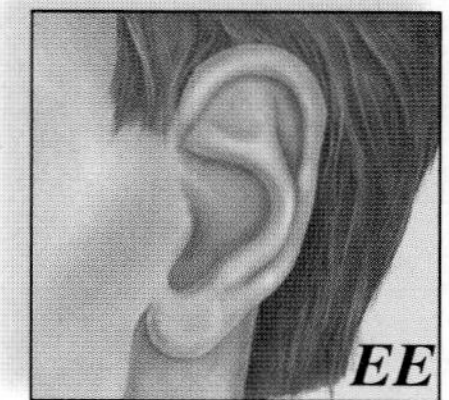

a. Unattached earlobe

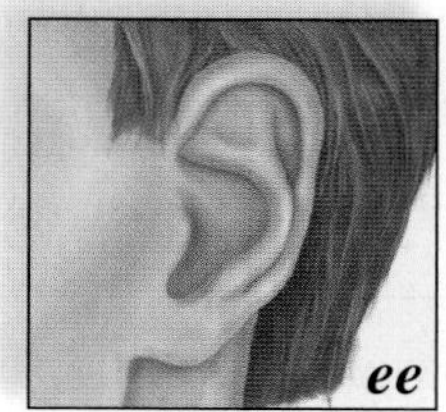

b. Attached earlobe

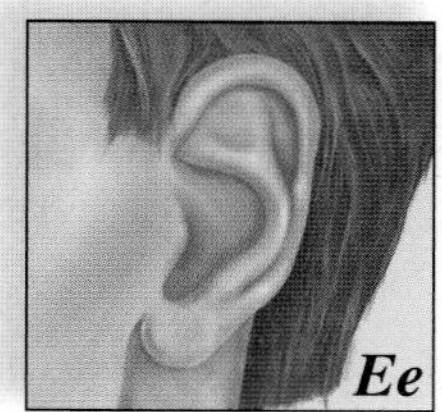

c. Unattached earlobe

Figure 11.1 Genotype versus phenotype. Unattached earlobes (*E*) are dominant over attached earlobes (*e*). **a.** Homozygous dominant individuals have unattached earlobes. **b.** Homozygous recessive individuals have attached earlobes. **c.** Heterozygous individuals have unattached earlobes.

11.1 Determining the Genotype

Humans inherit 46 chromosomes that occur in 23 pairs. Twenty-two of these pairs are called autosomes and one pair is the sex chromosomes. Autosomal traits are determined by alleles on the autosomal chromosomes.

Autosomal Dominant and Recessive Traits

Figure 11.2 shows a few human traits.

1. What is the homozygous dominant genotype for type of hairline? ______________ What is the phenotype? ______________________________

2. What is the homozygous recessive genotype for finger length? ______________ What is the phenotype? ______________________________

3. Why does the heterozygous individual *Ff* have freckles? ______________________________

Figure 11.2 Commonly inherited traits in humans. The alleles indicate which traits are dominant and which are recessive. (a) ©Superstock; (b) ©John Foxx/Stockbyte/Getty Images RF; (c–f) ©McGraw-Hill Education/Bob Coyle, photographer; (g) ©Hero/Corbis/Glow Images RF; (h) ©Glow Images RF

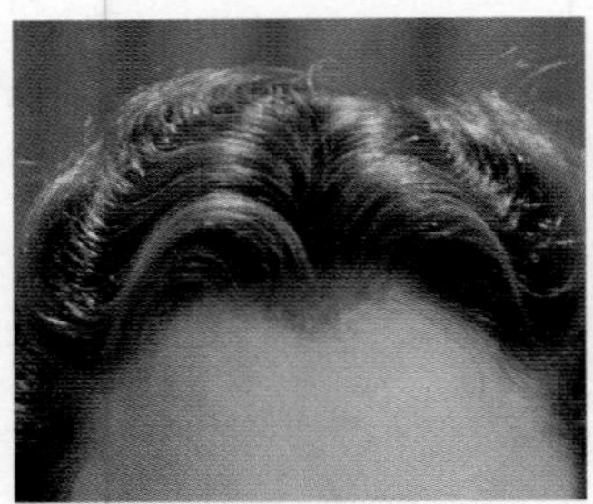

a. Widow's peak: *WW* or *Ww*

b. Straight hairline: *ww*

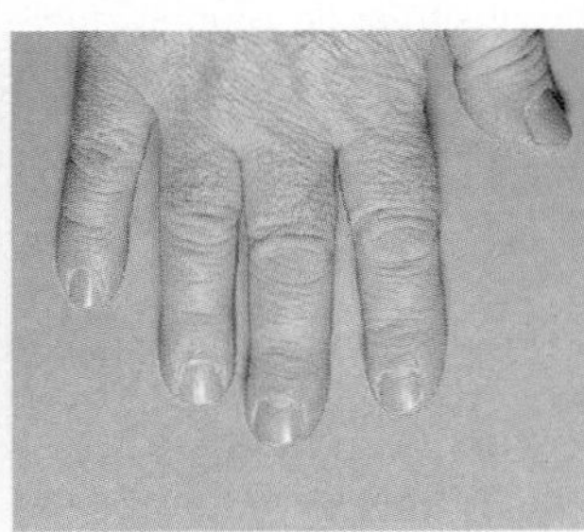

e. Short fingers: *SS* or *Ss*

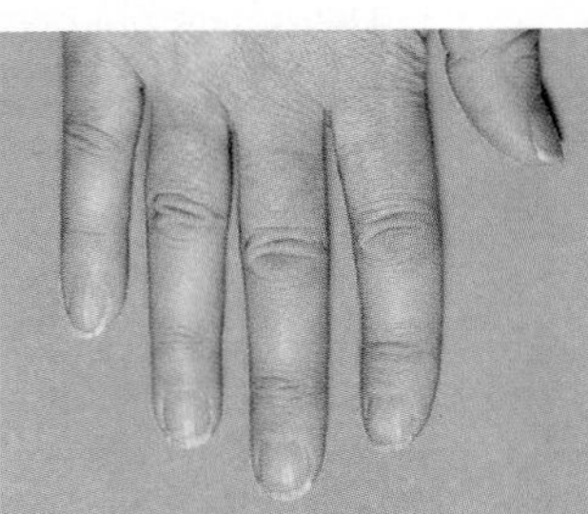

f. Long fingers: *ss*

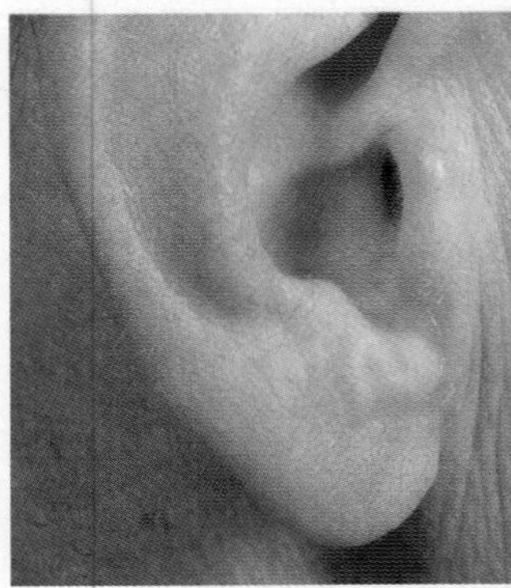

c. Unattached earlobes: *EE* or *Ee*

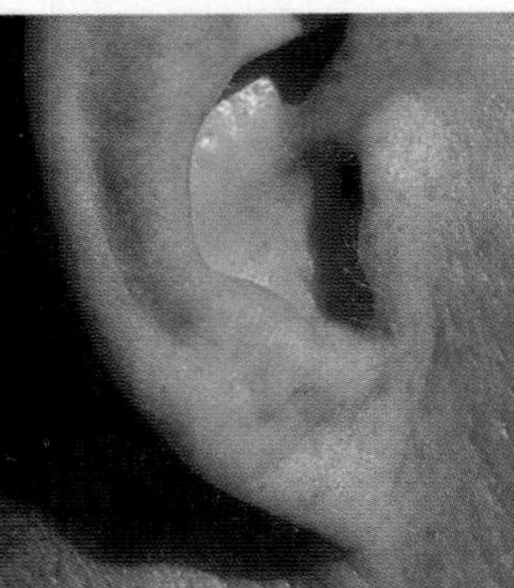

d. Attached earlobes: *ee*

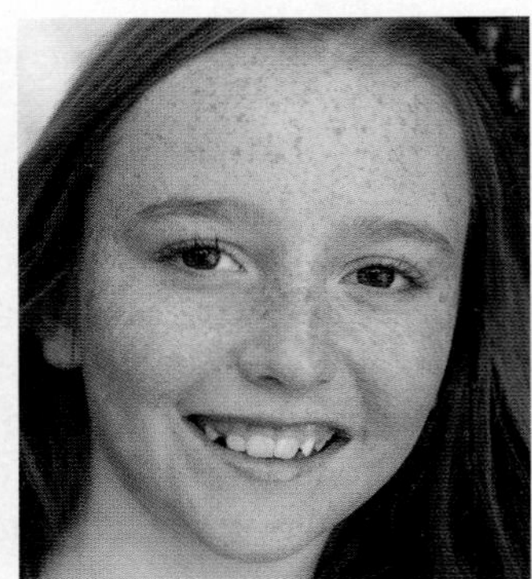

g. Freckles: *FF* or *Ff*

h. No freckles: *ff*

These genetic problems use the alleles from Figure 11.2 and Table 11.1.

4. Maria and the members of her immediate family have attached earlobes. What is Maria's genotype? ______________ Her maternal grandfather has unattached earlobes. Deduce the genotype of her maternal grandfather. ______________ Explain your answer. ______________________________

5. Moses does not have a bent little finger, but his parents do. Deduce the genotype of his parents. ______________ of Moses. ______________ Explain your answer. ______________________________

6. Manny is adopted. He has hair on the back of his hand. Could both of his parents have had hair on the back of the hand? ______________ Could both of his parents have had no hair on the back of the hand? ______________ Explain your answer. ______________________________

Experimental Procedure: Human Traits

1. For this Experimental Procedure, you will need a lab partner to help you determine your phenotype for the traits listed in the first column of Table 11.1.
2. Determine your probable genotype. If you have the recessive phenotype, you know your genotype. If you have the dominant phenotype, you may be able to decide whether you are homozygous dominant or heterozygous by recalling the phenotype of your parents, siblings, or children. Circle your probable genotype in the second column of Table 11.1.
3. Your instructor will tally the class's phenotypes for each trait so that you can complete the third column of Table 11.1.
4. Complete Table 11.1 by calculating the percentage of the class with each trait. Are dominant phenotypes always the most common in a population? ______________ Explain your answer. ____________________

____

Table 11.1 Autosomal Human Traits

Trait: d = Dominant r = Recessive	Probable Genotypes	Number in Class	Percentage of Class with Trait
Hairline:			
Widow's peak (d)	*WW* or *Ww*	________	________
Straight hairline (r)	*ww*	________	________
Earlobes:			
Unattached (d)	*UU* or *Uu*	________	________
Attached (r)	*uu*	________	________
Skin pigmentation:			
Freckles (d)	*FF* or *Ff*	________	________
No freckles (r)	*ff*	________	________
Hair on back of hand:			
Present (d)	*HH* or *Hh*	________	________
Absent (r)	*hh*	________	________
Thumb hyperextension—"hitchhiker's thumb":			
Last segment cannot be bent backward (d).	*TT* or *Tt*	________	________
Last segment can be bent back to 60° (r).	*tt*	________	________
Bent little finger:			
Little finger bends toward ring finger (d).	*LL* or *Ll*	________	________
Straight little finger (r)	*ll*	________	________
Interlacing of fingers:			
Left thumb over right (d)	*II* or *Ii*	________	________
Right thumb over left (r)	*ii*	________	________

11.2 Determining Inheritance

Recall that a Punnett square is a means to determine the genetic inheritance of offspring if the genotypes of both parents are known. In a **Punnett square,** all possible types of sperm are lined up vertically, and all possible types of eggs are lined up horizontally, or vice versa, so that every possible combination of gametes occurs within the square. Figure 11.3 shows how to construct a Punnett square when autosomal alleles are involved.

Figure 11.3 Punnett square.
In a Punnett square, all possible sperm are displayed vertically and all possible eggs are displayed horizontally, or vice versa. The genotypes of the offspring (in this case, also the phenotypes) are in the squares.

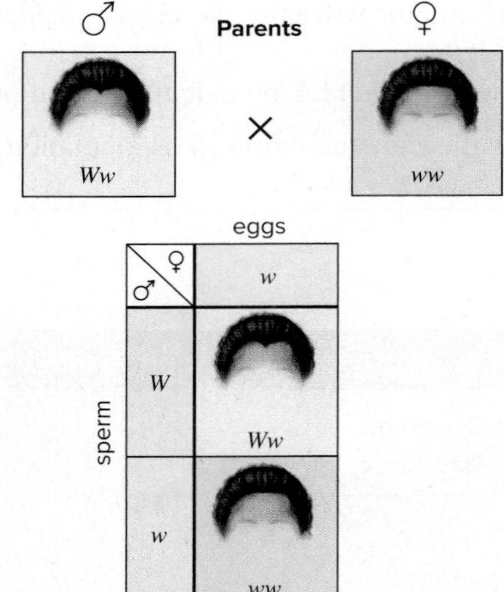

Results of Cross
Phenotypic ratio 1:1
☐ Chance of widow's peak $\frac{1}{2}$ = 50%
☐ Chance of straight hairline $\frac{1}{2}$ = 50%

Inheritance of Genetic Disorders

Figure 11.4 can be used to learn the chances that a particular phenotype will occur.

In Figure 11.4*a,*

$\frac{1}{4}$ of the offspring have the recessive phenotype = __________ % chance

$\frac{3}{4}$ of the offspring have the dominant phenotype = __________ % chance

In Figure 11.4*b,*

$\frac{1}{2}$ of the offspring have the recessive or the dominant phenotype = __________ % chance

In all the following genetic problems, use letters to fill in the parentheses with the genotype of the parents.

1. a. With reference to Figure 11.4*a,* if a genetic disorder is recessive and both parents are heterozygous (__________), what are the chances that an offspring will have the disorder? __________

b. With reference to Figure 11.4*a,* if a genetic disorder is dominant and both parents are heterozygous (__________), what are the chances that an offspring will have the disorder? __________

2. a. With reference to Figure 11.4*b,* if the parents are heterozygous (__________) by homozygous recessive (__________), and the genetic disorder is recessive, what are the chances that the offspring will have the disorder? __________

b. With reference to Figure 11.4*b,* if the parents are heterozygous (__________) by homozygous recessive (__________), and the genetic disorder is dominant, what are the chances that an offspring will have the disorder? __________

Figure 11.4 Two common patterns of autosomal inheritance in humans.
a. Both parents are heterozygous. **b.** One parent is heterozygous and the other is homozygous recessive. The letter *A* stands for any trait that is dominant and the letter *a* stands for any trait that is recessive. Substitute the correct alleles for the problem you are working on. For example, *C* = normal; *c* = cystic fibrosis.

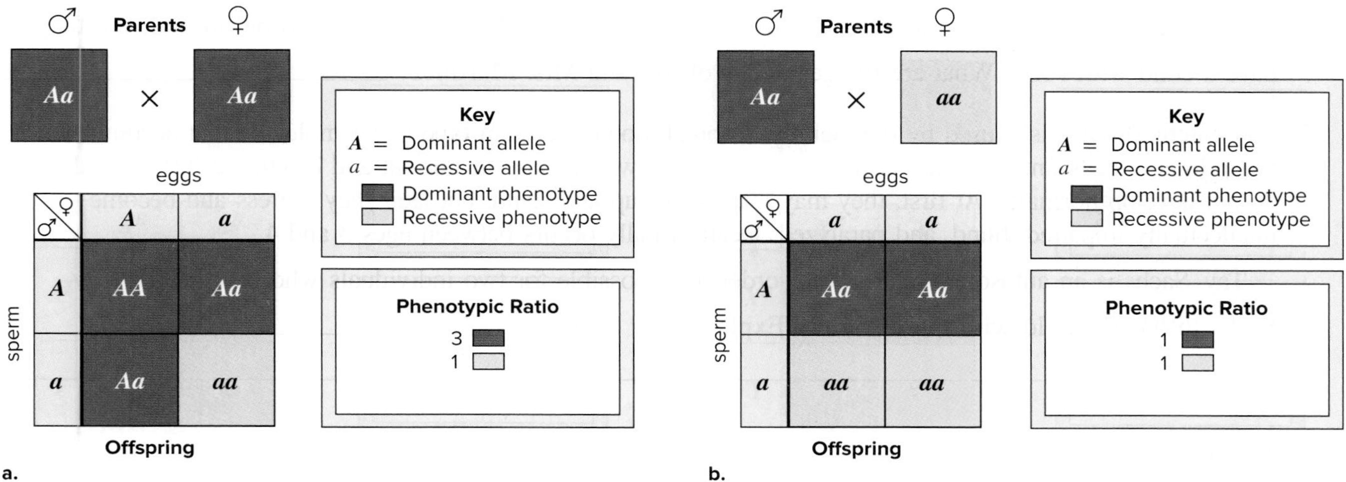

Autosomal Disorders

1. **Neurofibromatosis** (**NF**), sometimes called von Recklinghausen disease, is one of the most common genetic disorders. It affects roughly 1 in 3,000 people. It is seen equally in every racial and ethnic group throughout the world. At birth or later, the affected individual may have six or more large tan spots on the skin. Such spots may increase in size and number and become darker. Small benign tumors (lumps) called neurofibromas may occur under the skin or in the muscles. Neurofibromas are made up of nerve cells and other cell types.

   Neurofibromatosis is a dominant disorder. If a heterozygous woman reproduces with a homozygous normal man, what are the chances a child will have neurofibromatosis? ____________________

2. **Cystic fibrosis** is due to abnormal mucus-secreting tissues. At first, the infant may have difficulty regaining the birth weight despite good appetite and vigor. A cough associated with a rapid respiratory rate but no fever indicates lung involvement. Large, frequent, and foul-smelling stools are due to abnormal pancreatic secretions. Whereas children previously died in infancy due to infections, they now often survive because of antibiotic therapy.

   Cystic fibrosis is a recessive disorder. A **carrier** is an individual who appears to be normal but carries a recessive allele for a genetic disorder. A man and a woman are both carriers (__________) for cystic fibrosis. What are the chances a child will have cystic fibrosis? ____________________

3. **Huntington disease** does not appear until the 30s or early 40s. There is a progressive deterioration of the individual's nervous system that eventually leads to constant thrashing and writhing movements until insanity precedes death. Studies suggest that Huntington disease is due to a single faulty gene that has multiple effects, in which case there is now hope for a cure.

   People with Huntington disease seem to be more fertile than others. It is amazing that more than 1,000 of the cases in the United States in the past century can be traced to one man born in 1831.

   Huntington disease is a dominant disorder. Drina is 25 years old and as yet has no signs of Huntington disease. Her mother does have Huntington disease (__________), but her father is free (__________) of the disorder. What are the chances that Drina will develop Huntington disease? __________

   ____

Observation: Sex Chromosome Anomalies

A female with **Turner syndrome** (XO) has only one sex chromosome, an X chromosome; the O signifies the absence of the second sex chromosome. Because the ovaries never become functional, these females do not undergo puberty or menstruation, and their breasts do not develop. Generally, females with Turner syndrome have a short build, folds of skin on the back of the neck, difficulty recognizing various spatial patterns, and normal intelligence. With hormone supplements, they can lead fairly normal lives.

When an egg having two X chromosomes is fertilized by an X-bearing sperm, an individual with **poly-X syndrome** results. The body cells have three X chromosomes and therefore 47 chromosomes. Although they tend to have learning disabilities, poly-X females have no apparent physical anomalies, and many are fertile and have children with a normal chromosome count.

When an egg having two X chromosomes is fertilized by a Y-bearing sperm, a male with **Klinefelter syndrome** results. This individual is male in general appearance, but the testes are underdeveloped, and the breasts may be enlarged. The limbs of XXY males tend to be longer than average, muscular development is poor, body hair is sparse, and many XXY males have learning disabilities.

Jacob syndrome occurs in males who are usually taller than average, suffer from persistent acne, and tend to have speech and reading problems. At one time, it was suggested that XYY males were likely to be criminally aggressive, but the incidence of such behavior has been shown to be no greater than that among normal XY males.

Label each karyotype in Figure 11.6 as one of the syndromes just discussed. Explain your answers on the lines provided.

Figure 11.6 Sex chromosome anomalies.

(a–d) ©CNRI/SPL/Science Source

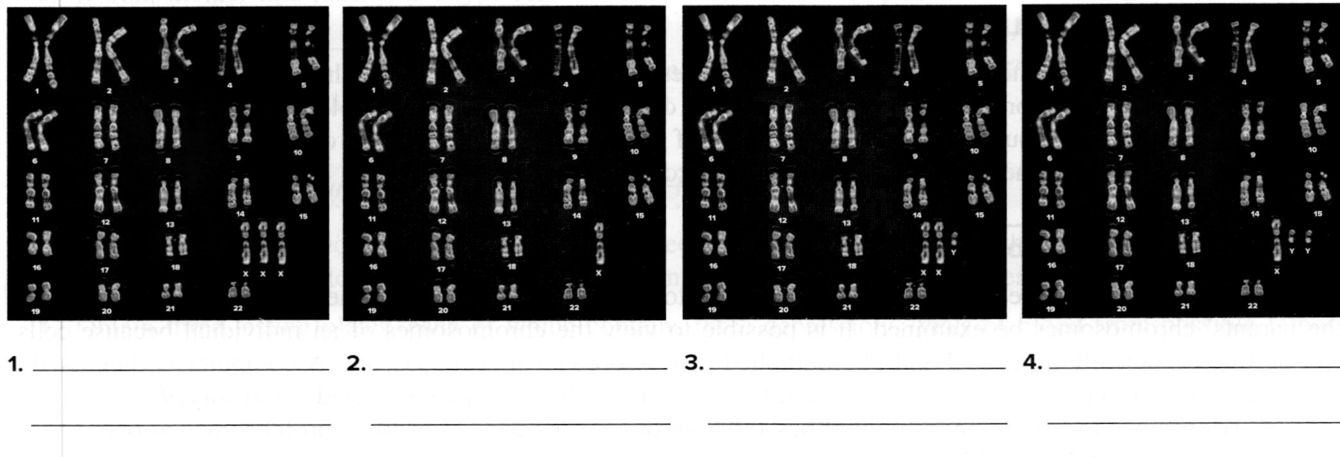

1. ______________________________
____
____

2. ______________________________
____
____

3. ______________________________
____
____

4. ______________________________
____
____

Determining the Pedigree

A pedigree shows the inheritance of a genetic disorder within a family and can help determine the inheritance pattern and whether any particular individual has an allele for that disorder. Then a Punnett square can be done to determine the chances of a couple producing an affected child.

The symbols used to indicate normal and affected males and females, reproductive partners, and siblings in a pedigree are shown in Figure 11.7.

For example, suppose you wanted to determine the inheritance pattern for straight hairline and you knew which members of a generational family had the trait (Fig. 11.8*a*).

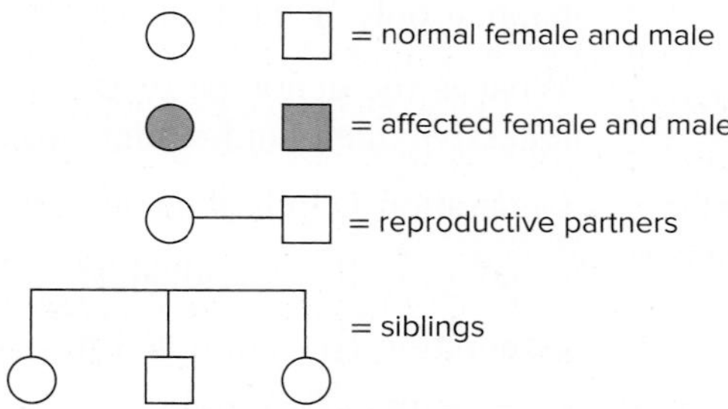

Figure 11.7 Pedigree symbols.

Figure 11.8 Autosomal pedigrees. **a.** Child with recessive phenotype can have parents without the recessive phenotype. **b.** Child with the dominant phenotype has parent(s) with the dominant phenotype; heterozygous parents can also have a child without the dominant phenotype.

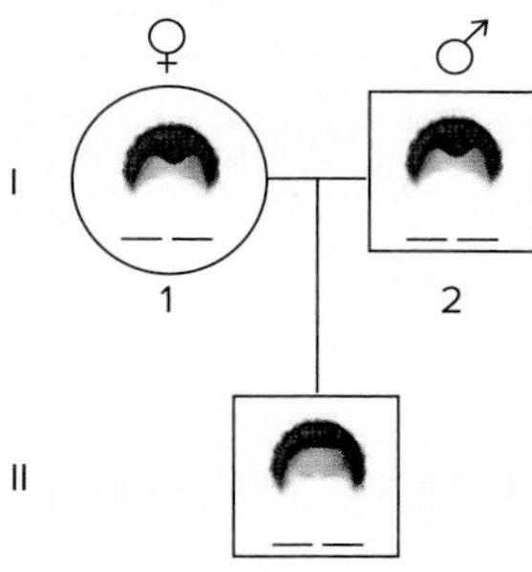

a. Straight hairline is recessive.

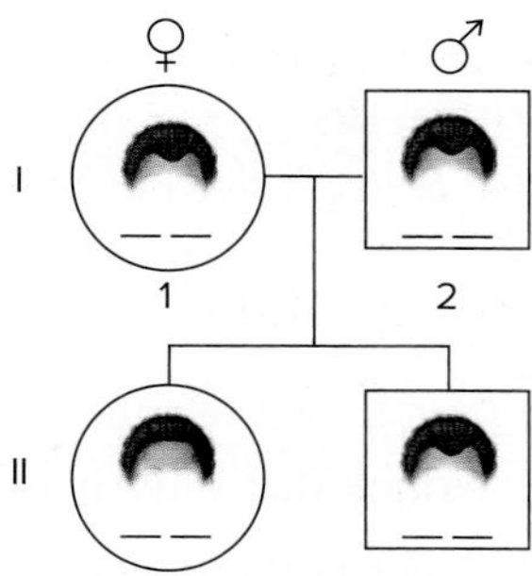

b. Widow's peak is dominant.

A pedigree allows you to determine that straight hairline is autosomal recessive because two parents without this phenotype have a child with the phenotype. This can happen only if the parents are heterozygous and straight hairline is recessive. Similarly, a pedigree allows you to determine that widow's peak is autosomal dominant (Fig. 11.8*b*): A child with this phenotype has at least one parent with the dominant phenotype, but again, heterozygous parents can produce a child without widow's peak. *Give each person in Figure 11.8*a *and* b *a genotype.*

Not shown is an X-linked recessive pedigree. An X-linked recessive phenotype occurs mainly in males, and it skips a generation because a female who inherits a recessive allele for the condition from her father may have a son with the condition.

Pedigree Analysis

For each of the following pedigrees, decide whether a trait is inherited as an autosomal dominant, autosomal recessive, or X-linked recessive. Then decide the genotype of particular individuals in the pedigree.

1. Study the following pedigree:

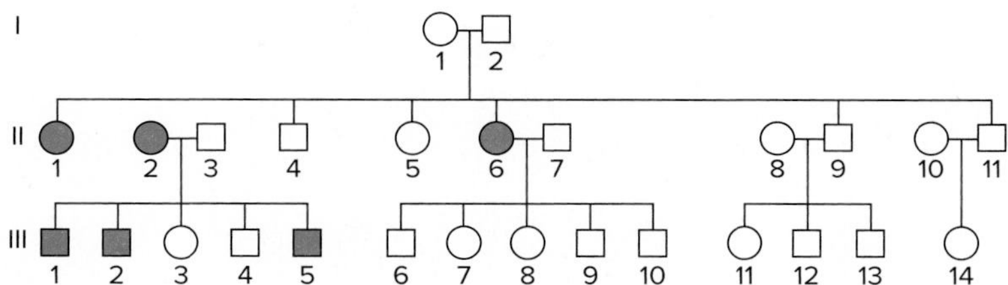

a. Notice that neither of the original parents is affected, but several children are affected. This could happen only if the trait were ____.

b. What is the genotype of the following individuals? Use *A* for the dominant allele and *a* for the recessive allele and explain your answer.

Generation I, individual 1: ____

Generation II, individual 1: ____

Generation III, individual 8: ____

2. Study the following pedigree:

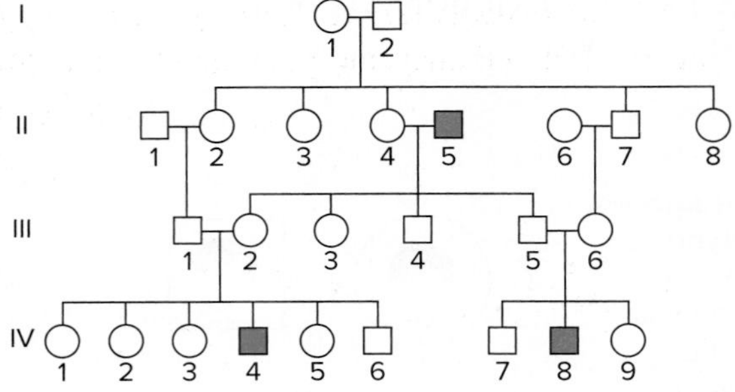

a. Notice that only males are affected. This could happen only if the trait were ____________________.

b. What is the genotype of the following individuals? Explain your answer.

Generation I, individual 1: ____

Generation II, individual 8: ____

Generation III, individual 1: ____

Construction of a Pedigree

You are a genetic counselor who has been given the following information from which you will construct a pedigree.

1. Your data: Henry has a double row of eyelashes, which is a dominant trait. Both his maternal grandfather and his mother have double eyelashes. Their spouses are normal. Henry is married to Isabella and their first child, Polly, has normal eyelashes. The couple wants to know the chances of any child having a double row of eyelashes.
2. What is your key for this trait?

   Key: __________ normal eyelashes __________ double row of eyelashes
3. *Construct two identical pedigrees with symbols (circles and squares) only for the underlined persons in #1.* The pedigrees start with the maternal grandfather and grandmother and end with Polly.

   Pedigree 1 **Pedigree 2**

4. Pedigree 1: Try out a pattern of autosomal dominant inheritance by assigning appropriate genotypes for an autosomal dominant pattern of inheritance to each person in this pedigree. Pedigree 2: Try out a pattern of X-linked dominant inheritance by assigning appropriate genotypes for this pattern of inheritance to each person in your pedigree. Which pattern is correct? ______________________
5. Use correct genotypes to show a cross between Henry and Isabella and from experience with crosses, state the expected phenotypic ratio among the offspring:

   Cross Henry Isabella **Phenotypic ratio:**

   __________ × __________ __________
6. What are the percentage chances that Henry and Isabella will have a child with double eyelashes? __________ "Chance has no memory," and each child has the same chance for double eyelashes. Explain why.

   ____

   ____

Laboratory Review 11

______________ **1.** What phrase describes an individual with two dominant alleles?

______________ **2.** What word refers to the alleles possessed by an individual?

______________ **3.** How many autosomes do humans have?

______________ **4.** An individual will have a widow's peak if the genotype is *WW* or *Ww*. Is widow's peak an autosomal dominant or autosomal recessive trait?

______________ **5.** What technique can be used to determine the genetic inheritance of offspring when the parents have known genotypes?

______________ **6.** Cystic fibrosis is an autosomal recessive disease. What word refers to an individual who has one recessive allele for cystic fibrosis?

______________ **7.** What is the genotype of individuals who will not develop Huntington disease (autosomal dominant)?

______________ **8.** From whom does a male child inherit color blindness—his mother or his father?

______________ **9.** What is the genotype of a carrier for hemophilia?

______________ **10.** How many dominant alleles does someone with AB blood type have?

______________ **11.** Can a child with AB blood type have a parent with O blood?

______________ **12.** What is the presence of three chromosomes (instead of two) called, and what human chromosome is most commonly affected in this fashion?

______________ **13.** What is created when chromosomes are paired by size and shape?

______________ **14.** What technique is used to determine the inheritance pattern of specific traits within a family?

Thought Questions

15. The blood types of parents and their two children are determined. The father has type A blood. The mother has type O blood. One child has type A blood and the other has type O blood. What are the genotypes of all people involved? Explain your reply.

16. What are the chances that a color-blind man will have a color-blind grandson (the child of the color-blind man's normal vision daughter and a normal vision male)? What are the genotypes of all people involved?

17. Recall what you learned about the fate of homologous chromosomes during meiosis I and the sister chromatids during meiosis II, and explain how trisomies like trisomy 21 or poly-X syndrome could occur.

LABORATORY

12

DNA Biology and Technology

Learning Outcomes

12.1 DNA Structure and Replication
- Explain how the structure of DNA facilitates replication.
- Explain how DNA replication is semiconservative.

12.2 RNA Structure
- List the ways in which RNA structure differs from DNA structure.

12.3 DNA and Protein Synthesis
- Describe how DNA is able to store information.
- State the function of transcription and translation during protein synthesis.

12.4 Isolation of DNA and Biotechnology
- Describe the procedure for isolating DNA.
- Describe the process of DNA gel electrophoresis.

12.5 Detecting Genetic Disorders
- Understand the relationship between an abnormal DNA base sequence and a genetic disorder.
- Suggest two ways to detect a genetic disorder.

Introduction

This laboratory pertains to molecular genetics and biotechnology. Molecular genetics is the study of the structure and function of **DNA** (**deoxyribonucleic acid**), the genetic material. Biotechnology refers to the use of natural biological systems to create a product or achieve some other end desired by humans. **Genetic engineering,** or modification of an organism's genes, is a form of biotechnology.

First we will study the structure of DNA and see how that structure facilitates DNA replication in the nucleus of cells. DNA replicates prior to mitosis; following mitosis, each daughter nucleus has a complete copy of the genetic material. DNA replication also precedes the production of gametes.

Then we will study the structure of **RNA** (**ribonucleic acid**) and how it differs from that of DNA before examining how DNA, with the help of RNA, specifies protein synthesis. The linear construction of DNA, in which nucleotide follows nucleotide, is paralleled by the linear construction of the primary structure of protein, in which amino acid follows amino acid. Essentially, we will see that the sequence of nucleotides in DNA codes for the sequence of amino acids in a protein. We will also review the role of three types of RNA in protein synthesis. DNA's code is passed to messenger RNA (mRNA), which moves to the ribosomes containing ribosomal RNA (rRNA). Transfer RNA (tRNA) brings the amino acids to the ribosomes, and they become linked by peptide bonds in the order directed by mRNA. In this way, a particular polypeptide forms.

We now understand that a mutated gene has an altered DNA base sequence, and altered sequences can cause genetic disorders. You will have an opportunity to carry out a laboratory procedure that detects whether an individual is normal, has sickle-cell disease, or is a carrier.

Translation

During translation, a polypeptide is made. DNA specifies the sequence of amino acids in a polypeptide because every three bases code for an amino acid. Therefore, DNA is said to have a **triplet code.** The bases in mRNA are complementary to the bases in DNA. Every three bases in mRNA are called a **codon.** One codon of mRNA represents one amino acid. Thus, the sequence of DNA bases serves as the blueprint for the sequence of amino acids assembled to make a protein. The correct sequence of amino acids in a polypeptide is the message that mRNA carries.

Messenger RNA leaves the nucleus and proceeds to the ribosomes, where protein synthesis occurs. As previously mentioned, transfer RNA (tRNA) molecules transfer amino acids to the ribosomes. Each RNA has one particular tRNA amino acid at one end and a specific **anticodon** at the other end (Fig. 12.5). *Label Figure 12.5,* where the amino acid is represented as a colored ball, the tRNA is green, and the anticodon is the sequence of three bases. (The anticodon is complementary to the mRNA codon.)

Figure 12.5 Transfer RNA (tRNA).
Each type of transfer RNA with a specific anticodon carries a particular amino acid to the ribosomes.

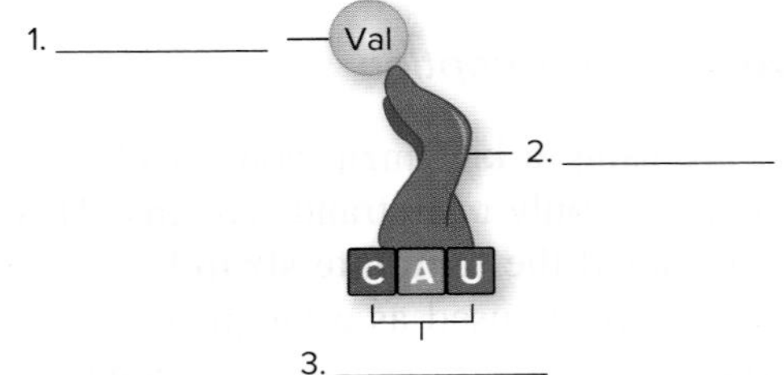

Observation: Translation

1. Figure 12.6 shows seven tRNA–amino acid complexes. Every amino acid has a name; in the figure, only the first three letters of the name are inside the ball.
2. If you are using a kit, arrange your tRNA–amino acid complexes in the order consistent with Table 12.7. Complete Table 12.7. Why are the codons and anticodons in groups of three? ____________________________
____

Figure 12.6 Transfer RNA diversity.
Each type of tRNA carries only one particular amino acid, designated here by the first three letters of its name.

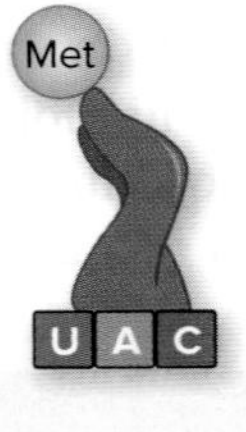

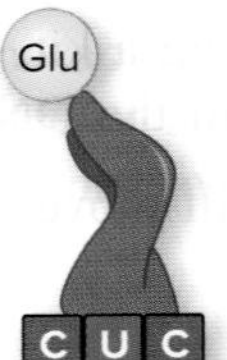

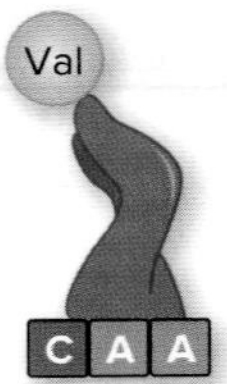

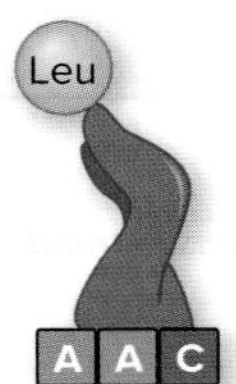

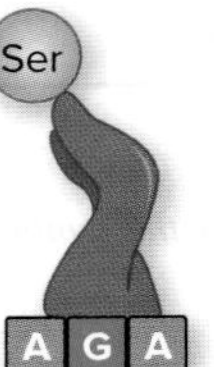

Table 12.7 Translation							
mRNA codons	AUG	CCC	GAG	GUU	GAU	UUG	UCU
tRNA anticodons							
Amino acid*							

*Use three letters only. See Table 12.8 for the full names of these amino acids.

Table 12.8 Names of Amino Acids

Abbreviation	Name
Met	methionine
Pro	proline
Asp	aspartate
Val	valine
Glu	glutamic acid
Leu	leucine
Ser	serine

3. Figure 12.7 shows the manner in which the polypeptide grows. A ribosome has three binding sites. They are the A (amino acid) site, the P (peptide) site, and the E (exit) site. A tRNA leaves from the E site after it has passed its amino acid or peptide to the newly arrived tRNA–amino acid complex. Then the ribosome moves forward, making room for the next tRNA–amino acid. This sequence of events occurs over and over until the entire polypeptide is borne by the last tRNA to come to the ribosome. Then a release factor releases the polypeptide chain from the ribosome. *In Figure 12.7, label the ribosome, the mRNA, and the peptide.*

Figure 12.7 Protein synthesis.
During protein synthesis, amino acids are added one at a time to the growing chain based on the mRNA codons. This process occurs at ribosomes in the cell.

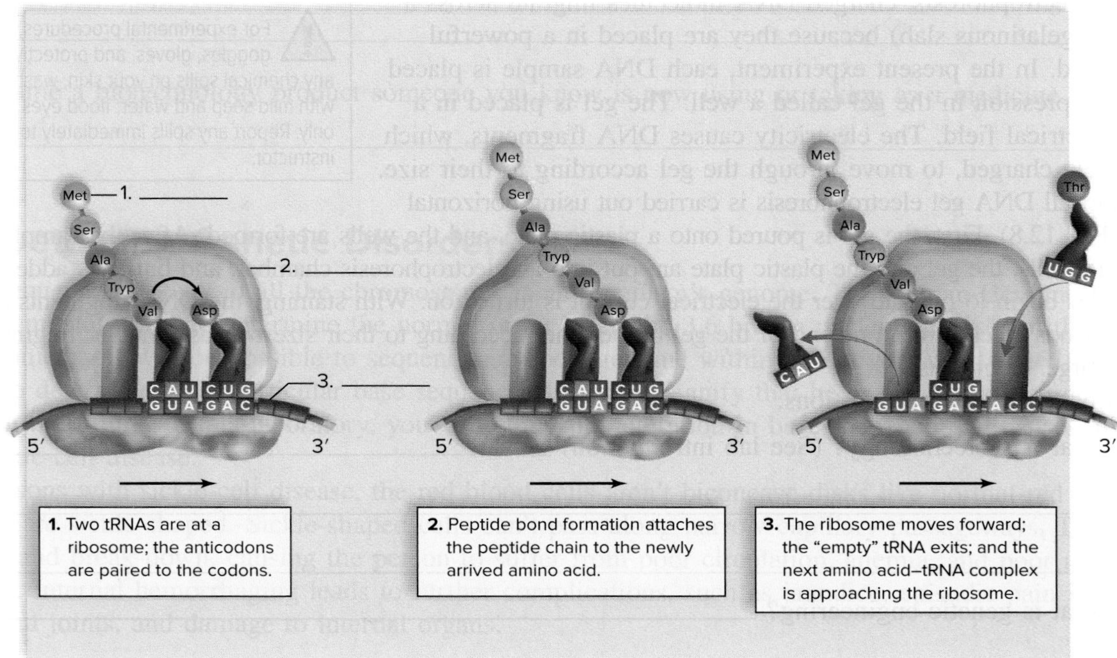

Met
Ser
Ala
Tryp
Val
Asp
Thr
1.
2.
3.
CAU CUG
GUA GAC
CUG
GUA GAC ACC
UGG
CAU
5′
3′
1. Two tRNAs are at a ribosome; the anticodons are paired to the codons.
2. Peptide bond formation attaches the peptide chain to the newly arrived amino acid.
3. The ribosome moves forward; the "empty" tRNA exits; and the next amino acid–tRNA complex is approaching the ribosome.

Sickle-shaped red blood cells are caused by an abnormal hemoglobin (Hb^S). Individuals with the Hb^AHb^A genotype are normal; those with the Hb^SHb^S genotype have sickle-cell disease, and those with the Hb^AHb^S have sickle-cell trait. Persons with sickle-cell trait do not usually have sickle-shaped cells unless they experience dehydration or mild oxygen deprivation.

Genetic Sequence for Sickle-Cell Disease

Examine Figure 12.9*a* and *b,* which show the DNA base sequence, the mRNA codons, and the amino acid sequence for a portion of the gene for Hb^A and the same portion for Hb^S.

1. In what one base does Hb^A differ from Hb^S? Hb^A ________ Hb^S ________

2. What are the codons that contain this base? Hb^A ________ Hb^S ________

3. What is the amino acid difference? Hb^A ________ Hb^S ________

This amino acid difference causes the polypeptide chain in sickle-cell hemoglobin to pile up as firm rods that push against the plasma membrane and deform the red blood cell into a sickle shape:

glutamic acid (polar *R* group)	valine (nonpolar *R* group)
$-CH_2-CH_2-C(=O)O^-$	$-CH_2-CH(CH_3)_2$

Figure 12.9 Sickle-cell disease.
a. When red blood cells are normal, the base sequence (in one location) for Hb^A alleles is CTC. **b.** In sickle-cell disease at these locations, it is CAC. (a–b) ©Bill Longcore/Science Source

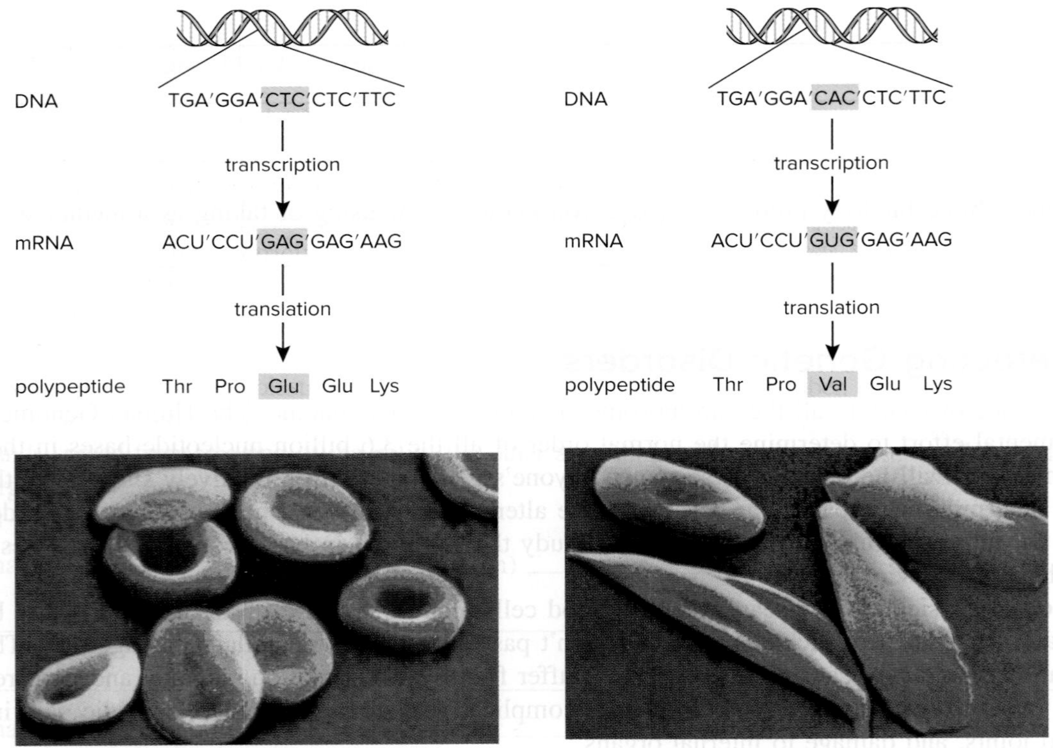

a. Normal red blood cells.

b. Sickle-shaped red blood cells.

Detection of Sickle-Cell Disease by Gel Electrophoresis

Three samples of hemoglobin have been subjected to protein gel electrophoresis. Protein gel electrophoresis (Fig. 12.10) is carried out in the same manner as DNA gel electrophoresis (see Fig. 12.8) except the gel has a different composition.

1. Sickle-cell hemoglobin (Hb^S) migrates more slowly toward the positive pole than normal hemoglobin (Hb^A) because the amino acid valine has no polar R groups, whereas the amino acid glutamic acid does have a polar R group.
2. In Figure 12.10, which lane contains only Hb^S, signifying that the individual is Hb^SHb^S?

   ____
3. Which lane contains only Hb^A, signifying that the individual is Hb^AHb^A? ________
4. Which lane contains both Hb^S and Hb^A, signifying that the individual is Hb^AHb^S? ________

Figure 12.10 Gel electrophoresis of hemoglobins.

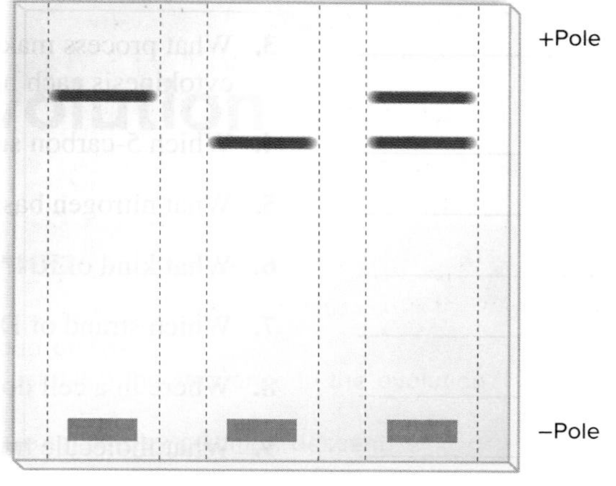

Detection by Genomic Sequencing

You are a genetic counselor. A young couple seeks your advice because sickle-cell disease occurs among the family members of each. You order DNA base sequencing to be done. The results come back that at one of the loci for normal hemoglobin, each has the abnormal sequence CAC instead of CTC. The other locus is normal. What are the chances that this couple will have a child with sickle-cell disease? ________________________________

Conclusion: Detecting Genetic Disorders

- What two methods of detecting sickle-cell disease were described in this section? ________________
  ____
- Which method is more direct and probably requires more expensive equipment to do? ________________
  ____
- Which method probably preceded the other method as a means to detect sickle-cell disease? ________________
  ____

13.2 Evidence from Comparative Anatomy

In the study of evolutionary relationships, parts of organisms are said to be **homologous** if they exhibit similar basic structures and embryonic origins. If parts of organisms are similar in function only, they are said to be **analogous.** Only homologous structures indicate an evolutionary relationship and are used to classify organisms.

Comparison of Adult Vertebrate Forelimbs

The limbs of vertebrates are homologous structures (Fig. 13.5). The similarity of homologous structures is explainable by descent from a common ancestor.

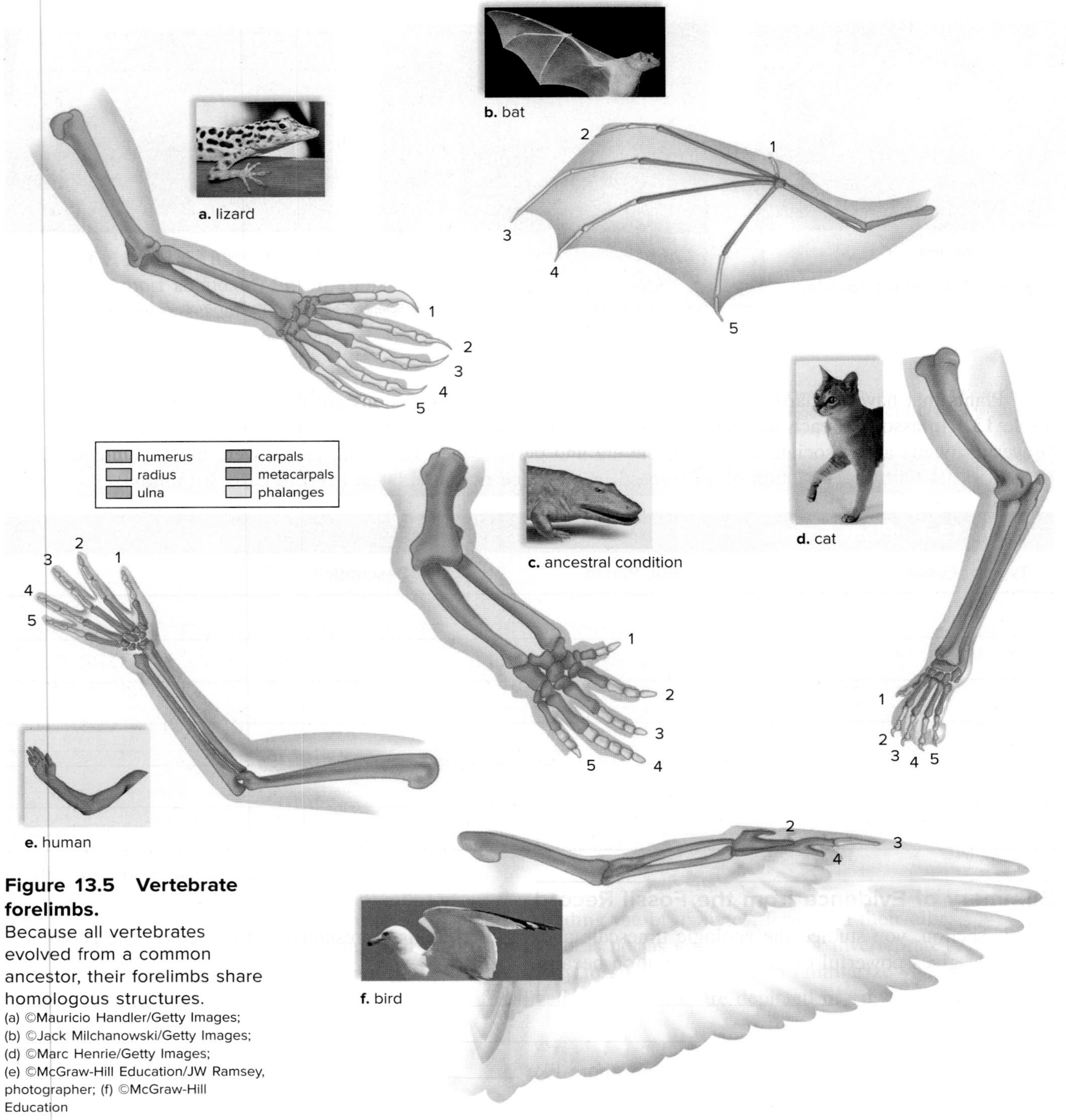

Figure 13.5 Vertebrate forelimbs.
Because all vertebrates evolved from a common ancestor, their forelimbs share homologous structures.
(a) ©Mauricio Handler/Getty Images; (b) ©Jack Milchanowski/Getty Images; (d) ©Marc Henrie/Getty Images; (e) ©McGraw-Hill Education/JW Ramsey, photographer; (f) ©McGraw-Hill Education

Observation: Vertebrate Forelimbs

1. Find the forelimb bones of the ancestral vertebrate in Figure 13.5. The basic components are the humerus (h), ulna (u), radius (r), carpals (c), metacarpals (m), and phalanges (p) in the five digits.
2. *Label the corresponding forelimb bones of the lizard, the bird, the bat, the cat, and the human.*
3. Fill in Table 13.5 to indicate which bones in each specimen appear to most resemble the ancestral condition and which most differ from the ancestral condition.
4. Adaptation to a way of life can explain the modifications that have occurred. Relate the change in bone structure to mode of locomotion in two examples.

   Example 1: ______________________________

   Example 2: ______________________________

Table 13.5 Comparison of Vertebrate Forelimbs

Animal	Bones That Resemble Common Ancestor	Bones That Differ from Common Ancestor
Lizard		
Bird		
Bat		
Cat		
Human		

Conclusion: Vertebrate Forelimbs

- Vertebrates are descended from a ____________________, but they are adapted to ____________________.

Comparison of Vertebrate Embryos

The anatomy shared by vertebrates extends to their embryological development. During early developmental stages, all animal embryos resemble each other closely, but as development proceeds the different types of vertebrates take on their own shape and form. In Figure 13.6, the reptile and bird embryo resemble each other more than either resembles a fish. What does that tell you about their evolutionary relationship? ______________________________

____

____

In this observation, you will see that as embryos all vertebrates have a postanal tail, a dorsal spinal cord, pharyngeal pouches, and various organs. In aquatic animals, pharyngeal pouches become functional gills (Fig. 13.6). In humans, the first pair of pouches becomes the cavity of the middle ear and auditory tube, the second pair becomes the tonsils, and the third and fourth pairs become the thymus and parathyroid glands.

Skull Features

Humans are omnivorous. A diet rich in meat does not require strong grinding teeth or well-developed facial muscles. Chimpanzees are herbivores, and a vegetarian diet requires strong teeth and strong facial muscles that attach to bony projections. Compare the skulls of the chimpanzee and the human in Figure 13.8 and answer the following questions:

1. **Supraorbital ridge:** For which skull is the supraorbital ridge (the region of frontal bone just above the eye socket) thicker? Record your observations in Table 13.7.
2. **Sagittal crest:** Which skull has a sagittal crest, a projection for muscle attachments that runs along the top of the skull? Record your observation in Table 13.7.
3. **Frontal bone:** Compare the slope of the frontal bones of the chimpanzee and human skulls. How are they different? Record your observations in Table 13.7.
4. **Teeth:** Examine the teeth of the adult chimpanzee and adult human skulls. Are the incisors (two front teeth) vertical or angled? Do the canines overlap the other teeth? Are the molars larger or moderate in size? Record your observations in Table 13.7.
5. **Chin:** What is the position of the mouth and chin in relation to the profile for each skull? Record your observations in Table 13.7.

Figure 13.8 Chimpanzee and human skulls.

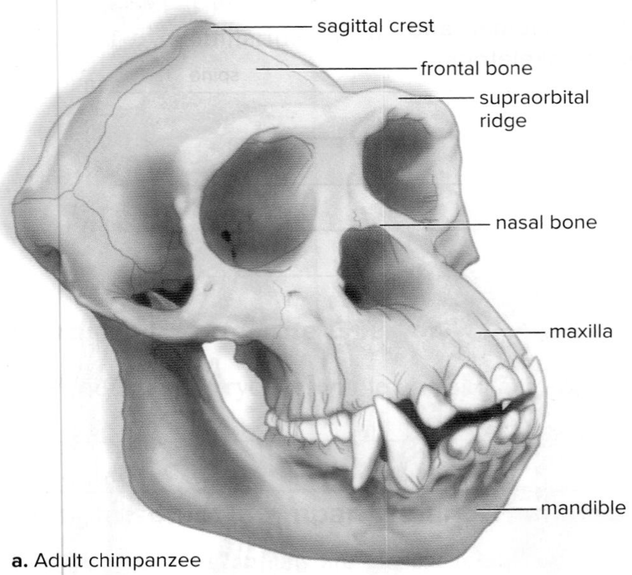

a. Adult chimpanzee

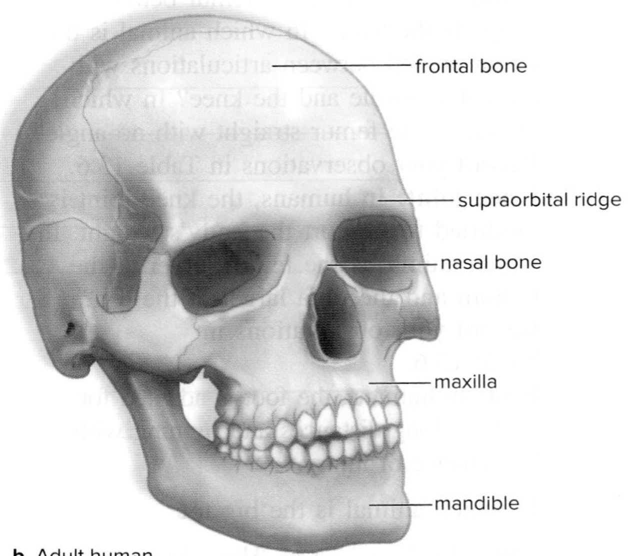

b. Adult human

Table 13.7 Skull Features of Chimpanzees and Humans

Feature	Chimpanzee	Human
1. Supraorbital ridge		
2. Sagittal crest		
3. Slope of frontal bone		
4. Teeth		
5. Chin		

Conclusion: Chimpanzee and Human Skeletons

- Do your observations show that the skeletal differences between chimpanzees and humans can be related to posture? ________ Explain your answer. ______________________________
- Do your observations show that diet can be related to the skull features of chimpanzees and humans? ______ Explain your answer. ______________________________

Comparison of Hominid Skulls

The designation *hominid* includes humans and primates that are humanlike. Paleontologists have uncovered many fossils representing the pathway of hominid evolution (Fig. 13.9). All past lineages either became extinct, were replaced by other species, or even interbred. At many points in history, multiple hominid species existed at once. Today, only *Homo sapiens* remains.

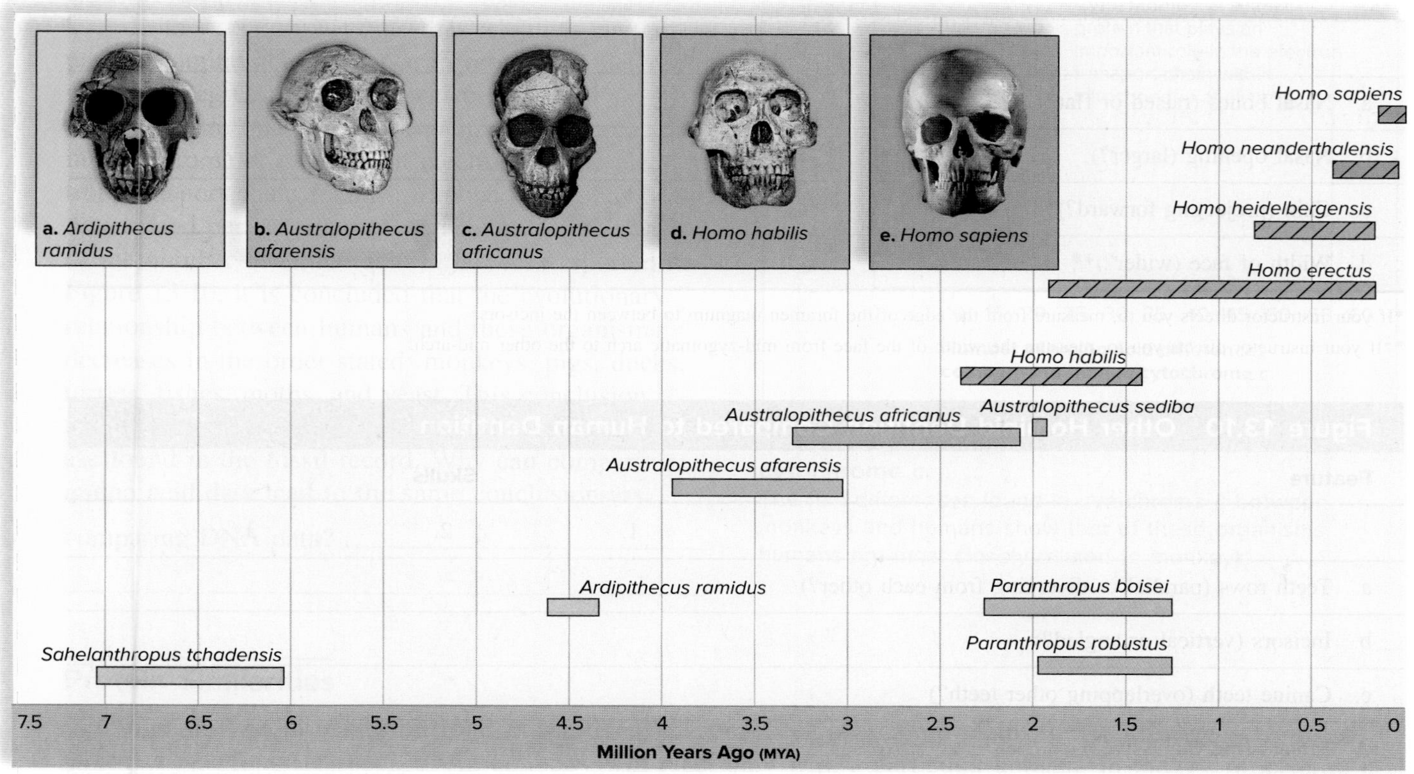

Figure 13.9 Human evolution.
(a) ©Richard T. Nowitz/Science Source; (b) ©Scott Camazine/Alamy; (c) ©Philippe Plailly/Science Source; (d) ©Kike Calvo VWPics/Superstock; (e) ©Kenneth Garrett/Getty Images

Observation of Hominid Skulls

Several of the skulls noted in Figure 13.9 may be on display. Use Tables 13.8, 13.9, and 13.10 to record data pertaining to the cranium (or braincase), the face, and the teeth. Compare the early skulls on display with a modern human skull. For example, is the forehead like or more flat than the human skull? List at least three skulls you examined for each feature.

Thought Questions

15. Analogous structures perform similar functions but do not arise from common ancestry. What explains the existence of analogous structures?

16. Why are there differences in the axial skeletons and skulls of chimpanzees and humans?

17. People with sickle-cell trait (heterozygotes $Hb^A\ Hb^S$) are immune to malaria, but people who are $Hb^S\ Hb^S$ will have sickle-cell disease.

    a. If two people with sickle-cell trait have offspring, what are the chances their offspring will have sickle-cell disease?

    b. Explain where, in the world, the Hb^S allele is most likely to be prevalent and why the allele persists in spite of the danger associated with having two Hb^S alleles.

LABORATORY

14

Natural Selection

Learning Outcomes

Introduction

- Explain the process of natural selection in terms of phenotype changes within a population.

14.1 Hardy-Weinberg Law

- State the Hardy-Weinberg law.
- Show how it is possible to use the Hardy-Weinberg equations to calculate the genotype frequencies in a population.

14.2 Natural Selection and Genetic Drift

- Show that during natural selection, a selective agent brings about gene pool frequency changes.
- Show that genetic drift results in gene pool frequency changes by chance.

Introduction

Natural selection is the mechanism, first described by Charles Darwin, that brings about adaptation to the environment as evolution occurs. Organisms are remarkably adapted to their way of life. Through natural selection, birds have become variously adapted to acquiring food (Fig. 14.1). The beak of a honeycreeper can catch an insect while the talons of an osprey can catch a fish, for example. Today, Charles Darwin's theory of natural selection can be described in this way:

1. The members of a population—all the members of a species living in one locale at the same time—have heritable variations (phenotypic traits) that can be passed from one generation to the next.
2. There is a competition for resources (such as food, mates, shelter) and members of the population with traits that allow them to better capture resources will reproduce to a greater extent than those that lack these traits.
3. Across generations, a larger proportion of the population will have these adaptive traits. In this way, the environment has selected how the genotype and phenotype makeup of the population will change over time.

a.

b.

Figure 14.1 Adaptations.
a. The beak of a European bee-eater is adapted to catching insects.
b. The talons of an osprey are adapted to catching fish.
(a) ©Juniors/Superstock; (b) ©Robert Burton/USFWS

14.1 Hardy-Weinberg Law

The **Hardy-Weinberg law** gives us a way to know when evolution has occurred. It states that normally, allele frequencies in the gene pool of a population stay the same from generation to generation. If they change, microevolution has occurred. The agents that can cause a change in gene pool frequencies are natural selection, genetic drift, mutation, gene flow (the movement of alleles between populations), and nonrandom mating (Fig. 14.2).

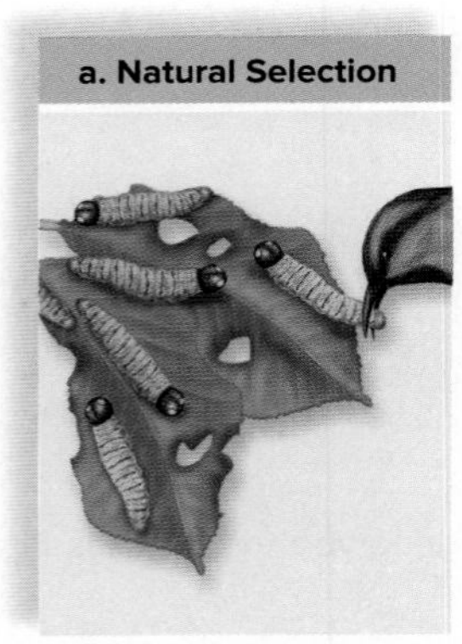

The only agent to result in adaptation to the environment.

Movement of new individuals or gametes into a population can cause gene pool frequency changes.

Chance results in gene pool frequency changes when population size decreases.

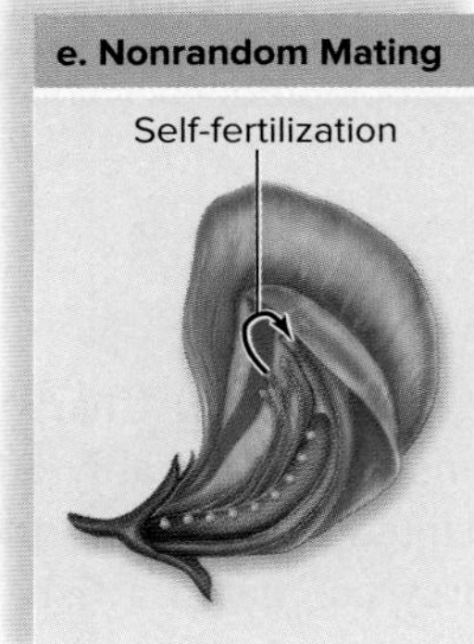

Inbreeding is a common reason for genotype frequency changes in a population.

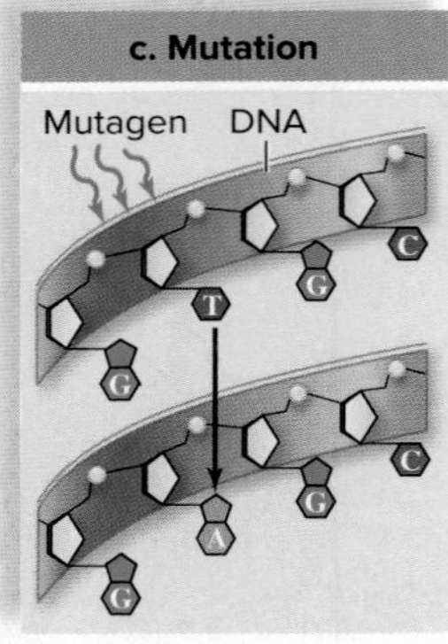

Mutation occurs infrequently but is the ultimate source of allele and phenotype variations.

Figure 14.2 The Hardy-Weinberg law states that there are five possible agents of evolutionary change: a. natural selection, **b.** genetic drift, **c.** mutation, **d.** gene flow, and **e.** nonrandom mating. Due to these agents, evolution evident by gene pool changes does occur.

In today's laboratory we are studying natural selection, genetic drift, and mutation as causes of microevolution.

In order to understand and use the law, understand these terms:

Population: all the members of a species living in the same locale at the same time. You and your fellow lab members will be the population in today's laboratory.

Gene pool: the frequency of the alleles and genotypes of all the individuals in the population.

Frequency: the proportion of alleles and genotypes relative to the total number of individuals in the population.

When we are dealing with a trait that has only two alleles, the Hardy-Weinberg equations allow us to calculate the gene pool frequencies in a population. The equations are:

$$p + q = 1$$
$$p^2 + 2pq + q^2 = 1$$

where
q = recessive allele frequency
q^2 = homozygous recessive frequency
p = dominant allele frequency
p^2 = homozygous dominant frequency
$2pq$ = heterzygous frequency

The Parent Population

In order to know if a change has occurred, we first need to know the allele and genotype frequencies for a particular trait in the gene pool of the present or **parent (*P*) population.** The trait under consideration will be the ability to taste PTC (phenylthiocarbamide). PTC is an antithyroid drug that prevents the thyroid gland from incorporating iodine into the thyroid hormone.

Observation: Gene Pool Frequencies of Parent Populations

Your Genotype

Taste a piece of paper impregnated with PTC. If you can taste the chemical, your genotype is either *TT* or *Tt*. If you cannot taste the chemical, your genotype is *tt*. What is your genotype? ____________

Homozygous Recessive Genotype (q^2) Frequency

Your instructor will determine what proportion of the class is *tt*. __________ = q^2 Frequencies are recorded as a decimal. For example, 25% becomes 0.25. Record the frequency of the *P* population that is homozygous recessive in Table 14.1.

Table 14.1 Parental (*P*) Generation

Genotypes	Genotype Frequencies
Homozygous recessive (q^2)	
Homozygous dominant (p^2)	
Heterozygous ($2pq$)	

Homozygous Dominant Genotype (p^2) Frequency

If q^2 = __________ , then q = __________ . Note that $q + p = 1$. Therefore, what is p? __________ What is p^2? __________ Record this number as the frequency of the *P* population that is homozygous dominant in Table 14.1.

Heterozygous Genotype (2pq) Frequency

As indicated, calculate the frequency of the heterozygous genotype in the *P* population by calculating $2pq$: $2 \times p \times q$ = __________ . Record this number as the frequency of the *P* population that is heterozygous in Table 14.1.

14.2 Natural Selection and Genetic Drift

15.1 Bacteria

Although two major groups of prokaryotes are now recognized, we will study the bacteria as a representative prokaryote. Most bacteria are **saprotrophic,** meaning that they send out digestive enzymes into the environment and thereafter take up the resulting nutrient molecules. Saprotrophic organisms are decomposers that play a major role in ecosystems by digesting the remains of dead organisms and returning inorganic nutrients to photosynthetic organisms. Some bacteria are **parasitic** and cause diseases, such as strep throat or gangrene. Other bacteria are **photosynthetic** or **chemosynthetic** and thus are able to make organic molecules utilizing inorganic molecules. Cyanobacteria are always photosynthetic. They contain chlorophyll, but the green color is often masked by other pigments. In fact, some cyanobacteria are red, brown, or even black.

Several techniques are used to identify bacteria; three methods are demonstrated in this laboratory: Gram stain, colony morphology, and shape of the bacterial cell.

Gram Stain

Most bacterial cells are protected by a **cell wall** that contains a unique molecule called **peptidoglycan.** Bacteria are commonly differentiated by using the Gram stain procedure, which distinguishes bacteria with a thick layer of peptidoglycan (Gram-positive) from those that have a thin layer of peptidoglycan (Gram-negative). Gram-positive bacteria retain a crystal violet-iodine complex and stain blue-purple, whereas Gram-negative bacteria decolorize and counterstain red-pink with safranin (Fig. 15.2).

Figure 15.2 Generalized structure of a bacterium.
a. Gram-positive cells have a thick layer of peptidoglycan. **b.** Gram-negative cells have a very thin layer of peptidoglycan. **c.** This difference causes Gram-positive cells to stain purple and Gram-negative cells to stain reddish-pink. (c) ©Science Source/Science Source

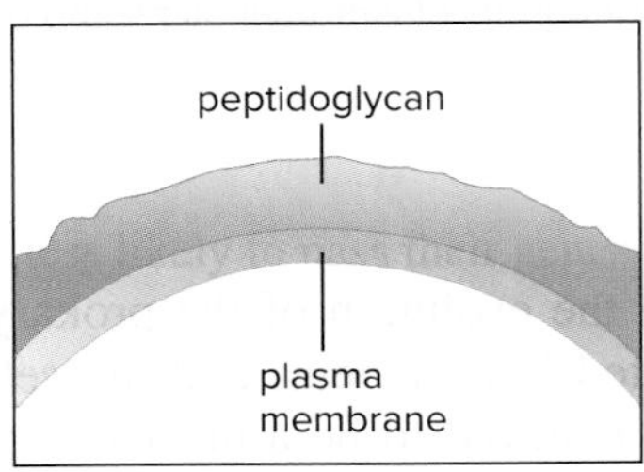

a. Gram-positive cell

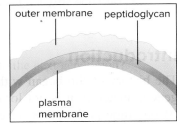

outer membrane
peptidoglycan
plasma membrane

b. Gram-negative cell

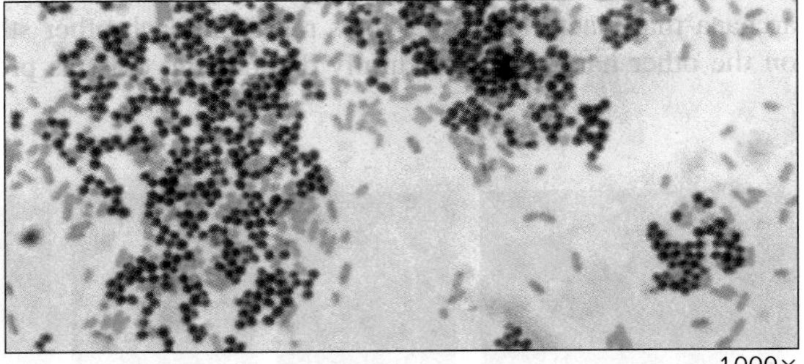

1,000×

c. Micrograph of Gram stained bacteria

Observation: Gram Staining

1. Examine prepared slides under high power. Note the color differences (Fig. 15.2*c*).
2. Study the directions for Gram staining on the next page.
3. Why do the Gram-positive cells in Figure 15.2*c* appear purple? ______________________

   Why do the Gram-negative cells appear reddish-pink? ______________________

Experimental Procedure: Gram Stain

Before you begin, fill in the first column of Table 15.1. Perform the Gram stain on at least two different bacteria, and complete the second column of Table 15.1.

Smear Preparation

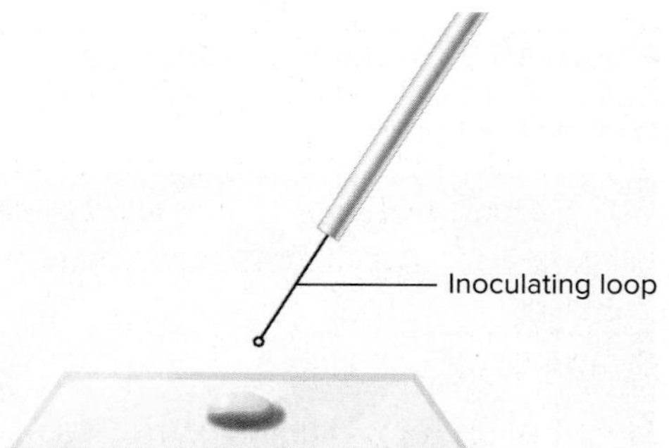

1. Obtain an inoculating loop and place one loopful of water on a slide.
2. Flame the loop and let cool briefly.
3. Touch the bacterial growth with the loop.
4. Briefly touch the loop to the drop, and move it back and forth one time.
5. Flame the loop.
6. Spread the cells to an area the size of a penny.
7. Reflame the loop.
8. Allow the slide to air dry.
9. Heat-fix the slide by passing the slide (smear side up) back and forth through the flame slowly three times.
10. Allow the slide to cool to the touch.

Gram Staining

1. To the fixed smear, apply several drops of crystal violet for 1–2 minutes.
2. Wash off excess stain under a stream of water.
3. Flood the slide with Gram's iodine, and let stand at least 1 minute.
4. Rinse with water, and shake off excess water.
5. Holding the slide at an angle, add decolorizing solution dropwise. When color is no longer removed (about 10 seconds), rinse with water.
6. Flood the slide with safranin, and let stand 2–3 minutes.
7. Drain excess and wash slide.
8. Carefully blot dry.
9. Examine with oil immersion lens. Gram-positive organisms will retain the crystal violet and appear purple. Gram-negative organisms will only be stained by safranin and will appear reddish-pink.
10. Complete Table 15.1.

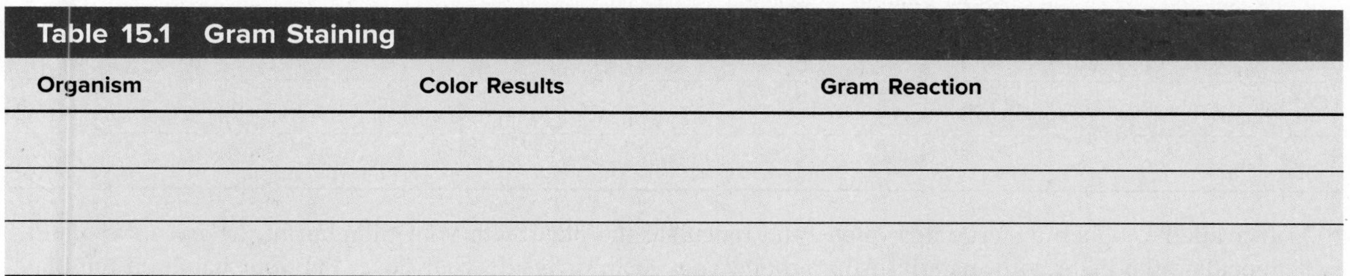

Table 15.1 Gram Staining

Organism	Color Results	Gram Reaction

Conclusions: Gram Stain

- What do you know about the cell wall of Gram-positive organisms?

  ____

- What do you know about the cell wall of Gram-negative organisms?

  ____

Colony Morphology

On a nutrient material called agar, bacteria grow as colonies (Fig. 15.3). A **colony** contains cells descended from one original cell. Sometimes, it is possible to identify the type of bacterium by the appearance of the colony.

Figure 15.3 Colony morphology.
Colonies of bacteria on agar plates. Note the variation that helps researchers identify the type of bacterium.
(both) ©Kathy Park Talaro

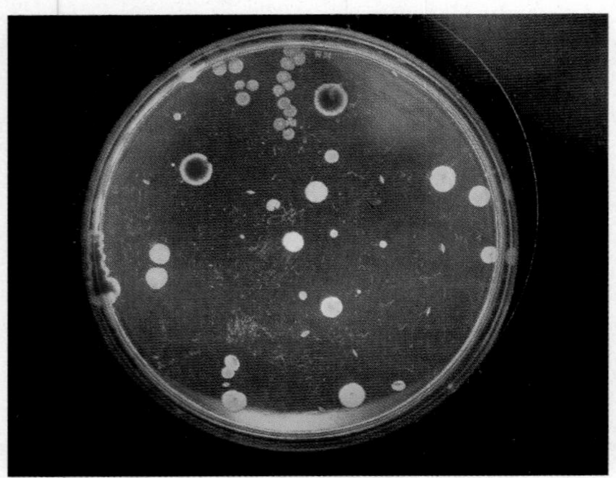

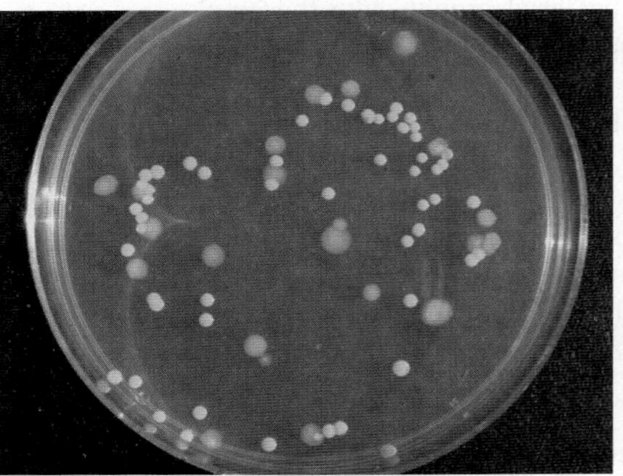

Observation: Colony Morphology

1. View agar plates that have been inoculated with bacteria and then incubated. Notice the "colonies" of bacteria growing on the plates.
2. Compare the colonies' color, surface, and margin, and note your observations in Table 15.2. It is not necessary to identify the type of bacterium.

Table 15.2 Agar Plates

Plate Number	Description of Colonies

3. If available, obtain a sterile agar plate, and inoculate the plate with your thumbprint, or use a swab and inoculate the plate with material from around your teeth or inside your nose. Put your name on the plate, and place it where directed by your instructor. Remember to view the plate next laboratory period. Describe your plate:

4. If available, obtain a sterile agar plate, and expose it briefly (at most for 10 minutes) anywhere you choose, such as in the library, your room, or your car. No matter where the plate is exposed, it subsequently will show bacterial colonies. Describe your plate:

Shape of Bacterial Cell

Most bacteria are found in three basic shapes: **spirillum** (spiral or helical), **bacillus** (rod), and **coccus** (round or spherical) (Fig. 15.4). Bacilli may form long filaments, and cocci may form clusters or chains. Some bacteria form endospores. An **endospore** contains a copy of the genetic material encased by heavy protective spore coats. Spores survive unfavorable conditions and germinate to form vegetative cells when conditions improve.

Observation: Shape of Bacterial Cell

1. View the microscope slides of bacteria on display. What magnification is required to view bacteria?
____
2. Using Figure 15.4 as a guide, identify the three different shapes of bacteria.
3. Do any of the slides on display show bacterial cells with endospores? __________ What is an endospore, and why does it have survival value? _______________

Figure 15.4 Diversity of bacteria.
a. Spirillum, a spiral-shaped bacterium. **b.** Bacilli, rod-shaped bacteria. **c.** Cocci, round bacteria.
(a) ©Michael Abbey/Science Source; (b) ©Eye of Science/Science Source; (c) ©SciMAT/Science Source

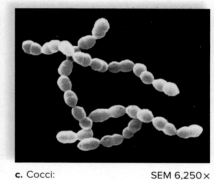

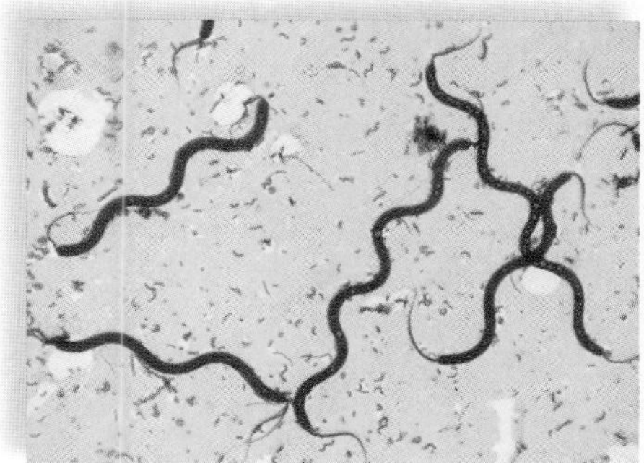

a. Spirillum: *Spirillum volutans* 250×

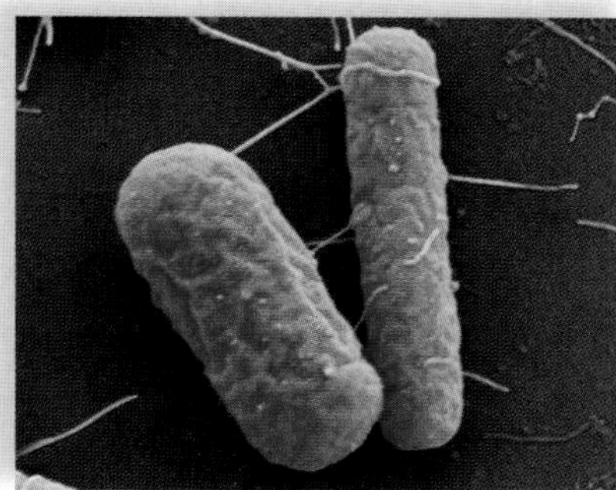

b. Bacilli: *Escherichia coli* SEM 13,300×

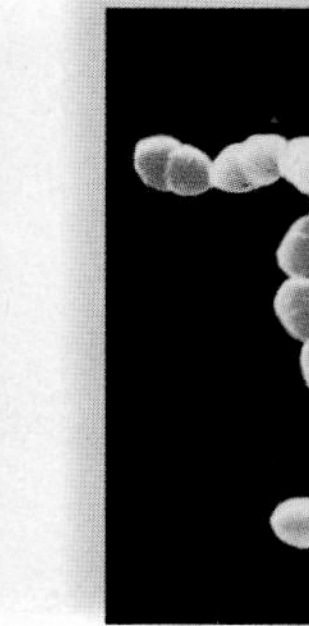

c. Cocci: *Streptococcus thermophilus* SEM 6,250×

Observation: Yersinia pestis

Commonly known as the Black Death or bubonic plague, *Yersinia pestis* is a Gram-negative, rod-shaped bacterium. It causes plague, normally a **zoonotic** (found in animals) disease of rodents (e.g., rats, mice, ground squirrels). Fleas that live on the rodents can sometimes pass the bacteria to human beings, who then suffer from the **bubonic** form of the plague (the presence of inflamed, tender swelling of lymph nodes, especially in the area of the armpit or groin). *Yersinia pestis* is a potential biological warfare bacterium if infected fleas were released or the bacterium itself can be released as an aerosol.

1. Examine the plague bacteria *Yersinia pestis* in Figure 15.5.
2. Try to identify the common "clothespin" shape of the rod-shaped cells.
3. Research the type(s) of antibiotics used to treat patients with bubonic or pneumonic plague in the United States.

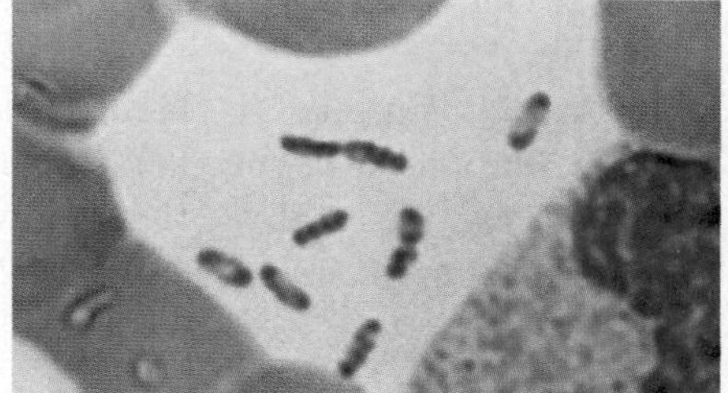

Figure 15.5 Plague bacteria in blood smear. ©Science Source/Science Source

Cyanobacteria

Cyanobacteria were formerly called blue-green algae because their general growth habit and appearance through a compound light microscope are similar to green algae. Electron microscopic study of cyanobacteria, however, revealed that they are structurally similar to other bacteria, particularly other photosynthetic bacteria. Although cyanobacteria do not have chloroplasts, they do have thylakoid membranes, where photosynthesis occurs.

Observation: Cyanobacteria

Gloeocapsa

1. Prepare a wet mount of a *Gloeocapsa* culture, if available, or examine a prepared slide, using high power (45×) or oil immersion (if available). The single cells adhere together because each is surrounded by a sticky, gelatinous sheath (Fig. 15.6).
2. What is the estimated size of a single cell? ____________________________________

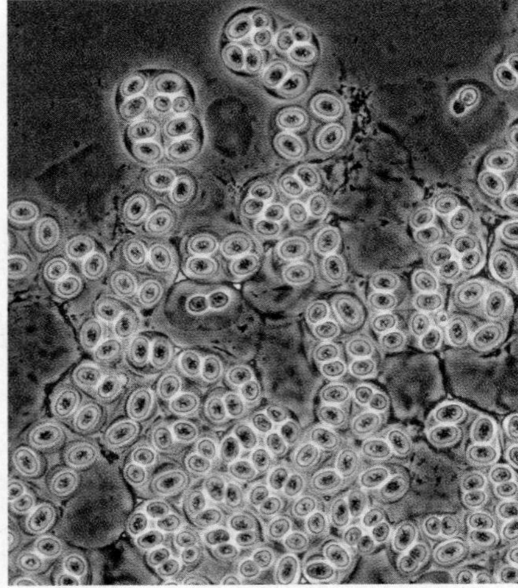

a. Micrograph at low magnification 100×

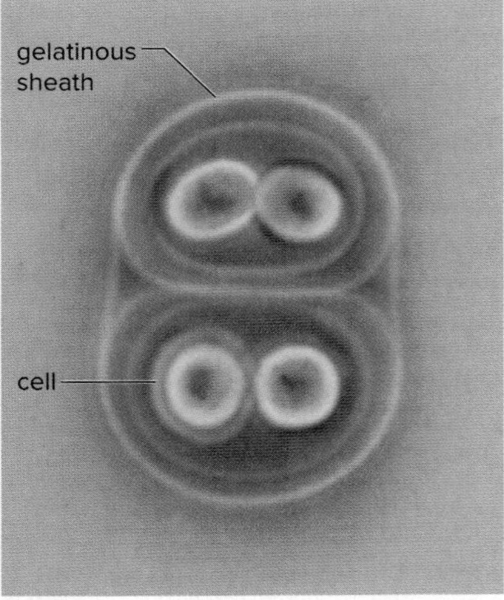

b. Micrograph at high magnification 500×

Figure 15.6 ***Gloeocapsa.*** (a) ©Michael Abbey/Science Source; (b) ©Biophoto Associates/Science Source

Oscillatoria

1. Prepare a wet mount of an *Oscillatoria* culture, if available, or examine a prepared slide, using high power (45×) or oil immersion (if available). This is a filamentous cyanobacterium with individual cells that resemble a stack of pennies (Fig. 15.7).
2. *Oscillatoria* takes its name from the characteristic oscillations that you may be able to see if your sample is alive. If you have a living culture, are oscillations visible?

____

250×

Figure 15.7 ***Oscillatoria.*** (right) ©M.I. Walker/Science Source

Anabaena

1. Prepare a wet mount of an *Anabaena* culture, if available, or examine a prepared slide, using high power (45×) or oil immersion (if available). This is also a filamentous cyanobacterium, although its individual cells are barrel-shaped (Fig. 15.8).
2. Note the thin nature of this strand. If you have a living culture, what is its color? ______________________

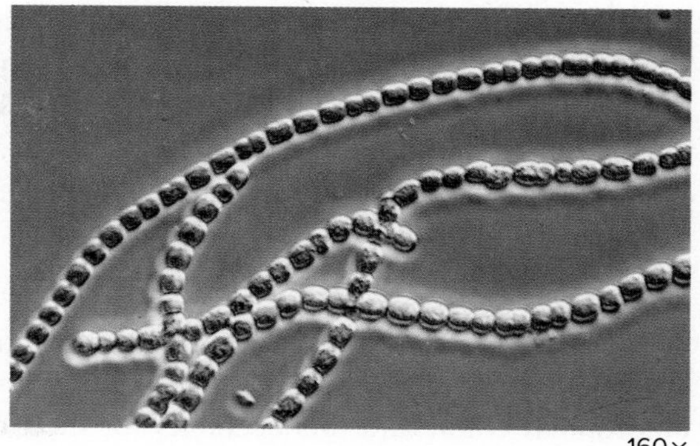

Figure 15.8 ***Anabaena.*** ©Robert Knauft/Biology Pics/Science Source

15.2 Protists

Protists were the first eukaryotes to evolve. Their diversity and complexity make it difficult to categorize them, but on the basis of molecular data, they are now placed in supergroups, which also include the animals, fungi, and land plants. See the accompanying table.

Protist Diversity

Supergroup		Members	Distinguishing Features
Archaeplastids	a. *Volvox*, colonial green alga with embedded daughter colonies	Green algae, red algae, charophytes **Other Members** Plants	Plastids; unicellular, colonial, and multicellular
SAR supergroup	b. Assorted fossilized diatoms	Stramenopiles: brown algae, diatoms, golden brown algae, water molds	Most with plastids; unicellular and multicellular
		Alveolates: ciliates, apicomplexans, dinoflagellates	Alveoli support plasma membrane; unicellular
	c. Radiolarians (assorted), produce a calcium carbonate shell	Rhizarians: foraminiferans, radiolarians	Thin pseudopods; some with tests; unicellular
Excavates	d. *Giardia*, a single-celled flagellated diplomonad	Euglenids, kinetoplastids, parabasalids, diplomonads	Feeding groove; unique flagella; unicellular
Amoebozoans	e. *Amoeba proteus*, a protozoan	Amoeboids, plasmodial and cellular slime molds	Pseudopods; unicellular
Opisthokonts	f. Choanoflagellate (unicellular), animal-like protist	Choanoflagellates, *nucleariids* **Other Members** Animals, fungi	Some with flagella; unicellular and colonial

(a) ©Stephen Durr RF; (b) ©M.I. Walker/Science Source; (c) ©Eye of Science/Science Source; (d) Source: CDC/Dr. Stan Erlandsen and Dr. Dennis Feely; (e) ©Melba/Media Bakery; (f) ©Image by David Patterson

Traditionally, protists were discussed according to their mode of nutrition, and this section will continue to do so while still noting the supergroup to which each type protist belongs. The photosynthetic algae photosynthesize in the same manner as land plants and certain of the green algae share a common ancestor with land plants. The heterotrophic (must acquire food from an external source) protists commonly called protozoans either ingest their food, as do animals, or they absorb it, as do fungi.

Photosynthetic Protists

Algae is a term that is used for aquatic organisms that photosynthesize in the same manner as land plants. All photosynthetic protists contain green chlorophyll, but they also may contain other pigments that mask the chlorophyll color, and this accounts for their common names. Commonly, we speak of the green algae, red algae, brown algae, and golden brown algae.

Observation: Green Algae

The green algae (supergroup Archaeplastids) are ancestral to the first land plants. Both groups possess chlorophylls *a* and *b,* both store reserve food as starch, and both have cell walls that contain cellulose.

If available, view a film loop showing the many forms of green algae. Notice that green algae can be single cells, filaments, colonies, or multicellular sheets. You will examine a filamentous form (*Spirogyra*) and a colonial form (*Volvox*). A **colony** is a loose association of cells.

Observation: Green Algae

If available, view a video showing the many forms of green algae. Notice that green algae can be single cells, filaments, colonies, or multicellular sheets. You will examine a filamentous form (*Spirogyra*) and a colonial form (*Volvox*). A **colony** is a loose association of cells.

Spirogyra

1. Make a wet mount of live *Spirogyra,* or observe a prepared slide. *Spirogyra* is a filamentous alga, lives in fresh water, and often is seen as a green scum on the surface of ponds and lakes. The most prominent feature of the cells is the spiral, ribbonlike chloroplast (Fig. 15.9*a*). How do you think *Spirogyra* got its name? ______________________________

____

Spirogyra's chloroplast contains a number of circular bodies, the **pyrenoids,** centers of starch polymerization. The nucleus is in the center of the cell, anchored by cytoplasmic strands.

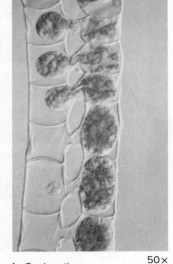

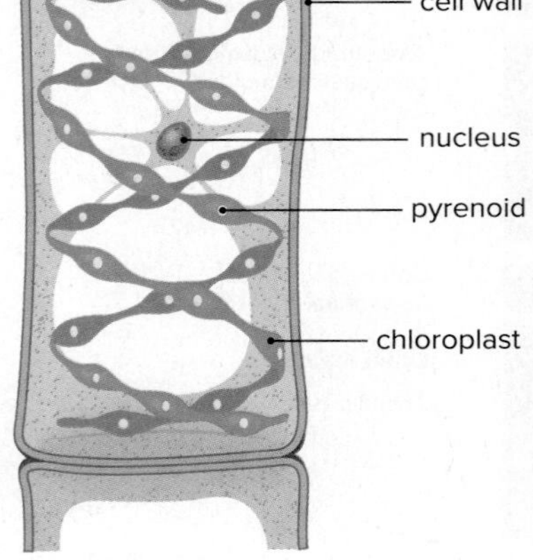

a. Drawing

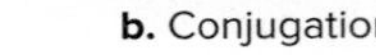

b. Conjugation 50×

Figure 15.9 ***Spirogyra.***
a. *Spirogyra* is a filamentous green alga, in which each cell has a ribbonlike chloroplast. **b.** During conjugation, the cell contents of one filament enter the cells of another filament. Zygote formation follows. (b) ©M.I. Walker/Science Source

2. Your slide may show **conjugation,** a sexual means of reproduction illustrated in Figure 15.9*b*. If it does not, obtain a slide that does show this process. Conjugation tubes form between two adjacent filaments, and the contents of one set of cells enter the other set. As the nuclei fuse, a zygote is formed. The zygote overwinters, and in the spring, meiosis and, subsequently, germination occur. The resulting adult protist is therefore haploid.

Volvox

1. Using a depression slide, prepare a wet mount of live *Volvox,* or study a prepared slide. *Volvox* is a green algal colony. It is motile (capable of locomotion) because the thousands of cells that make up the colony have flagella. These cells are connected by delicate cytoplasmic extensions (Fig. 15.10).

Volvox is capable of both asexual and sexual reproduction. Certain cells of the adult colony can divide to produce **daughter colonies** (Fig. 15.10) that reside for a time within the parental colony. A daughter colony escapes the parental colony by releasing an enzyme that dissolves away a portion of the matrix of the parental colony. During sexual reproduction, some colonies of *Volvox* have cells that produce sperm, and others have cells that produce eggs. The resulting zygote undergoes meiosis and the adult *Volvox* is haploid.

2. In Table 15.3, list the genus names of each of the green algae specimens available, and give a brief description.

Table 15.3 Green Algae Diversity

Specimen	Description

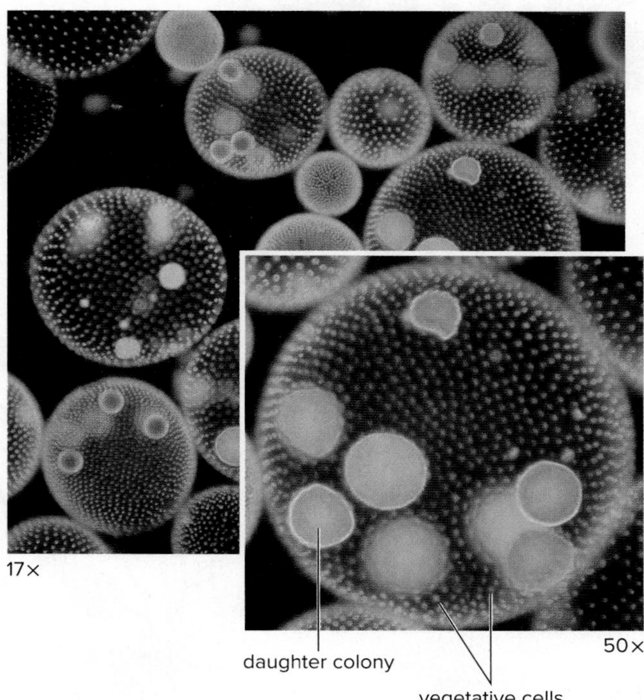

Figure 15.10 ***Volvox.*** *Volvox* is a colonial green alga. The adult *Volvox* colony often contains daughter colonies, asexually produced by special cells.
(both) ©Manfred Kage/Science Source

Observation: Brown Algae and Red Algae

Brown algae and red algae are commonly called *seaweed,* along with the multicellular green algae. Although red algae are also classified in the supergroup Archaeplastida, brown algae belong to the SAR supergroup within the subgroup of stramenopiles.

1. If available, study preserved specimens of brown algae (Fig. 15.11) that have brown pigments masking chlorophyll's green color. These algae are large and have specialized parts. *Laminaria* algae are called **kelps.** ***Fucus*** is called rockweed because it is seen attached to rocks at the seashore when the tide is out (Fig. 15.11). If available, view a preserved specimen. Note the dichotomously branched body plan, so called because the **stipe** repeatedly divides into two branches (Fig. 15.11). Note also the **holdfast** by which the alga anchors itself to the rock; the **air vesicles,** or bladders, that help hold the thallus erect in the water; and the **receptacles,** or swollen tips. The receptacles are covered by small, raised areas, each with a hole in the center. These areas are cavities in which the sex organs are located, with the gametes escaping to the outside through the holes. *Fucus* is unique among algae in that as an adult it is diploid (2n) and always reproduces sexually.
2. If available study preserved specimens of red algae. Like most brown algae, the red algae are multicellular, but they occur chiefly in warmer seawater, growing both in shallow waters and as deep as light penetrates. Some forms of red algae are filamentous, but more often, they are complexly branched with a feathery, flat, and expanded or ribbonlike appearance. Coralline algae are red algae that have cell walls impregnated with calcium carbonate ($CaCO_3$).
3. In Table 15.4, list the genus names of each of the brown and red algae specimens available, and give a brief description.

Figure 15.11 Brown algae.
Laminaria and *Fucus* are seaweeds known as kelps. They live along rocky coasts of the north temperate zone. The other brown algae featured, *Macrocystis* and *Nereocystis,* form spectacular underwater "forests" at sea.

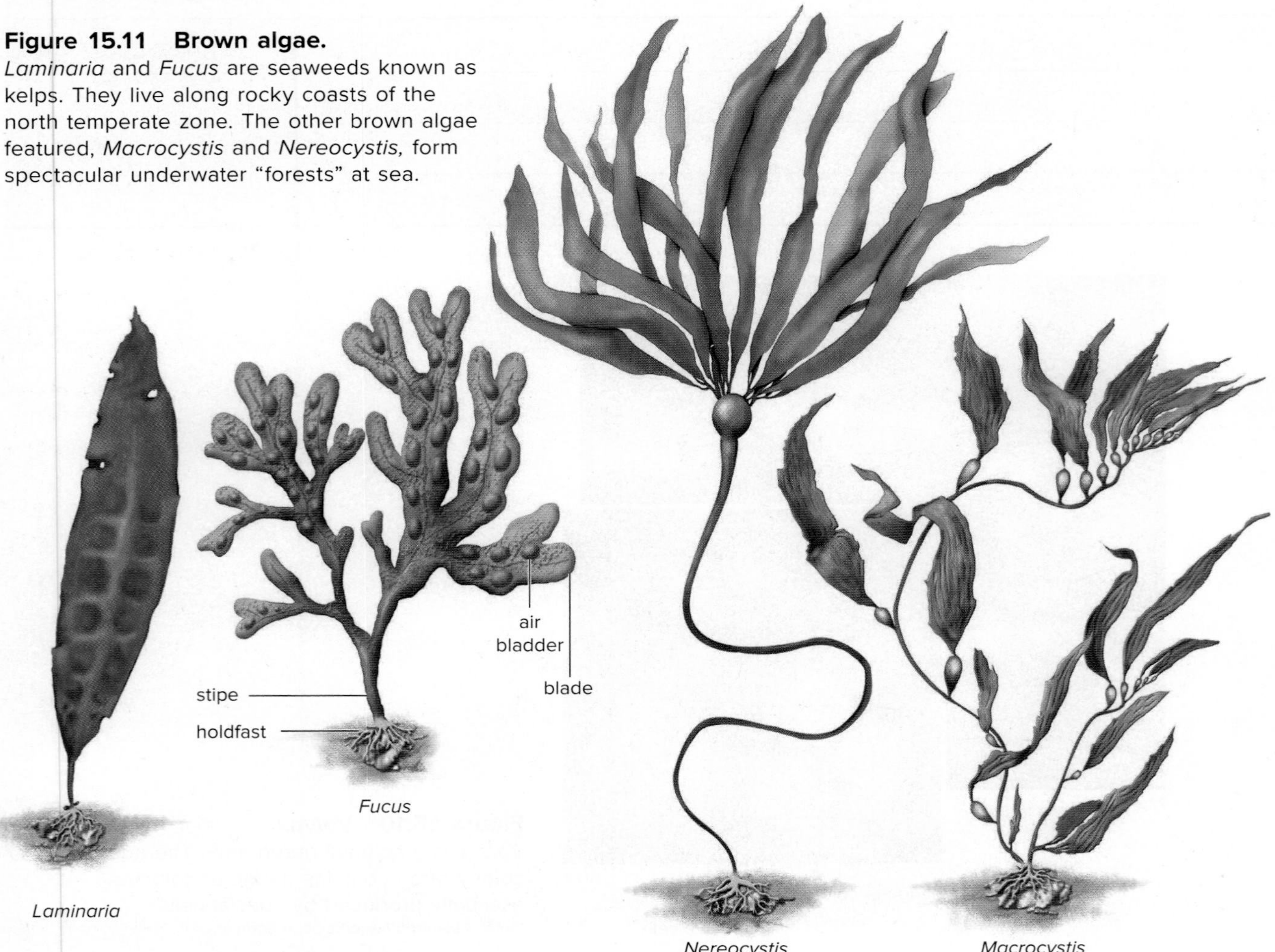

Table 15.4 Brown and Red Algae		
Specimen	**Genus**	**Description**
1		
2		
3		
4		

Observation: Diatoms and Dinoflagellates

Diatoms (golden-brown algae) are also stramenopiles in the SAR supergroup and have a yellow-brown pigment that, in addition to chlorophyll, gives them their color.

1. Make a wet mount of live diatoms, or view a prepared slide (Fig. 15.12). Describe what you see:________

____

____

____

____

The cell wall of **diatoms** is in two sections, with the larger one fitting over the smaller as a lid fits over a box. Since the cell wall is impregnated with silica, diatoms are said to "live in glass houses." The glass cell walls of diatoms do not decompose, so they accumulate in thick layers subsequently mined as diatomaceous earth and used in filters and as a natural insecticide. Diatoms, being photosynthetic and extremely abundant, are important food sources for the small heterotrophs (organisms that must acquire food from external sources) in both marine and freshwater environments.

2. Prepare a wet mount of live dinoflagellates or view a prepared slide (Fig. 15.13). Describe what you see.

____

____

____

____

130×

Figure 15.12 Diatoms. Diatoms, photosynthetic protists of the oceans.

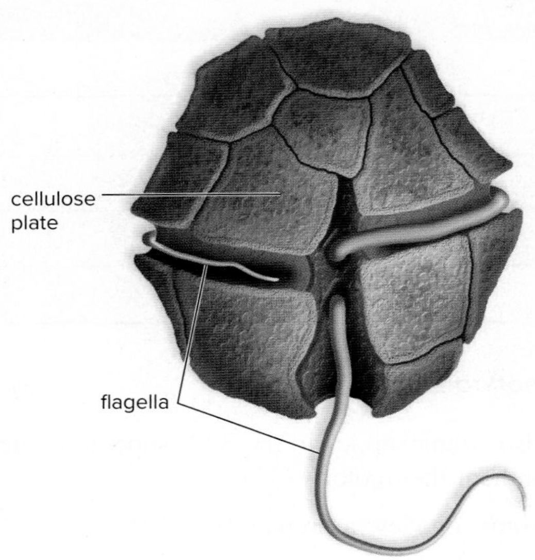

Figure 15.13 Dinoflagellates. Dinoflagellates such as *Gonyaulax* have cellulose plates.

Dinoflagellates (supergroup SAR) are photosynthetic, but they have two flagella; one is free, but the other is located in a transverse groove that encircles the animal. The beating of these flagella causes the organism to spin like a top. The cell wall, when present, is frequently divided into closely joined polygonal plates of cellulose. At times there are so many of these organisms in the ocean that they cause a condition called "red tide." The toxins given off in these red tides cause widespread fish kills and can cause paralysis in humans who eat shellfishes that have fed on the dinoflagellates.

Heterotrophic Protists

The term *protozoan* refers to unicellular eukaryotes and its use is often restricted to heterotrophic organisms that ingest food by forming **food vacuoles.** Other vacuoles, such as **contractile vacuoles** that rid the cell of excess water, are also typical. Usually protozoans have some form of locomotion and, as shown in Figure 15.14, some use **pseudopodia,** some move by **cilia,** and some use **flagella.**

Plasmodium vivax, which causes a common form of malaria is an apicomplexan. Apicomplexans have special organelles; one is an apicomplast at the base of their flagella and the other contains organic chemicals that assist in penetrating host tissues. *Plasmodium* spends a portion of its life cycle in mosquitoes (sexual phase) and the other part in human hosts (asexual phase). During the asexual phase of their life cycle, all apicomplexans exist as particulate spores, which accounts for why they are also called sporozoans. In general, how do sporozoans differ from the other protozoans shown in Figure 15.14? ______________________________

____

Plasmodium spores are the cause of malaria—they multiply in and rupture red blood cells.

Observation: Heterotrophic Protists

Individual Protozoans

1. Watch a video if available, and note the various forms of protozoans.
2. Prepare wet mounts or examine prepared slides of protozoans as directed by your instructor. You may already have had the opportunity to observe a protozoan such as *Euglena* or *Paramecium* in Laboratory 2. However, your instructor may want you to observe these organisms again. *Euglena* have flagella but many also have chloroplasts (see Fig. 2.13) and therefore they do not match our definition of a protozoan. But, then, some *Euglena* do not have chloroplasts.
3. Complete Table 15.5, listing the structures for locomotion in the types of protozoans you have observed.

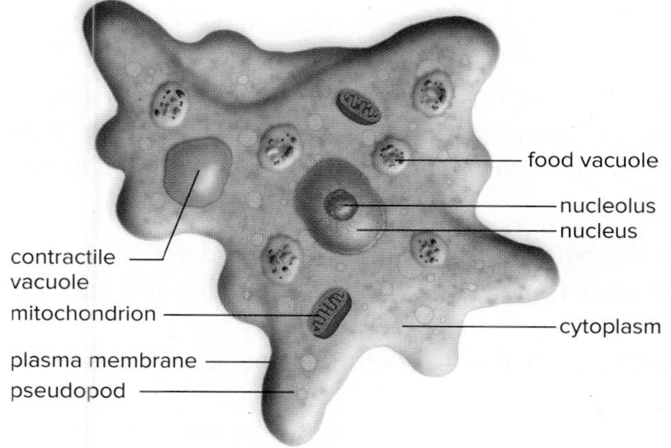

a. *Amoeba* moves by pseudopods

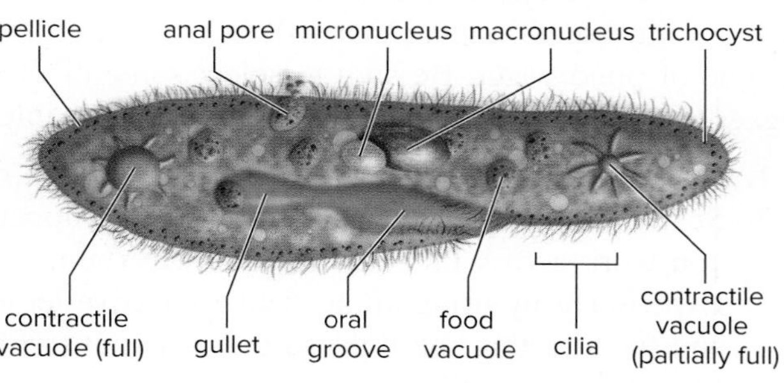

b. *Paramecium* moves by cilia

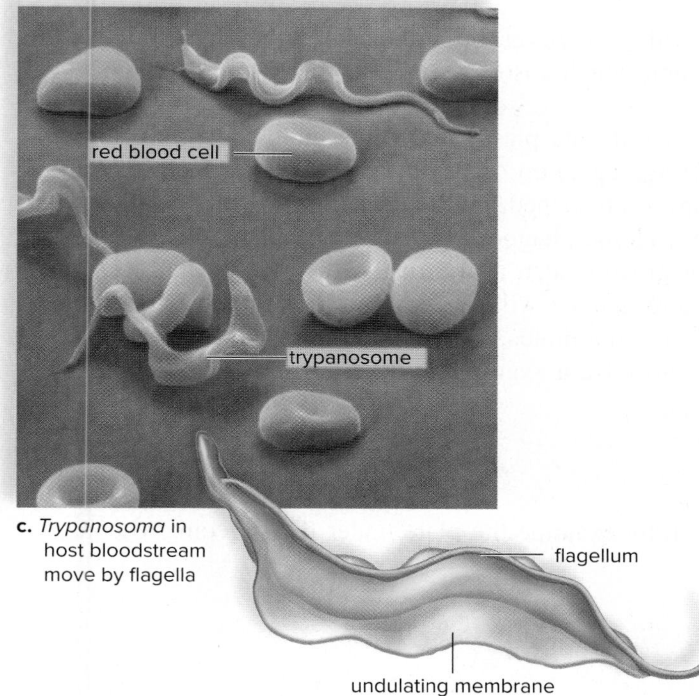

c. *Trypanosoma* in host bloodstream move by flagella

spore

red blood cell

32,400×

d. *Plasmodium* bursts forth as spores from infected red blood cells.

Figure 15.14 Protozoan diversity.
Many protozoans are motile. **a.** *Amoeba* (supergroup Amoebozoans) move by pseudopodia. **b.** *Paramecium* (supergroup Chromalveolates) move by cilia. **c.** *Trypanosoma* (kinetoplastid in supergroup Excavates) move by flagella. A blood infection is the cause of African sleeping sickness. **d.** *Plasmodium,* a cause of malarias, (alveolate in supergroup SAR) exist as nonmotile spores inside the red blood cells of humans. (c) ©Eye of Science/Science Source; (d) ©Dr. Gopal Murti/Science Source

Table 15.5 Heterotrophic Protists

Genus Name	Structures for Locomotion	Observations
Amoeba		
Paramecium		
Trypanosoma		

Pond Water

Pond water typically contains various examples of the protists studied in this laboratory. Your instructor may have an illustrated manual that will help you to identify the ones unfamiliar to you. Prepare a wet mount of a sample of pond water. Be sure to select some of the sediment on the bottom and a few strands of filamentous algae. Identify and classify them as fully as possible.

1. Make a wet mount of pond water by taking a drop from the bottom of a container of pond water.
2. Scan the slide for organisms: Start at the upper left-hand corner, and move the slide forward and back as you work across the slide from left to right.
3. Experiment by using all available objective lenses, by focusing up and down with the fine-adjustment knob, and by adjusting the light so that it is not too bright.
4. Identify the organisms you see by consulting Figure 2.13 on page 26, and use pictorial guides provided by your instructor.

Slime Molds

Slime molds (supergroup Amoebozoans) were once classified as fungi, but unlike fungi they have flagellated cells at one time during their life cycle. Also, unlike fungi, which absorb their food, slime molds phagocytize their food like amoeboids.

There are two types of slime molds: cellular slime molds and plasmodial slime molds. **Cellular slime molds** usually exist as individual amoeboid cells, which aggregate on occasion to form a pseudoplasmodium. **Plasmodial slime molds** usually exist as a **plasmodium,** a fan-shaped, multinucleated mass of cytoplasm.

The plasmodium of a plasmodial slime mold creeps along, phagocytizing decaying plant material in a forest or agricultural field. During times unfavorable for growth, such as a drought, the plasmodium develops many sporangia. A **sporangium** is a reproductive structure that we will see again among fungi, which produces wind-blown spores by meiosis. In some plasmodial slime molds, the spores become flagellated cells, and in others, they are amoeboid. In any case, they fuse to form a zygote that develops into a plasmodium (Fig. 15.15).

Observation: Plasmodial Slime Molds

1. Obtain a plate of *Physarum* growing on agar. Carefully examine the plate under the stereomicroscope.
2. Describe what you see.

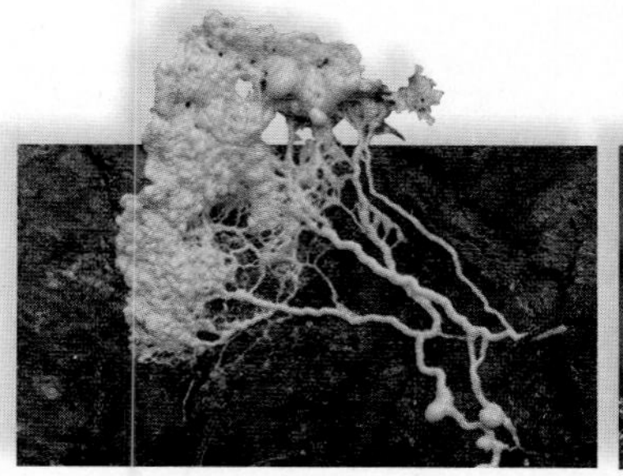

Plasmodium, *Physarum*

Sporangia, *Hemitrichia*

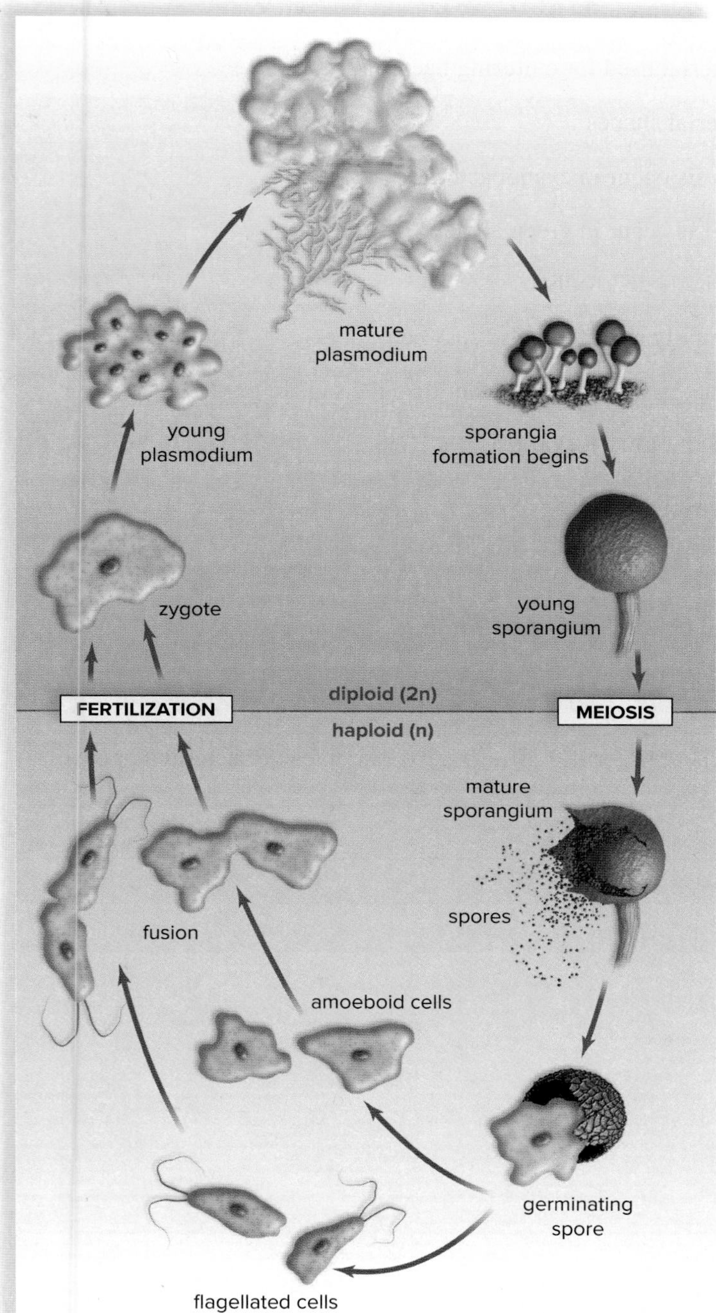

Figure 15.15 Plasmodial slime mold. The diploid adult forms sporangia during sexual reproduction, when conditions are unfavorable to growth. Wind-blown haploid spores germinate, releasing haploid amoeboid or flagellated cells that fuse. (left) ©NHPA/SuperStock; (right) ©NaturePL/SuperStock

Laboratory Review 15

_______________ **1.** Into which domain are protists, fungi, plants, and animals classified?

_______________ **2.** What molecule forms bacterial cell walls?

_______________ **3.** After Gram staining, what color do Gram-positive bacteria appear?

_______________ **4.** Which chemicals are used during the Gram stain procedure?

_______________ **5.** What is the term for a growth of cells descended from an original cell as observed on a culture plate?

_______________ **6.** What is the nutrient material used for culturing bacteria?

_______________ **7.** List three common bacterial shapes.

_______________ **8.** Where does photosynthesis occur in cyanobacteria?

_______________ **9.** Where does photosynthesis occur in green algae?

_______________ **10.** Which protist has silica in its cell walls?

_______________ **11.** Which protists are commonly referred to as "seaweed"?

_______________ **12.** Protists who gain their energy from sources other than the sun are termed _______________.

_______________ **13.** Which structures provide motility for protists?

_______________ **14.** Which protozoan moves using pseudopods?

Thought Questions

15. Describe the differences between Gram-positive and Gram-negative bacteria.

16. Many people have the misconception that all bacteria are "germs." Briefly give examples of bacteria that have beneficial roles.

17. Compare modes of locomotion among protists.

LABORATORY

16

Fungi

Learning Outcomes

16.1 Zygospore Fungi
- Identify the structures typical of black bread mold, and describe both the sexual and asexual life cycles.

16.2 Sac Fungi
- Identify types of sac fungi and both sexual and asexual reproductive structures.

16.3 Club Fungi
- Identify the parts of a mushroom and its sexual life cycle.

16.4 Fungal Diversity
- Explain the phylum names: Zygomycota, Ascomycota, Basidiomycota.

16.5 Fungi as Symbionts
- Explain the structure of a lichen and mycorrhizae.

Introduction

Fungi are multicellular heterotrophs that send out digestive juices into the environment and then absorb the resulting nutrients. In this way, fungi can be contrasted with animals that consume whole foods and then digest it. Fungal evolutionary relationships are being updated regularly as new molecular evidence is helping to clarify relationships among the phyla. Currently, seven phyla are recognized along with the informal "zygospore fungi" group (Fig. 16.1). Because of the rapidly changing phyla, fungi are commonly placed in these informal groups based on similar appearances or methods of reproduction. In this lab, you will study zygospore fungi, sac fungi, and club fungi—free-living terrestrial fungi that produce windblown spores during both asexual and sexual reproduction.

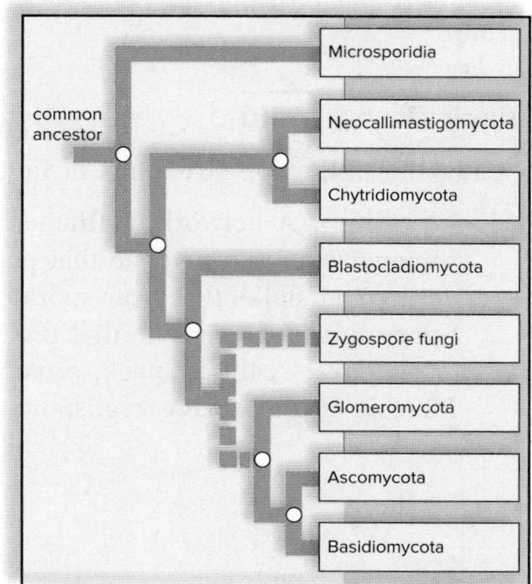

Figure 16.1 Evolutionary relationships among the fungi. The fungi branch of the eukaryote evolutionary tree with seven recognized phyla and the uncertain lineage of zygospore fungi. This evolutionary relationship represents a current hypothesis of fungal evolution.

Typically, the spores of a fungi develop directly into a **mycelium,** which is composed of hyphae. **Hyphae** are filaments of cells joined end to end. During sexual reproduction, hyphae from two different strains (called plus and minus) fuse and give rise to diploid nuclei. In zygospore fungi, the diploid nucleus undergoes meiosis and spore production occurs in a **sporangium.** Sac fungi and club fungi produce **fruiting bodies,** where many instances of meiosis are occurring. In sac fungi, the diploid nucleus undergoes meiosis and spore production occurs in asci, structures that are sac-shaped. In club fungi, the equivalent structures are club shaped and called basidia. As shown in the accompanying table, classification of fungi is based on means of sexual reproduction.

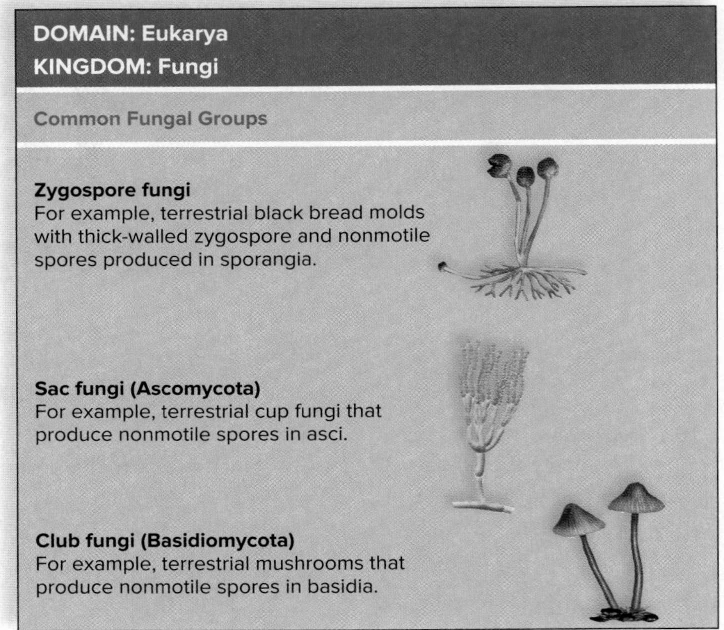

DOMAIN: Eukarya
KINGDOM: Fungi
Common Fungal Groups
Zygospore fungi
For example, terrestrial black bread molds with thick-walled zygospore and nonmotile spores produced in sporangia.
Sac fungi (Ascomycota)
For example, terrestrial cup fungi that produce nonmotile spores in asci.
Club fungi (Basidiomycota)
For example, terrestrial mushrooms that produce nonmotile spores in basidia.

16.1 Zygospore Fungi

Zygospore fungi, such as black bread mold, commonly grow on bread or any bakery goods kept at room temperature. Identify the sexual and the asexual portion of the life cycle of *Rhizopus,* the black mold shown in Figure 16.2.

Are the nuclei in the mycelium of *Rhizopus* haploid or diploid? __________ Where does meiosis occur in the *Rhizopus* life cycle? ____

Where are spores produced? ____

In zygosphore fungi, the hyphae are nonseptate (without cross walls); therefore, the hyphae are multinucleate.

Black Bread Mold

Identify the following structures in Figure 16.2:

1. **Mycelium:** A network of filaments called hyphae, of which there are three types: **Rhizoids** are somewhat rootlike hyphae that penetrate the bread, **stolons** are horizontal hyphae, and **sporangiophores** are hyphal stalks that bear sporangia.
2. **Sporangium:** A capsule that produces spores, which are black.
3. **Zygospore:** A thick, black, protective coat forms around the zygotes during sexual reproduction. Meiosis occurs during zygospore germination, and asexual sporangia are produced on sporangiophores.

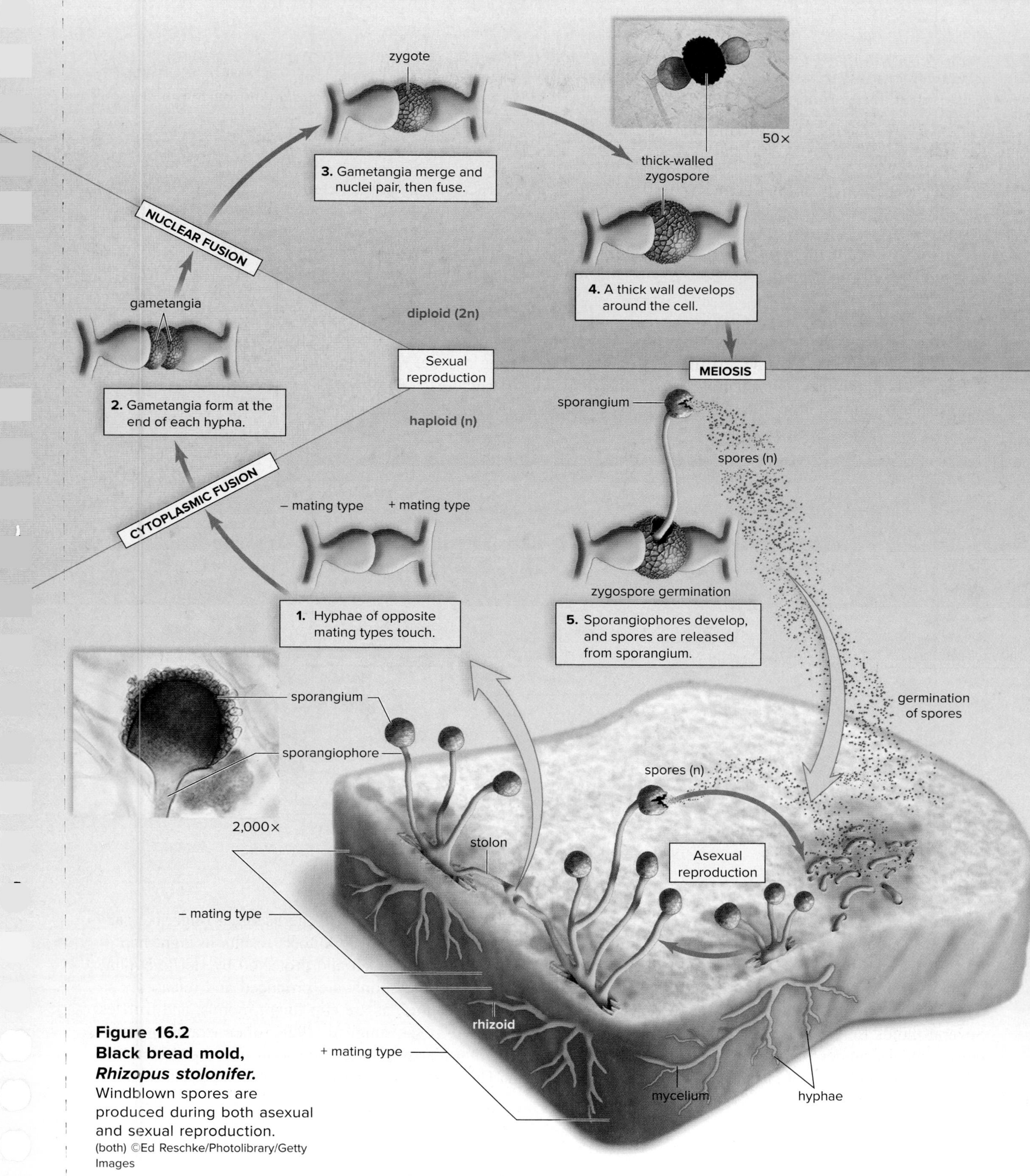

Figure 16.2
Black bread mold, *Rhizopus stolonifer*. Windblown spores are produced during both asexual and sexual reproduction.
(both) ©Ed Reschke/Photolibrary/Getty Images

Observation: Black Bread Mold

1. If available, examine bread that has become moldy. Do you recognize black bread mold on the bread? __________ Describe the mold you see. ____
____

2. Obtain a petri dish that contains living black bread mold. Observe with a stereomicroscope. Identify the three types of hyphae and the sporangia (black dots).______________________________________

3. View a prepared slide of *Rhizopus,* using both a stereomicroscope and the low-power setting of a light microscope. The absence of cross walls in the hyphae is an identifying feature of zygospore fungi. List the structures you can identify. ____
____

4. In the micrograph on the left, label structures seen during asexual reproduction: sporangiophore, rhizoid, and sporangium. In the micrograph on the right, label structures seen during sexual reproduction: stolon, gametangium, and zygospore.

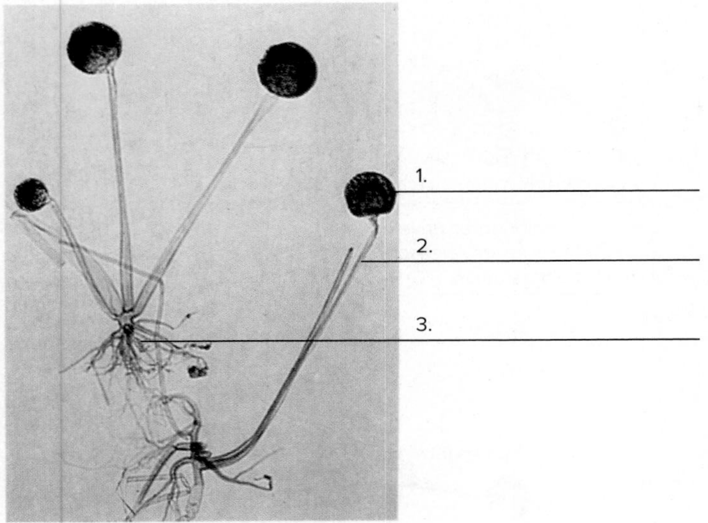

©Carolina Biological Supply Company/Phototake

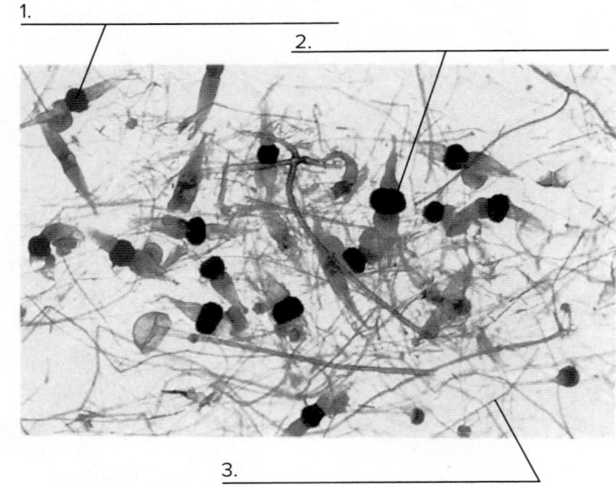

©Carolina Biological Supply Company/Phototake

16.2 Sac Fungi

Sac fungi are composed of septate hyphae. The phylum name Ascomycota refers to the **ascus,** a fingerlike sac that develops during sexual reproduction. Within each ascus, meiosis followed by mitosis results in eight haploid nuclei, and these become eight ascospores. The asci are usually surrounded and protected by sterile hyphae within a fruiting body. A **fruiting body** is a reproductive structure where spores are produced and released.

Various molds, including red bread mold (*Neurospora*), are sac fungi, as are cup fungi, morels, and truffles. Several molds (e.g., *Penicillium* and *Aspergillus*) are now classified as sac fungi. In 1928, when examining a petri dish, Alexander Fleming observed an area around *Penicillium* that was free of bacteria because the mold produced an antibacterial substance he later called penicillin. This discovery changed medicine and has saved countless lives. Other sac fungi have great medical importance also. For example, *Candida albicans* is a yeast normally present on the skin and mucous membranes of the mouth, vagina, or rectum. When it overgrows in the mouth, the medical condition is called thrush.

Some sac fungi, such as Dutch elm disease, ergot, and powdery mildews, cause serious plant diseases. Ergot contains LSD, and this hallucinogenic substance sometimes has been ingested accidentally.

Yeasts

Yeasts differ from most other fungi because they are not composed of hyphae. Even though yeasts are unicellular, they form an ascus during sexual reproduction. Usually, however, yeasts reproduce asexually, either by mitosis and cell division or by **budding.** Budding involves an unequal distribution of cytoplasm during cytokinesis.

Yeasts, such as *Saccharomyces,* are used during the production of wine and beer because of the alcohol they produce when fermenting. Yeasts are also helpful in that the carbon dioxide they give off makes bread rise.

Observation: Yeast

1. Obtain a sample of *Saccharomyces* (yeast) culture, and make a wet mount.
2. Observe, using all three objectives of your microscope. Add a drop of methylene blue stain if you cannot see the cells well. See methylene blue caution on page 199.
3. Label the vegetative cell, bud, ascus, and ascospore in the following diagram of *Saccharomyces:*

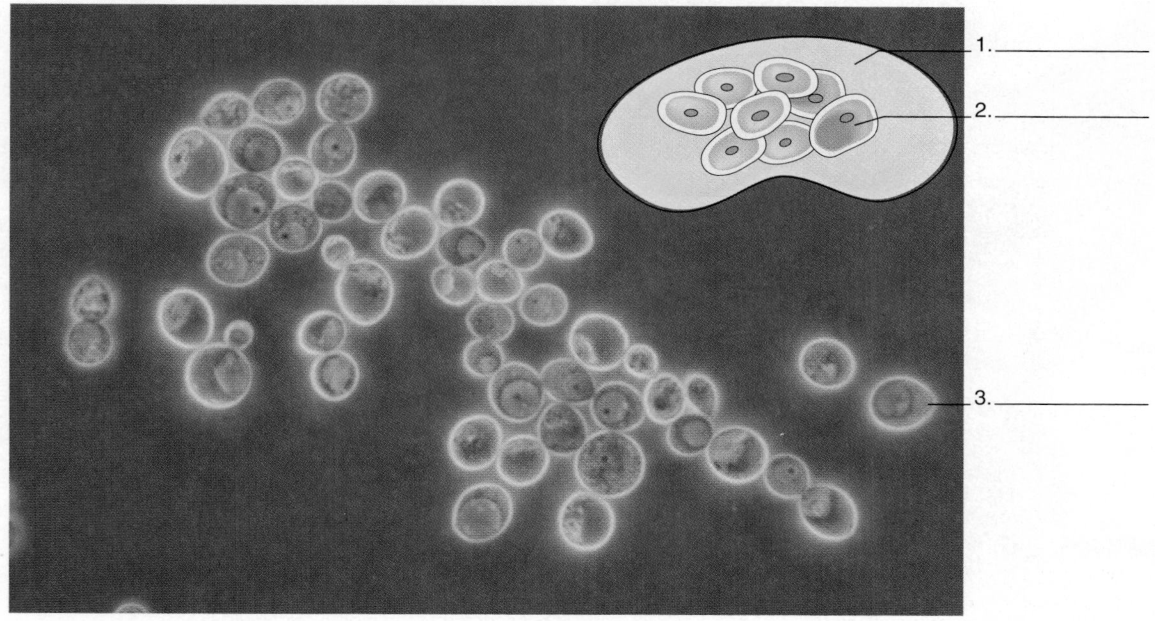

©Biophoto Associates/Science Source

4. View a prepared slide of *Schizosaccharomyces.* Identify an ascus and ascospores. How many ascospores are in each ascus? ______________________

Cup Fungi

Cup fungi are representative of sac fungi because they produce an **ascocarp** (fruiting body) in which saclike asci develop (Fig. 16.3*a*). The fruiting body contains a mass of sterile hyphae intertwined with reproductive hyphae. The sterile hyphae protect the asci. A diploid (2n) nucleus is produced in each ascus when two haploid nuclei unite. This zygote nucleus undergoes meiosis and then divides mitotically to produce **ascospores.** In some cup fungi, the ascocarp is cup shaped (Fig. 16.3*a*). In morels, the ascocarp is stalked and crowned by bell-shaped, convoluted tissue that bears the asci (Fig. 16.3*b*).

Figure 16.3 Sac fungi.
a. Cup fungus such as *Sarcoscypha* has a fruiting body shaped like a cup. **b.** A morel such as *Morchella* has a pitted, bell-shaped fruiting body. Ascospores line the pits. (left) ©Carol Wolfe RF; (center) ©Robert Marien/Corbis/Getty Images RF; (right) ©Carolina Biological Supply Company/Phototake

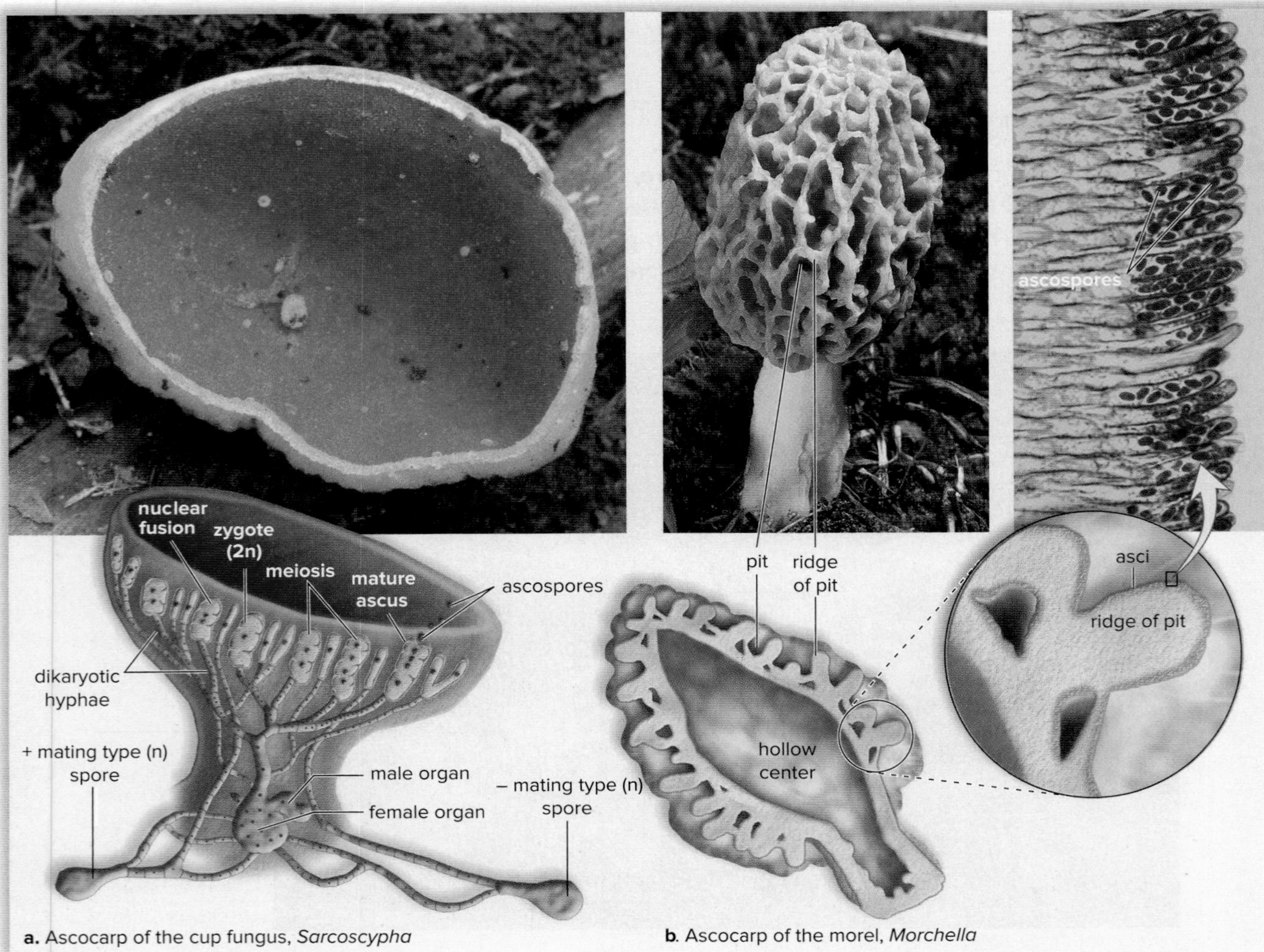

Observation: Cup Fungi

View a prepared slide of *Peziza,* and identify the ascocarp, asci, ascospores, and sterile hyphae. Are the ascospores inside or outside the asci? ______________________________

If available, obtain and view samples of morels and other cup fungi. The "pits" of a morel are lined with ____________________.

Conidiospores

When sac fungi reproduce asexually, they produce **conidiospores** (conidia) on upright hyphae known as conidiophores.

Observation: Conidiophores and Conidiospores

1. Obtain a petri dish of an *Aspergillus* culture. Use a dropper bottle of *methylene blue* to make a streak of stain 2.5 cm long on a microscope slide. Place the center of a 4 cm piece of clear tape over the *Aspergillus* culture so that a sample of the mold sticks to the tape. Put this mold sample on the microscope slide on which methylene blue stain has been placed. Observe with all three objectives of your microscope. What do you see?

   ___

> **Methylene blue** Avoid ingestion, inhalation, and contact with skin, eyes, and mucous membranes. Exercise care in using this chemical. If any should spill on your skin, wash the area with mild soap and water. Methylene blue will also stain clothing. Follow your instructor's directions for its disposal.

2. Obtain a prepared slide of *Penicillium.* Locate the periphery of the mass under low power, and then switch to high power. You should now be able to see conidiophores.
3. View a prepared slide of *Aspergillus,* and observe how the conidiophore arrangement differs from that of *Penicillium.*
4. Label the following diagram of *Aspergillus:*

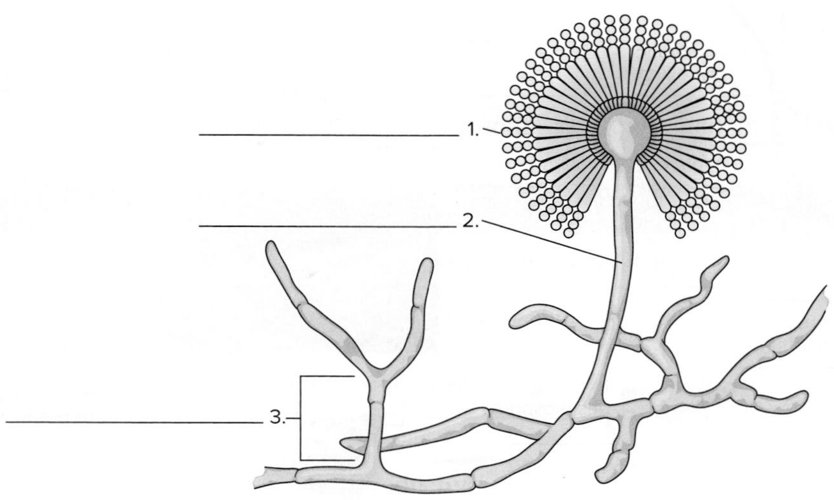

16.3 Club Fungi

Club fungi are composed of septate hyphae. The phylum name Basidiomycota refers to the **basidium,** a club-shaped structure that develops during sexual reproduction. Within each basidium, meiosis results in four basidiospores, which project from the basidium.

Study Figure 16.4, and note that when monokaryotic (n) hyphae from two different strains meet, they fuse to produce a dikaryotic (n + n) mycelium that, in turn, sometimes produces a fruiting body called a **basidiocarp.** A mushroom is a commonly known basidiocarp. In gill mushrooms, a button-shaped structure expands to become a full-sized mushroom, which consists of a stalk or stipe, and a terminal cap or pileus with gills on the underside. The gills are thin plates (lamellae) on which the basidia develop. Each basidium produces four basidiospores. Pore mushrooms have pores (tubes) instead of gills (Fig. 16.4).

Laboratory Review 16

_______________ **1.** Are fungi autotrophs or heterotrophs?

_______________ **2.** A mass of hyphae filaments is called a ______.

_______________ **3.** Black bread mold is an example of ______ fungi.

_______________ **4.** Which structure produces spores in zygospore fungi?

_______________ **5.** In which structure does meiosis occur in zygospore fungi?

_______________ **6.** What type of unique structure do sac fungi produce during sexual reproduction?

_______________ **7.** What is the name of a reproductive structure in sac fungi where spores are produced and released?

_______________ **8.** Which fungi produces an antibacterial substance discovered by Alexander Fleming?

_______________ **9.** In which two ways can yeast reproduce asexually?

_______________ **10.** What metabolic process do yeast perform that is used in alcohol production?

_______________ **11.** What is the common name for club fungi?

_______________ **12.** Where are basidia located on cup fungi?

_______________ **13.** An association between fungi and a cyanobacterium or green alga is called a ____.

_______________ **14.** Which type of symbiotic relationship is represented by mycorrhizae?

Thought Questions

15. Contrast fungi and animals in terms of their method to gather nutrients.

16. Why is it advantageous for fungi to reproduce sexually and asexually?

17. Why does Fig. 16.6 show the relationship between plants and mycorrhizae is mutualistic rather than parasitic?

LABORATORY

17

Nonvascular Plants and Seedless Vascular Plants

Learning Outcomes

17.1 The Evolution and Diversity of Land Plants

- Understand the relationship of charophytes to land plants and the main events in the evolution of plants as they became adapted to life on land.
- Describe the plant life cycle and the concept of the dominant generation.
- Contrast the role of meiosis in the plant and animal life cycles.

17.2 Nonvascular Plants

- Describe, in general, the nonvascular plants and how they are adapted to reproducing on land.
- Describe the life cycle of a moss, including the appearance of its two generations.
- Describe the liverworts, another group of nonvascular plants.

17.3 Seedless Vascular Plants

- Describe, in general, the seedless vascular plants and how they are adapted to reproducing on land.
- Describe the lycophytes, whisk ferns, and horsetails.
- Describe the life cycle of a fern and the appearance of its two generations.

Introduction

Plants are multicellar photosynthetic eukaryotes; their evolution is marked by adaptations to a land existence. Among aquatic green algae, the **charophytes** are most closely related to the plants that now live on land. The charophytes have several features that would have promoted the evolution of land plants, including retention of and care of the zygote. The most successful land plants are those that protect all phases of reproduction (sperm, egg, zygote, and embryo) from drying out and that have an efficient means of dispersing offspring on land.

In this laboratory, you will have the opportunity to examine the various adaptations of plants to living on land. Although this lab will concentrate on land plant reproduction, you will also see that much of a land plant's body is covered by a **waxy cuticle** that prevents water loss, and that land plants require structural support to oppose the force of gravity and to lift their leaves up toward the sun. **Vascular tissue** offers this support and transports water to, and nutrients from, the leaves.

Also, you will examine the adaptations of nonvascular plants and the seedless vascular plants to a land existence, and you will see that reproduction in these plants requires an outside source of water.

17.1 The Evolution and Diversity of Land Plants

Figure 17.1 shows the evolution of land plants. *Circle the evolutionary events that led to the adaptation of plants to a land existence.*

Figure 17.1 Evolutionary relationships among the plants.

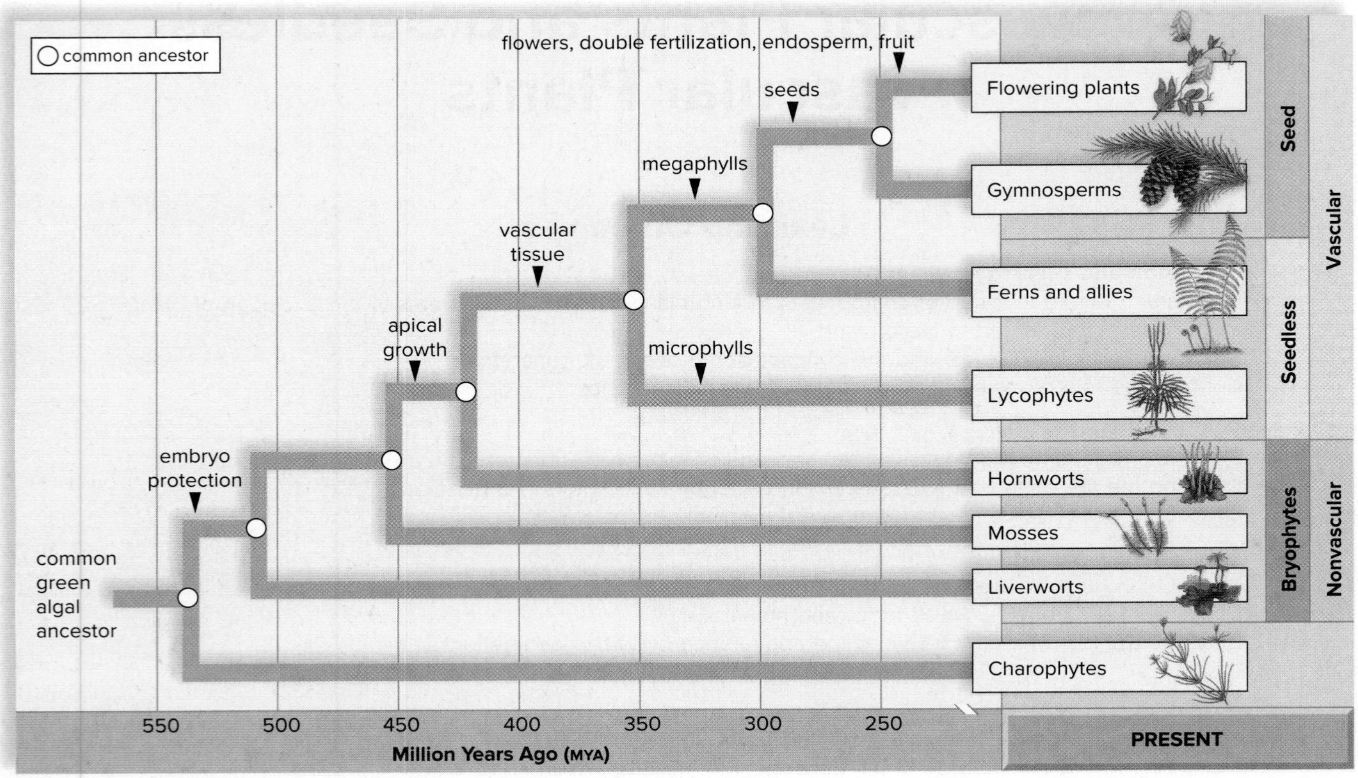

common ancestor

flowers, double fertilization, endosperm, fruit

seeds

megaphylls

vascular tissue

apical growth

microphylls

embryo protection

common green algal ancestor

Flowering plants

Gymnosperms

Ferns and allies

Lycophytes

Hornworts

Mosses

Liverworts

Charophytes

Seed

Seedless

Vascular

Bryophytes

Nonvascular

550 500 450 400 350 300 250

Million Years Ago (MYA)

PRESENT

Algal Ancestor of Land Plants

In the evolutionary tree, the charophytes are green algae that share a common ancestor with land plants. This common ancestor may have resembled a Charales, such as *Chara,* which can live in warm, shallow ponds that occasionally dry up. Adaptation to such periodic desiccation may have facilitated the ability of certain members of the common ancestor population to invade land, and with time, become the first land plants.

Observation: Chara

Examine a living *Chara* (Fig. 17.2). How does it superficially resemble a land plant? ____________________

____

Chara is a filamentous green alga that consists of a primary branch and many side branches (Fig. 17.2). Each branch has a series of very long cells. Note where one cell ends and the other begins. Measure the length of one cell. ____________________ Gently pick up and handle *Chara* while you are examining it. What does it feel like? ____________________ Its cell walls are covered with calcium carbonate deposit.

Figure 17.2 ***Chara*.**
Chara is an example of a stonewort, the type of green alga believed to be most closely related to the land plants. (left) ©Bob Gibbons/Alamy Stock Photo; (right) ©Kingsley Stern

Conclusion: Chara

- What characteristics cause *Chara* to resemble land plants? ______________________
____
- Why are *Chara* called stoneworts? ______________________

Alternation of Generations

Land plants have a two-generation life cycle called **alternation of generations.**

1. The **sporophyte** (diploid) **generation** produces haploid spores by meiosis. Spores develop into a haploid generation by mitosis. *Label the appropriate arrow "mitosis" in the first part of Figure 17.3.*
2. The **gametophyte** (haploid) **generation** produces **gametes** (eggs and sperm) by mitosis. The gametes then unite to form a diploid zygote. The zygote becomes the sporophyte by mitosis. *Label the appropriate arrow "mitosis" in the first part of Figure 17.3.*

In this life cycle, the two generations are dissimilar, and one dominates the other. The dominant generation is larger and exists for a longer period. Figure 17.3 contrasts the plant life cycle (alternation of generations) with the animal life cycle (**diploid life cycle**).

1. In the plant life cycle, meiosis occurs during the production of ______________________.
2. In the human life cycle, meiosis occurs during the production of ______________________.
3. In the plant life cycle, the generation that produces gametes is (n or 2n) ______________________.
4. In the human life cycle, the individual that produces gametes is (n or 2n) ______________________.

Figure 17.3 Plant and animal life cycles.

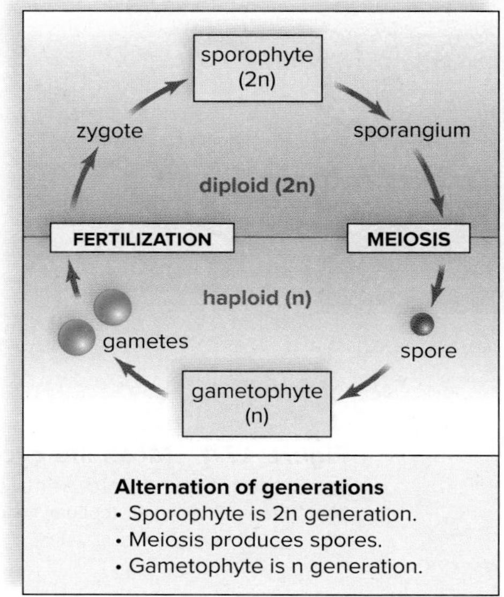

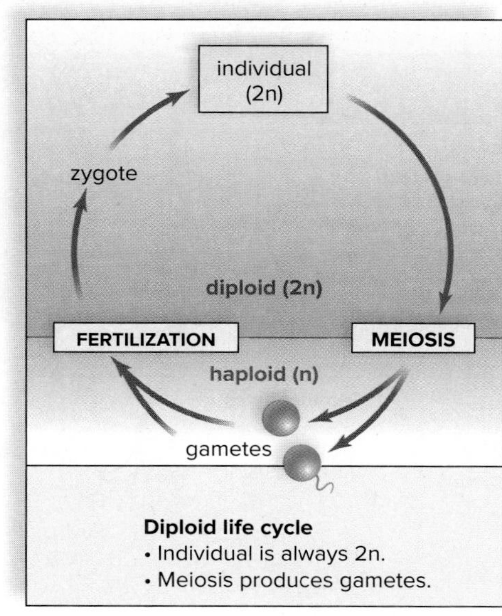

17.2 Nonvascular Plants

The nonvascular plants include the **mosses** (phylum Bryophyta) and **liverworts** (phylum Hepaticophyta). The gametophyte is dominant in all nonvascular plants. The gametophyte produces eggs within **archegonia** and flagellated sperm in **antheridia.** The sperm swim to the egg, and the embryo develops within archegonia. The nonvascular sporophyte grows out of the archegonium. The sporophyte is dependent on the female gametophyte (Fig. 17.4) for nourishment.

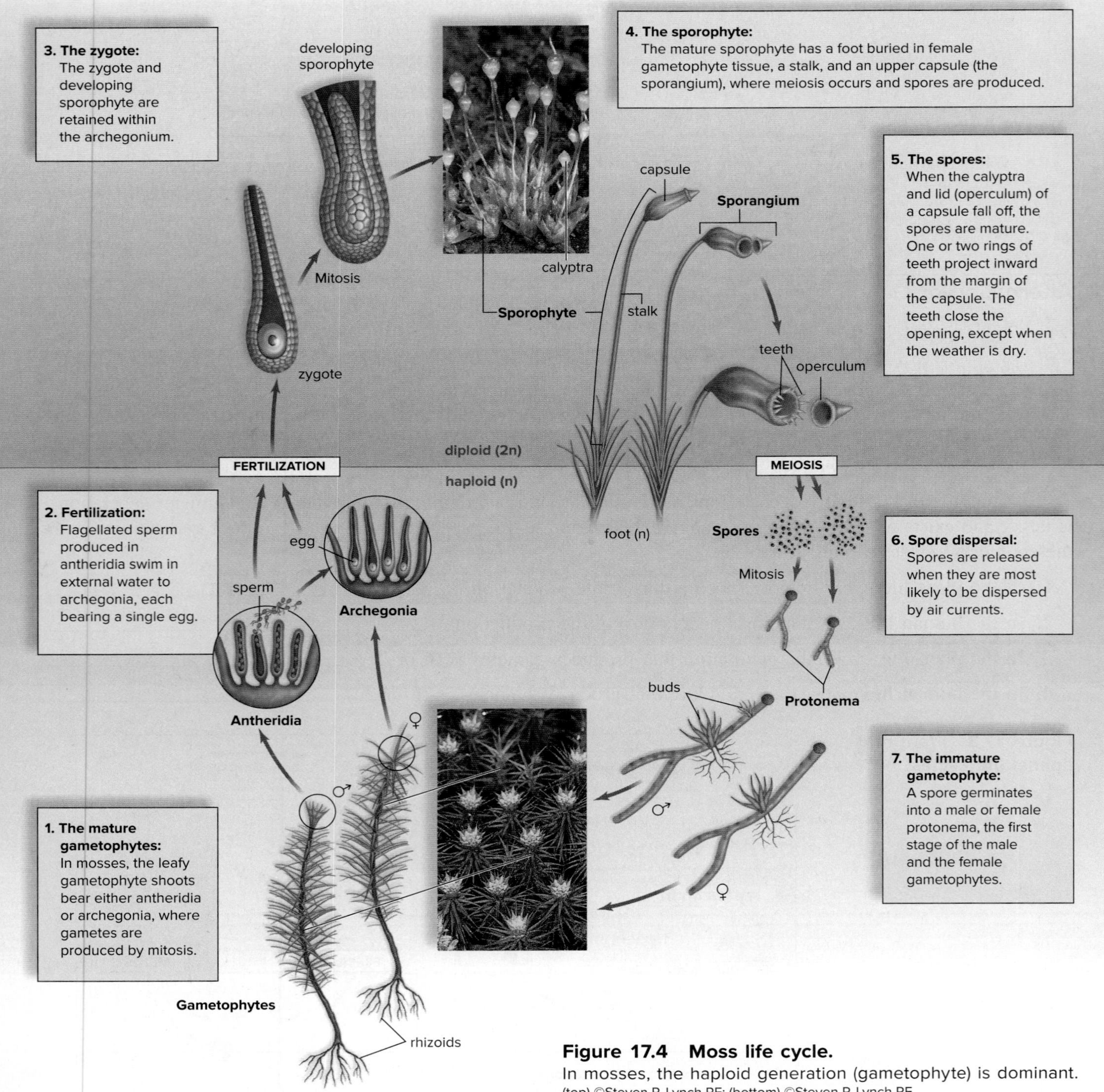

3. The zygote:
The zygote and developing sporophyte are retained within the archegonium.
developing sporophyte
Mitosis
zygote
4. The sporophyte:
The mature sporophyte has a foot buried in female gametophyte tissue, a stalk, and an upper capsule (the sporangium), where meiosis occurs and spores are produced.
calyptra
Sporophyte
capsule
Sporangium
stalk
teeth
operculum
5. The spores:
When the calyptra and lid (operculum) of a capsule fall off, the spores are mature. One or two rings of teeth project inward from the margin of the capsule. The teeth close the opening, except when the weather is dry.
FERTILIZATION
diploid (2n)
haploid (n)
MEIOSIS
2. Fertilization:
Flagellated sperm produced in antheridia swim in external water to archegonia, each bearing a single egg.
egg
sperm
Archegonia
Antheridia
foot (n)
Spores
Mitosis
6. Spore dispersal:
Spores are released when they are most likely to be dispersed by air currents.
buds
Protonema
♀
♂
7. The immature gametophyte:
A spore germinates into a male or female protonema, the first stage of the male and the female gametophytes.
1. The mature gametophytes:
In mosses, the leafy gametophyte shoots bear either antheridia or archegonia, where gametes are produced by mitosis.
Gametophytes
rhizoids

Figure 17.4 Moss life cycle.
In mosses, the haploid generation (gametophyte) is dominant.

1. Put a check mark beside the phrases that describe nonvascular plants:

I	II
________ No vascular tissue to transport water	________ Vascular tissue to transport water
________ Flagellated sperm that swim to egg	________ Sperm protected from drying out
________ Dominant gametophyte	________ Dominant sporophyte

2. Which listing of features (**I** or **II**) would you expect to find in a plant fully adapted to a land environment? ________ Explain. ____

3. In nonvascular plants, windblown spores are dispersal agents, and some species forcefully expel their spores. How are windblown spores an adaptation to reproduction on land? ____

____

Observation: Moss Gametophyte

Living or Plastomount

Obtain a living moss gametophyte or a plastomount of this generation. Describe its appearance. ____________

____

The leafy green shoots of a moss are said to lack true roots, stems, and leaves because, by definition, roots, stems, and leaves are structures that contain vascular tissue.

Microscope Slide

1. Study a slide of the top of a male moss shoot that contains antheridia, the reproductive structures where sperm are produced (Fig. 17.5). What is the chromosome number (choose 2n or n) of the sperm (see Fig. 17.4)? ________ Are the surrounding cells haploid or diploid? ________________________

2. Study a slide of the top of a female moss shoot that contains archegonia, the reproductive structures where eggs are produced (Fig. 17.6). What is the chromosome number of the egg? ____________ Are the surrounding cells haploid or diploid? ____________ When sperm swim from the antheridia to the archegonia, a zygote results. The zygote develops into the sporophyte. Is the sporophyte haploid or diploid? ________

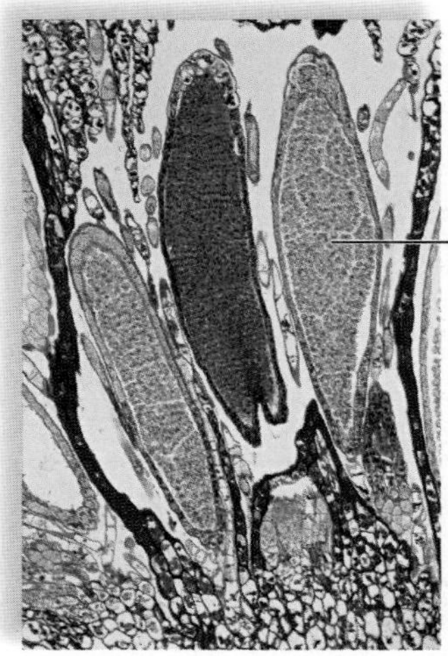

Figure 17.5 Moss antheridia. Flagellated sperm are produced in antheridia.
©Ed Reschke/Photolibrary/Getty Images

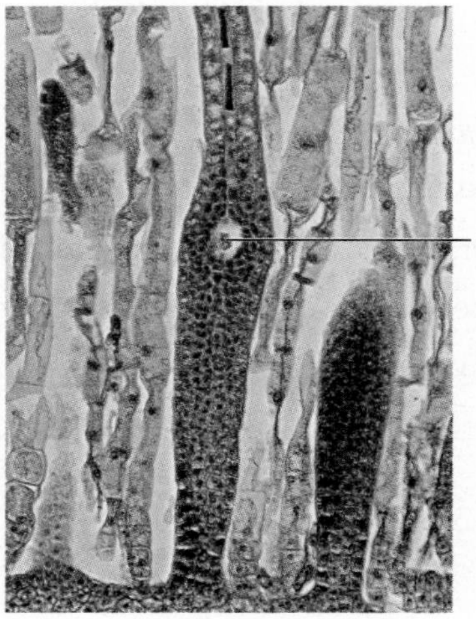

Figure 17.6 Moss archegonia. Eggs are produced in archegonia.
©Robert Knauft/Biology Pix/Science Source

Observation: Moss Sporophyte

Living Sporophyte

1. Examine the living sporophyte of a moss in a minimarsh, or obtain a plastomount of a female shoot with the sporophyte attached. Identify the **sporangium,** a capsule where spores are produced and released.
2. *Bracket and label the gametophyte and sporophyte in Figure 17.7*a. *Place an* n *beside the gametophyte and a* 2n *beside the sporophyte.*

Figure 17.7 Moss sporophyte.
a. The moss sporophyte is dependent on the female gametophyte. **b.** The sporophyte produces spores by meiosis.
(b) ©Ed Reschke

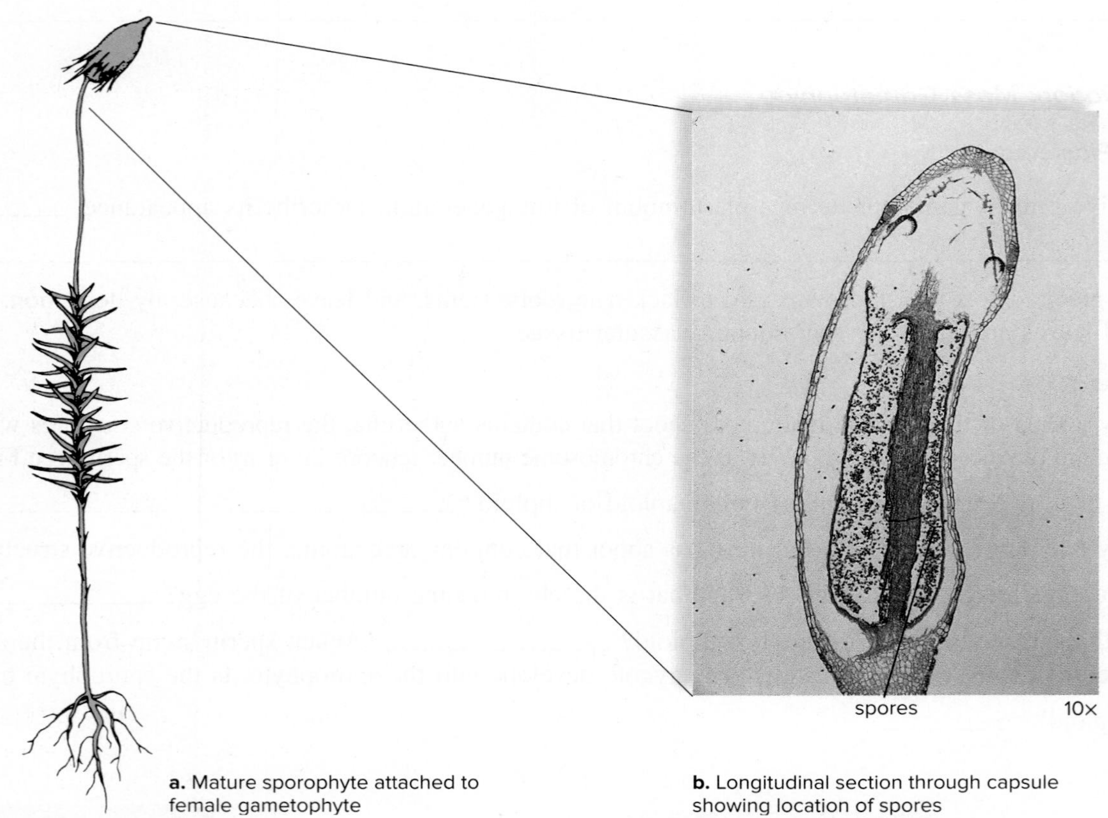

a. Mature sporophyte attached to female gametophyte

b. Longitudinal section through capsule showing location of spores

Microscope Slide

Examine a slide of a longitudinal section through a moss sporophyte (Fig. 17.7*b*). Identify the stalk and the sporangium, where spores are being produced. By what process are the spores being produced? ____________

When spores germinate, what generation begins to develop? ____________

Why is it proper to say that spores are dispersal agents? ____________

____

Observation: Liverworts

Obtain a living sample of the liverwort *Marchantia* (Fig. 17.8) thallus, body of the plant. Examine it under the stereomicroscope. What generation is this sample? __________ Identify the following:

1. The gametophyte consisting of lobes. Each lobe is about a centimeter or so in length; the upper surface is smooth, and the lower surface bears numerous **rhizoids** (rootlike hairs).
2. **Gemma cups** on the upper surface of the thallus. These contain groups of cells called **gemmae** that can asexually start new plants.
3. Disk-headed stalks bear antheridia, where flagellated sperm are produced.
4. Umbrella-headed stalks bear archegonia, where eggs are produced. Following fertilization, tiny sporophytes arise from the archegonia.

Figure 17.8 Liverwort, *Marchantia*.
a. Gemmae can detach and start new plants. **b.** Antheridia are present in disk-shaped structures. **c.** Archegonia are present in umbrella-shaped structures. (a) ©Ed Reschke/Photolibrary/Getty Images; (b) ©J.M. Conrarder/Nat'l Audubon Society/Science Source; (c) ©Ed Reschke/Oxford Scientific/Getty Images

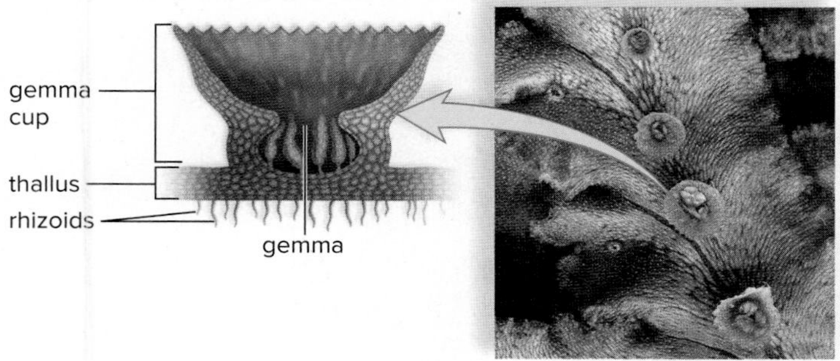

a. Thallus with gemma cups

b. Male gametophytes bear antheridia

c. Female gametophytes bear archegonia

17.3 Seedless Vascular Plants

Seedless vascular plants include the **lycophytes** (club mosses) and **pteridophytes**—ferns and their allies, the whisk ferns and horsetails. These plants were prevalent and quite large during the swampy forest Carboniferous period. At that time, the coal deposits still used today were formed.

The sporophyte is dominant in the seedless vascular plants. The dominant sporophyte has adaptations for living on land; it has vascular tissue and produces windblown spores. The spores develop into a separate gametophyte generation that is very small (less than 1 cm). The gametophyte lacks vascular tissue and produces flagellated sperm.

1. Place a check mark beside the phrases that describe seedless vascular plants:

I	II
__________ Independent gametophyte	__________ Gametophyte dependent on sporophyte, which has vascular tissue
__________ Flagellated sperm	__________ Sperm protected from drying out

2. Which listing (**I** or **II**) would you expect to find in a plant fully adapted to a land environment? ________

   Explain. ____

   ____

   Are seedless vascular plants fully adapted to living on land? Explain your answer. ______________________

   ____

Observation: Lycophytes

Lycophytes (phylum Lycophyta) are commonly called **club mosses.** Lycophytes are representative of the first vascular plants. They have an aerial stem and a horizontal root (rhizome with attached rhizoids) both of which have vascular tissue. The leaves are called **microphylls** because they have only one strand of vascular tissue.

Ground Pines

1. Examine a living or preserved specimen of *Lycopodium* (Fig. 17.9).
2. Note the shape and the size of the microphylls and the branches of the stems.
3. Note the terminal clusters of leaves, called **strobili,** that are club-shaped and bear sporangia.
4. *Label strobili, leaves, stem, and rhizoids in Figure 17.9.*
5. Examine a prepared slide of a *Lycopodium* that shows the sporangia with spores inside. The spore develops into a tiny microscopic gametophyte that remains in the soil.

Figure 17.9 ***Lycopodium.***
In the club moss *Lycopodium,* green photosynthetic stems are covered by scalelike leaves, and spore-bearing leaves are clustered in strobili. ©Steve Solum/Bruce Coleman/Photoshot

Observation: Pteridophytes

Molecular (DNA) studies tell us that the whisk ferns, horsetails, and ferns are closely related.

Whisk Ferns

Psilotum is representative of **whisk ferns,** named for their resemblance to whisk brooms.

1. Examine a preserved specimen of *Psilotum,* and note that it has no leaves. The underground stem, called a **rhizome,** gives off upright, aerial stems with a dichotomous branching pattern, where bulbous sporangia are located (Fig. 17.10).

   What generation are you examining? ____________________

2. *Label a sporangium, the stem, and the rhizome in Figure 17.10*b.

Figure 17.10 Whisk fern, *Psilotum*.
This whisk fern has no roots or leaves—the branches carry on photosynthesis. The sporangia are yellow.
©Carolina Biological Supply Company/Phototake

Horsetails

In **horsetails,** a rhizome produces aerial stems that stand about 1.3 meters tall.

1. Examine *Equisetum,* a horsetail, and note the minute, scalelike leaves (Fig. 17.11).
2. Feel the stem. *Equisetum* contains a large amount of silica in its stem. For this reason, these plants are sometimes called scouring rushes and may be used by campers for scouring pots.
3. Strobili appear at the tips of the stems, or else special buff-colored stems bear the strobili. Sporangia are in the strobili.
4. *Label the strobilus, branches, leaves, and rhizome of the horsetail in Figure 17.11.*

Figure 17.11 Horsetail.
In the horsetail *Equisetum,* whorls of branches or tiny leaves appear in the joints of the stem. The sporangia are borne in strobili. ©Robert P. Carr/Bruce Coleman/Photoshot

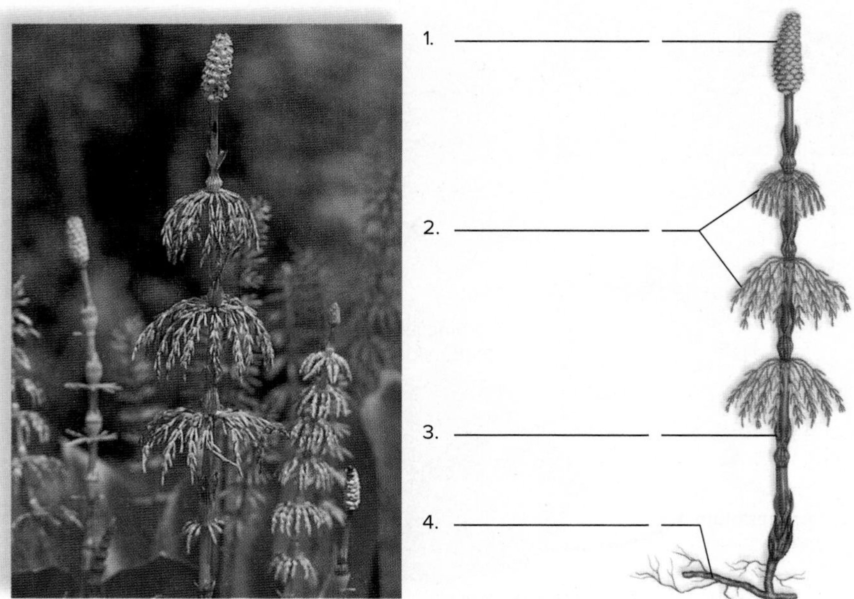

Ferns

Ferns are quite diverse and range in size from those that are low growing and resemble mosses to those that are as tall as trees. The rhizome grows horizontally, which allows ferns to spread without sexual reproduction (Fig. 17.12). The gametophyte (called a **prothallus**) is small (about 0.5 cm) and usually heart-shaped. The prothallus contains both archegonia and antheridia. Ferns are largely restricted to moist, shady habitats because sexual reproduction requires adequate moisture. Why? ______________________________________

____

Figure 17.12 Fern life cycle.
The sporophyte is the frond, and the gametophyte is the heart-shaped prothallus.

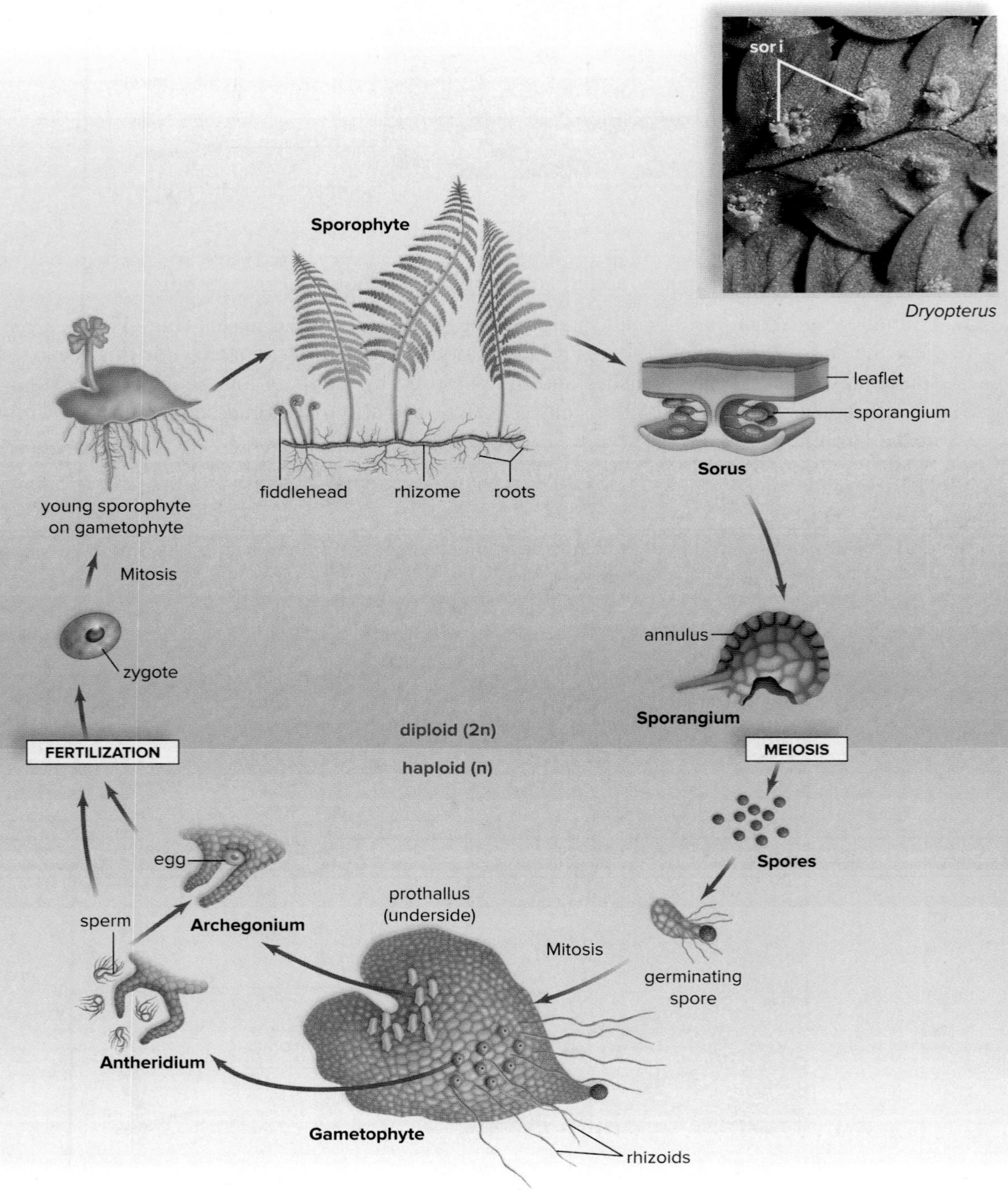

Fern spores are produced by meiosis in sporangia, which in many species occur on the underside of large leaves called **fronds.** These leaves are called **megaphylls** because they are broad leaves with several strands of vascular tissue. The spores are released when special cells of the **annulus** (a line or ring of thickened cells on the outside of the sporangium) dry out, and the sporangium opens. How do ferns move to new locations? _____

____

Observe the ferns on display, and then complete Table 17.1.

Table 17.1 Fern Diversity

Type of Fern	Description of Frond

Observation: Fern Sporophyte

Study the life cycle of the fern (Fig. 17.12), and find the sporophyte generation. This large, complexly divided leaf is known as a frond. Fronds arise from an underground stem called a rhizome.

Living or Preserved Frond

Examine a living or preserved specimen of a frond, and on the underside, notice a brownish clump called a **sorus** (pl., sori), each a cluster of many sporangia (Fig. 17.13).

What is being produced in the sporangia? ______________________________

Given that this is the generation we call the fern, what generation is dominant in ferns? ______________

Figure 17.13 Underside of frond leaflets.
Sori occur on the underside of frond leaflets.

Microscope Slide of Sorus

1. Examine a prepared slide of a cross section of a frond leaflet. Using Figure 17.14 as a guide, locate the fern leaf above and the sorus below.

Figure 17.14 Micrograph of cross section of a frond leaflet.
Micrograph of the internal anatomy of a sorus depicts many sporangia, where spores are produced. ©Ed Reschke/Photolibrary/Getty Images

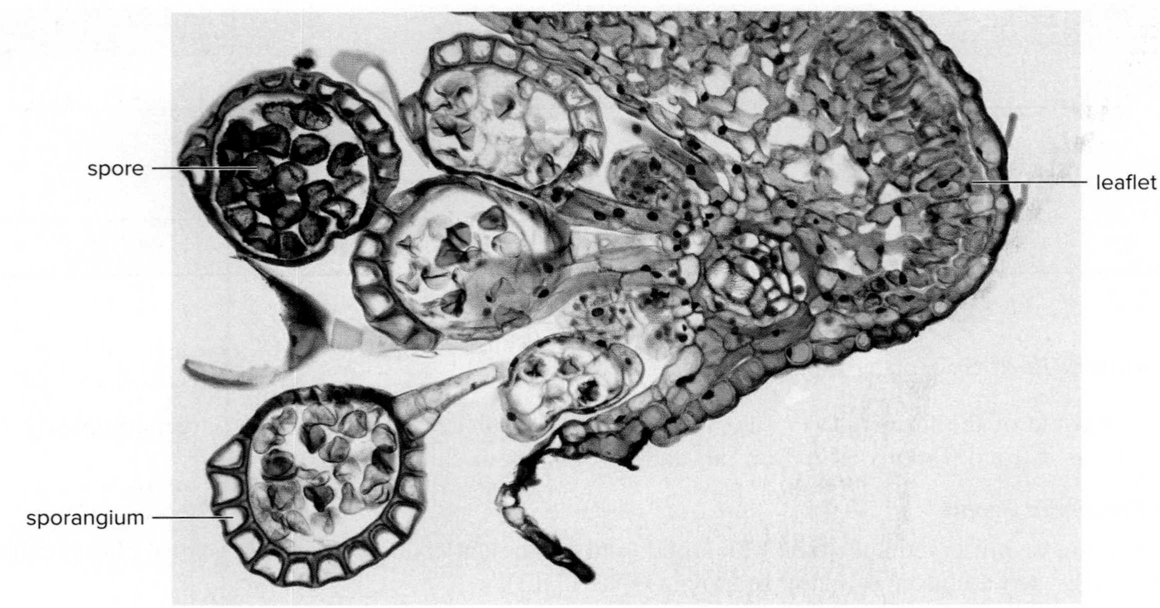

2. Within the sorus, find the sporangia and spores. Look for an **indusium** (not present in all species), a shelflike structure that protects the sporangia until they are mature. Does this fern have an indusium? __________

Observation: Fern Gametophyte

Plastomount

1. Examine a plastomount showing the fern life cycle.
2. Notice the prothallus, a small, heart-shaped structure. Can you find this structure in your fern minimarsh (if available)? __________ The prothallus is the gametophyte generation of the fern. Most people do not realize that this structure exists as a part of the fern life cycle. What is the function of this structure? _____
____

Microscope Slide

1. Examine whole-mount slides of fern prothallium-archegonia and fern prothallium-antheridia (Fig. 17.15).
2. If you focus up and down very carefully on an archegonium, you may be able to see an egg inside. What is being produced inside an antheridium? __________ When sperm produced in an antheridium swim to the archegonium in a film of water, what results? ______________________________ This structure develops into what generation? ______________________________

Conclusions: Ferns

- How are ferns dispersed from one area to another? ______________________________
- Is either generation in the fern dependent for any length of time on the other generation? __________ Explain. ______________________________

Figure 17.15 Fern prothallus.
The underside of the heart-shaped fern prothallus contains archegonia, where eggs are produced, and antheridia, where flagellated sperm are produced. (right side, top) ©Robert Knauft/Biology Pix/Science Source; (right side, bottom) ©Carolina Biological Supply Company/Phototake

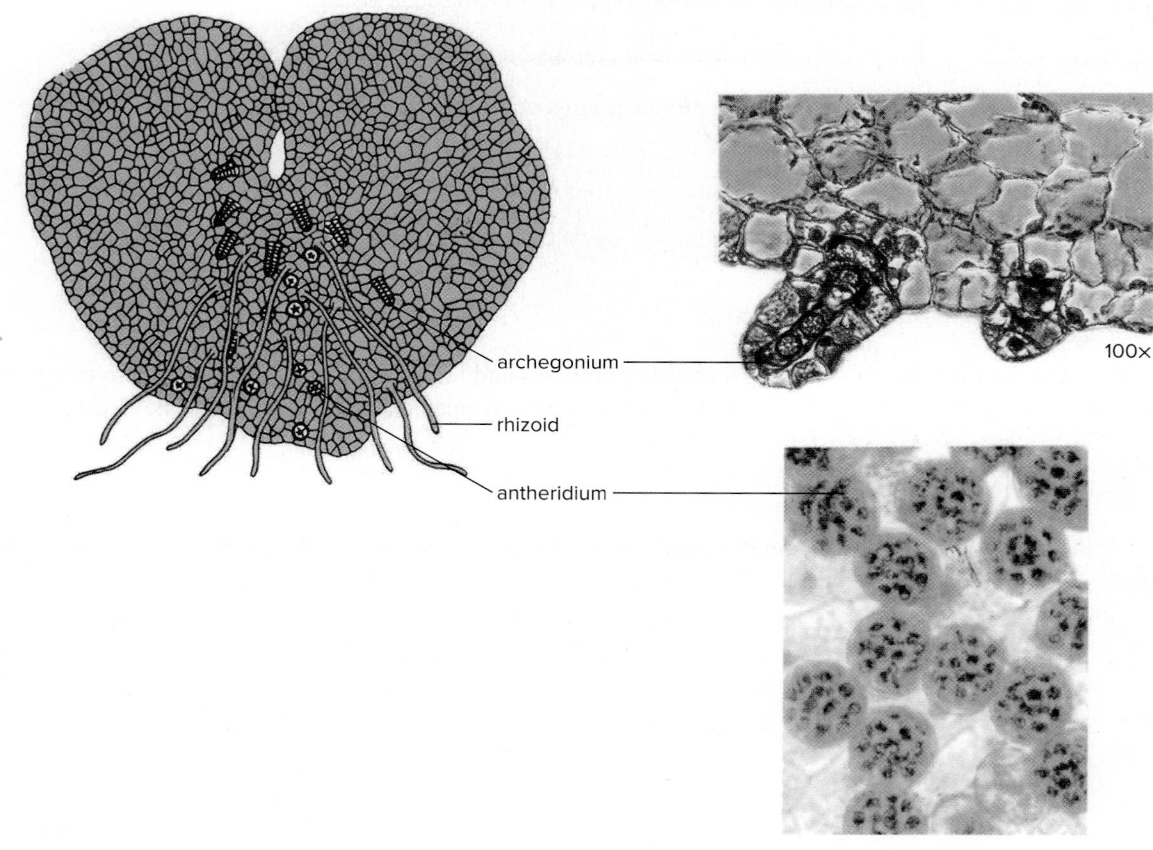

Laboratory Review 17

________________ **1.** What kind of tissue provides support and moves water to, and nutrients from, the leaves?

________________ **2.** What generation of the plant life cycle undergoes meiosis, sporophyte or gametophyte?

________________ **3.** What occurs, meiosis or mitosis, when spores develop into the gametophytes?

________________ **4.** When two gametes fuse, what plant generation is formed?

________________ **5.** When a plant identified as moss is observed, what generation of the moss life cycle is it?

________________ **6.** What kind of gamete is produced by moss antheridia?

________________ **7.** Do the rhizoids that anchor moss in the soil contain vascular tissue?

________________ **8.** How are mosses dispersed to new locations?

________________ **9.** Do seedless vascular plants have dominant sporophytes or gametophytes?

________________ **10.** Which generation in the life cycle of ferns lacks vascular tissue?

________________ **11.** Are fern zygotes dispersed to new locations?

________________ **12.** Do fern roots and rhizomes contain vascular tissue?

________________ **13.** What are the brownish clumps on the underside of a fern frond called?

________________ **14.** What structure produces eggs in the life cycle of ferns?

Thought Questions

15. Is there anything comparable to a spore in the life cycle of animals? Justify your reply.

16. Compare and contrast the sporophyte generation of mosses and ferns. Your reply could take into consideration vascular tissue, dominant form, diploid or haploid, and what happens to it as the life cycle advances.

17. During the life cycle of animals, gametes are produced when meiosis occurs. Why must gametes be produced by mitosis during the plant life cycle?

LABORATORY

18

Seed Plants

Learning Outcomes

18.1 Life Cycle of Seed Plants
- Compare the life cycle of nonseed and seed plants.
- List significant innovations of seed plants.

18.2 Gymnosperms
- Name the four groups of gymnosperms and give the distinguishing features of each group.
- Distinguish between pollination and fertilization in the life cycle of a pine tree.
- Identify the structure and function of pollen cones and seed cones, and tell where you would find the male and female gametophytes and seeds.

18.3 Angiosperms
- Identify the parts of a flower and compare two major groups of flowering plants.
- Identify the innovations that make the life cycle of a flowering plant different from that of a pine tree.
- Identify the male and female gametophytes and their roles in the life cycle of flowering plants.

18.4 Comparison of Gymnosperms and Angiosperms
- Compare the adaptations of flowering plants to those of conifers.
- Contrast the life cycle of gymnosperms and angiosperms.

Introduction

Planning Ahead Your instructor may suggest that you prepare the pollen grain slide at the start of the laboratory session.

Among plants, **gymnosperms** and **angiosperms** are seed plants. Review the plant evolutionary tree (see Fig. 17.1) and note that the gymnosperms and angiosperms share a common ancestor, which produced **seeds,** a structure that contains the next sporophyte generation. We shall see that the use of seeds to disperse the next generation requires a major overhaul of the alternation of generations life cycle. Nonseed plants disperse the gametophyte, the haploid (n) generation, by production of spores. Seed plants disperse the sporophyte, the diploid (2n) generation. Which generation—gametophyte or sporophyte—is better adapted to a land environment because it contains vascular tissue? __________

Nonseed plants utilize an archegonium to protect the egg; and following fertilization, the sporophyte develops immediately within the archegonium. Seed plants protect the entire female gametophyte within an **ovule.** Following fertilization the ovule develops into a sporophyte-containing seed. You will see that the formation of the ovule and also the **pollen grains** (male gametophytes) are radical innovations in the life cycle of seed plants. In gymnosperms, pollen grains are windblown, but many flowering plants have a mutualistic relationship with animals, particularly flying insects, which disperse their pollen. Various animals also help disperse the seeds of many flowering plants. Relationships with animals help explain why flowering plants are so diverse and widespread today.

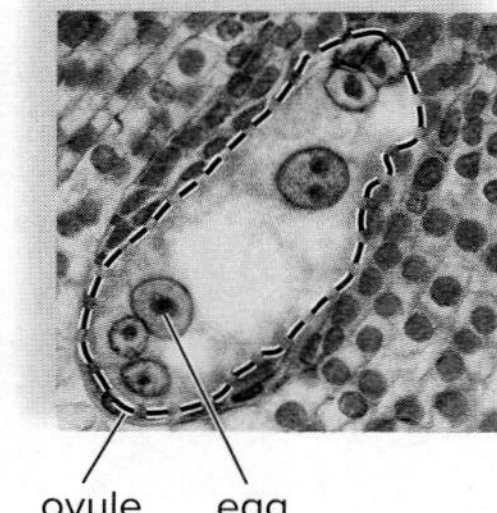

Female gametophyte of angiosperms
©W. P. Armstrong 2004

18.1 Life Cycle of Seed Plants

Figure 18.1 shows alternation of generations as it occurs in nonseed plants, and Figure 18.2 shows alternation of generations as it occurs in seed plants. Note which structures are haploid and which are diploid in these diagrams.

1. In which life cycle, nonseed or seed, do you note pollen sacs (microsporangia) and ovules (megasporangia)? __________ In which life cycle, nonseed or seed, do you note two types of spores, microspores, and megaspores? __________ The formation of **heterospores** (unlike spores) is an innovation that leads to the production of pollen grains and the formation of ovules in seed plants. *Label heterospores where appropriate in Figure 18.2.* In which life cycle do you note male gametophyte (pollen grain) and female gametophyte (embryo sac in ovule)? __________
2. In nonseed plants, flagellated sperm must swim in external water to the egg. In seed plants, **pollination** (the transport of male gametophytes by wind or pollen carrier to the vicinity of female gametophytes) does not require external water. This is an innovation in seed plants. *Label pollination where appropriate in Figure 18.2.*
3. In which life cycle does a **seed** appear between the zygote and the sporophyte? ____________________ What generation is present in a seed? ____________________ Formation of a seed from an ovule following fertilization is an innovation in seed plants.

Figure 18.2 Alternation of generations in flowering plants, which are seed plants.

Figure 18.1 Alternation of generations in nonseed plants.

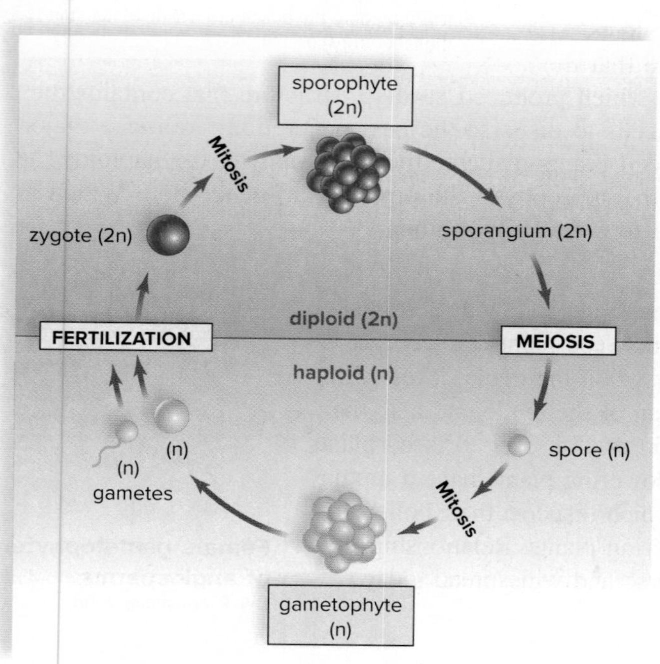

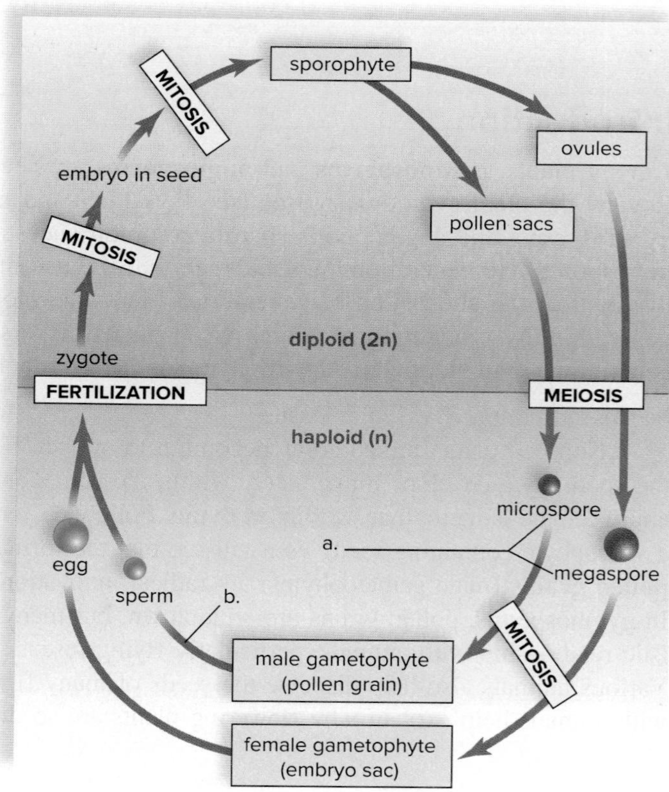

18.2 Gymnosperms

The term *gymnosperm* means "naked seed." The ovules and seeds of gymnosperm are exposed on a cone scale or some other comparable structure. Most gymnosperms do produce cones. The four groups of gymnosperms are:

Cycadophyta: cycads	Gnetophyta: gnetophytes
Ginkgophyta: ginkgoes	Coniferophyta: conifers

Cycads

Cycads are cone-bearing, palmlike plants found today mainly in tropical and subtropical regions. Cycads flourished during the Mesozoic era. They were so numerous that this era is sometimes referred to as the Age of Cycads and Dinosaurs. Herbivorous dinosaurs most likely fed on cycad seeds and foliage.

Observation: Cycads

Among cycads, male and female plants are separate. If available, examine cycad specimens, including *Zamia* and *Cycas*. If you are examining *Zamia,* note the male and female cones. If you are examining *Cycas,* note that only the male plant bears cones; the female reproductive structure is much more leaflike.

After examining cycads and the following photograph, give three characteristics you could use to recognize a cycad.

1. ______________________________
2. ______________________________
3. ______________________________

©FLPA/David Hosking/age fotostock

Ginkgoes

Only one species of **ginkgo**—the maidenhair tree—survives today. The maidenhair tree was largely restricted to ornamental gardens in China until it was found to do quite well in polluted areas. Since the female trees produce rather smelly seeds, usually only male trees are used as ornamentals, such as those planted in city parks.

Observation: The Maidenhair Tree

The maidenhair tree can be quite tall (12–25 meters). A few of the lower branches may be pendulous. The fan-shaped leaves turn a vivid yellow in the fall. The tree is deciduous. The male tree bears cones, and the female tree produces the seeds later surrounded by a fleshy covering.

After examining ginkgo leaves and the following photograph, give three characteristics you could use to recognize a male ginkgo tree.

1. ______________________________

2. ______________________________

3. ______________________________

©suchi187/123RF

©Hadrian/Shutterstock RF

©Vladimir Wrangel/Shutterstock

Gnetophytes

Gnetophytes are seed-bearing plants that can grow as shrubs, trees, or vines and share similarities with both gymnosperms and angiosperms. Three genera survive today: *Ephedra* is a shrub found in arid regions of the southwestern United States. It has small, scalelike leaves. *Gnetum* is a vine or tree found in tropical rain forests of Asia, South America, and Africa. Its large, leathery leaves resemble those of angiosperms. *Welwitschia* is found only in South Africa. A woody central disk has only two enormous straplike leaves that split lengthwise with age. Cone-bearing branches arise from the margin of the disk. Most of the plant is buried in sandy soil. Molecular data supports the long-held belief that these plants are related.

Observation: Gnetophytes

If available, examine gnetophyte specimens and then *label the following photographs on the lines provided.*

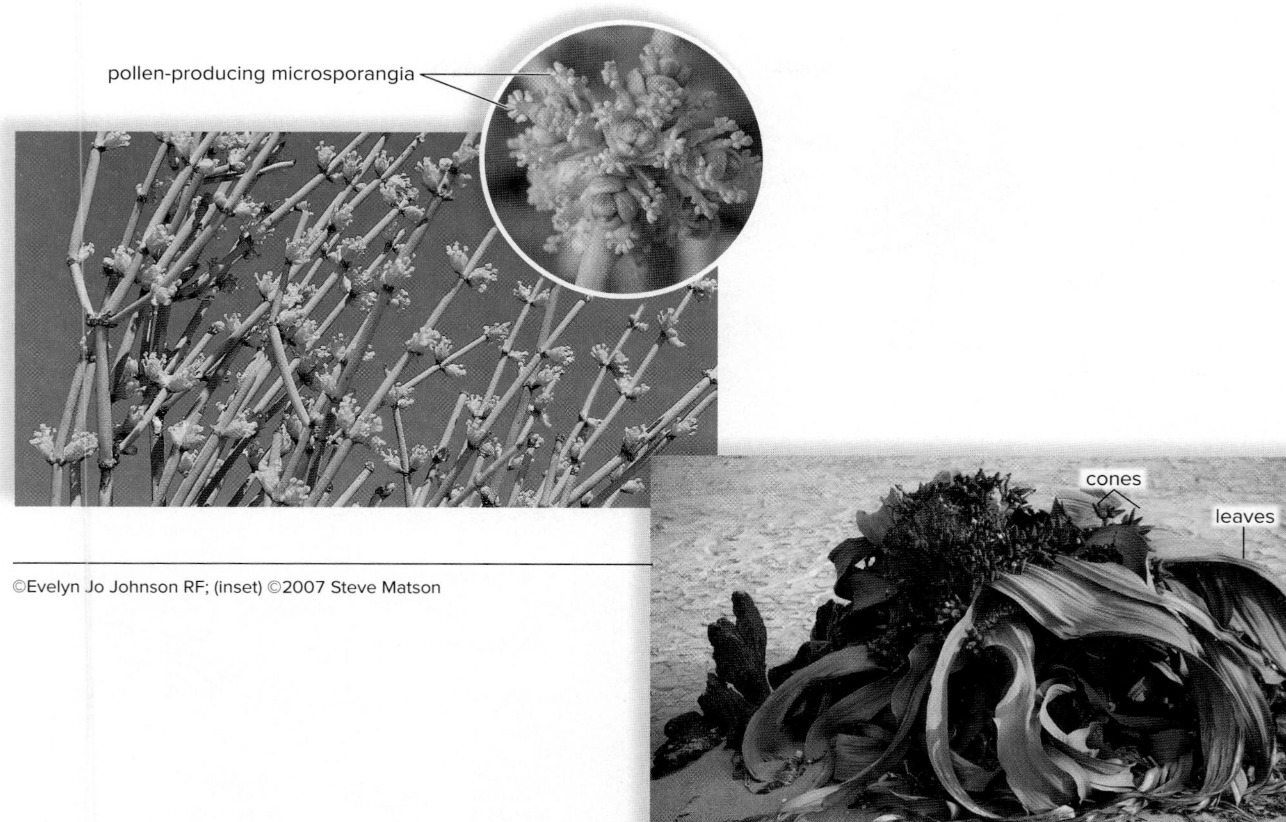

©Evelyn Jo Johnson RF; (inset) ©2007 Steve Matson

©Steve Robinson/NHPA/Photoshot

Conifers

The **conifers** are by far the largest group of gymnosperms. Pines, hemlocks, and spruces are evergreen conifers because their leaves remain on the tree through all seasons. The cypress tree and larch (tamarack) are examples of conifers that are not evergreen.

The gymnosperms have the first real development of **wood,** which is derived from dead transport tissue. Pines in particular are grown to serve as sources for wood and paper.

A northern coniferous forest of evergreen trees
©Creatas/Jupiterimages RF

Cones of lodgepole pine, *Pinus contorta*
©Kathy Merrifield/Science Source

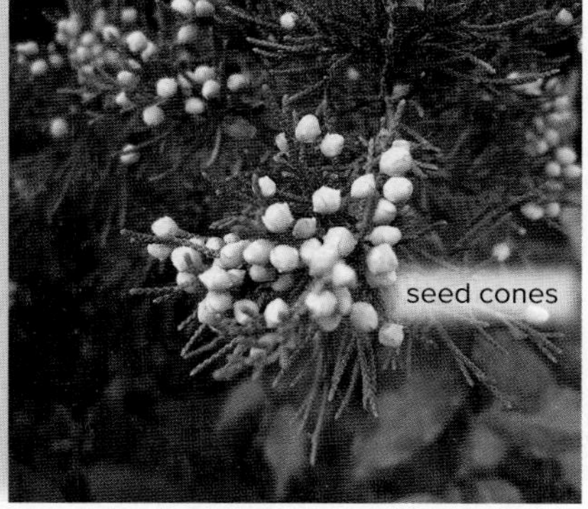

Fleshy seed cones of juniper, *Juniperus*
©McGraw-Hill Education/Evelyn Jo Johnson, photographer

Figure 18.3 Pine life cycle.
The sporophyte is the tree. The male gametophytes are windblown pollen grains shed by pollen cones. The female gametophytes are retained within ovules on seed cones. The ovules develop into windblown seeds.

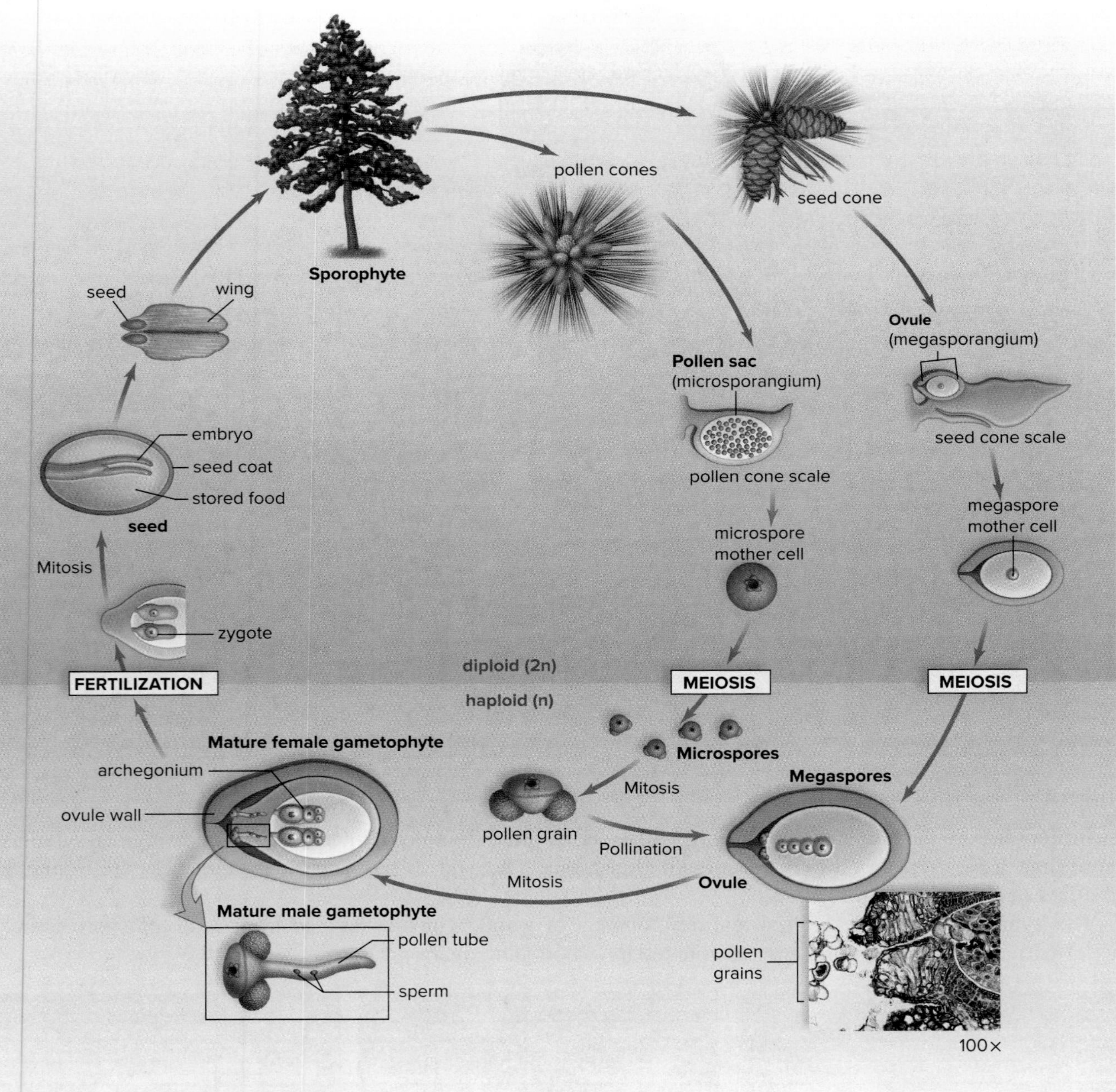

Pine Trees

The pine tree is the dominant sporophyte generation (Fig. 18.3) in the life cycle of a pine. Vascular tissue extends from the roots, through the stem, to the leaves. Pine tree leaves are needlelike; leathery; and covered with a waxy, resinous cuticle. The **stomata,** openings in the leaves for gas exchange, are sunken. The structure of the leaf and the leaf's internal anatomy are adaptive to a drier climate.

In pine trees, the **pollen sacs** and **ovules** are located in cones. Pollen sacs on pollen cone scales contain microspore mother cells, which undergo meiosis to produce **microspores.** Microspores (n) become sperm-bearing **male gametophytes** (pollen grains). Seed cone scales bear ovules where megaspore mother cells undergo meiosis to produce **megaspores.** A megaspore (n) becomes an egg-bearing **female gametophyte.**

Pollination occurs when pollen grains are windblown to the seed cones. After **fertilization,** the egg becomes a sporophyte (2n) embryo enclosed within the ovule, which develops a seed coat. The seeds are winged and are dispersed by the wind.

1. Which part of the pine life cycle is an adult sporophyte? ______________________________
2. Which part of the pine life cycle is the male gametophyte? The female gametophyte?

   ____

   ____
3. Where does fertilization occur and what structure becomes a seed? ____________________

   ____

Observation: Pine Leaf

Obtain a cluster of pine leaves (needles). A very short woody stem is at its base. Each type of pine has a typical number of leaves in a cluster (Fig. 18.4). How many leaves are in the cluster you are examining? __________ What is the common name of your specimen? ______________________________

Figure 18.4 Pine leaves (needles). **a.** The needles of white pines are in clusters of five. **b.** The needles of pitch pines are in clusters of three. **c.** The needles of red pines are in clusters of two.

Observation: Pine Cones

Preserved Cones

1. Compare a pine pollen cone to a pine seed cone. ______________________________

____

____

Note the size and texture of the pollen cones relative to the larger, woody, seed cones.

2. Pollen cones. Remove a single scale (sporophyll) from the pollen cone and examine with a stereomicroscope. Note the two pollen sacs on the lower surface of each scale (Fig. 18.5*a*). What do the pollen sacs produce? ______

____

____

Figure 18.5 Pine cones.
a. The scales of pollen cones bear pollen sacs where microspores become pollen grains. **b.** The scales of seed cones bear ovules that develop into winged seeds.

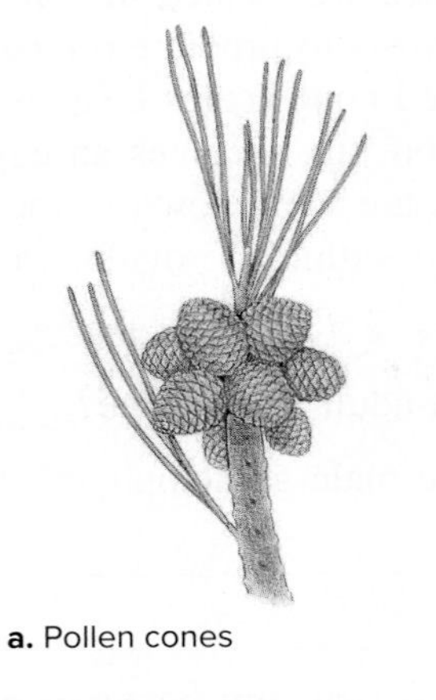

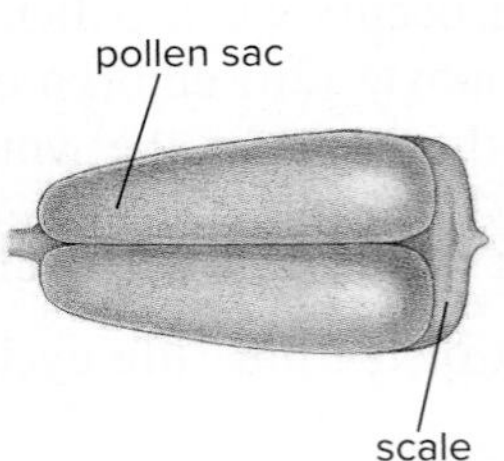

a. Pollen cones

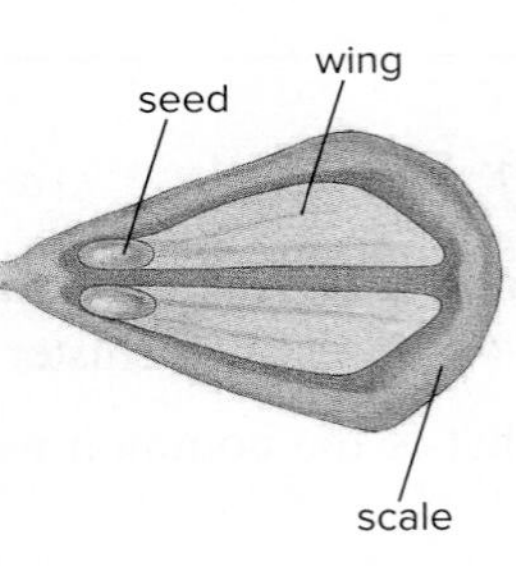

b. Seed cone

3. Seed cones.
   - **a.** Seed cones of three distinct ages are present. First-year cones are about 1 cm wide. Second-year cones are about 10 cm long. They are green and the scales are tightly closed together. In third-year cones the scales have opened, revealing the mature seeds at the base of the cone scale.
   - **b.** If available, examine a first-year cone cut lengthwise. The ovules are visible as small, milky-white domes at the base of the cone scale. Ovules will hold the female gametophyte generation, and then they become seeds.
   - **c.** Examine mature seed cones. See if any seeds are present (Fig. 18.5*b*). Where are they located? ______________________________ Note the hardness of the seed coat.

     What is the function of the seed coat? ______________________________

     Are the seeds covered by tissue donated by the original sporophyte? ______________________________

     What does gymnosperm mean? ______________________________

     If instructed to do so, use tweezers to pull out a seed and note the wing. What is the wing for?

     ____

     Replace the seed in the cone when you are finished.

4. Pine seeds. If available, examine a pine seed (called a pine nut) that has the seed coat removed. These seeds can be used for cooking foods such as pesto. Carefully cut the seed lengthwise and examine. Can you find an embryo inside? ______________________________

Microscope Slides of Pine Cones

1. Examine a prepared slide of a longitudinal section through a mature pine pollen cone. *Label a pollen sac in Figure 18.6*a *and a pollen grain in Figure 18.6*b. A pollen grain (male gametophyte) has a central body and two attached hollow bladders. How do these help in the dispersal of pine pollen? ___________

   ____

   The central body has two cells. One cell will divide to become two nonflagellated sperm, one of which fertilizes the egg after pollination. The other cell forms the **pollen tube** through which a sperm travels to the egg.

2. Examine a prepared slide of a longitudinal section through an immature pine seed cone. As mentioned, seed cone scales bear ovules. The ovule contains a megaspore mother cell, which undergoes meiosis to produce four megaspores, three of which disintegrate. This megaspore (n) becomes an egg-bearing female gametophyte. *Label the ovule and the megaspore mother cell in Figure 18.7. Also, label the pollen grains that you can see just outside the ovule.*

Figure 18.6 Pine pollen cone.
Pollen cones bear **a.** pollen sacs (microsporangia) in which microspores develop into pollen grains. **b.** Enlargement of pollen grains (male gametophytes).

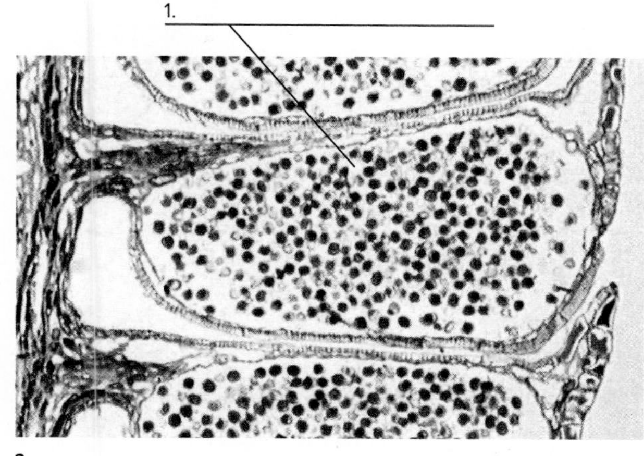

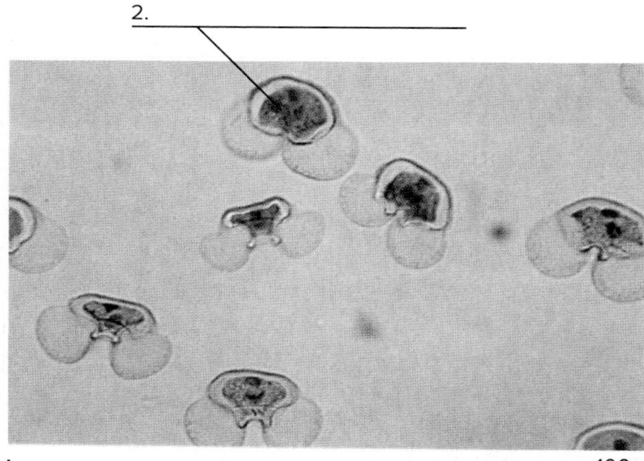

Figure 18.7 Seed cone.
Seed cones bear ovules, each of which will contain a female gametophyte. Note pollen grains near the entrance.

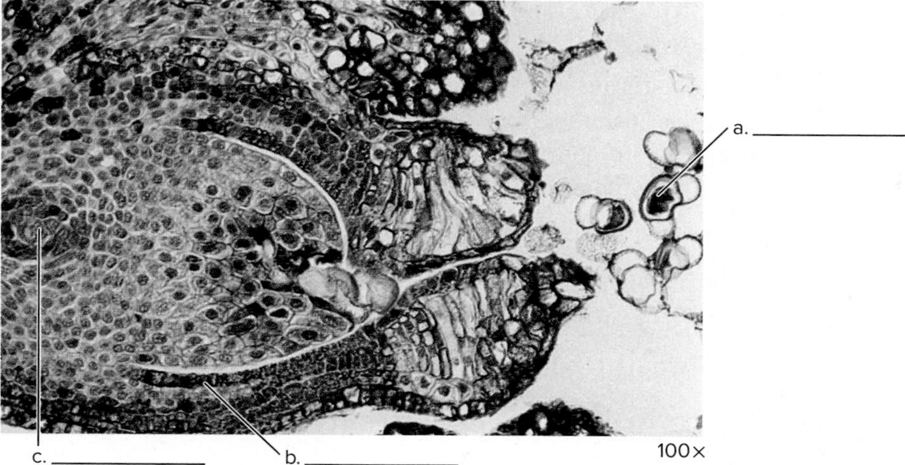

18.3 Angiosperms

Flowering plants are the dominant plants today. They occur as trees, shrubs, vines, and garden plants. At some point in their life cycle, all flowering plants bear flowers (Fig. 18.8).

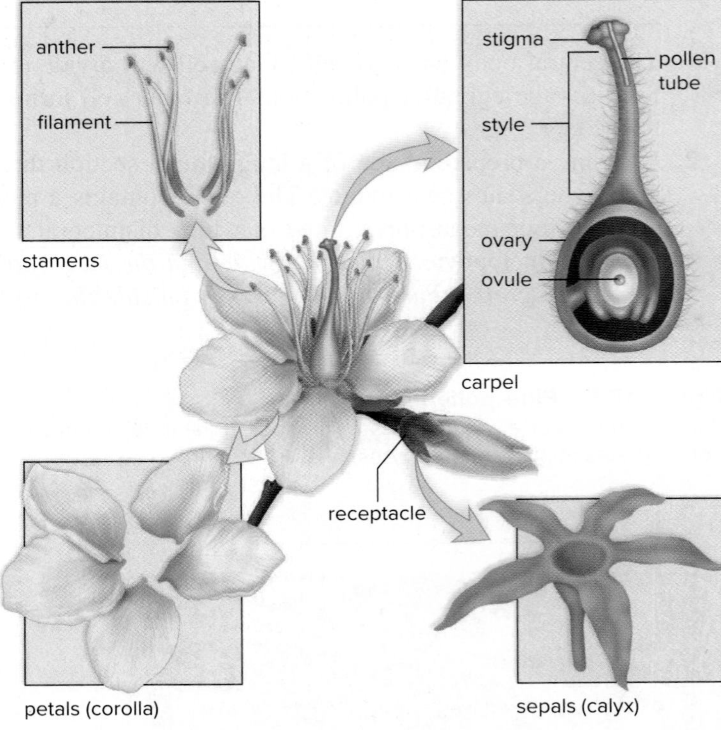

Figure 18.8 Generalized flower.
A flower has four main kinds of parts: sepals, petals, stamens, and a carpel. A stamen has an anther and filament. A carpel has a stigma, style, and ovary. An ovary contains ovules.

Observation: A Flower

1. With the help of Figure 18.8, identify the following structures on a model of a flower:
   a. **Receptacle:** The portion of a stalk to which the flower parts are attached.
   b. **Sepals:** An outermost whorl of modified leaves, collectively called the **calyx.** Sepals are green in most flowers. They protect a bud before it opens.
   c. **Petals:** Usually colored leaves that collectively constitute the **corolla.**
   d. **Stamen:** A swollen terminal **anther** and the slender supporting **filament.** The anther contains two pollen sacs, where microspores develop into male gametophytes (pollen grains).
   e. **Carpel:** A modified sporophyll consisting of a swollen basal ovary; a long, slender **style** (stalk); and a terminal **stigma** (sticky knob).
   f. **Ovary:** The enlarged part of the carpel that develops into a fruit.
   g. **Ovule:** The structure within the ovary where a megaspore develops into a female gametophyte (embryo sac). The ovule becomes a seed.
2. Carefully inspect a fresh flower. What is the common name of your flower? ______________________
3. Remove the sepals and petals by breaking them off at the base. How many sepals and petals are there? ______________________
4. Are the stamens taller than the carpel? ______________________
5. Remove a stamen, and touch the anther to a drop of water on a slide. If nothing comes off in the water, crush the anther a little to release some of its contents. Place a coverslip on the drop, and observe with low- and high-power magnification. What are you observing? ______________________

6. Remove the carpel by cutting it free just below the base. Make a series of thin cross sections through the ovary. The ovary is hollow, and you can see nearly spherical bodies inside. What are these bodies? ______________________________

____

7. Flowering plants are divided into two groups, called **monocots** and **eudicots.** In the meantime, Table 18.1 lists significant differences between the two classes of plants. Is your flower a monocot or eudicot? _____

____

____

Table 18.1 Monocots and Eudicots

Monocots	Eudicots
One cotyledon	Two cotyledons
Flower parts in threes or multiples of three	Flower parts in fours or fives or multiples of four or five
Usually herbaceous	Woody or herbaceous
Usually parallel venation	Usually net venation
Scattered bundles in stem	Vascular bundles in a ring
Never woody	Can be woody

Life Cycle of Flowering Plants

The life cycle of a flowering plant (Fig. 18.9) is like that of the pine tree except for these innovations:

- In angiosperms, the often brightly colored flower contains the pollen sacs and ovules. Locate these structures in Figure 18.9, and trace the life cycle of flowering plants from the sporophyte generation (the tree) through the various stages to the sporophyte generation once again.
- **Pollination** in flowering plants—when pollen is delivered to the stigma of the carpel—is sometimes accomplished by wind but more likely by the assistance of an animal pollinator. The pollinator acquires nutrients (e.g., nectar) from the flower and inadvertantly collects pollen, which it takes to the next flower.
- Notice that flowering plants, unlike gymnosperms, practice **double fertilization.** A mature pollen grain contains two sperm; one fertilizes the egg, and the other joins with the two polar nuclei to form **endosperm** (3n), which serves as food for the developing embryo.
- Also, flowering plants produce seeds the same as gymnosperm, but their seeds are enclosed within fruits. **Fruits** protect the seeds and aid in seed dispersal. Sometimes, animals eat the fruits, and after the digestion process, the seeds are deposited far away from the parent plant. The term *angiosperm* means "covered seeds." The seeds of angiosperm are found in fruits, which develop from parts of the flower, particularly the ovary.

Figure 18.9 Flowering plant life cycle.
The parts of the flower involved in reproduction are the anthers of stamens and the ovules in the ovary of a carpel.

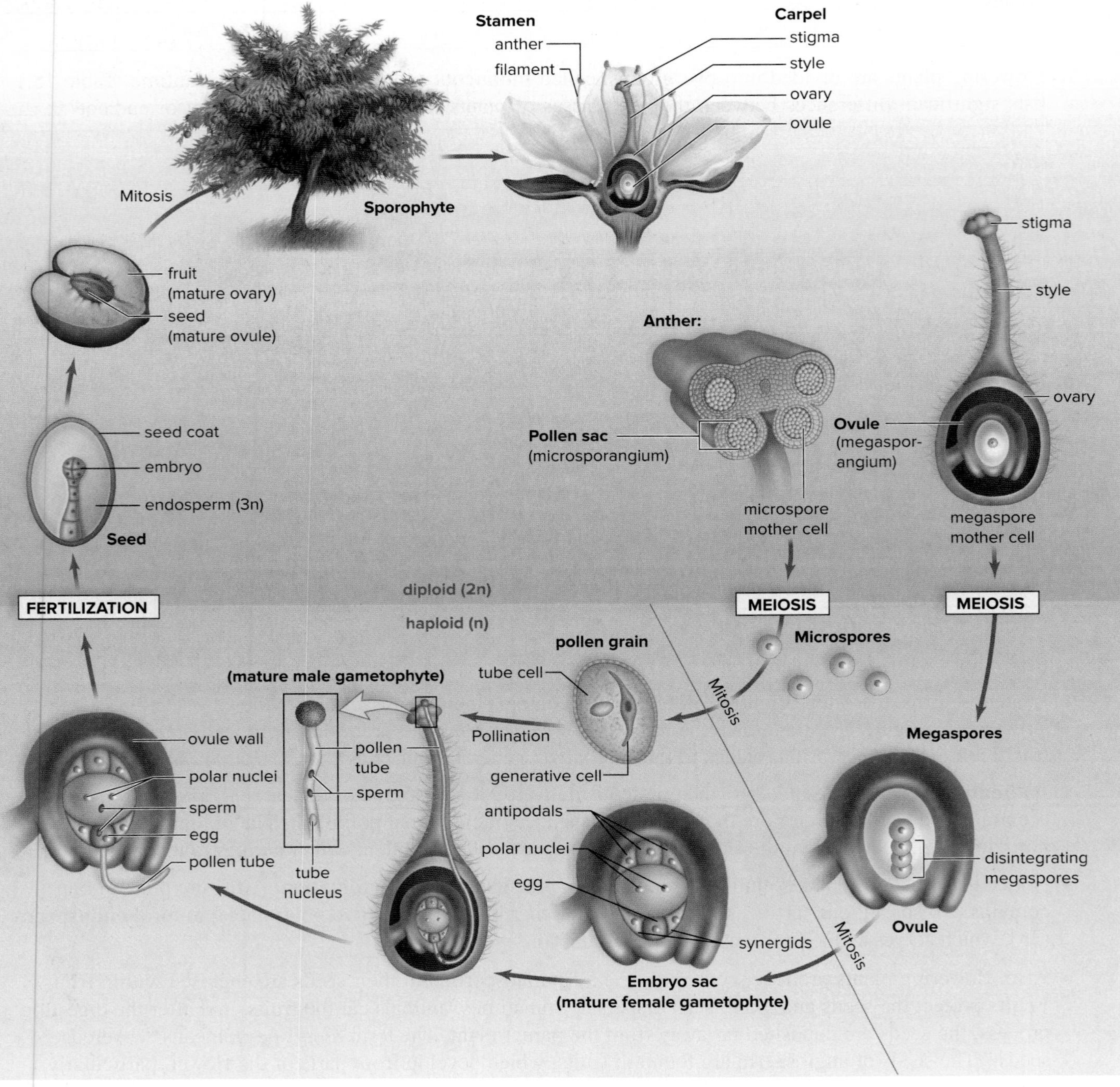

The Male Gametophyte

Notice that in the life cycle of flowering plants (Fig. 18.9), meiosis in pollen sacs produces four microspores, each of which will become a two-celled pollen grain. In flowering plants, ______________ is the transfer of the pollen grain from the anther to the stigma, where the pollen grain germinates and becomes the mature male gametophyte. By what means is pollination accomplished in the flowering plant life cycle? ____

Observation: Pollen Grain Slide

1. Examine a pollen grain slide before germination. In this observation, you will observe nongerminated pollen grains and germinated pollen grains. A pollen grain newly released from the anther has two cells. The larger of the two cells is the **tube cell,** and the smaller is the **generative cell.** Obtain and examine a prepared slide of pollen grains. Identify the tube cell and the generative cell. Sketch your observation here.

2. Examine a pollen grain slide following germination. During germination, the pollen grain's tube cell gives rise to the pollen tube. As it grows, the pollen tube passes through the stigma and grows through the style of the carpel and into the ovary. Two sperm cells produced by division of the pollen grain's generative cell migrate through the pollen tube into the embryo sac. Obtain and examine a prepared slide of germinated pollen grains with pollen tubes. You should be able to see the tube nucleus and two sperm cells. What portion of a germinated pollen grain tells you that it is the mature male gametophyte?

____

**Experimental Procedure: Pollen Grains*

Inoculate an agar-coated microscope slide with pollen grains. Invert the slide, and place it onto a pair of wooden supports in a covered petri dish. Leave the inverted slide in the covered petri dish for one hour. Then remove the slide, and examine it with the compound light microscope. Have any of the pollen grains germinated? ____

If so, describe. ____

____

*Source: Carter, Richard, and Faircloth, Wayne R. General Laboratory Studies, Dubuque: IA: Kendall/Hunt Publishing, 1991.

The Female Gametophyte

In the ovule, the surviving megaspore undergoes three mitotic divisions to produce a seven-celled (eight-nuclei) female gametophyte called an **embryo sac** (Fig. 18.10). One of these cells is an egg cell. The largest cell contains two **polar nuclei.** The embryo sac is the mature female gametophyte.

Figure 18.10 Embryo sac in a lily ovule.
An embryo sac is the female gametophyte of flowering plants. It contains seven cells, one of which is the egg. The fate of each cell is noted. ©W. P. Armstrong 2004

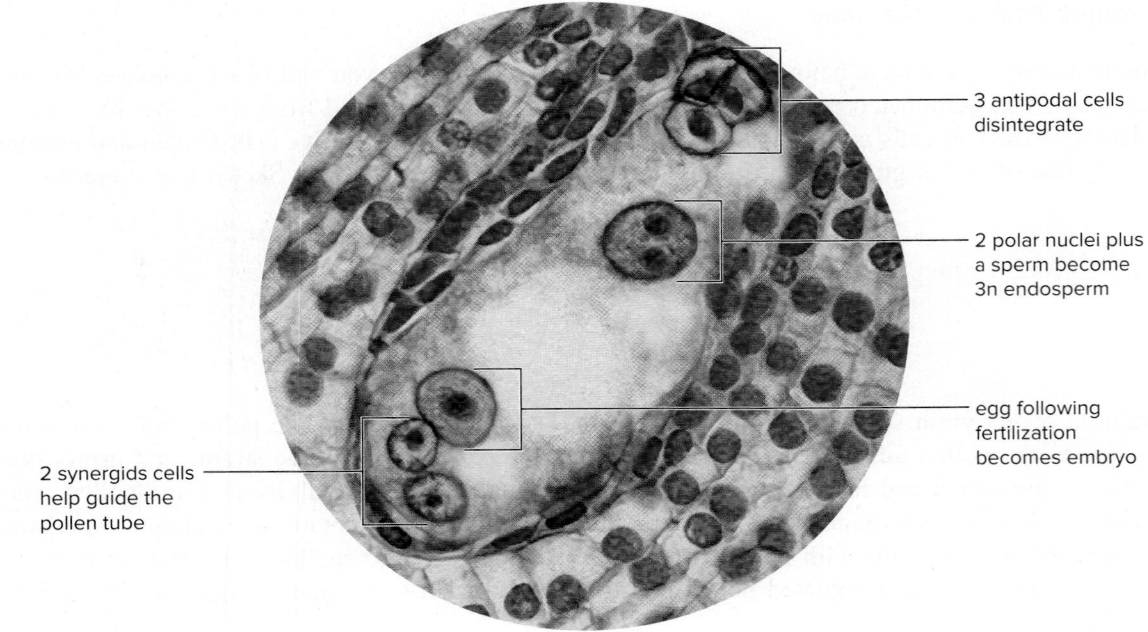

1. What portion of the embryo sac tells you that it is the female gametophyte? ______________________
2. When the pollen tube delivers sperm to the embryo sac, double fertilization, as defined previously, occurs. Due to the first fertilization, what happens to the egg? ______________________
____

   Due to the second fertilization, what happens to the polar nuclei? ______________________
____
3. Following fertilization, the ovules develop into seeds, and the ovary develops into the fruit. What are the three parts of a seed? ______________________
____

Observation: Embryo Sac Slide

Examine the demonstration slide of the mature embryo sac of *Lilium.* Identify the egg labeled in Figure 18.10.

18.4 Comparison of Gymnosperms and Angiosperms

1. Beneath the photos, list ways to tell a gymnosperm from an angiosperm.

©Evelyn Jo Johnson

©Evelyn Jo Johnson

a. ____
____
____

b. ____
____
____

2. Complete Table 18.2 using "yes" or "no" to compare gymnosperms to angiosperms.

Table 18.2 Comparison of Gymnosperms and Angiosperms

	Heterospores	Pollen grains/ Ovule	Cones	Flower	Fruit
Gymnosperms					
Angiosperms					

3. What structure in gymnosperms and angiosperms delivers sperm from pollen sacs to the vicinity of the egg? ____

   Does delivery require external water? ____

4. What structure in gymnosperms and angiosperms becomes a seed? ______________________________

5. Is a sporophyte or gametophyte embryo in a seed? ______________________________

6. What innovation in angiosperms led to the seeds being covered by fruit? ______________________________

Laboratory Review 18

_______________ 1. Which generation, gametophyte or sporophyte, is dispersed by seed plants?

_______________ 2. What protects the entire female gametophyte in seed plants?

_______________ 3. When meiosis occurs in a seed plant, what results?

_______________ 4. What is the transfer of the seed plants' male gametophyte by wind or a carrier called?

_______________ 5. What is the name given to plants that produce "naked seeds"?

_______________ 6. How are the seeds of a pine tree dispersed?

_______________ 7. In the live cycle of a pine, what is the sporophyte?

_______________ 8. In the life cycle of a pine, where are the ovules located?

_______________ 9. What flower structure is formed by the anther and filament?

_______________ 10. What does mitotic division of the generative cell in a pollen grain produce?

_______________ 11. What structure do the sperm move through to reach an angiosperm's female gametophyte?

_______________ 12. What is produced when double fertilization occurs in angiosperms?

_______________ 13. What is the name of the female gametophyte of flowering plants?

_______________ 14. What flowers structure forms the covering around the seed of an angiosperm?

Thought Questions

15. a. How are the gametophytes of mosses and flowering plants different from one another?

b. How are the gametophytes of mosses and flowering plants similar to one another?

16. Why do pines produce clusters of pollen cones, and why are the pollen cones located at the tips of branches instead of closer to the trunk of the tree?

17. Angiosperm means "covered seed." What covers the seeds of angiosperms, and what role does this covering have in ensuring the success of the embryos in the seeds?

L A B O R A T O R Y

19

Organization of Flowering Plants

Learning Outcomes

19.1 External Anatomy of a Flowering Plant
- Distinguish between the root system and the shoot system of a plant.
- Identify the external anatomical features of a flowering plant.

19.2 Major Tissues of Roots, Stems, and Leaves
- List and give functions for the major tissues in the organs of a plant.
- List five differences between monocots and eudicots.

19.3 Root System
- Name and describe the zones of a root tip.
- Identify a cross section and the specific tissues of a eudicot root and a monocot root.
- Identify various types of root diversity.

19.4 Stems
- Identify a cross section and the specific tissues of eudicot herbaceous and monocot herbaceous stems.
- Identify various types of stem diversity.
- Distinguish between primary and secondary growth of a stem in two ways.
- Explain the structure of a winter twig and compare its structure to that of a living nonwoody plant.
- Explain the occurrence of annual rings, and determine the age of a tree from a trunk cross section.

19.5 Leaves
- Identify a cross section and the specific tissues of a leaf.
- Distinguish between simple leaves and two types of compound leaves.
- Distinguish between various arrangements of leaves on a stem..

Introduction

Despite their great diversity in size and shape, all flowering plants have three vegetative organs that have nothing to do with reproduction: the root, the stem, and the leaf (Fig. 19.1). Roots anchor a plant and absorb water and minerals from the soil. A stem usually supports the leaves so that they are exposed to sunlight. Leaves carry on photosynthesis and thereby produce the nutrients that sustain a plant and allow it to grow.

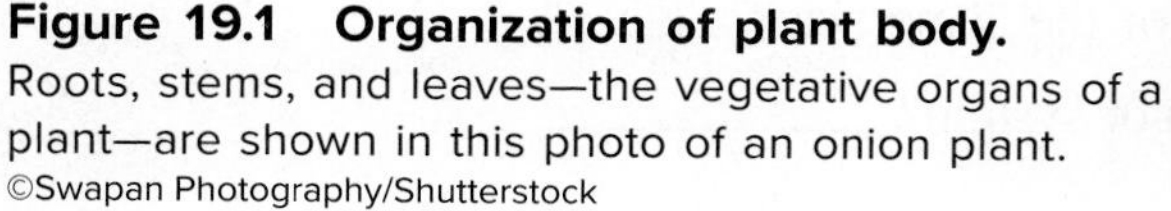

Figure 19.1 Organization of plant body. Roots, stems, and leaves—the vegetative organs of a plant—are shown in this photo of an onion plant.

19.1 External Anatomy of a Flowering Plant

Figure 19.2 shows that a plant has a root system and a shoot system. The **root system** consists of the roots. The **shoot system** consists of the stem and the leaves.

Observation: A Living Plant

Shoot System

What structures are in the shoot system?

____

____

____

The Leaves

Leaves carry on photosynthesis. Which part of a leaf (blade or petiole) is the most expansive part?

____

____

The Stem

1. In the **stem,** locate a **node** and an **internode.**
2. Measure the length of the internode in the middle of the stem. Does the internode get larger or smaller toward the apex of the stem? ____________ Toward the roots? ________ Based on the fact that a stem elongates as it grows, explain your observation. ____________

   ____

   ____

   ____

3. Where is the **terminal bud** (also called the shoot tip) of a stem?____________

   ____

   Where are axillary buds? ____________

   ____

Root System

Observe the root system of a living plant if the root system is exposed. Does this plant have a strong primary root or many roots of the same size? ________

What structures are in the root system? ____________

____

____

____

Where is the root tip? ____________

Figure 19.2 Organization of a plant.
Roots, stems, and leaves are vegetative organs.

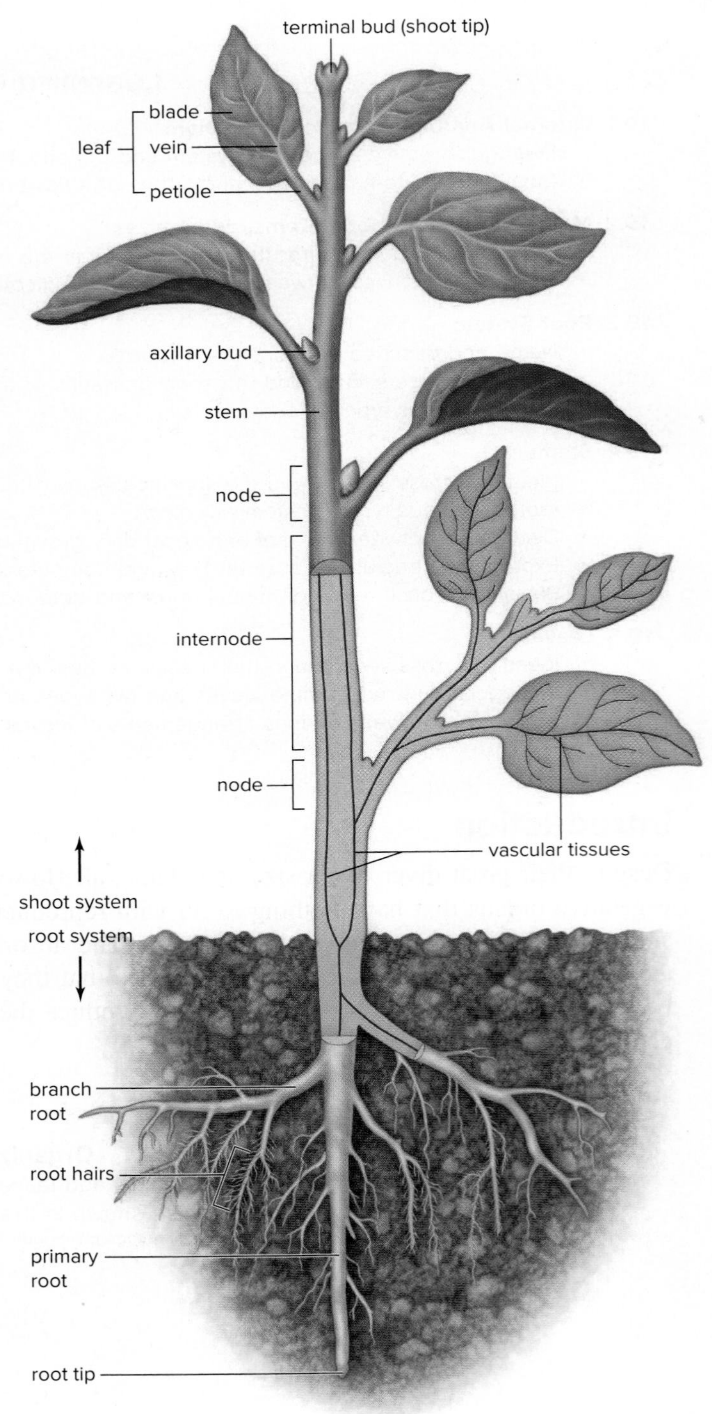

19.2 Major Tissues of Roots, Stems, and Leaves

Unlike humans, flowering plants grow in size their entire life because they have an immature tissue called **meristematic tissue** composed of cells that divide. **Apical meristem** is located at the terminal end of the stem, the branches, and at the root tip and the root branches. When apical meristem cells divide, some of the cells differentiate into the mature tissues of a plant:

Dermal tissue: Forms the outer protective covering of a plant organ

Ground tissue: Fills the interior of a plant organ; photosynthesizes and stores the products of photosynthesis

Vascular tissue: Transports water and sugar, the product of photosynthesis, in a plant and provides support

Note in Table 19.1 that roots, stems, and leaves have all three tissues, but they are given different specific names.

Table 19.1 Mature Tissues of Vegetative Organs

Tissue Type	Roots	Stems	Leaves
1. Dermal tissue (epidermis)	Protects inner tissues Root hairs absorb water and minerals.	Protects inner tissues	Protects inner tissues Cuticle prevents H_2O loss. Stomata carry on gas exchange.
2. Ground tissue	Cortex: Stores products of photosynthesis Pith: Stores products of photosynthesis	Cortex: Carries on photosynthesis, if green Pith: Stores products of photosynthesis	Mesophyll: Photosynthesis
3. Vascular tissue (xylem and phloem)	Vascular cylinder: Transports water and nutrients	Vascular bundle: Transports water and nutrients	Leaf vein: Transports water and nutrients

Observation: Tissues of Roots, Stems, and Leaves

Meristematic tissue. A slide showing the apical meristem in a shoot tip and another showing the apical meristem in a root tip are on demonstration. Meristematic cells are spherical and stain well when they are dense and have thin cell walls (Fig. 19.3). What is the function of meristematic tissue? ______________________

____

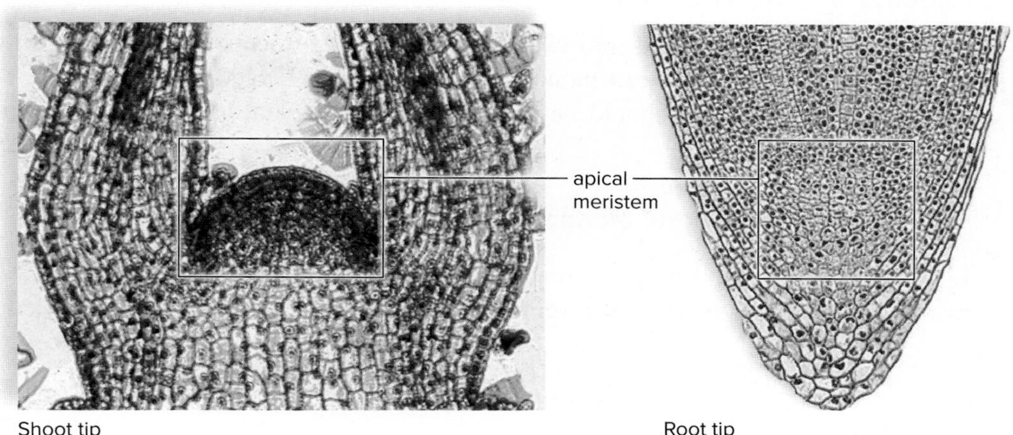

Figure 19.3 Apical meristem. A shoot tip and a root tip contain meristem tissue, which allows them to grow longer the entire life of a plant.

(left) ©Steven P. Lynch RF; (right) ©Ray F. Evert, University of Wisconsin

Mature Tissues

1. **Dermal tissue.** In a cross section of a leaf (Fig. 19.4), for example, focus only on the upper or lower epidermis. Epidermal cells tend to be square or rectangular in shape. In a leaf, the epidermis is interrupted by openings called **stomata** (sing., stoma). Later you will have an opportunity to see the epidermis in roots, stems, and leaves. What is a function of epidermis in all three organs (see Table 19.1)? ____________

____

____

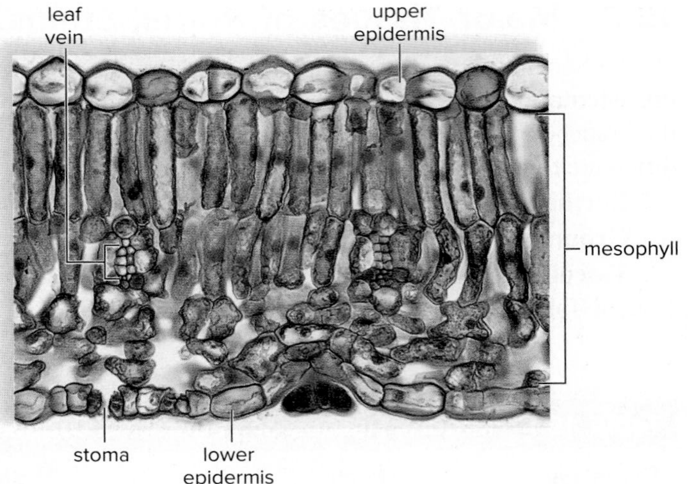

Figure 19.4 Microscopic leaf structure.
Like the stem and root, a leaf contains epidermal tissue, vascular tissue (leaf vein), and ground tissue (mesophyll).

2. **Ground tissue.** The ground tissue fills the space between epidermis in roots, stems, and leaves. In leaves, for example, ground tissue is called mesophyll (Fig. 19.4). Ground tissue largely contains parenchyma cells and sclerenchyma cells. **Parenchyma cells** can be of different sizes and vary from fairly circular to oval. Those that contain chloroplasts carry on photosynthesis. Those that contain leucoplasts store starches and oils. **Sclerenchyma cells** are usually elongate and have thick walls impregnated with lignin. These dead cells appear hollow, and the presence of lignin means that they stain a red color. Sclerenchyma cells are strong and provide support. Which type of cell (parenchyma or sclerenchyma) carries on photosynthesis or stores the products of photosynthesis in a leaf? ________

   Which one lends strength to ground tissue in roots and stems? ____________________________
3. **Vascular tissue.** In a leaf, strands of vascular tissue are called leaf veins (Fig. 19.4). There are two types of vascular tissue, called xylem and phloem. **Xylem** contains hollow dead cells that transport water. The presence of lignin makes the cell walls strong, stains red, and makes xylem easy to spot. **Phloem** contains thin-walled, smaller living cells that transport sugars in a plant. Phloem is harder to locate than xylem, but it is always found in association with xylem. Which type of tissue (xylem or phloem) transports sugars in a plant? ____________ Which type of tissue transports water? ____________________

Monocots Versus Eudicots

Flowering plants are classified into two major groups: **monocots** and **eudicots.** In this laboratory, you will be studying the differences between monocots and eudicots as noted in Figure 19.5. The arrangement of tissues is distinctive enough that you should be able to identify the plant as a monocot or eudicot when examining a slide of a root, stem, or leaf.

Experimental Procedure: Monocot Versus Eudicot

1. Examine the live plant again (see Fig. 19.2). The leaf vein pattern—that is, whether the veins run parallel to one another or whether the veins spread out from a central location (called the net pattern)—indicates

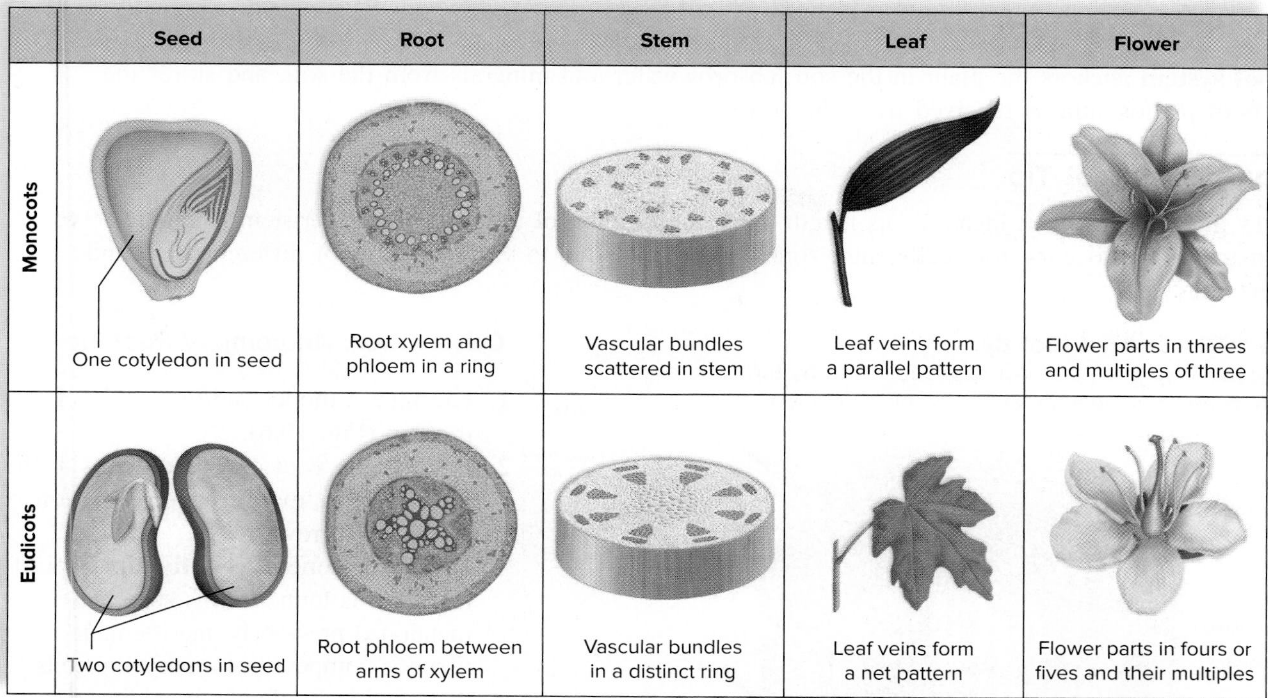

Seed
Root
Stem
Leaf
Flower
Monocots
One cotyledon in seed
Root xylem and phloem in a ring
Vascular bundles scattered in stem
Leaf veins form a parallel pattern
Flower parts in threes and multiples of three
Eudicots
Two cotyledons in seed
Root phloem between arms of xylem
Vascular bundles in a distinct ring
Leaf veins form a net pattern
Flower parts in fours or fives and their multiples

Figure 19.5 Monocots versus eudicots.
The five features illustrated here are used to distinguish monocots from eudicots.

that a plant is either a monocot or a eudicot. Is the plant in Figure 19.2 a monocot or eudicot? ________

The leaf pattern in Figure 19.4 appears to be parallel. Is this the leaf of a monocot or eudicot? ________

2. Observe any other available plants, and record in Table 19.2 the name of the plant and note the leaf vein pattern. On the basis of the leaf vein pattern, decide if this plant is a monocot or a eudicot.
3. Aside from leaf vein pattern, other external features indicate whether a plant is a monocot or a eudicot. For example, open a peanut if available. The two halves you see are cotyledons. Is the plant that produced the peanut a monocot or eudicot? ________
If available, examine a flower. If a flower has three petals or six petals, or any multiple of three, is the plant that produced this flower a monocot or a eudicot? ________

In this laboratory you will have the opportunity to examine the cross sections of roots and stems microscopically; the arrangement of vascular tissue roots and stems also indicates whether a plant is a monocot or a eudicot.

Table 19.2 Monocots Versus Eudicots

Name of Plant	Organization of Leaf Veins	Monocot or Eudicot?

19.3 Root System

The **root system** anchors the plant in the soil, absorbs water and minerals from the soil, and stores the products of photosynthesis received from the leaves.

Anatomy of a Root Tip

Primary growth of a plant increases its length. Note the location of the root apical meristem in Figure 19.6. As primary growth occurs, root cells enter zones that correspond to various stages of differentiation and specialization.

Figure 19.6 Eudicot root tip.
In longitudinal section, the root cap is followed by the zone of cell division, zone of elongation, and zone of maturation.

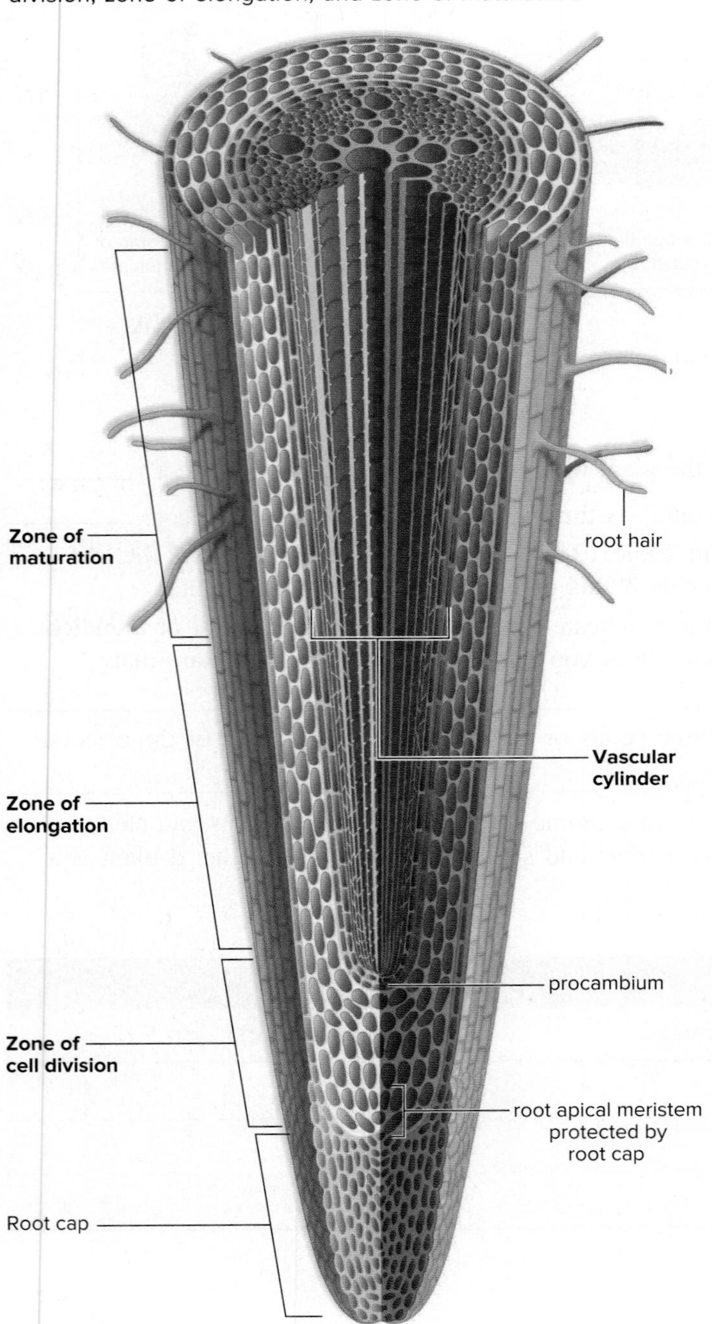

Observation: Anatomy of Root Tip

1. Examine a model and/or a slide of a root tip (Fig. 19.6).
2. Identify the **root cap** (dead cells at the tip of a plant that provide protection as the root grows).
3. Locate the **zone of cell division.** Apical meristem is found in this zone. As mentioned previously, meristematic tissue is composed of embryonic cells that continually divide, providing new cells for root growth.
4. Find the **zone of elongation.** In this zone, the newly produced cells elongate as they begin to grow larger.
5. Identify the **zone of maturation.** In this zone, the cells become differentiated into particular cell types. When epidermal cells differentiate, they produce **root hairs,** small extensions that absorb water and minerals. Also noticeable are the cells that make up the xylem and phloem of vascular tissue.

Anatomy of Eudicot and Monocot Roots

Eudicot and monocot roots differ in the arrangement of their vascular tissue.

Observation: Cross-Section Anatomy of Eudicot and Monocot Roots

Eudicot Root

1. Obtain a prepared cross-section slide of a buttercup (*Ranunculus*) root. Use both low power and high power to identify the **epidermis** (the outermost layer of small cells that gives rise to root hairs). The epidermis protects inner tissues and absorbs water and minerals.
2. Locate the **cortex,** which consists of several layers of thin-walled cells (Fig. 19.7*a, b*). In Figure 19.7*b,* note the many stained starch grains in the cortex cells. The cortex is ground tissue that functions in food storage.
3. Find the **endodermis,** a single layer of cells whose walls are thickened by a layer of waxy material known as the **Casparian strip.** (It is as though these cells are glued together with a waxy glue.) Because of the Casparian strip, the only access to the xylem is through the living endodermal cells. The endodermis regulates what materials that enter a plant through the root? ______________

   ____

   Use this illustration to trace the path of water and minerals from the root hairs to xylem. ______________________________
4. Identify the **pericycle,** a layer one or two cells thick just inside the endodermis. Branch roots originate from this tissue.
5. Locate the **xylem** in the vascular cylinder of the root. Xylem has several "arms" that extend like the spokes of a wheel. This tissue conducts water and minerals from the roots to the stem.
6. Find the **phloem,** located between the arms of the xylem. Phloem conducts organic nutrients from the leaves to the roots and other parts of the plant.

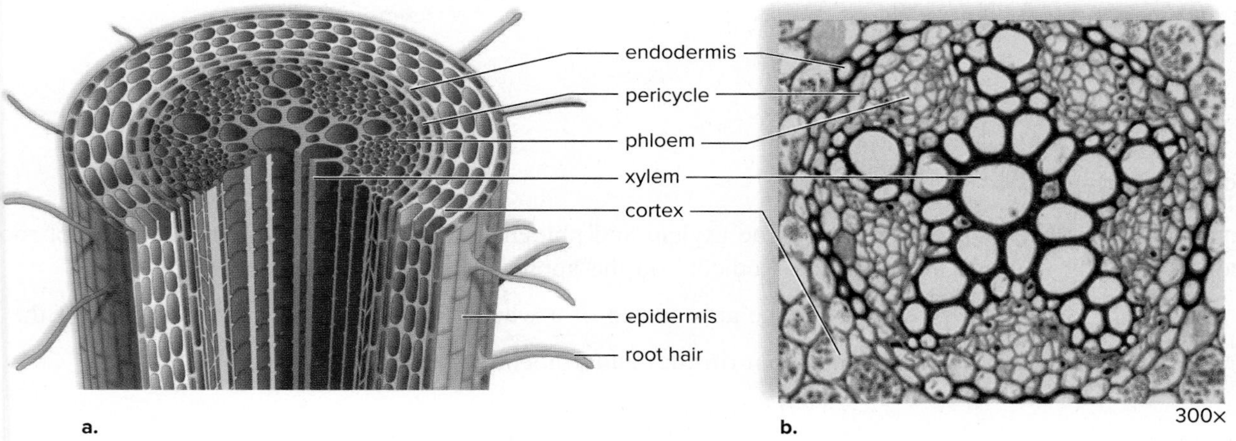

Figure 19.7 Eudicot root cross section.
The vascular cylinder of a dicot root contains the vascular tissue. Xylem is typically star-shaped, and phloem lies between the points of the star. **a.** Drawing. **b.** Micrograph.

Monocot Root

1. Obtain a prepared cross-section slide of a corn (*Zea mays*) root (Fig. 19.8*a*, *b*). Use both low power and high power to identify the six tissues mentioned for the eudicot root.
2. In addition, identify the **pith,** a centrally located ground tissue that functions in food storage.

Figure 19.8 Monocot root cross section.
a. Micrograph of a monocot root cross section. **b.** An enlarged portion.
(a) ©Dr. Keith Wheeler/Science Source; (b) ©George Ellmore, Tufts University

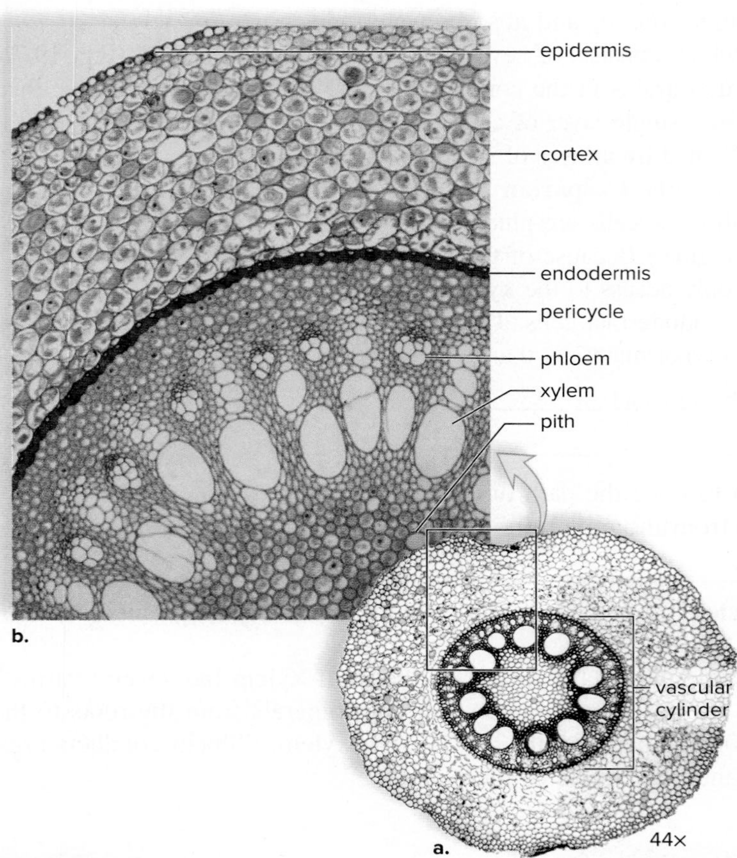

Comparison

Contrast the arrangement of vascular tissue (xylem and phloem) in the vascular cylinder of monocot roots and eudicot roots by writing "monocot" or "eudicot" on the appropriate line.

__________________ Xylem has the appearance of a wheel. Phloem is between the spokes of the wheel.

__________________ Ring of xylem (inside) and phloem (outside) surrounds pith.

Root Diversity

Roots are quite diverse, and we will take this opportunity to become acquainted with only a few select types.

Observation: Root Diversity

Taproots and fibrous roots. Most plants have either a taproot or a fibrous root. Note in Figure 19.9*a* that carrots have a **taproot.** The main root is many times larger than the branch roots. Grasses such as the type shown in Figure 19.9*b* have a **fibrous root:** All the roots are approximately the same size.

Examine the *taproots on display,* and name one or two in which the taproot is enlarged for storage. ______

____

The *dandelion on display* has a taproot. Describe. ______________________

____

Taproots in particular function in food storage.

Adventitious roots. Some plants have **adventitious** roots. Roots that develop from nonroot tissues, such as nodes of stems, are called adventitious roots. Examples include the prop roots of corn (Fig. 19.9*c*) and the aerial roots of ivy that attach this plant to structures such as stone walls.

Which plants on display have adventitious roots? ______________________

Other types of roots. Mangroves and other swamp-dwelling trees have roots called pneumatophores that rise above the waterline. Pneumatophores have numerous lenticels, which are openings that allow gas exchange to occur.

What root modifications not noted here are on display in your laboratory? ______________

____

Figure 19.9 Root diversity.
a. Carrots have a taproot. **b.** Grass has a fibrous root system. **c.** A corn plant has prop roots, and (**d**) black mangroves have pneumatophores to increase their intake of oxygen.

a. Taproot system

b. Fibrous root system

c. Prop roots, a type of adventitious root

d. Pneumatophores of black mangrove trees

19.4 Stems

Stems are usually found aboveground where they provide support for leaves and flowers. Vascular tissue extends from the roots through the stem and its branches to the leaves. Therefore, what function do botanists assign to stems in addition to support for branches and leaves? ______________________________ ______________________________ Explain why a branch cannot live if severed from the rest of the plant. ______________________________

Stems that do not contain wood are called **herbaceous,** or nonwoody, stems. Usually, monocots remain herbaceous throughout their lives. Some eudicots, such as those that live a single season, are also herbaceous. Other eudicots, namely trees, become woody as they mature.

Anatomy of Herbaceous Stems

Herbaceous stems undergo primary growth. **Primary growth** results in an increase in length due to the activity of the apical meristem located in the terminal bud (see Fig. 19.3) of the shoot system.

Observation: Anatomy of Eudicot and Monocot Herbaceous Stems

Eudicot Herbaceous Stem

1. Examine a prepared slide of a eudicot herbaceous stem (Fig. 19.10), and identify the **epidermis** (the outer protective layer). *Label the epidermis in Figure 19.10*a.
2. Locate the **cortex,** which may photosynthesize or store nutrients.
3. Find a **vascular bundle,** which transports water and organic nutrients. The vascular bundles in a eudicot herbaceous stem occur in a ring pattern. *Label the vascular bundle in Figure 19.10*a. Which vascular tissue (xylem or phloem) is closer to the surface? ______________________________
4. Label the central **pith,** which stores organic nutrients. Both cortex and pith are composed of which tissue type listed in Table 19.1? ______________________________

Figure 19.10 Eudicot herbaceous stem.
The vascular bundles are in a definite ring in this photomicrograph of a eudicot herbaceous stem. Complete the labeling as directed by the Observation.
(a) ©Ed Reschke; (b) ©Ray F. Evert, University of Wisconsin

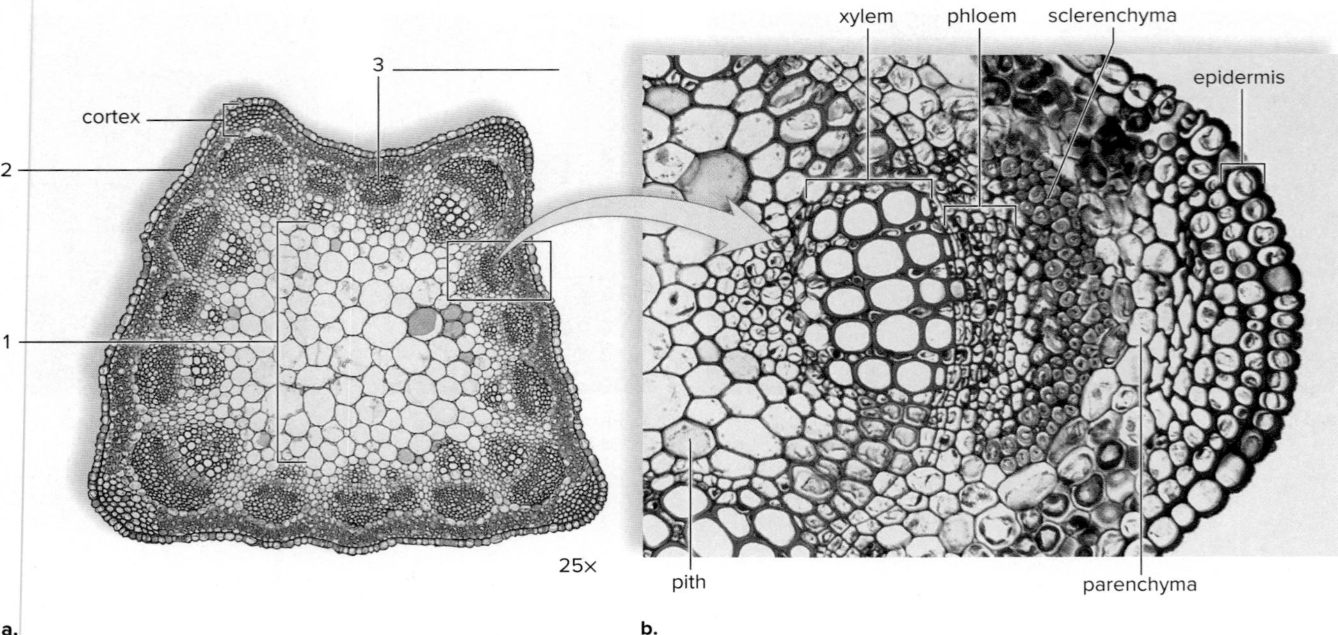

Monocot Herbaceous Stem

1. Examine a prepared slide of a monocot herbaceous stem (Fig. 19.11). Locate the epidermal, ground, and vascular tissues in the stem.
2. The vascular bundles in a monocot herbaceous stem are said to be scattered. Explain. ____

Figure 19.11 Monocot stem.
The vascular bundles, one of which is enlarged, are scattered in this photomicrograph of a monocot herbaceous stem.
(left) ©Carolina Biological Supply Company/Phototake; (right) ©Kingsley Stern

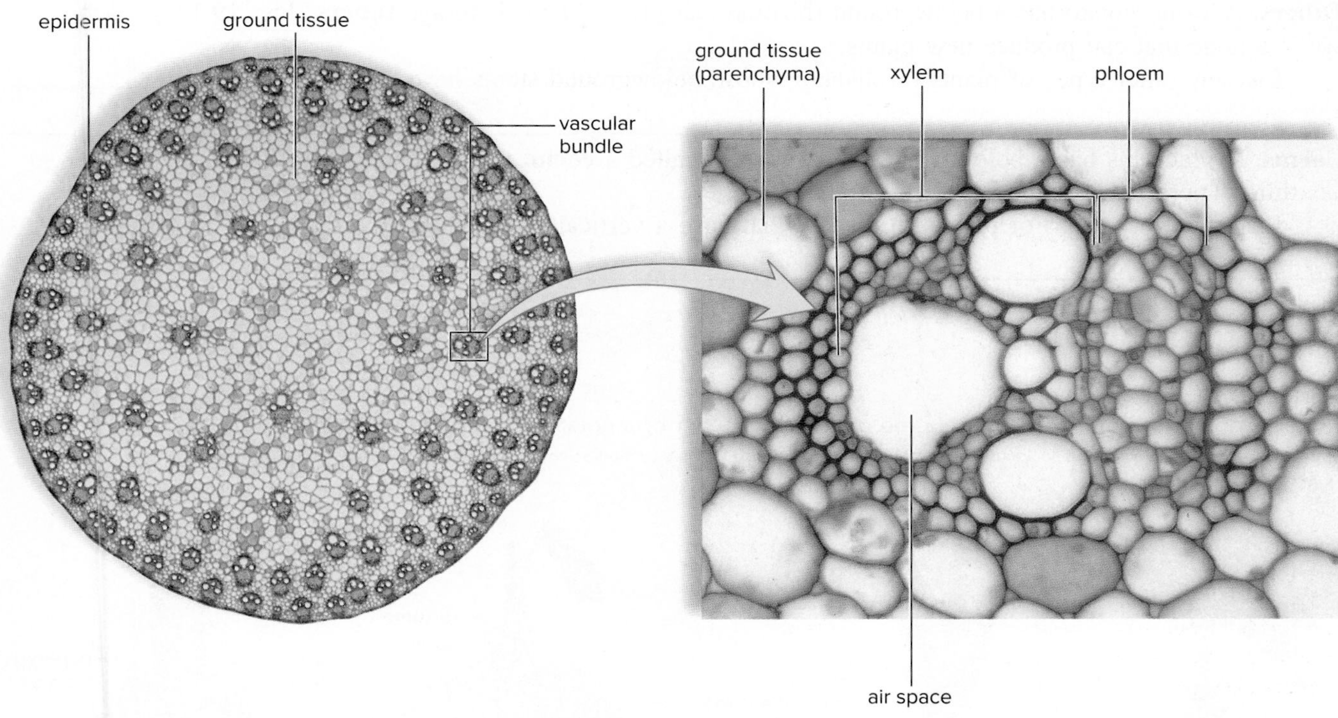

Comparison

1. Compare the arrangement of ground tissue in eudicot and monocot stems. ____
2. Compare the arrangement of vascular bundles in the stems of eudicots and monocots. ____

Stem Diversity

Stems can be quite diverse as well. We take this opportunity to become acquainted with those that allow a plant to accomplish vegetative reproduction and/or function in food storage. Several of these and other types of stems may be on display in the laboratory.

Observation: Stem Diversity

Stolons. The strawberry plant in Figure 19.12*a* has a horizontal aboveground stem called a runner or **stolon.** The stolon produces adventitious roots and new shoots at nodes. *Label an adventitious root and a new shoot in Figure 19.12*a.

List any other plants on display that spread and produce new shoots by sending out stolons. ______________

____

Rhizomes. An iris has a belowground horizontal stem called a **rhizome,** which functions as a fleshy food storage organ (Fig. 19.12*b*). New plants can grow from a single piece of the rhizome.

List any other plants on display that have the same belowground horizontal stems (rhizomes) as the iris. ____

Tubers. A white potato has a belowground rhizome that gives off food storage **tubers** (Fig. 19.12*c*). Each eye is a node that can produce new plants.

List any other types of plants on display whose belowground stems have tubers. ______________

____

Corms. A gladiolus has a belowground vertical stem called a **corm,** which functions in food storage and has thin, papery leaves (Fig. 19.12*d*).

List any other types of plants on display that have a vertical stem called a corm. ______________

____

Figure 19.12 Stem diversity.
a. Stolon of a strawberry plant. **b.** Rhizome of an iris. **c.** Tuber of a potato. **d.** Corm of a gladiolus.
(a) ©McGraw-Hill Education/Evelyn Jo Johnson, photographer; (b) ©Science Pictures Limited/Science Source; (c, d) ©McGraw-Hill Education/Carlyn Iverson, photographer

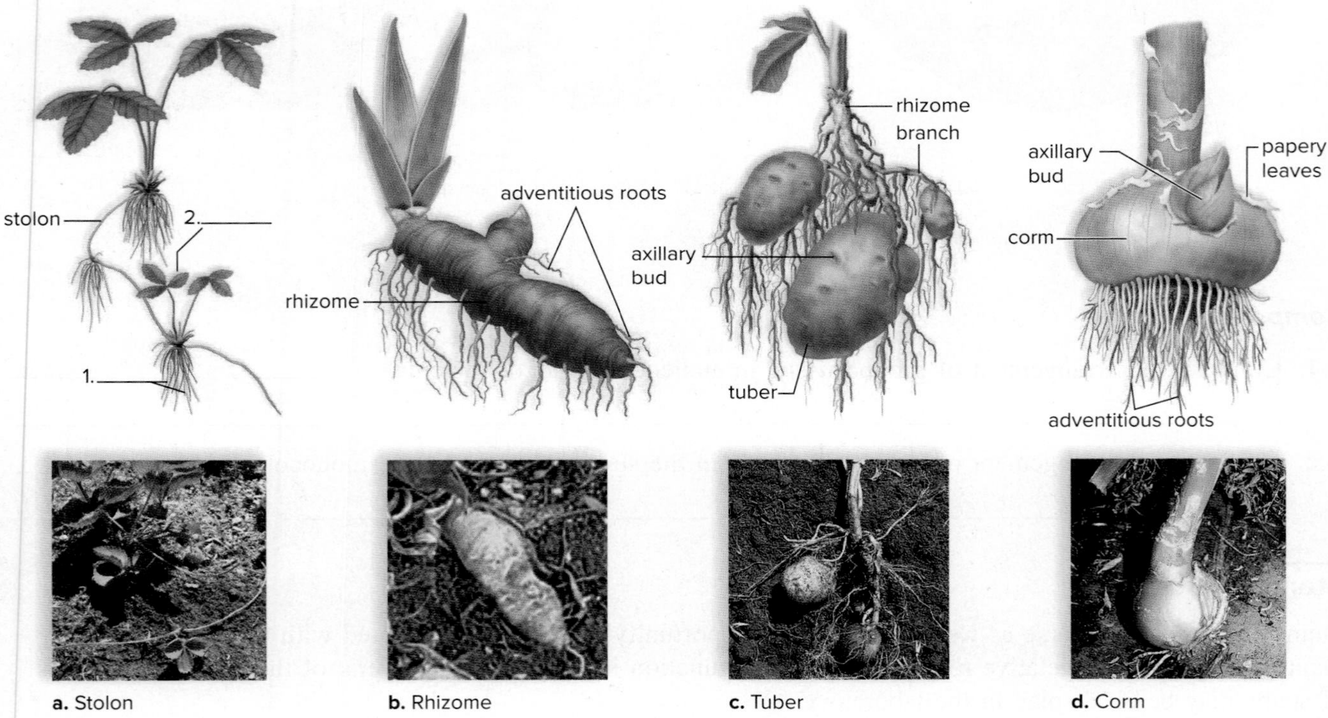

Anatomy of Woody Stems

Woody stems undergo both primary growth (increase in length) and secondary growth (increase in girth). When *primary growth* occurs, the apical meristem within a terminal bud is active. When *secondary growth* occurs, the vascular cambium is active. **Vascular cambium** is meristem tissue, which produces new xylem and phloem called **secondary xylem** and **phloem** each year. The buildup of secondary xylem year after year is called **wood.** Complete Table 19.3 to distinguish between primary growth and secondary growth of a stem.

Table 19.3 Primary Growth Versus Secondary Growth

	Primary Growth	Secondary Growth
Active meristem		
Result		

Observation: Anatomy of a Winter Twig

1. A winter twig typically shows several past years' primary growth. Examine several examples of winter twigs (Fig. 19.13), and identify the **terminal bud** located at the tip of the twig. This is where new primary growth will originate. During the next growing season, the terminal bud produces new tissues including vascular bundles and leaves.
2. Locate a **terminal bud scar.** These are marks left on a stem from terminal bud scales (modified leaves protecting the bud). The distance between two adjacent terminal bud scars equals one year's primary growth.
3. Find a **leaf scar.** Mark where a leaf was attached to the stem.
4. Note the **bundle scars.** Complete this sentence: Marks left in the leaf scar where the vascular tissue ______________________________.
5. Identify a **node.** This is the region where you find leaf scars and bundle scars. The region between nodes is called an ______________________________.
6. Locate an **axillary bud.** This is where new branch growth can occur.
7. Note the numerous lenticels, breaks in the outer surface where gas exchange can occur.

Figure 19.13 External structure of a winter twig.
Counting the terminal bud scars tells the age of a particular branch.

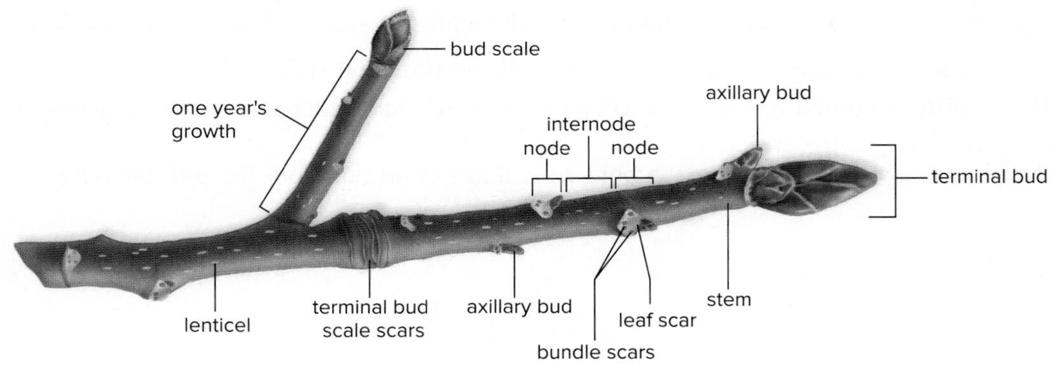

Observation: Anatomy of Woody Stem

1. Examine a prepared slide of a cross section of a woody stem (Fig. 19.14), and identify the **bark** (the dark outer area), which contains **cork,** a protective outer layer; **cortex,** which stores nutrients; and **phloem,** which transports organic nutrients.

Figure 19.14 Woody eudicot stem cross section.
Because xylem builds up year after year, it is possible to count the annual rings to determine the age of a tree. This tree is 3 years old.
(b) ©Carolina Biological Supply Company/Phototake

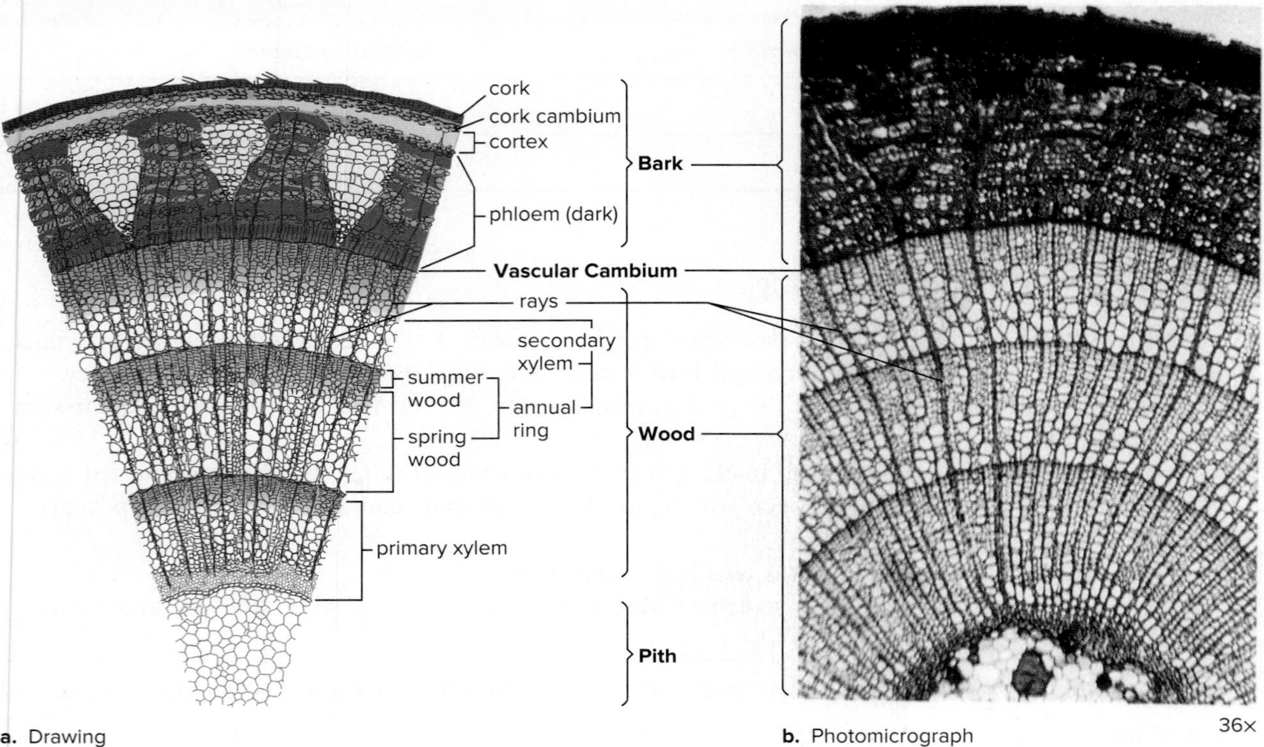

a. Drawing
b. Photomicrograph

2. Locate the **vascular cambium** at the inner edge of the bark, between the bark and the wood. Vascular cambium is meristem tissue whose activity accounts for secondary growth, which causes increased girth of a tree. Secondary phloem (which disappears) and secondary xylem (which builds up) are produced by vascular cambium each growing season.
3. Find the **wood,** which contains annual rings. An **annual ring** is the amount of xylem added to the plant during one growing season. Rings appear to be present because spring wood has large xylem vessels and looks light in color, while summer wood has much smaller vessels and appears much darker. How old is the stem you are observing? ____________ Are all the rings the same width? ____________
4. Identify the **pith,** a ground tissue at the center of a woody stem that stores organic nutrients and may disappear.
5. Locate **rays,** groups of small, almost cuboid cells that extend out from the pith laterally.

19.5 Leaves

A **leaf** is the organ that produces food for the plant by carrying on photosynthesis. Leaves are generally broad and quite thin. An expansive surface facilitates the capture of solar energy and gas exchange. Water and nutrients are transported to the cells of a leaf by leaf veins, extensions of the vascular bundles from the stem.

Anatomy of Leaves

Observation: Anatomy of Leaves

1. Examine a model of a leaf. With the help of Figure 19.15, identify the waxy **cuticle,** the outermost layer that protects the leaf and prevents water loss.
2. Locate the **upper epidermis** and **lower epidermis,** single layers of cells at the upper and lower surfaces. Trichomes are hairs that grow from the upper epidermis and help protect the leaf from insects and water loss.
3. Find the leaf veins in your model. The bundle sheath is the outer boundary of a vein; its cells surround and protect the vascular tissue. If this is a model of a monocot, all the leaf veins will be ____________ ________________________. If this is a model of a eudicot, some leaf veins will be circular and some will be oval. Why? ____
4. Identify the **palisade mesophyll,** located near the upper epidermis. These cells contain chloroplasts and carry on most of the plant's photosynthesis. Locate the **spongy mesophyll,** located near the lower epidermis. These cells have air spaces that facilitate the exchange of gases across the plasma membrane. *Label the layers of mesophyll in Figure 19.15.* Collectively, the mesophyll represents which of the three types of tissue found in all parts of a plant (see Table 19.1)? ____________________
5. *Label the two layers of epidermis in Figure 19.15.* Find a **stoma** (pl., stomata), an opening through which gas exchange occurs and water escapes. Stomata are more numerous in the lower epidermis. A stoma has two guard cells that regulate its opening and closing.

Figure 19.15 Leaf anatomy.

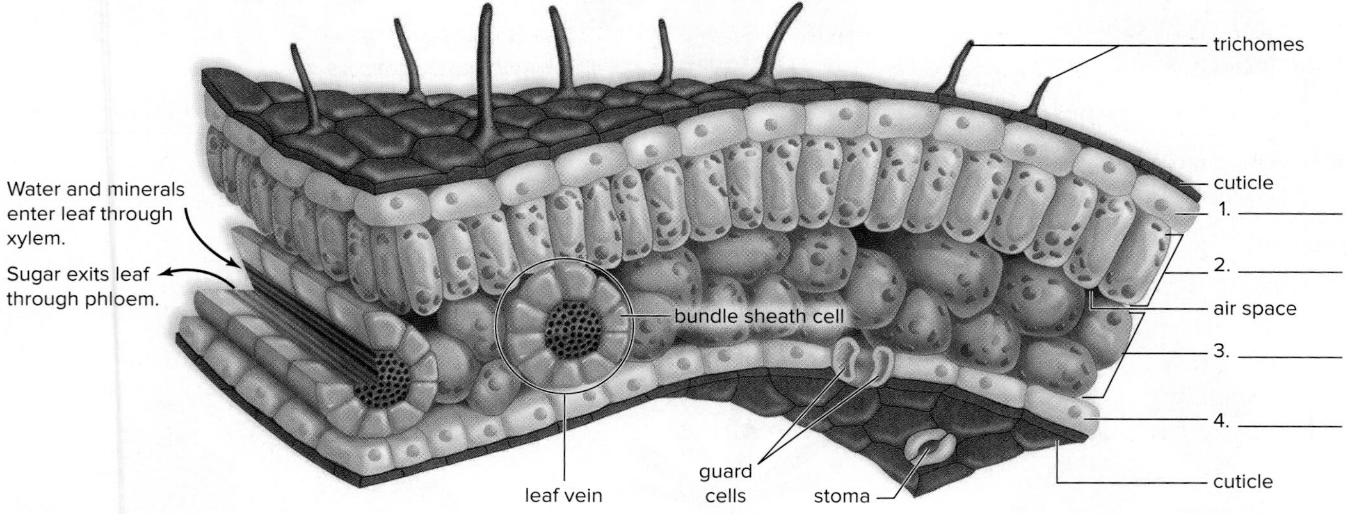

Leaf Diversity

A eudicot leaf consists of a flat blade and a stalk, called the petiole. An axillary bud appears at the point where the petiole attaches a leaf to the stem. In other words, an axillary bud is a tip-off that you are looking at a single leaf.

Observation: Leaf Diversity

Several types of eudicot leaves will be on display in the laboratory. Examine them using these directions.

- A leaf may be **simple,** in which case it consists of a single blade; or a leaf may be **compound,** meaning its single blade is divided into leaflets. *In Figure 19.16*a *write "simple" or "compound" before the word* leaf *in 1–3.* (Among the leaves on display, find one that is simple and one that is compound.)
- A compound leaf can be **palmately** compound, meaning the leaflets are spread out from one point. Which leaf in Figure 19.16*a* is palmately compound? *Add "palmately" in front of "compound" where appropriate.* (See if you can find a palmately compound leaf among those on display.)
- A compound leaf can be **pinnately** compound, meaning the leaflets are attached at intervals along the petiole. Which leaf in Figure 19.16*a* is pinnately compound? *Add "pinnately" in front of "compound" where appropriate.* (See if you can find a pinnately compound leaf among those on display.)
- As shown in Figure 19.16*b,* leaves can be in various positions on a stem. On which stem in Figure 19.16*b,* 4–6, are the leaves **opposite** one another? *Write the word "opposite" where appropriate.* On which stem do the leaves **alternate** along the stem? *Write the word "alternate" where appropriate.* On which stem do the leaves **whorl** about a node? *Add the word "whorl" where appropriate in Figure 19.16*b. (See if you can find different arrangements of leaves among the stems on display.)

Figure 19.16 Classification of leaves.

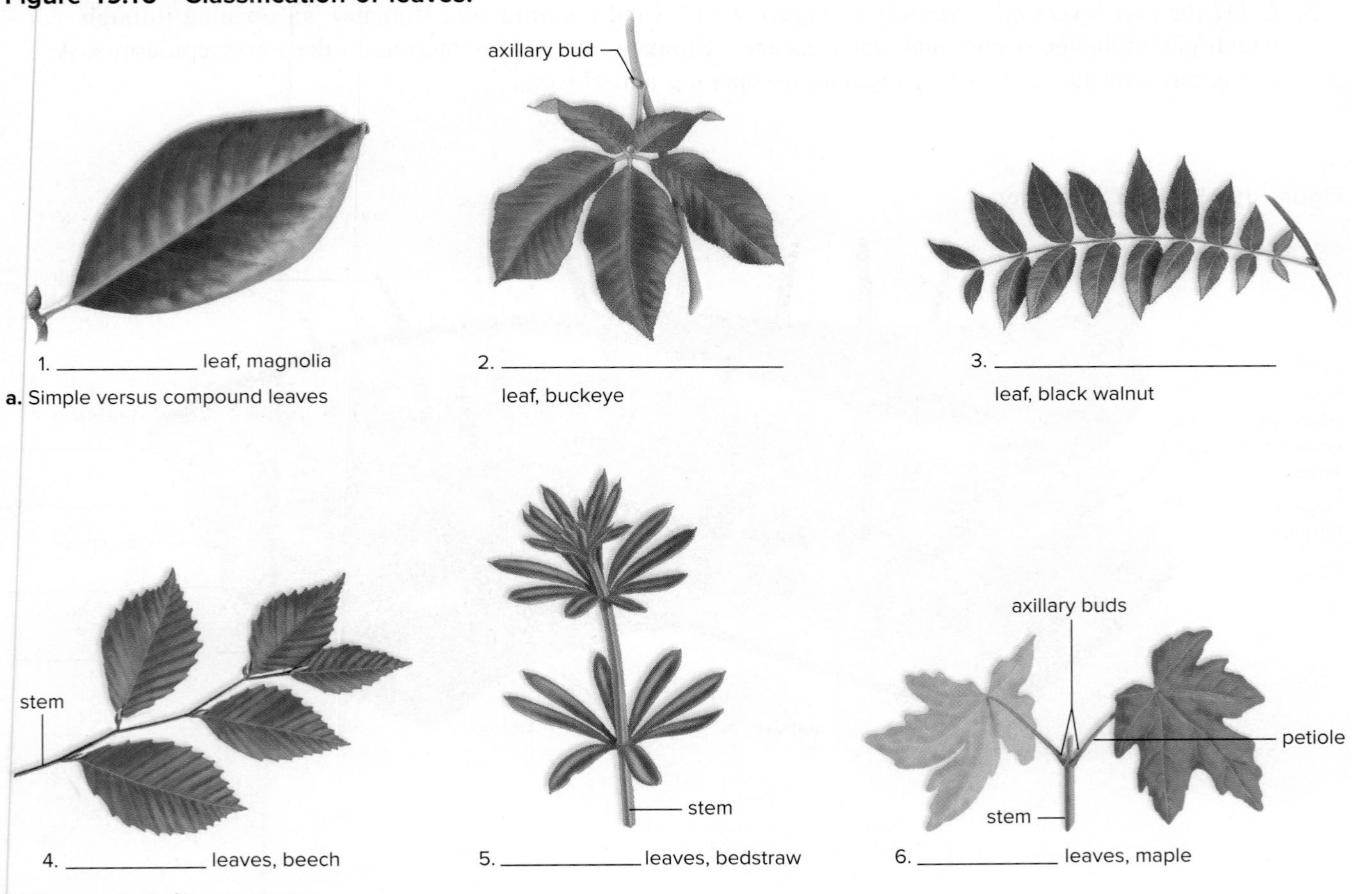

Laboratory Review 19

________________ 1. What kind of plant tissue is composed of cells that divide?

________________ 2. What term is used to identify ground tissue in a leaf?

________________ 3. What kind of plant, monocot or eudicot, has the xylem and phloem in its leaves arranged in a parallel pattern?

________________ 4. In what zone of a eudicot root will tracheids and vessel elements be found?

________________ 5. What type of tissue gives rise to root hairs?

________________ 6. How is xylem typically arranged in a eudicot root cross section?

________________ 7. What term refers to stems that are nonwoody?

________________ 8. What kind of plant, monocot or eudicot, has stem vascular bundles that occur in a ring?

________________ 9. What kind of growth do woody plants undergo when the diameter of their trunk increases?

________________ 10. Where does new primary growth occur on a winter twig?

________________ 11. What type of vascular tissue forms the annual rings in a woody stem?

________________ 12. What is the outermost layer of a leaf that protects and prevents water loss?

________________ 13. What cells are on either side of a stoma and regulate whether it is open or closed?

________________ 14. If its leaf veins form a net pattern and there is one blade attached to a petiole, is the leaf simple or compound, and is it from a monocot or a eudicot?

Thought Questions

15. Where is glucose produced in a leaf? What type of tissue carries glucose away from a leaf, and where is the glucose stored in a plant like a beet?

16. Stomata are more numerous in the lower epidermis of leaves. Explain why it would be detrimental to the plant to have more stomata in the upper epidermis of leaves.

17. Compare and contrast monocots and eudicots by identifying two similarities and two differences between them.

LABORATORY

20

Water Absorption and Transport in Plants

Learning Outcomes

20.1 Water Absorption by Root Hairs

- Explain how tonicity affects water uptake at the roots.

20.2 The Water Column in Stems

- Explain how the structure of xylem and the properties of water bring about a continuous water column in stems.

20.3 Transpirational Pull at Leaves

- Hypothesize the rate of transpiration under varied conditions.

20.4 Stomata and Their Role in Transport of Water

- Explain the relationship between stomata aperture and transpiration rate.
- Explain the relationship between stomata aperture and water stress.

Introduction

Water from the soil is absorbed into a plant through extensions of epidermal root cells, called **root hairs** (Fig. 20.1, *bottom*). It enters **xylem,** the vascular tissue that transports water up through the stem to the leaves. Water forms a continuous column in xylem for two reasons due to the presence of hydrogen bonding between water molecules: First, water molecules exhibit **adhesion** because they momentarily stick to the sides of the xylem vessels. Second,

Water evaporates, pulling the water column from the roots to the leaves.
Water molecules cling together and adhere to sides of vessels in stems.
H_2O
Water enters a plant at root hairs.

Figure 20.1 Water transport in a tree. When water evaporates from the leaves, the water column in xylem is pulled upward due to the clinging together (cohesion) of water molecules with one another and the sticking (adhesion) of water molecules to the sides of the vessels.

water molecules exhibit **cohesion** because they cling together and do not ordinarily break apart.

The water column moves upward due to **transpiration** (evaporation of water from the leaves at the stomata). As water molecules exit at the leaves, they are replaced by other water molecules from leaf veins. Therefore, transpiration exerts a tension that pulls the water column in the xylem up from the roots to the leaves. This explanation for xylem transport is called the cohesion-tension model of xylem transport.

Planning Ahead Section 20.3 requires students to take readings for 40 minutes; therefore, you may wish to start this procedure early during the laboratory session. Alternately, you may wish to have students do the virutal lab, "Plant Transpiration," also described in Section 20.3.

20.1 Water Absorption by Root Hairs

The tip of a root consists of a root cap, the zone of cell division, the zone of elongation, and the zone of maturation, where root hairs appear (Fig. 20.2). Root hairs are extensions of epidermal cells. Root hairs increase the surface area for absorption of water that moves through the cortex by way of osmosis and enters the vascular cylinder where xylem is located.

Figure 20.2 Root hairs. Root epidermis has root hairs to absorb water. ©Evelyn Jo Johnson RF

Observation: Root Hairs

1. Obtain a young, germinated *seedling,* and float it in some water in a petri dish while you examine it. The root is rapidly elongating and therefore has grown some distance beyond the seed coat.
2. Using a stereomicroscope, locate the root cap, which covers the root tip, and the region where the root hairs have formed on the root surface.
3. Remove the root from the seedling, and make a wet mount of the root, using water and/or a solution of 0.1% neutral red.
4. Observe your slide under the microscope. Does every epidermal cell have a root hair? ________________

   How does the structure of a root hair aid absorption? ________________________________

   ____

Experimental Procedure: Absorption of Water by Osmosis

This Experimental Procedure uses potato cylinders to determine under which osmotic conditions water is likely to enter root hairs. Recall from your previous study of tonicity that **isotonic** refers to equal tonicity, where the concentration of solute in two solutions is the same. **Hypertonic** refers to solutions that contain more solute particles (and less water) than the one to which it is being compared. **Hypotonic** refers to a solution that contains fewer solute particles (more water) than the one to which it is being compared.

1. Obtain five test tubes and a rack or beaker to hold the tubes. The tubes must be large enough to hold a small piece of potato (see step 3).
2. Number the test tubes and fill them with enough solution to cover a piece of potato. Record the contents in Table 20.1.

   Fill tube 1 with a solution of 0.05% sucrose.
   Fill tube 2 with a solution of 0.07% sucrose.
   Fill tube 3 with a solution of 0.09% sucrose.
   Fill tube 4 with a solution of 0.12% sucrose.
   Fill tube 5 with a solution of 0.14% sucrose.

3. Using a cork borer, remove cylinders of tissue from a large potato. With a scalpel, cut the cylinders into uniform 2 cm lengths until you have five sections, placing them under a damp paper towel as you cut them. Cut all sections from the *same* potato.
4. Weigh each potato piece before placing it in one of the numbered test tubes. Record the pretreatment weights of each in Table 20.1. (*Note:* You must know which potato piece you put in which tube.)
5. After one hour; remove, dry with a paper towel, and weigh each potato piece and record its post-treatment weight in Table 20.1. (*Note:* You must know which potato piece goes with which tube.)
6. Complete the last two columns in Table 20.1.

Table 20.1 Absorption of Water by Osmosis

Test Tube	Sucrose %	Pretreatment Weight	Post-treatment Weight	Movement of Water (into or out of potato)	Tonicity of Original Solution Compared to Potato
1					
2					
3					
4					
5					

Conclusions: Absorption of Water by Osmosis

- Since potato cells (modified stems) are believed to behave much as root hairs do, this experiment indicates that only if groundwater is ________________ to cytoplasm in root hairs will root hairs be able to absorb water.
- If, by chance, a plant's roots are surrounded by a hypertonic solution, the roots will ________________ water.

20.2 The Water Column in Stems

The water column in stems is continuous and moves up the stem for three reasons: (1) Xylem (Fig. 20.3) contains two types of nonliving and hollow conducting cells: vessel elements and tracheids. A **vessel element** has performation plates as end walls. The plate may have a single, large opening or it may have a series of openings. The vessel elements align one on top of the other; and the performation plates allow water and minerals to freely flow from one to the other. **Tracheids** are nonliving and hollow but they are longer and more narrow than a vessel element. Water can move from one tracheid to the other because they have pitted walls. Pits are depressions where the secondary cell wall does not form; making it easier for water to flow from one tracheid to the other. (2) The column of water in xylem is continuous because water molecules are **cohesive** (they cling together) and because water molecules **adhere** to the sides of xylem cells. (3) Water evaporation from stomata creates a force that pulls the water column upward.

Figure 20.3 Xylem structure.
Photomicrograph of xylem vascular tissue (left) and drawings (right) showing general organization of xylem tissue.
(left) ©Garry DeLong/Science Source

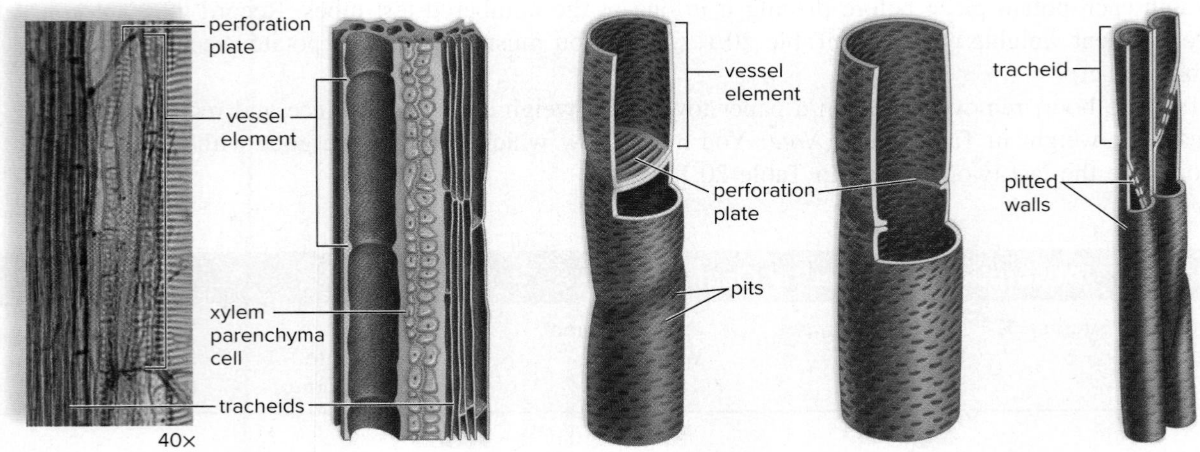

Experimental Procedure: The Water Column in Stems

1. Place a small amount of red-colored water in two beakers. Label one beaker "wet" and the other beaker "dry."
2. Transfer a stalk of celery (which was cut and then immediately placed in a container of water) into the "wet" beaker so that the large end is in the colored water.
3. Transfer a stalk of celery of approximately the same length and width (but that was kept in the air after being cut) into the "dry" beaker so that the large end is in the colored water.
4. With scissors, cut off the top end of each stalk, leaving about 10 cm total length.
5. Time how long it takes for the red-colored water to reach the top of each stalk, and record these data in Table 20.2.

Table 20.2 The Water Column in Stems

Stalk	Speed of Dye (Minutes)	Conclusion
Cut end placed in water prior to experiment		
Cut end kept in air prior to experiment		

Conclusions: The Water Column in Stems

- Exposing a celery stalk to air breaks the water column. Is a continuous water column helpful to the conduction of water in plants? __________ Explain on the basis of your results. ______________________

  ____
- Conclude why speed of conduction was faster/slower, and write this conclusion in the last column in Table 20.2.

20.3 Transpirational Pull at Leaves

Evaporation of water from leaves by way of stomata is called **transpiration.** As transpiration occurs, the continuous water column is pulled upward—first within the leaf, then from the stem, and finally from the roots.

Experimental Procedure: Transpiration

Assembling a Transpirometer

1. Tightly fit one end of a 4 cm piece of rubber tubing over one end of a 15–20 cm glass pipet (Fig. 20.4*a*).
2. Immerse the assembled glass and rubber tube in a large tub of *water* so that it will fill. Check that there are no air bubbles in the tube.
3. Cut off a portion of a geranium plant (stem with five to seven leaves), and place the cut end of this stem into the tub of water, but keep the leaves dry (Fig. 20.4*a*). With sharp scissors held under water, cut off a 1.5 cm piece of the stem at an oblique angle. (If this were done in the air, a vacuum would occur in the vascular system.)
4. Keeping the leaves dry and the stem end submerged, fit the cut stem end into the rubber tubing that has been filled with water. Squeeze out all of the air bubbles. If necessary, tightly wind a rubber band or tie a string around the juncture (while still submerged) for added tightness, but do not crush the stem.
5. Remove the assembled apparatus from the water. Hold it so that the geranium stem cutting is upright.
6. Clamp the assembly to a ring stand (Fig. 20.4*b*). The water will not run out of the tube due to water's adhesive and cohesive properties.

Determining Transpiration Rate Under Normal Conditions

1. Wait 5 minutes, and then mark with a wax pencil the water level at the lower end of the transpirometer.
2. Then, every 10 minutes for the next 40 minutes, mark the water level, and measure (in ml) the amount the water has moved. Record your data in the first two columns of Table 20.3. These figures indicate the ml of water transpired.

Figure 20.4 Assembling a transpirometer.
Complete directions are given for **a.** assembling the transpirometer apparatus, and **b.** clamping the finished assembly to a ring stand.

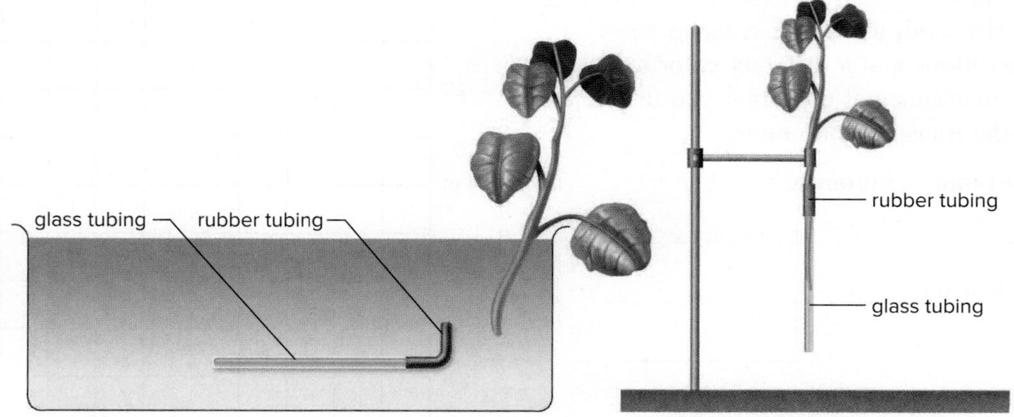

a. Assemble the transpirometer apparatus under water.

b. Clamp the finished assembly to a ring stand.

Determining Transpiration Rates Under Varied Environmental Conditions

1. Your instructor will assign you one of these conditions. Repeat the experiment using the assigned condition.

a. Focus a light source on the plant (to simulate heat). The plant should be located at least 25 cm away from the light source. Hypothesize how an increase in *temperature* will affect the rate of transpiration.

____

b. Spray the plant and the inside of a plastic bag with water (to simulate a rise in humidity). Put the plastic bag over the plant, and use string to draw it closed around the tubing. Hypothesize how *humidity* will affect the rate of transpiration.

____

c. Use a small fan to gently blow air across the plant (to simulate wind). Hypothesize how *wind* will affect the rate of transpiration.

____

2. Remove your previous marks from the glass pipet. Again wait 5 minutes. Then, as before, every 10 minutes for the next 40 minutes mark the water level and measure (in ml) the amount the water has moved. Record your data in Table 20.3.

Table 20.3 Effect of [Temperature, Humidity, Wind]* on Transpiration Rate

	Normal Conditions		Test Conditions	
Time	**Reading (in ml)**	**Total Change (in ml)**	**Reading (in ml)**	**Total Change (in ml)**
After 10 minutes				
After 20 minutes				
After 30 minutes				
After 40 minutes				

*Circle the condition you tested.

3. Plot the results of your two experiments on the graph provided, using one color to show normal conditions and a different color to show the environmental condition you tested. Calculate the transpiration rate.

Total ml/40 min = ml/min

Control: _______________________

Condition tested (_______________________):

____

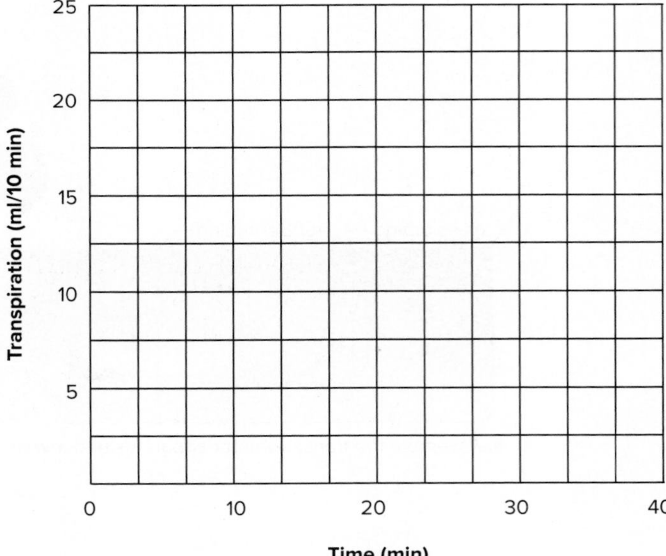

Conclusion: Transpiration

Conclude one of the following according to your data. Conclude the other two according to data collected by other laboratory groups.

1. Effect of temperature on transpiration rate. Did the results support or not support your hypothesis? Explain. ______________________________

____

2. Effect of humidity on transpiration rate. Did the results support or not support your hypothesis? Explain. ______________________________

____

3. Effect of wind on transpiration rate. Did the results support or not support your hypothesis? Explain. ______________________________

____

20.4 Stomata and Their Role in Transport of Water

In leaves, the lower epidermis in particular has openings called **stomata.** The opening and closing of a stoma is regulated by **guard cells** on either side of a stoma. When stomata are open, gas exchange occurs; also, water evaporates from the leaves.

The more stomata present and the more they are open (degree of aperture) the greater the rate of transpiration.

Observation: Number of Stomata

1. Calculate the area of the high-power microscopic field with the following formula:

$$\text{Area} = \pi r^2$$

where r (radius) of the field = 0.2 mm (or as determined in Laboratory 2) and $\pi = 3.14$.

Area = πr^2 = ________ mm^2

2. Obtain a leaf from a plant designated by your instructor, and have a slide with a drop of distilled water ready.
3. Using the technique shown in Figure 20.5*a*, obtain a strip of epidermis from the underside of the leaf, and put it in the drop of water on the slide, outer side up. Add a coverslip, and examine it microscopically, using both low power and high power.
4. Count the number of stomata (Fig. 20.5*b*) you see in the high-power field. __________ stomata
5. Divide the number of stomata by the area of the field calculated in step 1. This will tell you the number of stomata in 1 mm^2. __________ stomata/mm^2
6. If the underside surface area of your leaf were 400 mm^2, how many stomata would be present on its surface? __________ Such a large number accounts for the large amount of water lost by transpiration.

Figure 20.5 Stomata.
a. Method of obtaining a strip of epidermis from the underside of leaf. **b.** False-colored scanning electron micrograph of leaf surface. (b) ©Andrew Syred/SPL/Science Source

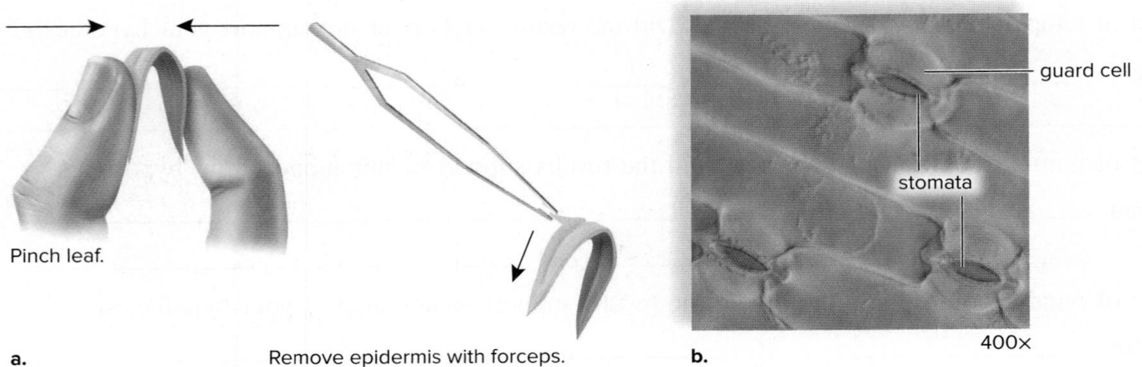

Experimental Procedure: Open Versus Closed Stomata

Stomata are expected to close when plants are water stressed because closure prevents further water loss. In this Experimental Procedure, we will simulate water stress by placing a strip of epidermis in a salty environment.

1. Prepare two slides, one of which has a drop of distilled water and the other a drop or two of 5% salt solution. Label the slides *W* (for water) and *S* (for salts).
2. Obtain two strips of epidermis (Fig. 20.5*a*) and place one on each slide.
3. Wait 5 minutes.
4. Observe each slide in turn. Which slide contains open stomata, and which slide contains closed stomata? ______________________________
5. Explain your results. ______________________________
____

Summary

Complete Table 20.4 to explain the mechanism of water transport in plants.

Table 20.4 Water Transport in Plants

Process	Where	Mechanism
Absorption of water		
Formation of water column		
Transpirational pull		

Laboratory Review 20

______________ 1. Which type of vascular tissue transports water in plants?

______________ 2. Which structure on a plant root absorbs water from the soil?

______________ 3. Name two forces that allow water to be transported through a plant.

______________ 4. What type of chemical bonding aids in the transport of water through xylem?

______________ 5. How does a plant lose water?

______________ 6. Will roots gain or lose water if they are hypertonic to groundwater?

______________ 7. Are plants likely to lose or gain water if they are placed in salty soil?

______________ 8. Which two cells form xylem?

______________ 9. Which force of tension is lost if air is introduced into xylem?

______________ 10. What conditions can increase the rate of transpiration from leaves?

______________ 11. Where does water exit leaves?

______________ 12. How is water loss through transpiration regulated on leaves?

______________ 13. On which part of a leaf are stomata more abundant?

______________ 14. In addition to water, what else is transported through stomata?

Thought Questions

15. Explain the structural and chemical characteristics of roots that allow for increased water absorption.

16. How are plants adapted to survive in dry conditions?

17. How must a plant balance transpiration and photosynthesis?

LABORATORY

21

Control of Plant Growth and Responses

Learning Outcomes

Introduction

- Distinguish between a positive tropism and a negative tropism.

21.1 Gravitropism

- Contrast the gravitropism seen in stems with that seen in roots, and explain on the basis of auxin distribution.

21.2 Phototropism

- Explain the positive phototropism seen in stems on the basis of response to blue light and auxin distribution.

21.3 Gibberellins and Stem Elongation

- Explain how dwarf plants respond to the hormone gibberellin.

21.4 Etiolation

- Explain the phenomenon of etiolation, and relate it to the pigment phytochrome.

Figure 21.1 Examples of tropism. Stems exhibit negative gravitropism. Roots exhibit positive gravitropism.

Introduction

Plant growth and development are regulated in part by temperature, nutrients, and water availability. Plants respond to certain environmental stimuli, such as light and gravity, by utilizing hormones that affect cell growth. A **hormone** is a chemical messenger produced by one set of cells that affects the behavior of a different set of cells. Growth of a plant or plant part in the direction of the stimulus is said to be a **positive tropism.** Growth of a plant or plant part away from a stimulus is a **negative tropism.** In Figure 21.1 the root system of a plant exhibits **positive gravitropism.** Why? Because the force of gravity is toward the soil, and roots grow into the soil. On the other hand, the shoot system of a plant exhibits **negative gravitropism.** Why? Because stems grow up in the opposite direction to the force of gravity.

Planning Ahead You may wish to start the Response of Seedlings experiment in section 21.1 at the beginning of the laboratory session to allow time for a response to occur.

21.1 Gravitropism

The negative gravitropism of stems can be demonstrated by laying a potted plant on its side and noticing that it will eventually bend upward. Cell elongation on the lower side of the stem causes the stem to bend up (Fig. 21.2). Plant cells contain **amyloplasts** (plastids that store starch), a type of plastid, and gravity causes amyloplasts to settle in the lower part of the cell. The settling of amyloplasts is believed to lead to an increase in the concentration of a plant hormone called **auxin** in the lower side of a stem. The presence of auxin causes cell elongation on the lower side and a bending of the stem upward.

To explain the reaction of roots to the presence of gravity, it is hypothesized that cell growth is inhibited, rather than stimulated by the presence of so much auxin on the lower side of a root.

Figure 21.2 Negative gravitropism.
After 24 hours, the stem of a coleus plant had this appearance after the plant was placed on its side. ©Kingsley Stern

Experimental Procedure: Gravitropism

Our experimental material will be *Brassica rapa* plants, which are related to plants in the mustard and cabbage family. They have been bred by Wisconsin Fast Plants™ to produce plants that mature in only 35 days.

Do you predict that light will have any effect on gravitropism? ______________________

Why or why not? ______________________

Knowing that gravitropism is dependent on the settling of plastids (that store starch) on the lower portion of the stem, hypothesize how light might affect gravitropism. ______________________

____

Figure 21.3 Gravitropism of mature stem.
Follow the directions in the text to measure the bend in a stem that has been placed on its side. Keeping the plant in the dark removes phototropism as a variable.

Response of Mature Stem

1. Observe two *Brassica rapa* plants less than 14 days old that have been turned on their sides at least two hours ago. One plant is in the open; the other plant is under a box so that it is in the dark.
2. Use a protractor to determine the angle of stem bending in both plants (Fig. 21.3). The straight edge of the protractor is held along the main stem, while the point of bend in the stem is held under the protractor's 90° point. The tip of the stem will point to the degree of bend.
3. Record the angle of bending for each plant in Table 21.1.

Table 21.1 Gravitropism of Stems

Mature Plant	Angle of Bending/Degrees	Gravitropism of Stem	Conclusion
In the light			
In the dark			

Conclusions: Response of Mature Stem

- Complete Table 21.1. Do your data suggest that the presence of light/dark affects the degree of gravitropism in stems? ____________________
- If your prediction was not supported, suggest a new model (scenario) that might explain your results. ____________________
- Why is it adaptive for seedlings to exhibit negative gravitropism? ____________________

Response of Seedlings

1. Examine a petri dish in which *Brassica rapa* seeds germinated a few days ago (Fig. 21.4). (Do not remove the dish from the water reservoir.)

Figure 21.4 Gravitropism of seedlings.
After *Brassica rapa* seeds germinate in a petri dish, determine the effect of gravity on the orientation of the stem and root.

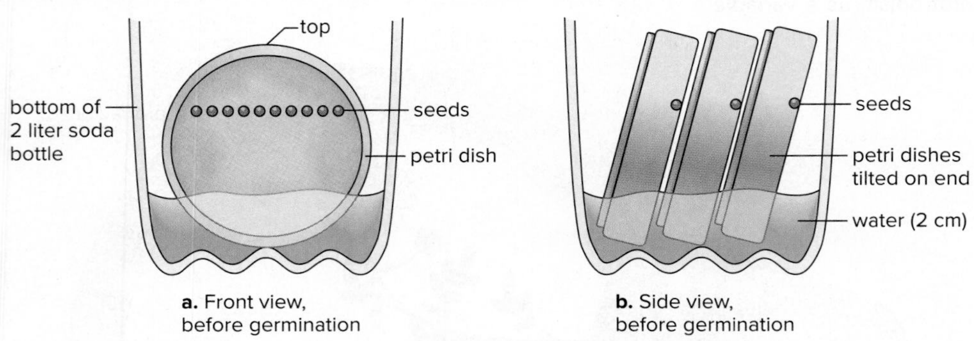

2. Use the space provided to sketch the position of the seedlings. This is time t_0. In Table 21.2, note the position of the root and stem for a plant grown in the light.
3. Rotate the petri dish in the water so that the former top is now at the bottom. You are testing for gravitropism only, so these plants need to be put under a box to make sure they are not influenced by light while they are being tested.
4. Observe the seedlings again after 45 minutes. Compare them with your sketch and with an unturned petri dish on display. Use the space provided to sketch the seedlings' position. This is time t_1. Describe the position of the root and stem in Table 21.2.

Table 21.2 Gravitropism in Seedlings

Time	Position of Root	Position of Stem	Explanation
t_0			
t_1			

Conclusions: Response of Seedlings

Based on the response of the mature stem (in the dark and in the light), and on the response of seedlings

- Do roots exhibit negative or positive gravitropism? ____________________________
- Do stems exhibit negative or positive gravitropism? ____________________________
- Fill in the last column in Table 21.2.

21.2 Phototropism

Phototropism is the bending of plants in response to unidirectional light. When a plant is exposed to unidirectional light, the hormone auxin migrates from the bright side of a stem to the shady side. The cells on that side elongate faster than those on the bright side, causing the stem to curve toward the light (Fig. 21.5).

As you will see, blue light brings about phototropism. Therefore, it is hypothesized that a plant pigment capable of absorbing blue light initiates phototropism.

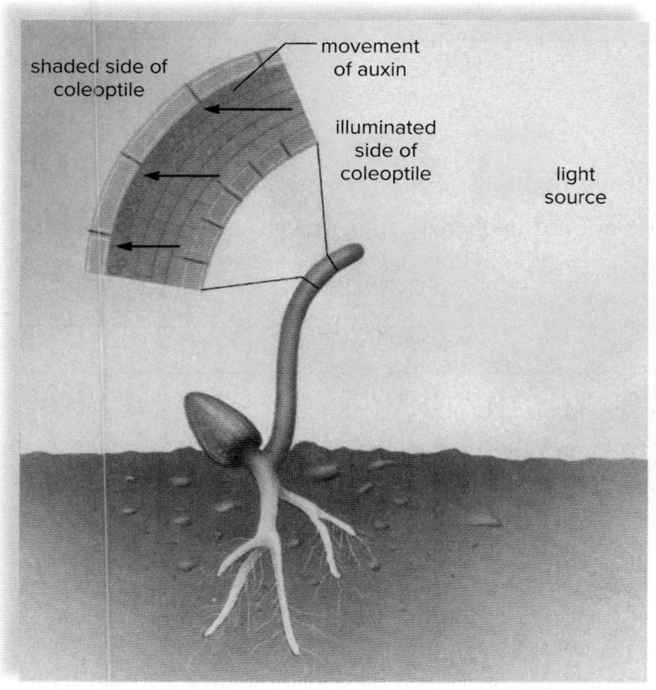

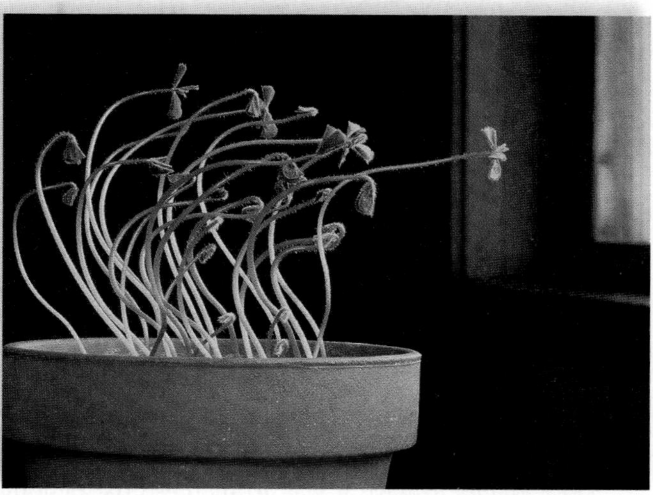

Figure 21.5 Phototropism in a seedling and plant. **a.** Experiments with seedlings show that when a seedling bends toward the light, auxin (arrows) migrates from the bright side to the shady side. Cell elongation follows and this causes bending toward the light. **b.** Adult plants also respond to light by bending toward it.

Experimental Procedure: Phototropism

Phototropism chambers can be made with film canisters (Fig. 21.6). Three windows are punched into the sides of the film canister, and three *Brassica rapa* seeds are placed on wet blotting paper. Each window of the canister then is covered with a red, blue, or green acetate square. A control canister has windows covered with clear acetate.

1. Obtain a control canister and a canister with colored windows.
2. Open the canisters and observe the growth of the seedlings placed there by your instructor. Reclose the canisters, being sure that the orientation of the colored windows is the same as before.
3. Record the response of the seedlings in Table 21.3.

Figure 21.6 Color of light and phototropism. Seeds are exposed to different colors of light in a phototropism chamber.

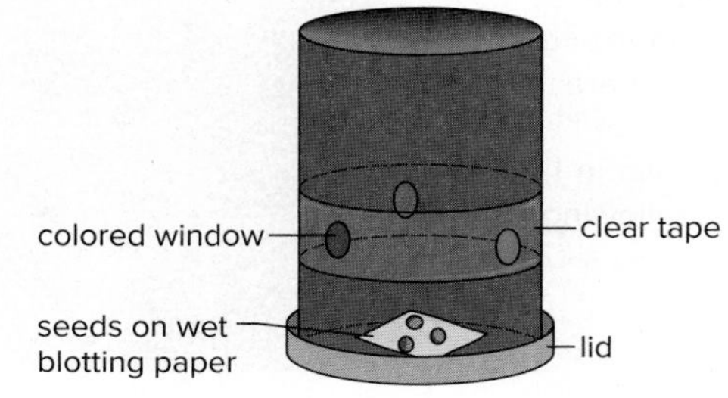

Table 21.3 Phototropism

Type of Canister	Response of Seedlings	Conclusion
Clear windows (control)		
Colored window (red)		
Colored window (blue)		
Colored window (green)		

Conclusions: Phototropism

- Record the results of this experiment and your conclusion in Table 21.3.
- Do your results support the model that blue light (but not red and green light) reception is involved in positive phototropism of stems? ______________________________
  Explain how you came to this conclusion. ______________________________
  ____
- Investigators have found that the activation of a plasma membrane photoreceptor for blue light, now called phototropin, leads to the binding of auxin by cells and the bending of stems. Why is it adaptive for plants to have a way to increase the bending of stems in reponse to unidirectional light? __________

21.3 Gibberellins and Stem Elongation

Gibberellins are plant hormones that cause stem elongation (Fig. 21.7). The presence of gibberellin in a cell activates a gene that codes for amylase. Amylase breaks down starch, providing an energy source for wall construction during elongation.

Experimental Procedure: Stem Elongation

Do you hypothesize that gibberellins would cause dwarf plants (called dwarf plants because the stem cells do not usually elongate) to grow taller? __________ Explain. ______________________________
____

a.

b.

Figure 21.7 Gibberellins cause stem elongation.
a. The plant on the right was treated with gibberellins; the plant on the left was not treated. **b.** The grapes are larger on the right because gibberellins caused an increase in the space between the grapes, allowing them to grow larger. (a) ©Science Source; (b) ©Amnon Lichter, The Volcani Center

Normal Plants

1. Observe three normal *Brassica rapa* plants of increasing age. Your instructor will tell you how many days each plant has been growing. Record the ages of the plants in Table 21.4.
2. See the boxed instructions on how to measure an internode. Choose the two longest internodes on each plant. Carefully measure (and record) their lengths, and determine the average internode distance for these two measurements. Record your data in Table 21.4.
3. Return the plants to the location specified by your instructor.

Untreated Dwarf Plants

1. Observe three dwarf plants (a mutant strain of *Brassica rapa*) of increasing age. Your instructor will tell you how many days each plant has been growing. Record the ages of the plants in Table 21.4.
2. See the boxed instructions on how to measure an internode. Choose the two longest internodes on each plant. Carefully measure (and record) their lengths, and determine the average internode distance for these two measurements. Record your data in Table 21.4.
3. Return the plants to the location specified by your instructor.

Treated Dwarf Plants

1. Observe three dwarf plants of increasing age that have been sprayed with *gibberellin* since germination. Your instructor will tell you how many days each plant has been growing. Record the ages of the plants in Table 21.4.
2. See the boxed instructions on how to measure an internode. Choose the two longest internodes on each plant. Carefully measure (and record) their lengths, and determine the average internode distance for these two measurements. Record your data in Table 21.4.
3. Return the plants to the location specified by your instructor.

Measuring an Internode

A **node** is where a leaf is attached to the stem and an **internode** is the region of a stem between nodes. To arrive at the length of an internode, use a metric ruler to measure from the **base of the petiole** (leaf stalk) of the upper leaf to the **leaf axil** of the lower leaf.

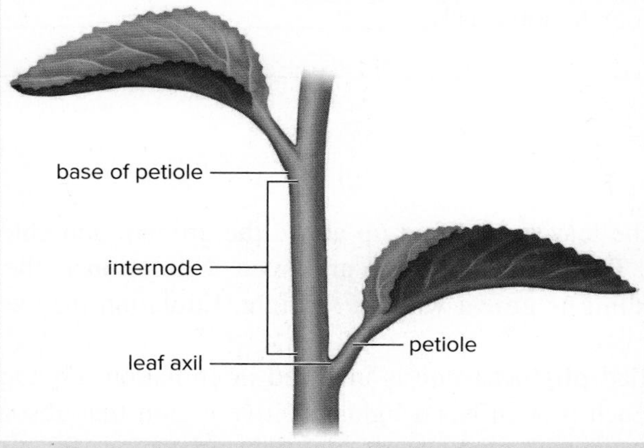

Table 21.4 Effect of Gibberellin

	Internode 1 Measurement	Internode 2 Measurement	Average Internode Distance
Normal Plants			
Age			
Age			
Age			
Untreated Dwarf Plants			
Age			
Age			
Age			
Treated Dwarf Plants			
Age			
Age			
Age			

Conclusions: Stem Elongation

- How do the three groups of plants differ from each other? ______________________________

  ____

  ____

- Did your data support your hypothesis? ______________________________
- If not, why not? ______________________________

21.4 Etiolation

In a plant grown in the light, the leaves are lifted up above the ground, and chlorophyll is synthesized so that photosynthesis may begin. Therefore, the leaves are green. **Etiolation** is the summation of altered growth patterns that result when a seedling is grown without sunlight. Etiolation may assist the seedling in reaching sunlight.

A blue-green pigment called **phytochrome** is involved in etiolation. Phytochrome is composed of two identical proteins (Fig. 21.8). Each protein has a light-sensitive region that absorbs red light during the day and far-red light at night. When phytochrome absorbs red light it becomes Pfr, the active form of phytochrome, which promotes normal growth; the leaves expand and become green. This form of phytochrome is known as Pfr because it absorbs far-red light. After doing so, it becomes Pr, the inactive form of phytochrome, which is so called because it absorbs red light during the day. As long as Pr is present and not Pfr, a plant will undergo etiolation.

Figure 21.8 Phytochrome conversion cycle.

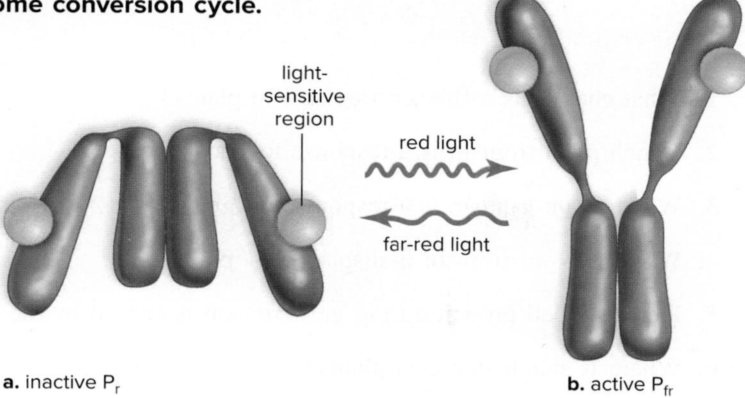

Experimental Procedure: Bean Seedling Etiolation

1. Compare the appearance of a bean seedling that germinated and grew in the dark with one that germinated and grew in the light (Fig. 21.9). Record your data in Table 21.5.
2. Complete Table 21.5 by explaining why the etiolated plant differs from the normal plant in the characteristics listed.

Figure 21.9 Etiolation.
Bean seedlings on the left are etiolated because they were grown in the dark. Bean seedlings on the right are much shorter with well-formed leaves because they were grown in the light. ©Nigel Cattlin/ Science Source

Table 21.5 Comparison of Normal and Etiolated Plants

	Normal Plant	Etiolated Plant	Explanation
Color of stem			
Color of leaf			
Length of stem			
Size of leaf			
Stiffness of stem (examine gently)			
Other comments, if any			

Laboratory Review 21

_______________ **1.** What chemicals influence tropisms in plants?

_______________ **2.** Which plant tropism is a response to gravity?

_______________ **3.** Which plant tropism is a response to light?

_______________ **4.** Which type of tropism is displayed by plant roots?

_______________ **5.** Increased cell growth during gravitropism is caused by the hormone __________.

_______________ **6.** Where is starch stored in plants?

_______________ **7.** When seedlings first germinate, are they affected by phototropism or gravitropism?

_______________ **8.** What color of light triggers phototropism?

_______________ **9.** Phototropism is influenced by the photoreceptor __________.

_______________ **10.** Stem elongation is caused by the hormone __________.

_______________ **11.** What is the role of amylase in a plant?

_______________ **12.** A plant grown in the dark will exhibit __________.

_______________ **13.** Are etoliated plants taller or shorter than plants grown in the light?

_______________ **14.** Etoliation is influenced by the pigment __________.

Thought Questions

15. What purposes do tropisms serve in plants?

16. Why is it advantageous for plants to respond to gravity and light?

17. How might etoliation allow plants to better survive?

LABORATORY

22

Reproduction in Flowering Plants

Learning Outcomes

22.1 The Flower
- Identify the parts of a flower.

22.2 The Flowering Plant Life Cycle
- Describe the alternation-of-generation life cycle of flowering plants.
- Identify the sporophyte and gametophyes in a flowering plant life cycle.

22.3 Pollination and Development of the Embryo
- Describe pollination in flowering plants and hypothesize what is the particular pollinator for various plants.
- Describe the developmental stages of a eudicot embryo.

22.4 Fruits
- Classify simple fruits as fleshy or dry, and then use a dichotomous key to identify the specific fruit type.

22.5 Seeds
- Identify the parts of a seed and of an embryonic plant. Distinguish between eudicot and monocot seeds.
- Describe the germination of eudicot and monocot seeds.

Introduction

The evolution of the flower and the involvement of animals in the plant life cycle helps account for the great success of flowering plants, also called **angiosperms.** The flower is the center of sexual reproduction for angiosperms and is involved in the production of male and female gametophytes, which are gametes and embryos enclosed within seeds. The flower also gives rise to fruits that cover the seeds.

Flowering plants are stationary, but they have evolved a successful mechanism for pollination that helps them complete their life cycle. Motile animals, often insects, visit flowers and collect the pollen. Pollen moves sperm that fertilizes an egg, which triggers the development of a seed. The relationship between the pollinator and the flower (Fig. 22.1) is mutualistic because the pollinator acquires nectar, a nutrient substance produced by the flower. Animals also help flowering plants disperse their seeds. When animals feed on fruits, they inadvertently take the seeds to new locations.

Figure 22.1 Pollinator on a landing platform. This composite flower contains many individual flowers. The bee is attracted to this flower because of its bright color and sweet smell, and the bee uses the flower's reproductive parts as a place to land. Pollen will be collected by the bee when it feeds on nectar.

22.1 The Flower

The structure of a flower allows a flowering plant to produce seeds whose dispersal accounts for the success of these plants.

Structure of a Flower

The flower contains the reproductive structures necessary to the life cycle of a flowering plant.

Observation: Structure of a Flower

1. Examine a model and identify these parts of a flower (Fig. 22.2).

Figure 22.2 Flower structure.

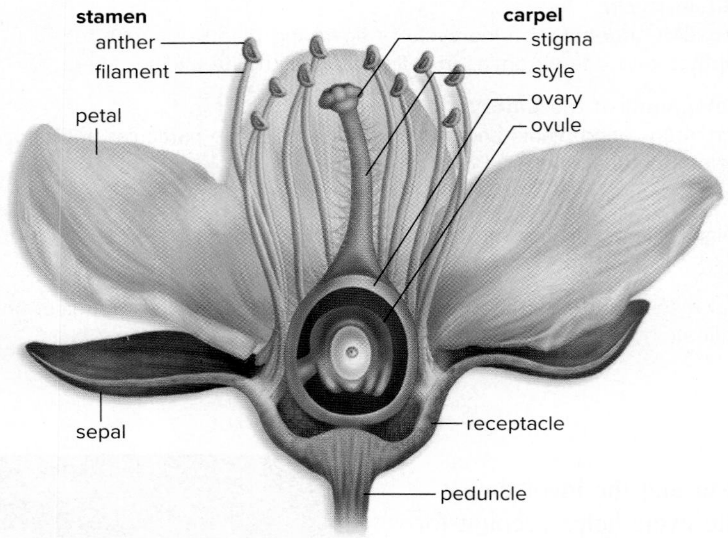

Peduncle: The flower stalk.

Receptacle: The portion of a flower that bears the other parts.

Sepals: Collectively called the **calyx,** the **sepals** protect the flower bud before it opens. The sepals may drop off or may be colored like the petals. Usually, sepals are green and remain attached to the receptacle.

Petals: Collectively called the **corolla,** the **petals** are quite diverse in size, shape, and color. Their color or arrangement can attract a pollinator.

Stamens: Each **stamen** consists of two parts—the **anther,** and the **filament,** a slender stalk. Pollen grains develop in the anther.

Carpel: A vaselike structure with three major regions from top to bottom: the **stigma,** an enlarged, sticky knob; the **style,** a slender stalk; and the **ovary,** an enlarged base that contains one or more ovules. **Ovules** develop into seeds. The ovule wall becomes the seed coat, which protects an embryonic plant and its stored food. An ovary becomes a fruit. Fruit is instrumental in the distribution of seeds. For example, after a bird eats a berry, it will probably fly to a new location, where it defecates the seed.

Flower parts in threes and multiples of three

Table 22.1 Monocots and Eudicots

Monocots	Eudicots
One cotyledon in seed	Two cotyledons in seed
Flower parts in threes or multiples of three	Flower parts in fours or fives or their multiples
Usually herbaceous	Woody or herbaceous
Usually parallel venation in leaves	Usually net venation in leaves
Scattered vascular bundles in stem	Ring of vascular bundles in stem

Flower parts in fours or fives and their multiples

2. As we studied in the previous lab, some flowering plants are monocots (monocotyledons) and some are eudicots (eudicotyledons) as described in Table 22.1. Note the difference between monocot and eudicot flowers. Is the flower model you have been examining a monocot or a eudicot? ____________________

Observation: Live Flower

1. Obtain a live flower similar to the one in Figure 22.3. It should be possible to see the sepals, petals, stamens, and carpel(s) of this flower as in Figure 22.3. Sketch the overall appearance of the whole flower here and label as much as possible using the terms in Figure 22.2.

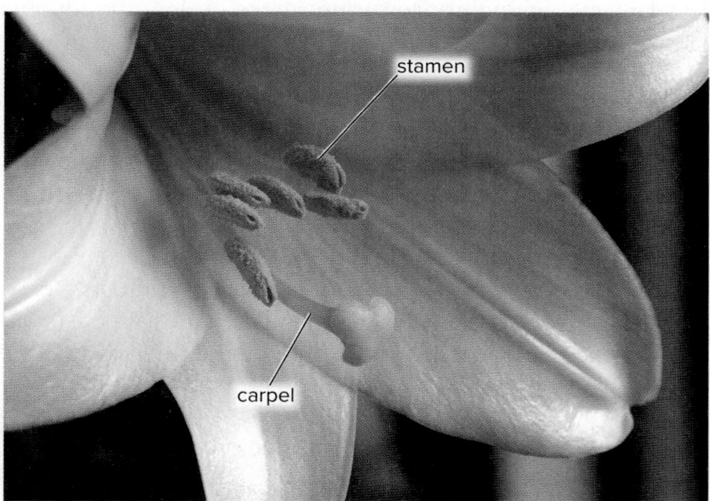

Figure 22.3 Lily (*Lilium longiflorum*).

2. Describe the color and scent of your flower, if any. ____________________
____

3. Count the number of petals and sepals (see Table 22.1). Is this flower a monocot or a eudicot? __________
____________________ Explain. ____________________

4. Determine where the petals and sepals are attached and remove them. What color are the sepals? __________
__________ Do they resemble the petals in any way? ____________________ If yes, how?
____

5. Count the number of stamens. Is this the same as the petal number or a multiple of the petal number? ___________ Remove the stamens and examine one to locate the anther and filament. Place a stamen on a slide and use a scalpel to remove and open the anther to disperse pollen on the slide. Remove all but a few pollen grains from the slide. Add a coverslip and examine microscopically, using high power. Can you find two different nuclei? ___________ At this point, a pollen grain contains two cells. One of these cells will divide to produce the sperm and the other will produce the pollen tube through which the sperm will travel to the embryo sac in an ovule of an ovary.
6. Before removing the carpel, identify the stigma, style, and ovary. Remove a carpel and place it on a slide. Use a scalpel to slit it longitudinally. Do you see any ovules? ___________
Describe. ___________
Use the scalpel to make a cross section of the ovary. Does this ovary contain chambers, each with ovules? ___________
7. If available, examine the fruit produced by this type of flower. The exterior wall came from what part of the carpel? ___________ The seeds came from what structure? ___________

22.2 The Flowering Plant Life Cycle

Land plants, including flowering plants, follow the **alternation-of-generation life cycle** (Fig. 22.4). Just as your arms are shaped differently from your legs, land plants have two generations that look different and have different functions:

1. The **sporophyte** (diploid generation) produces haploid spores in a structure called a sporangium by meiosis. **Spores** undergo mitosis and develop into the haploid generation.
2. The **gametophyte** (haploid generation) produces gametes (eggs and sperm) by mitosis. If fertilization occurs, the gametes unite to form a diploid zygote (a young sporophyte).

In humans, meiosis is involved in the production of ___________. What does meiosis produce in plants? ___________

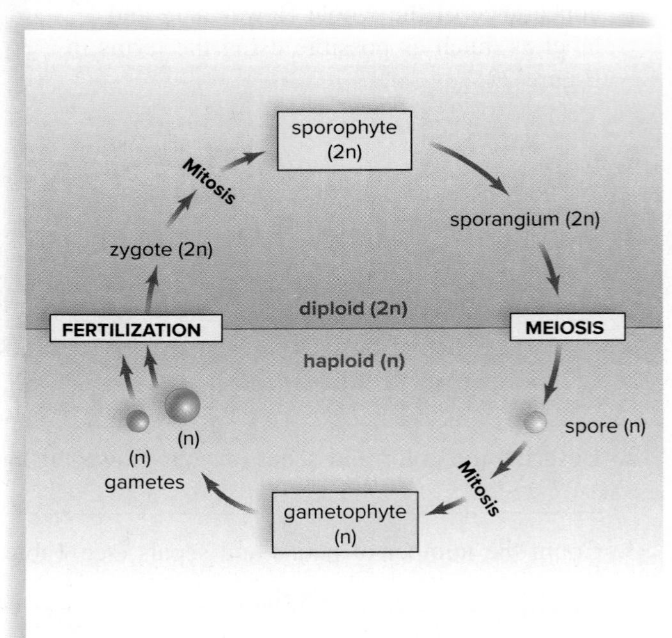

Figure 22.4 Alternation of generations.

Alternation of Generation in Flowering Plants

What do the sporophyte and gametophyte of a flowering plant look like? The sporophyte is the dominant generation in flowering plants. It is the generation we can easily see because we recognize it as the plant. An oak tree, a rose bush, a daffodil plant—each is the sporophyte generation.

A flowering plant produces spores in sporangia, but they are microscopic and reside in the body of the dominant generation. Further the life cycle is modified. Flowering plants have two types of sporangia (Fig. 22.5). A megasporangium inside the ovule produces a megaspore, and microsporangia inside pollen sacs produce microspores. A microspore develops into a pollen grain, the male gametophyte, which produces two sperm. The **megaspore** within an ovule develops into a **female gametophyte** (embryo sac), which contains an egg.

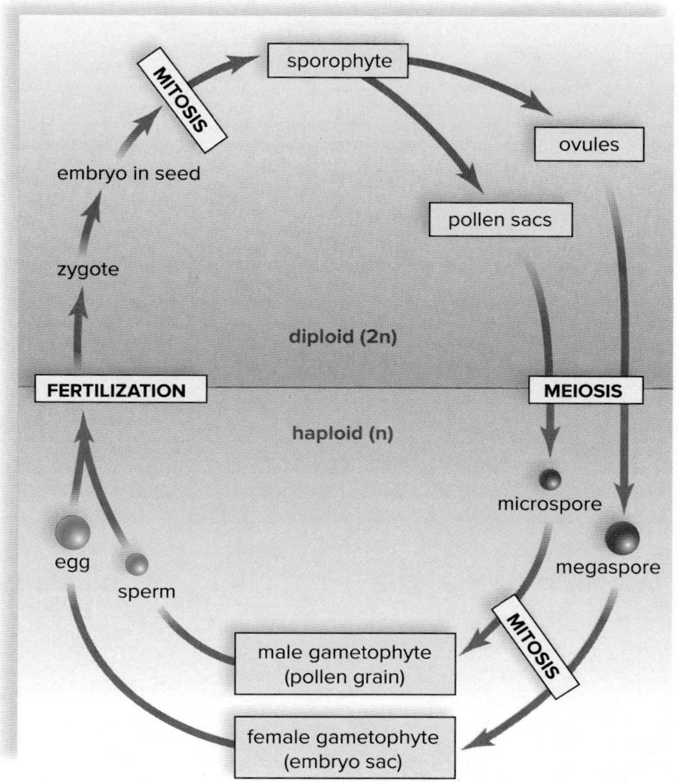

Figure 22.5 Alternation of generations in flowering plants, which are seed plants.

Flowering plants practice **double fertilization.** When a pollen grain of the correct species lands on a stigma, it develops a **pollen tube** that carries the sperm to the embryo sac in the ovule, where one sperm fertilizes the egg and the other participates in development of the **endosperm,** which is food for the developing embryo. An ovule develops into a seed.

Explain why a pollen grain is called the *male* gametophyte. ________________________________

____

Explain why an embryo sac is called the *female* gametophyte. ________________________________

____

A seed contains which generation? ________________________________

Figure 22.6 Flowering plant life cycle.

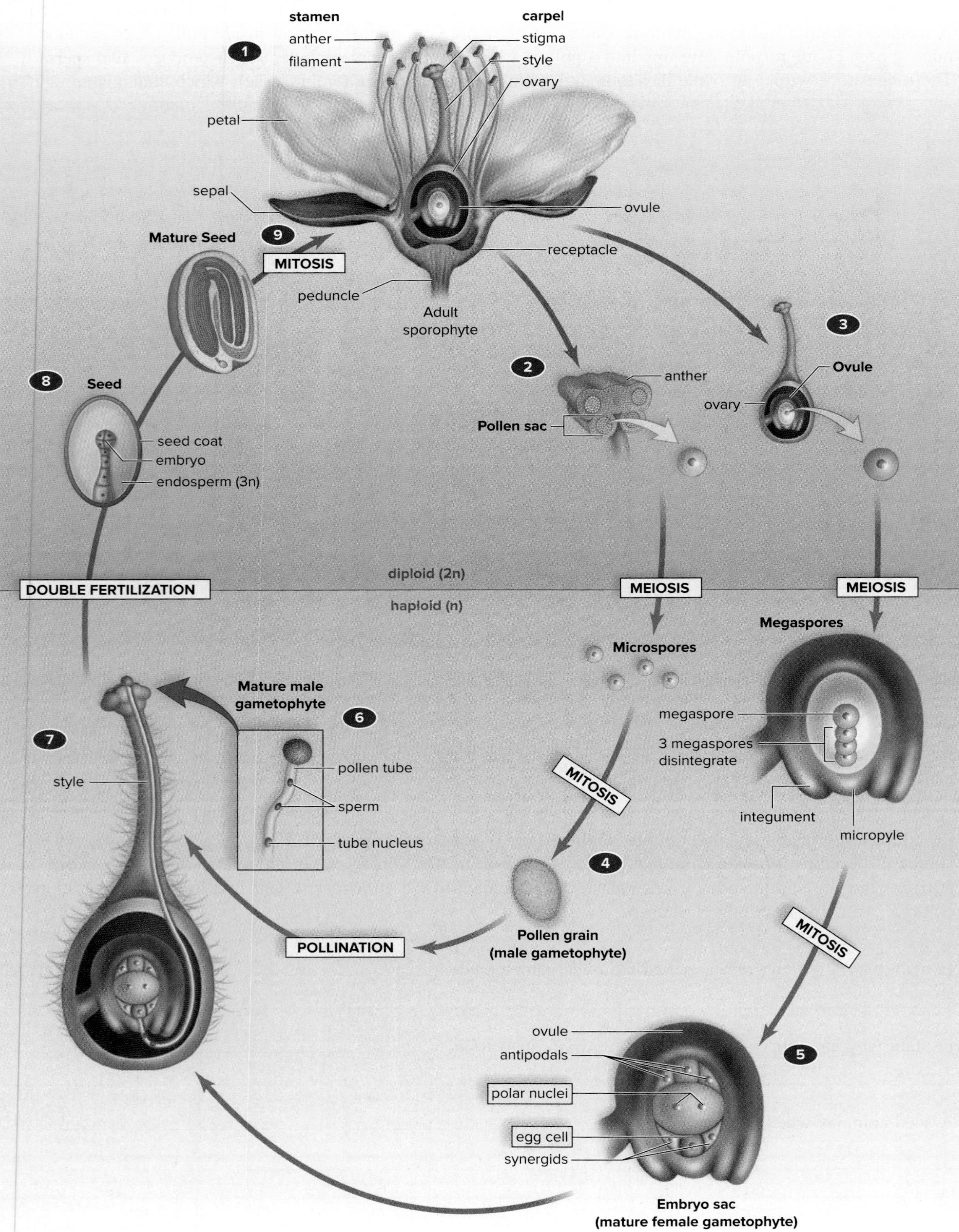

stamen
anther
filament
carpel
stigma
style
ovary
1
petal
sepal
ovule
Mature Seed
9
MITOSIS
receptacle
peduncle
Adult
sporophyte
3
Ovule
2
anther
ovary
Pollen sac
8
Seed
seed coat
embryo
endosperm (3n)
diploid (2n)
DOUBLE FERTILIZATION
MEIOSIS
MEIOSIS
haploid (n)
Megaspores
Microspores
Mature male
gametophyte
6
7
megaspore
3 megaspores
disintegrate
pollen tube
MITOSIS
style
sperm
integument
micropyle
tube nucleus
4
MITOSIS
Pollen grain
(male gametophyte)
POLLINATION
ovule
antipodals
5
polar nuclei
egg cell
synergids
Embryo sac
(mature female gametophyte)

Observation: Flowering Plant Life Cycle

Use Figure 22.6 as a guide to describe the life cycle of flowering plants.

1. The parts of the flower involved in reproduction are the ______________ and the ________________.
2. The anther at the top of the stamen has two ________________ sacs, which produce ________________ by meiosis.
3. Within an ovule, a megaspore mother cell undergoes meiosis to produce four ________________, three of which die.
4. A microspore undergoes mitosis and becomes a ________________, the male gametophyte.
5. One megaspore undergoes mitosis and develops into a(n) ________________, which contains, in addition to other cells, two ________________ nuclei and the ________________ cell. (These terms are shaded.)
6. **Pollination** is the transport of pollen by wind or by animals such as insects (see Fig. 22.1) from the pollen sacs to the stigmas of carpels. After a pollen grain lands on the stigma of a carpel, it develops a pollen tube that passes down the style and takes ________________ sperm to the embryo sac.
7. During double fertilization, one sperm from the pollen tube fertilizes the egg within the embryo sac, and the other joins with the two ________________ nuclei.
8. The fertilized egg becomes an ________________, and the joining of polar nuclei and sperm becomes the triploid (3n) **endosperm.** The ovule wall becomes the seed coat. A **seed** contains a sporophyte embryo, stored food, and a seed coat. In angiosperms, seeds are enclosed by ________________ (not shown). A fruit assists in the dispersal of seeds (e.g., when animals eat fleshy fruits, they may ingest the seeds and disperse them by defecation sometime later).
9. After dispersal of the seed, it germinates and begins to grow, eventually becoming an adult sporophyte that bears flowers.

22.3 Pollination and Development of the Embryo

Pollination, the movement of pollen from the pollen sacs to the stigmas of carpels, is an important event in flowering plants. In some plants, wind accomplishes pollination and these plants produce copious amounts of pollen (Fig. 22.7). Why would you expect wind to carry out pollination for a tree and not for a garden plant? ________________________

____

____

____

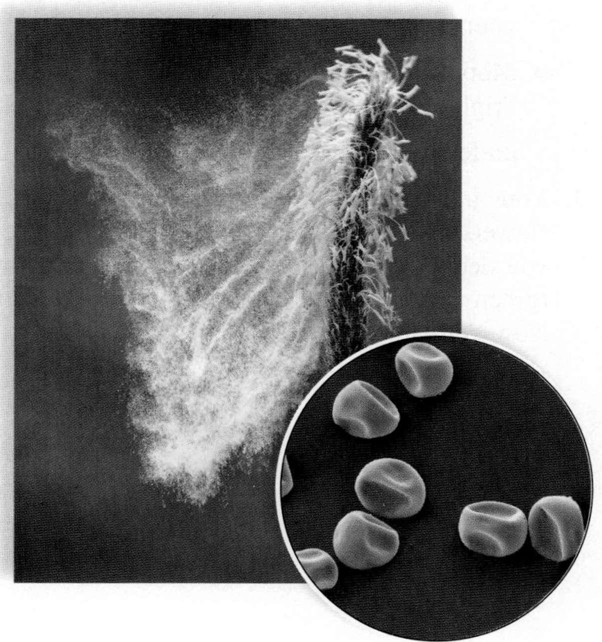

Figure 22.7 Wind pollination of a grass, with SEM of pollen grains.
©Tim Gainey/Alamy Stock Photo; (inset) ©Eye of Science/Science Source

Plants and Their Pollinators

Through the course of evolution, many plants—even trees—have come to depend on animals (e.g., beetles, bats, insects) called **pollinators** to carry out pollination. The relationship is mutualistic because both the plant and the pollinator benefit from the relationship. **Nectar** is a sugary substance produced by the plant for the pollinator. The pollinator goes from plant to plant to gather food (nectar and even pollen) and in the process carries pollen from flower to flower.

It might seem as if a pollinator such as a bee goes to all types of flowers, but it doesn't. Bees visit only certain flowers—the ones that provide them with nectar, a sugary liquid that serves as their food. Bee-pollinated flowers have a shape and appearance preferred by bees, even a particular type of bee. Particular bees and particular flowers are suited to one another because there has been interaction between them over time that caused coevolution to occur.

How is this advantageous to both the plant and the bee? ______________________________

____

____

Observation: Plants and Their Pollinators

1. The description you gave in the live flower observation (section 22.1) can help you decide what type of pollinator would be attracted to your flower. See also Figure 22.8.
   - Bees and moths can smell. But bees like a delicate, sweet smell, while moths prefer a strong smell that can allow them to find flowers in the dark. Moths are nocturnal and feed at night. If the flower has a smell, which of these two pollinators might pollinate your flower? ______________________________
   - Bees and butterflies generally need a landing platform. Bees can land on a small petal, but butterflies typically walk around on a cluster of flowers.
   - Moths and hummingbirds do not need a landing platform because they hover (flap their wings to stay in one place). Moths and hummingbirds have a long tongue to reach nectar at the bottom of the floral tube. Based on this information, which type of pollinator mentioned so far might pollinate your flower? ______________________________
   - Moth-pollinated flowers are typically white; bees can't see red but can see yellow; butterflies like brightly colored flowers; and hummingbirds prefer the color red. Which of these pollinators might prefer your flower? ______________________________
2. Your instructor may have suggested that you bring other flowers to lab. Examine several available flowers and tell how each of the following features help decide the pollinator. These features will help you decide the pollinator for each type of flower. If a flower is the same color as the rest of the plant (green or brown) and hangs down from the end of a branch, it likely produces pollen that is windblown. Now complete Table 22.2.

   How does each of these flower features help decide the pollinator?

   Landing platform ______________________________

   Color ______________________________

   Smell ______________________________

   Shape of flower ______________________________

a.

b.

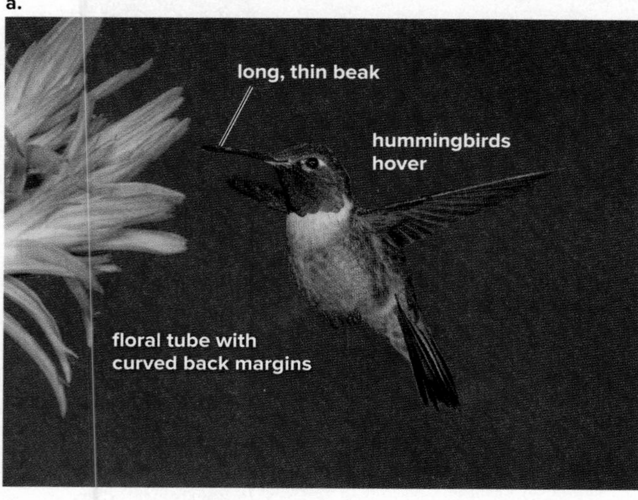

c.

Figure 22.8 Pollinators.
a. The butterfly-pollinated flower is often a brightly colored composite containing many individual flowers. The broad exposure provides room for the butterfly to land. **b.** A moth-pollinated flower is usually light in color; a moth depends on scent to find the flower it prefers in the dark. **c.** Hummingbird-pollinated flowers are curved back, allowing the bird to insert its beak to reach the rich supply of nectar.
(a) ©Juk Atrasat/Shutterstock; (b) ©Andrew Darrington/Alamy Stock Photo; (c) ©Anthony Mercieca/Science Source

Table 22.2 Plants and Their Pollinators

Description of Flower	Possible Pollinator	Explanation
1		
2		
3		
4		
5		
6		

Development of Eudicot Embryo

Stages in the development of a eudicot embryo are shown in Figure 22.9. During development, the **suspensor** anchors the embryo and transfers nutrients to it from the mature plant. The **cotyledons** store nutrients that the embryo uses as nourishment. An embryo consists of the **epicotyl,** which becomes the leaves; the **hypocotyl,** which becomes the stem; and the **radicle,** which becomes the roots.

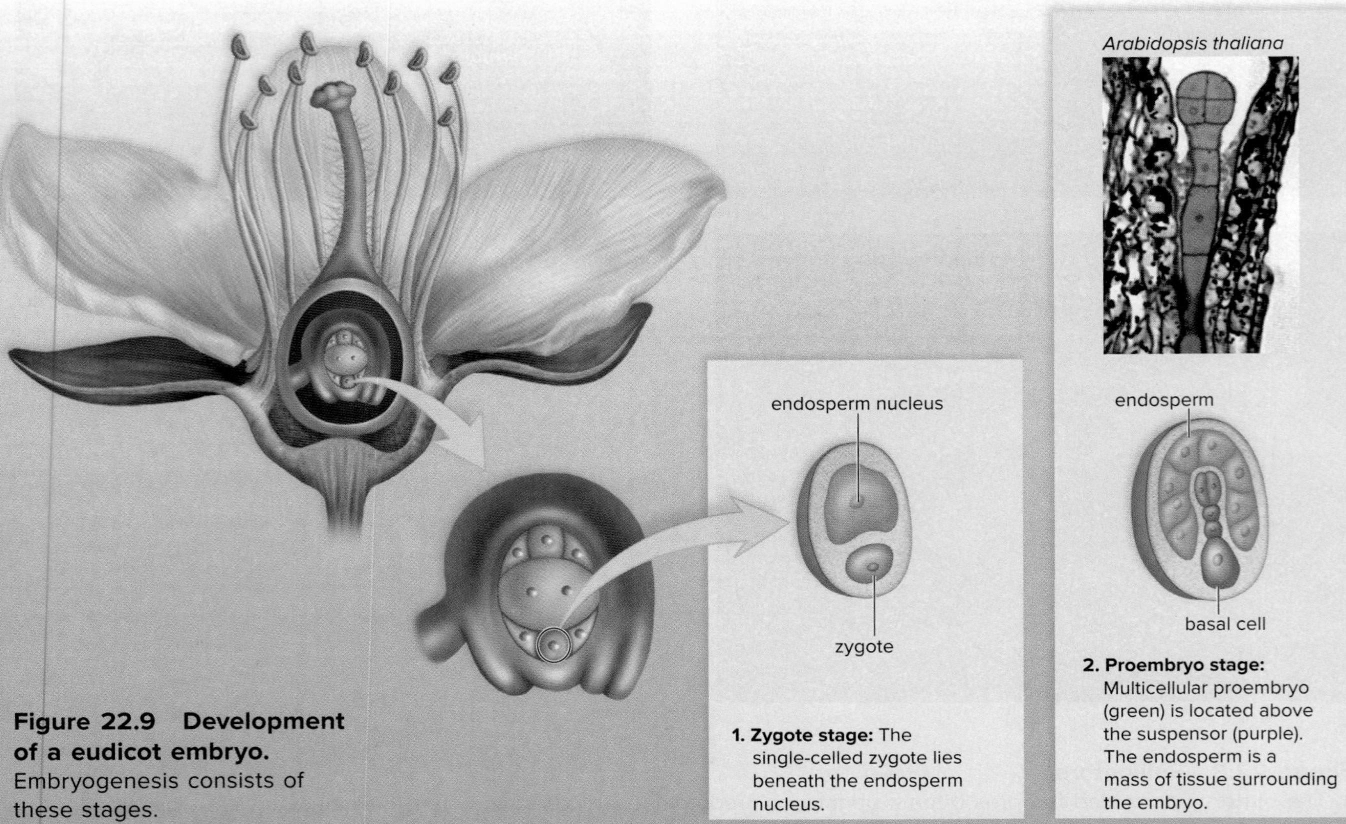

1. Zygote stage: The single-celled zygote lies beneath the endosperm nucleus.

2. Proembryo stage: Multicellular proembryo (green) is located above the suspensor (purple). The endosperm is a mass of tissue surrounding the embryo.

Figure 22.9 Development of a eudicot embryo. Embryogenesis consists of these stages.

(top right) ©Dr. Chun-Ming Liu

Observation: Development of Eudicot Embryo

Study prepared slides and identify the stages described in Figure 22.9. List the stages you were able to identify. ______________________________

____

____

____

____

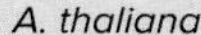

A. thaliana

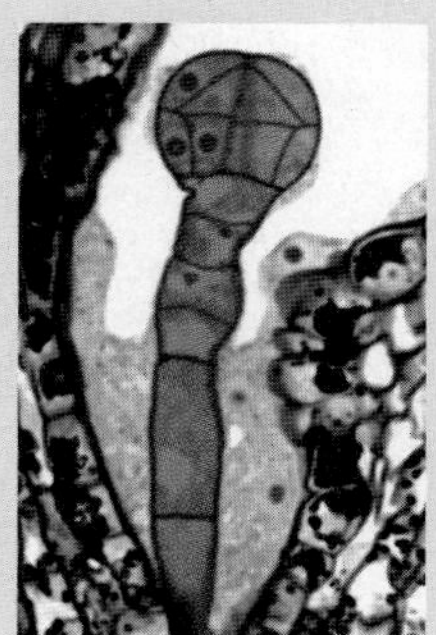

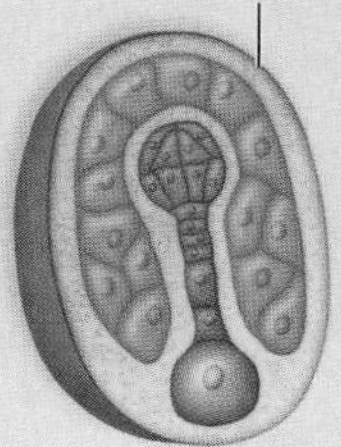

3. Globular stage: As cell division continues, the proembryo (green) becomes globe-shaped. The stalklike suspensor (purple) anchors the embryo.

A. thaliana

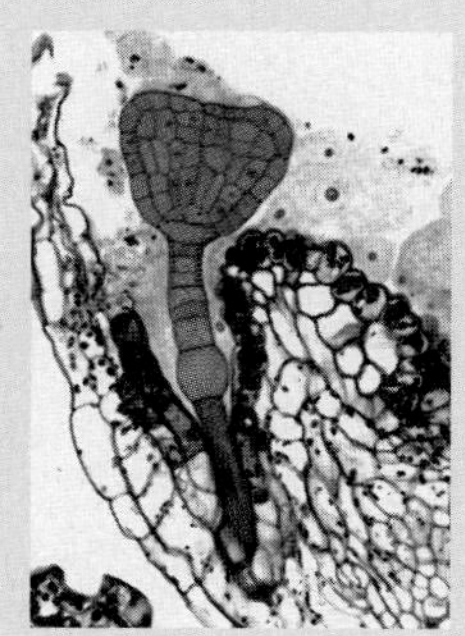

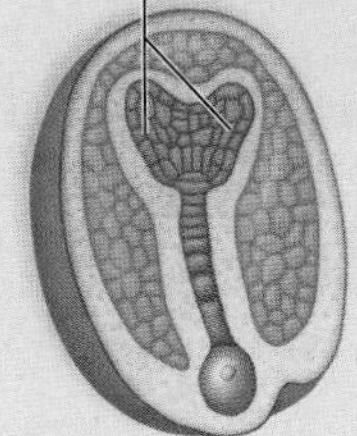

4. Heart stage: The embryo becomes heart-shaped as the cotyledons begin to appear.

Capsella

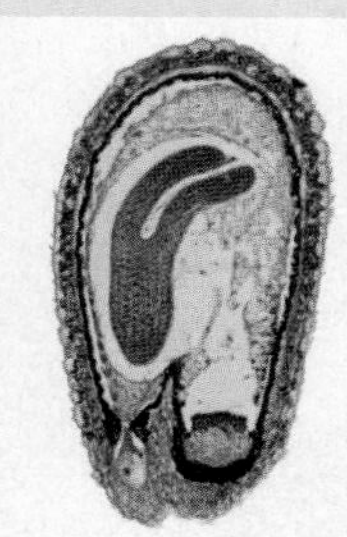

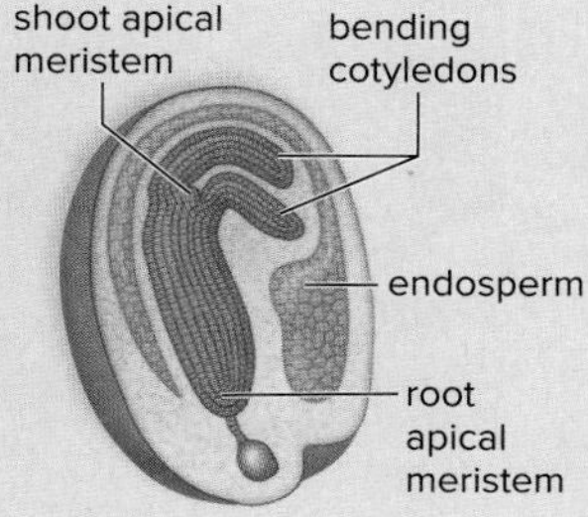

5. Torpedo stage: The embryo becomes torpedo-shaped as the cotyledons enlarge. The endosperm lessens, and tissues become differentiated.

Capsella

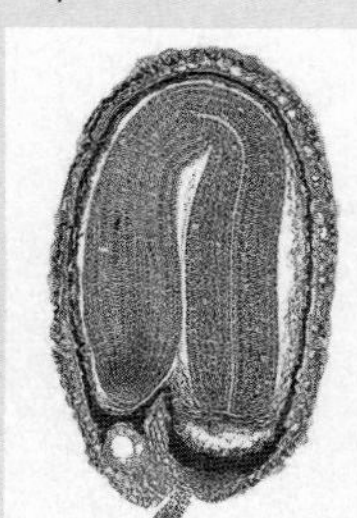

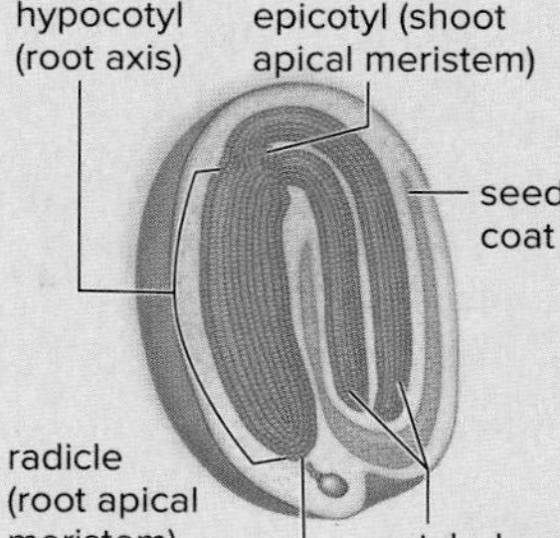

6. Mature embryo stage: The embryo consists of the epicotyl (represented here by the shoot apex), the hypocotyl, and the radicle (which contains the root apex).

22.4 Fruits

Because fruits are defined as the mature ovary, what people normally call a vegetable can be a fruit to a botanist. For example, in addition to a blackberry, an almond, a pea pod, and a tomato are fruits (Fig. 22.10).

As a fruit develops from an ovary, the ovary wall thickens to become the **pericarp.** The pericarp can have as many as three layers: exocarp, mesocarp, and endocarp. The **exocarp** forms the outermost skin of a fruit. The **mesocarp** between the exocarp and endocarp can be fleshy. The receptacle of a flower often

Figure 22.10 Fruit diversity.
a. The almond fruit is fleshy, with a single seed enclosed by a hard covering. **b.** The tomato is derived from a compound ovary. **c.** Dry fruit of the pea plant develops from a simple ovary. **d.** The blackberry fruit is an aggregate fruit derived from a flower that had many ovaries. Each ovary produces one of the "berries" in the aggregate fruit.

contributes to the flesh of a fruit also. The **endocarp** serves as the boundary around the seed(s). The endocarp may be fleshy (as in tomatoes), hard (as in peach pits), or papery (as in apples). Botanists classify fruits in the manner described in the following key. This is a simplified key, but it does contain some of the most common types of fruits.

Dichotomous Key to Major Types of Fruit

I. Fleshy fruits
- **A.** Simple fruits (i.e., from a single ovary)
  - **1.** Flesh mostly of ovary tissue, particularly the mesocarp
    - **a)** Endocarp is hard and stony; ovary superior and single-seeded (cherry, olive, coconut): **drupe**
    - **b)** Endocarp is fleshy or slimy; ovary usually many-seeded (tomato, grape, green pepper): **berry**
  - **2.** Flesh mostly of receptacle tissue (apple, pear, quince): **pome**
- **B.** Complex fruits (i.e., from more than one ovary)
  - **1.** Fruit from many carpels on a single flower (strawberry, raspberry, blackberry): **aggregate fruit**
  - **2.** Fruit from carpels of many flowers fused together (pineapple, mulberry): **multiple fruit**

II. Dry fruits
- **A.** Fruits that split open at maturity (usually more than one seed)
  - **1.** Split occurs along two seams in the ovary. Seeds borne on one of the halves of the split ovary (pea and bean pods, peanuts): **legume**
  - **2.** Seeds released through pores or multiple seams (poppies, irises, lilies): **capsule**
- **B.** Fruits that do not split open at maturity (usually one seed)
  - **1.** Pericarp hard and thick, with a cup at its base (acorn, chestnut, hickory): **nut**
  - **2.** Pericarp thin and winged (maple, ash, elm): **samara**
  - **3.** Pericarp thin and not winged (sunflower, buttercup): **achene** (cereal grains): **caryopsis**

Source: Vodopich and Moore: *Biology Laboratory Manual* 8/e, p. 344.

To use this key understand that at each step you have at least two choices and you pick one of these in order to proceed further. Therefore, first determine if the fruit is fleshy or dry. If fleshy, proceed further with I and if dry, proceed further with II. Next choose either A or B and so forth, until you arrive at the particular type of fruit. The last column in Table 22.3 will be one of the boldface terms from the key.

Observation: Fruits

1. Again examine the fruit of your flower. Use the key to determine the fruit type and list this fruit as an example in Table 22.3.
2. Examine an apple that has been sliced to give a longitudinal and a cross-section view of the interior. The flesh of an apple is from the receptacle and only the core of an apple is from the ovary. Locate the outer limit of the pericarp and the limit of the endocarp. An ovary can be simple (have one chamber) or can be compound (have more than one chamber). What type of ovary does an apple have? ______________

   How could an animal, such as a deer, help disperse the seeds of an apple? ______________

   ____

   Use the apple as an example of a fruit in Table 22.3.

3. Examine the pod of a string bean or pea plant. How many seeds (beans or peas) are in the pod? ____________________ Would it help disperse the seeds of a pea plant if an animal were to eat the peas? ____________ Why or why not? ____

Split the pea or bean and look for the embryo. Is this plant a monocot or eudicot plant? __________

How do you know? ____

Use a pea pod as an example of a fruit in Table 22.3.

4. Examine a sunflower fruit and remove the seed. The outer coat of a sunflower seed is actually ________ ________________________. How can examining the seed tell you that the sunflower plant is a eudicot? ____

Except for the apple, all the fruits you have examined are dry fruits. What does this mean? __________ ____

Add sunflower fruit to Table 22.3.

5. Examine other available fruits and complete Table 22.3.

Table 22.3 Identification of Simple Fruits

Common Name	Fleshy or Dry?	Eaten as a Vegetable, Fruit, Other?	Type of Fruit (from Dichotomous Key)
1			
2			
3			
4			
5			
6			
7			
8			
9			
10			

22.5 Seeds

The seeds of flowering plants develop from ovules. As noted previously in Figures 22.6 and 22.9, a seed contains an embryonic plant, stored food, and a seed coat. Monocot seeds have one **cotyledon** (seed leaf); eudicots have two cotyledons.

Observation: Eudicot and Monocot Seeds

Bean Seed

1. Obtain a presoaked bean seed (eudicot). Carefully dissect it, using Figure 22.11 to help you identify the following:

   a. Seed coat: The outer covering. Remove the seed coat with your fingernail.

   b. Cotyledons: Food storage organs. The endosperm was absorbed by the cotyledons during development. What is the function of cotyledons? ____

   ____

   c. Epicotyl: The small portion of the embryo located above the attachment of the cotyledons. The first true leaves (**plumules**) develop from the epicotyl.

   d. Hypocotyl: The small portion of embryo located below the attachment of the cotyledons. The lower end develops into the embryonic root, or **radicle.**

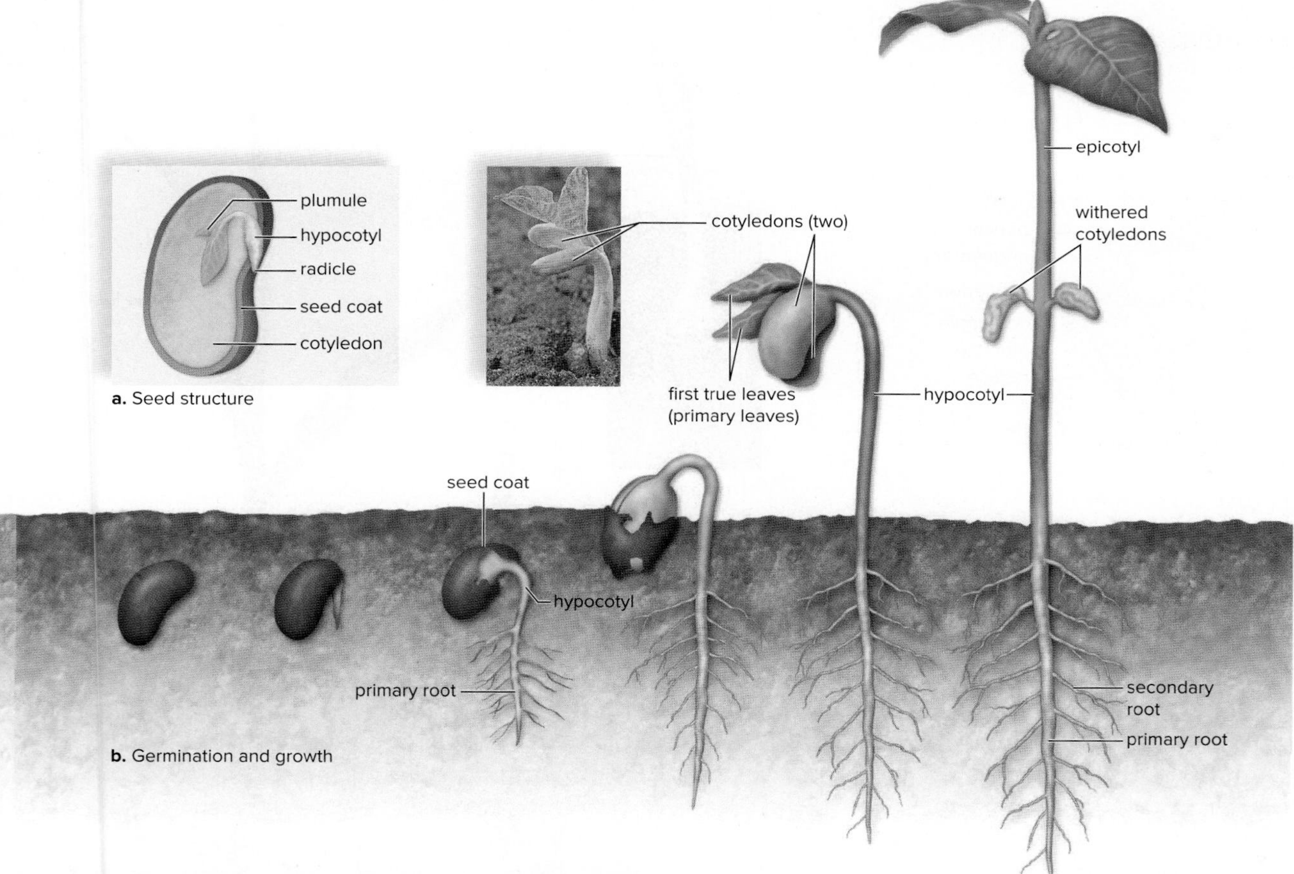

Figure 22.11 Eudicot seed structure and germination.
a. A seed contains an embryo, stored food, and a seed coat as exemplified by a bean seed. A eudicot seed has two cotyledons. Following (**b**) germination, a bean seedling grows to become a mature plant.
(top center) ©Ed Reschke

2. As observed in Figure 22.11, which organ emerges first from a dicot seed—the plumule or the radicle (i.e., root)? ______________________________
Of what advantage is this to the plant? ______________________________
____

3. The hypocotyl is the first part to emerge from the soil. What is the advantage of the hypocotyl pulling the plumule up out of the ground instead of pushing it up through the ground? ______________________________
____

4. Do cotyledons stay beneath the ground in eudicots? ______________________________

Corn Kernel

1. Obtain a presoaked corn kernel (monocot). Lay the seed flat, and with a razor blade or scalpel, carefully slice it in half. A corn kernel is a fruit, and the seed coat is tightly attached to the pericarp (Fig. 22.12).
2. Identify the cotyledon, plumule, and radicle. In addition, identify the
   a. **Endosperm:** Stored food for the embryo; the nutrients pass into the cotyledon as the seedling grows.
   b. **Coleoptile:** A sheath that covers the emerging leaves.
3. As observed in Figure 22.12, does the cotyledon of a corn seed, our example of a monocot, stay beneath the ground? ______________________________

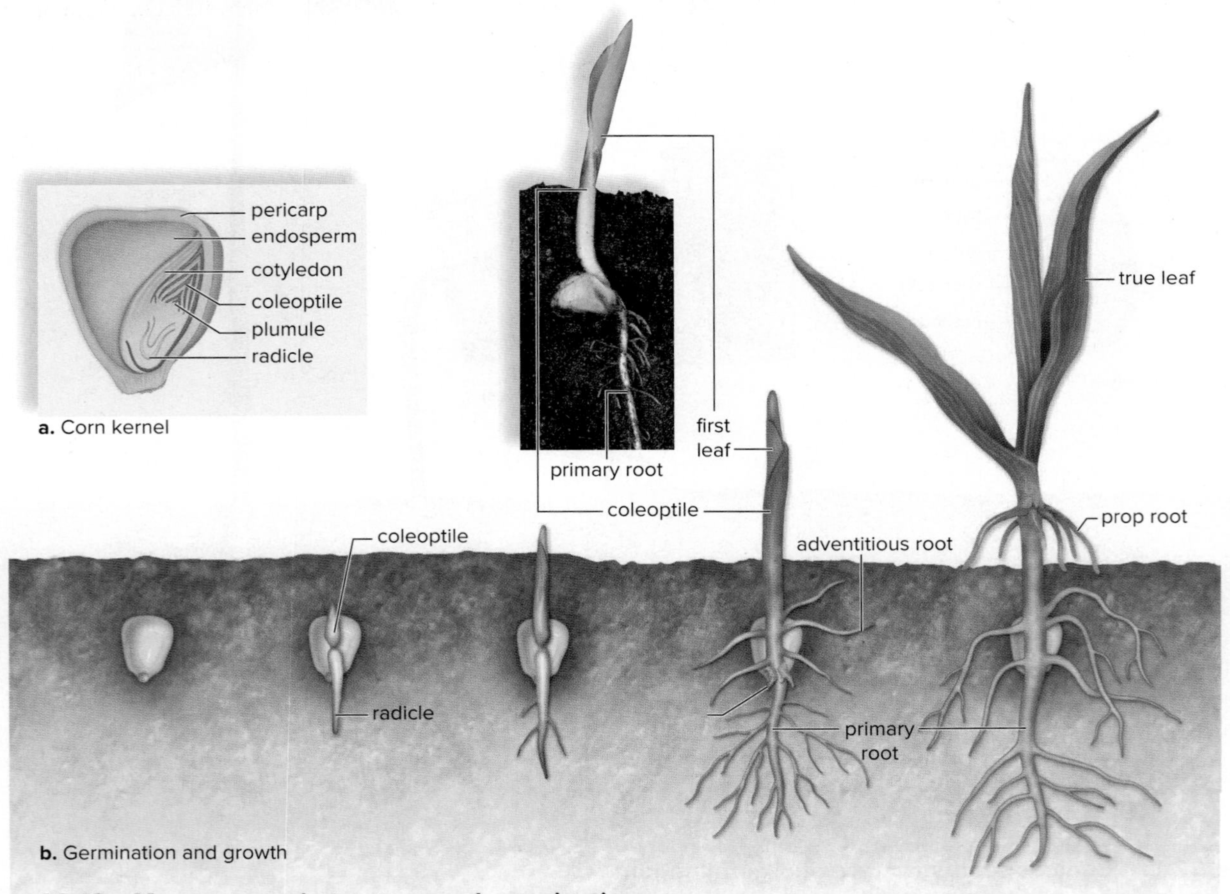

Figure 22.12 Monocot seed structure and germination.
a. A monocot seed has only one cotyledon as exemplified by a corn kernel. A corn kernel is a fruit—the seed is covered by a pericarp. Following (**b**) germination, a corn seedling grows to become a mature plant.
(top center) ©Ed Reschke/Getty Images

Seed Germination

Mature seeds contain an embryo that does not resume growth until after germination, which requires the proper environmental conditions. Mature seeds are dry, and for germination to begin, the dry tissues must take up water in a process called **imbibition.** After water has been imbibed, enzymes break down the food source into small molecules that can provide energy or be used as building blocks until the seedling is ready to photosynthesize.

Experimental Procedure: Seed Germination

Your instructor has placed sunflower seeds in five containers and watered them. The seedlings are in increasing stages of growth.

1. Order the seedlings in a series according to increasing stages of growth. What criteria did you use to order the seedlings? ______________________________

____

2. Is this plant a monocot or a eudicot? ______________ What criteria did you use to decide? __________

____

3. Can you see the cotyledons? ____________ Explain why the cotyledons of a eudicot seedling shrivel as the seedling grows. ______________________________

4. Use the space below to draw two contrasting stages of growth and add these labels to your drawings: hypocotyl, cotyledons, epicotyl, leaves, stem, terminal bud, node.

Laboratory Review 22

_______________ 1. In what flower structures are pollen grains produced?

_______________ 2. What do the stigma, style, and ovary of a flower collectively form?

_______________ 3. What do a flower's ovule and ovary become?

_______________ 4. What kind of angiosperm makes flowers with parts in threes or multiples of three?

_______________ 5. What is the haploid generation of the alternation of generations called?

_______________ 6. What process occurs to forms gametes in plants?

_______________ 7. What is the name of the female gametophyte of flowering plants?

_______________ 8. What process occurs to form microspores and megaspores in flowering plants?

_______________ 9. What transports sperm from the flower's stigma to the embryo sac?

_______________ 10. What does the fusion of the polar nuclei and sperm form, and how many n is it?

_______________ 11. What is the likely pollinator for a night blooming, white flower with a strong scent?

_______________ 12. What is an example of a simple fruit that has a fleshy endocarp?

_______________ 13. What term describes a fruit that develops from more than one ovary?

_______________ 14. What is the name for the part of the embryo above the attachment of the cotyledons?

Thought Questions

15. If a flower has five united petals, five anthers, and is red in color without a scent, is it made by a monocot or a eudicot plant, and what is the mostly likely pollinator for its flowers? Justify your choice.

16. Why is the relationship between a flowering plant and its pollinator considered mutualistic?

17. How do angiosperms increase the likelihood that their embryos are moved away from the parent plant? Why does moving the embryos away from the parent plant increase the success of the embryo?

LABORATORY

23

Introduction to Invertebrates

Learning Outcomes

Introduction

- Distinguish between invertebrate and vertebrate animals.

23.1 Evolution of Animals

- Discuss the evolution of animals in terms of symmetry, number of tissue layers, and pattern of development.

23.2 Sponges

- Identify a sponge and describe the anatomy of a sponge.
- Show that the anatomy and behavior of a sponge aids its survival and ability to reproduce.

23.3 Cnidarians

- Identify various types of cnidarians, and describe the anatomy of *Hydra* in particular.
- Show that the anatomy and behavior of *Hydra* aid its survival and ability to reproduce.

23.4 Flatworms

- Identify various flatworms, both free-living and parasitic forms. Contrast their lifestyles.
- Contrast the structure of a planarian with that of *Hydra,* and show that each is adapted to a particular way of life.

23.5 Roundworms

- Identify a roundworm, and contrast the anatomy of a roundworm with that of a planarian.
- Show that roundworms have successfully exploited various ways of life, including the parasitic way of life.

23.6 Rotifers

- Identify rotifers and describe their anatomy and behavior.

Introduction

In our survey of the animal kingdom, we will see that animals are very diverse in structure. Even so, all animals are multicellular and **heterotrophic,** which means their food consists of organic molecules made by other organisms. Consistent with the need to acquire food, animals achieve locomotion by the use of muscle fibers. Animals are always diploid, and during sexual reproduction, the embryo undergoes specific developmental stages.

While we tend to think of animals in terms of **vertebrates** (e.g., dogs, fishes, squirrels), which have a backbone, most animal species are those that lack a backbone, commonly known as **invertebrates.** In this laboratory, we will examine those invertebrates that lack a true body cavity, called a **coelom.** A survey of the rest of the animal kingdom is covered in other labs.

Planning Ahead To see hydra and planarians feed, have students observe the animals at the start of lab, add food, and then check frequently until food engulfment occurs.

23.1 Evolution of Animals

Today, molecular data is used to trace the evolutionary history of animals. These data tell us, as shown in the phylogenetic (evolutionary) tree (Fig. 23.1), that all animals share a common ancestor. This common ancestor was most likely a colonial protist consisting of flagellated cells. All but one of the phyla depicted in the tree consist of only invertebrates—the phylum Chordata contains a few invertebrates and also the vertebrates.

Figure 23.1 Evolutionary tree of animals.

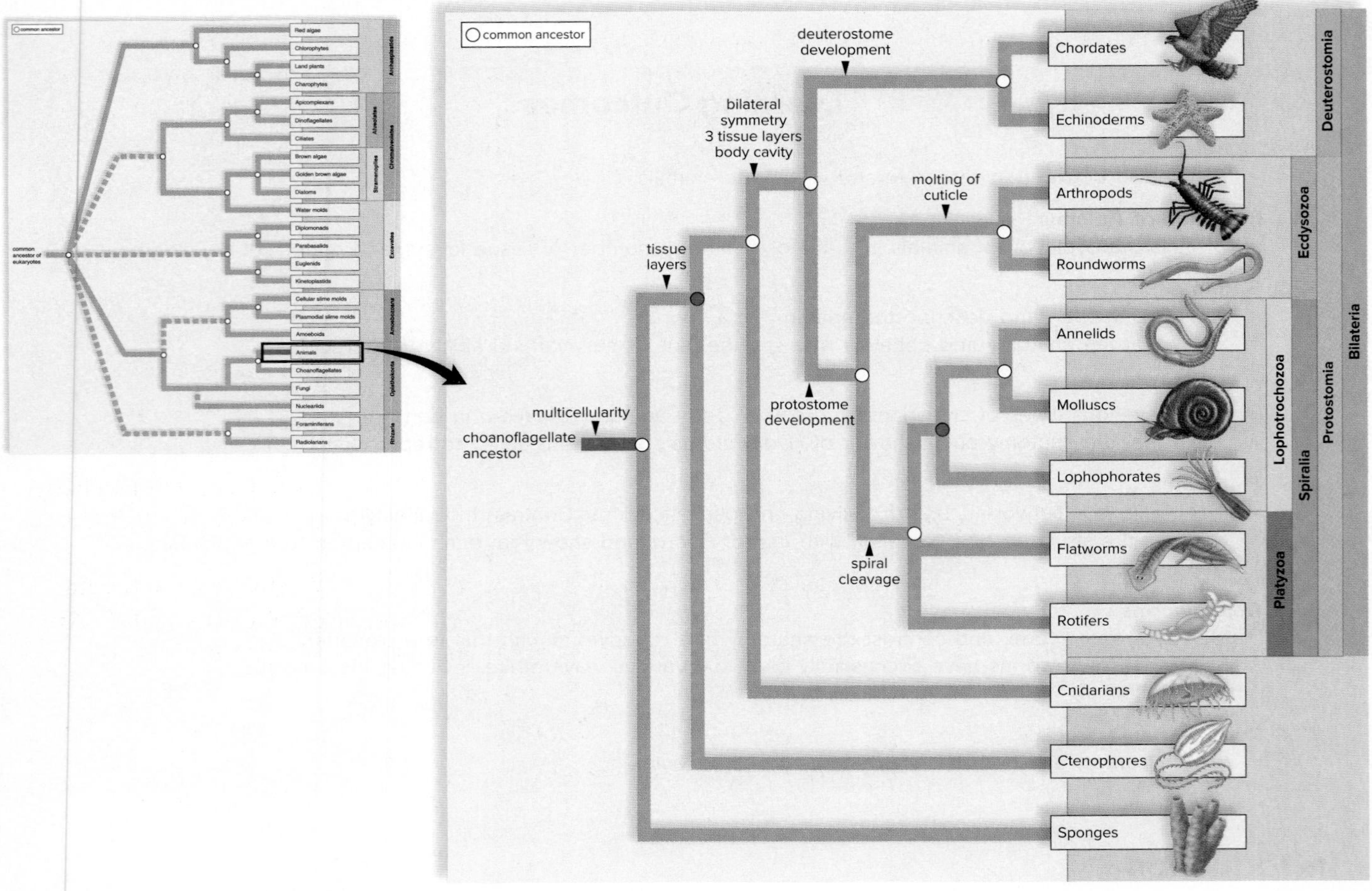

Certain anatomical features of animals are used in the tree. One feature of interest is formation of tissue layers. Sponges have no true tissue layers. Which phyla in the tree have only two tissue layers? ______________________________ The other phyla have three tissue layers.

Another feature of interest is **symmetry. Asymmetry** means the animal has no particular symmetry. Which phyla have **radial symmetry,** in which, as in a wheel, two identical halves are obtained no matter how the animal is longitudinally sliced? ______________________________ The other phyla have **bilateral symmetry,** which means the animals have one plane of symmetry with a definite right half and left half. Animals in phylum Echinodermata are bilateral in the larval form, but they have radial symmetry as adults.

Finally, complex animals are either **protostomes** or **deuterostomes.** Protostomes have a pattern of development in which the first opening of the embryo forms the mouth. The second opening of the embryo forms the mouth in deuterostomes; the first opening becomes the anus. Which pattern of development do the flatworms, rotifers, and roundworms (animals included in this laboratory) have? ______________________________

23.2 Sponges

Sponges (phylum Porifera) live in water, mostly marine, attached to rocks, shells, and other solid objects. An individual sponge is typically shaped like a tube, cup, or barrel. Sponges grow singly or in colonies whose overall appearances vary widely. A single sponge can become a colony by asexual budding.

Anatomy of Sponges

Sponges consist of loosely organized cells and have no well-defined tissues. They are asymmetrical or radially symmetrical and **sessile** (immotile). They can reproduce asexually by budding or fragmentation, but they also reproduce sexually by producing eggs and sperm.

Sponges have a few types of specialized cells. Most notably they have flagellated **collar cells (choanocytes)**. The movement of their flagella keeps water moving through the pores into the central cavity and out the osculum of a sponge (Fig. 23.2). Collar cells also take in suspended food particles from the water and digest them for the benefit of all the other cells in a sponge.

Observation: Anatomy of Sponges

Preserved Sponge

1. Examine a preserved sponge (Fig. 23.2*a*). Note the main excurrent opening (**osculum**) and the multiple incurrent pores. Water is constantly flowing in through the pores and out the osculum.
*Label the arrows in Figure 23.2*a *to indicate the flow of water.* Use the labels *water out* and *water in* through pores.
2. Examine a sponge specimen cut in half. Note the central cavity and the sponge wall. The wall is convoluted in some sponges, and the pores line small canals. Does this particular sponge have pore-lined canals? ______
3. You may be able to see **spicules,** fine projections over the body and especially encircling the osculum. Does this sponge have spicules? ______________________________

Figure 23.2 Sponge anatomy.
a. Movement of water through pores into the central cavity and out the osculum is noted. **b.** Collar cells line the central cavity of a sponge, and the movement of their flagella keeps the water moving through the sponge. **c.** Draw an enlargement of spicules here.

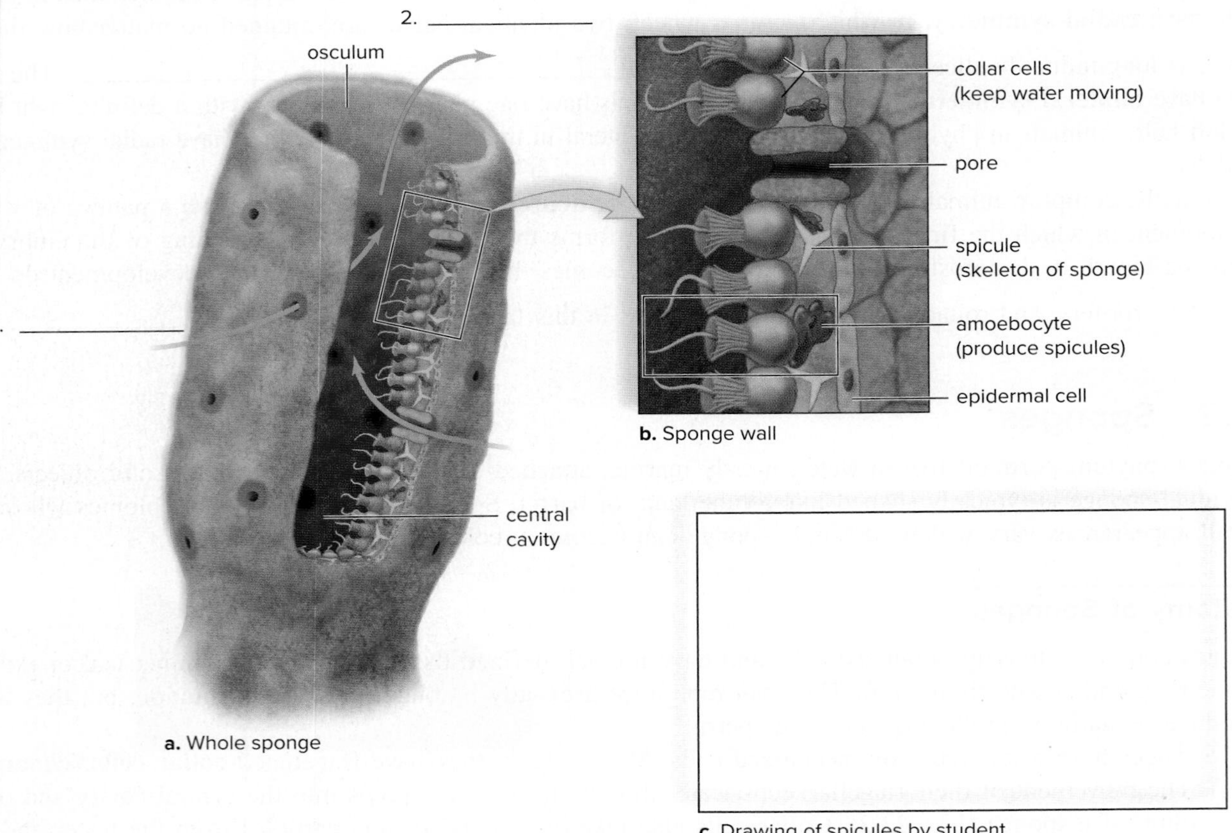

a. Whole sponge

b. Sponge wall

c. Drawing of spicules by student

Prepared Slides

1. Examine a prepared slide of *Grantia*.
   a. Find the collar cells that line the interior (Fig. 23.2*b*). A sponge is a **sessile filter feeder.** Collar cells phagocytize (engulf) tiny bits of food that come through the pores along with the water flowing through the sponge. They then digest the food in food vacuoles. Explain the expression *sessile filter feeder.* ____________
   ____
   b. Do you see any spicules? Do they project from the wall of a sponge? ____________
   c. Depending on the sponge, spicules are made of either calcium carbonate, silica (glass), or protein. Calcium carbonate and silica produce hard, sharp spicules. Name two possible advantages of spicules to a sponge. ____________
   ____
2. Examine a prepared slide of sponge spicules. What do you see? ____________
   *Draw a sketch of four spicules, each having a different appearance, in the space provided in Figure 23.2*c.

Diversity of Sponges

Sponges are very diverse and come in many shapes and sizes. Some sponges live in fresh water, although most live in the sea and are a prominent part of coral reefs, areas of abundant sea life discussed in the next section. Zoologists have described over 5,000 species of sponges, which are grouped according to the type of spicule (Fig. 23.3).

a. Calcareous sponge, *Clathrina canariensis*

b. Bath sponge, *Xestospongia testudinaria*

c. Glass sponge, *Euplectella aspergillum*

Figure 23.3 Diversity of sponges.
a. Calcareous (chalk) sponges have spicules of calcium carbonate. **b.** Bath sponges have a skeleton of spongin. **c.** Glass sponges have glassy spicules.

Conclusions: Anatomy of Sponges

- The anatomy and behavior of a sponge aid its survival and its ability to reproduce. How does a sponge
  - **a.** protect itself from predators? ______________________________
  - **b.** acquire and digest food? ______________________________
  - **c.** reproduce asexually and sexually? ______________________________

23.3 Cidarians

Cnidarians (phylum Cnidaria) are tubular or bell-shaped animals that live in shallow coastal waters, except for the oceanic jellyfishes. Two basic body forms are seen among cnidarians. The mouth of a **polyp** is directed upward, while the mouth of a jellyfish, or **medusa,** is directed downward. At one time, both body forms may have been a part of the life cycle of all cnidarians. When both are present, the sessile polyp stage produces medusae, and this motile stage produces egg and sperm (Fig. 23.4). Today in some cnidarians, one stage is dominant and the other is reduced; in other species, one form is absent altogether. Regardless, all cnidarians are radially symmetrical. How can radial symmetry be a benefit to an animal? ______________________________

____

How can a life cycle that involves two forms, called **polymorphism,** be of benefit to an animal, especially if one stage is sessile (stationary)? ______________________________

____

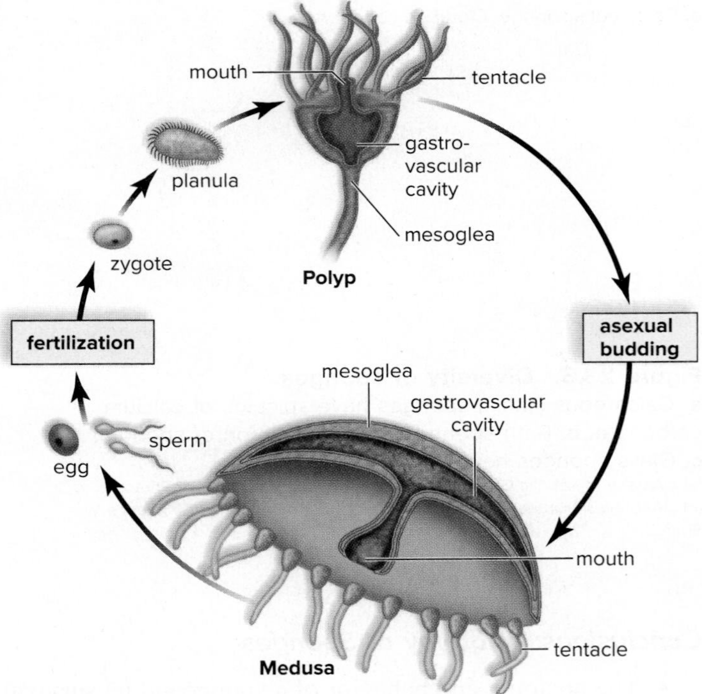

Figure 23.4 The life cycle of a cnidarian. Some cnidarians have both a polyp stage and a medusa stage; in others, one stage may be dominant or absent altogether.

Anatomy of *Hydra*

Figure 23.5 shows the anatomy of a hydra, which will be studied as a typical cnidarian. Hydras exist only as sessile polyps; there is no alternate stage. Note the **tentacles** that surround the **mouth,** the large **gastrovascular cavity,** and the basal disk. A gastrovascular cavity has a single opening that is used as both an entrance for food and an exit for wastes.

Figure 23.5 Anatomy of *Hydra*.
Hydra typifies the anatomy of a cnidarian.
(left) ©NHPA/M.I. Walker/Photoshot RF

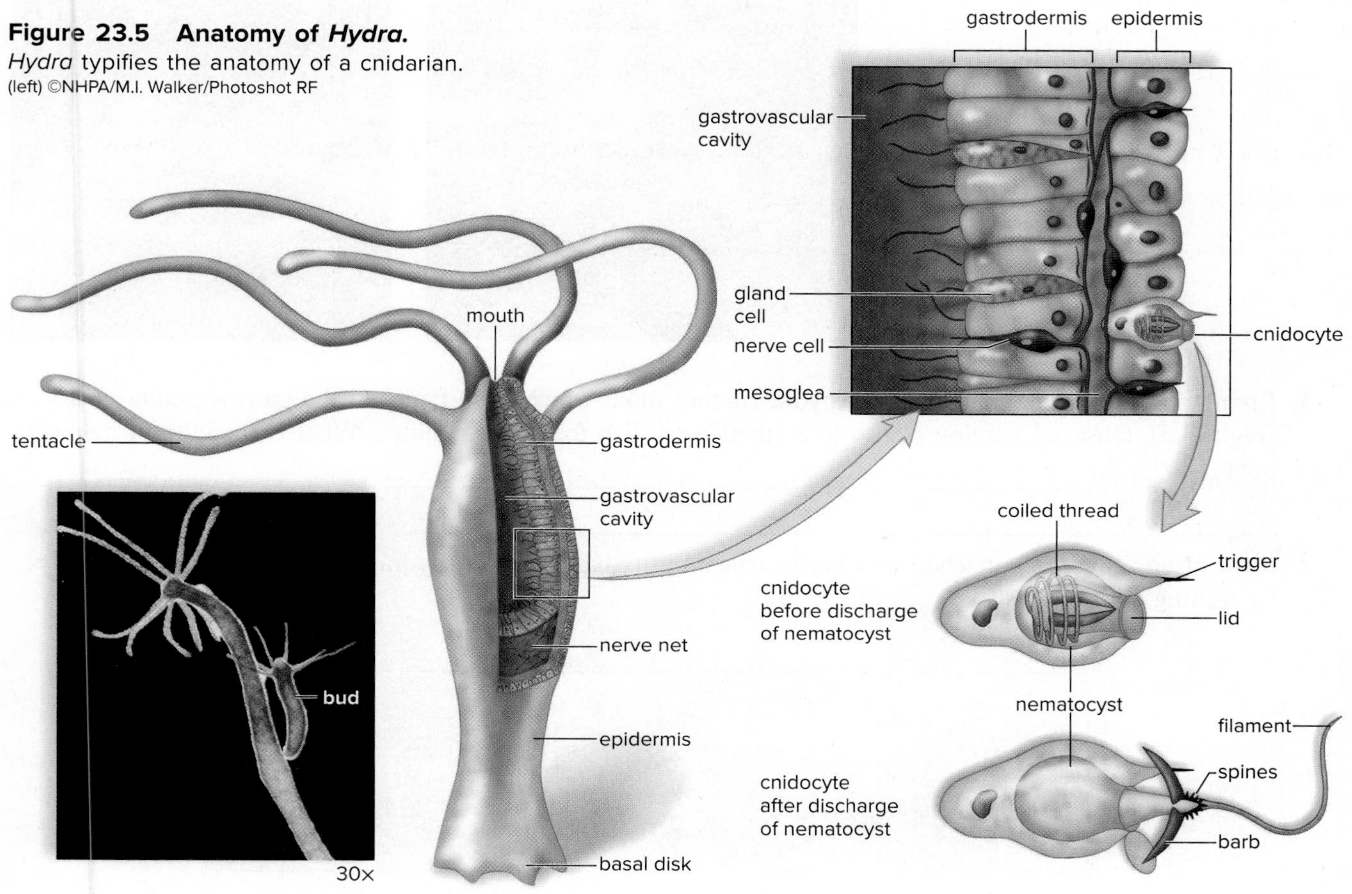

Observation: Anatomy of Hydra

1. Preserved *Hydra.* With the aid of a hand lens, examine preserved specimens of *Hydra.* Hydras typically reproduce asexually by budding. Do you see any evidence of buds that are developing directly into small hydras? __________ The body wall can also produce ovaries and testes that produce eggs and sperm. The testes are generally located near the attachment of the tentacles; the ovaries appear farther down on the trunk, toward the basal disk.
2. Slide of *Hydra.* Examine prepared slides of cross and longitudinal sections of *Hydra.* With the help of Figure 23.5, note the epidermis, the mesoglea (a gelatinous material between the two tissue layers), and the gastrodermis, which lines the gastrovascular cavity. Switch to high power. Do you find any cells? __________ Describe them. ____

____

Figure 23.6 Hydra feeding.
(both) ©Roland Birke/Getty Images

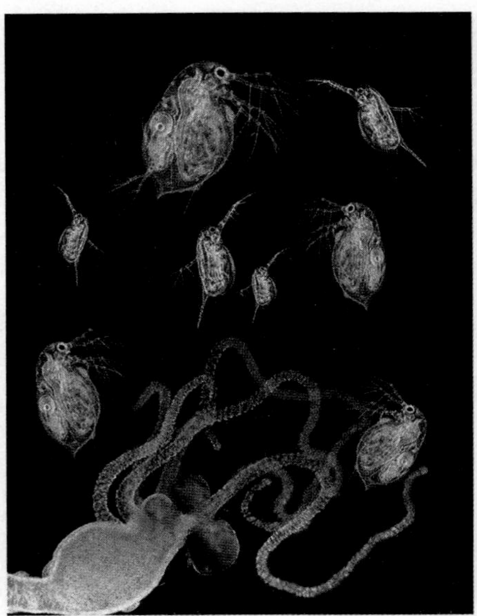

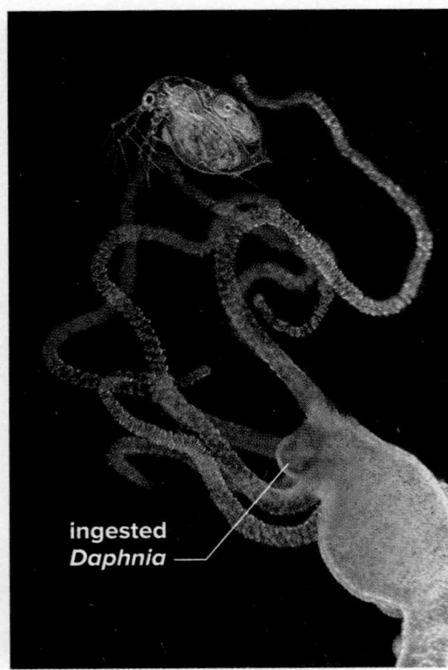

3. Living specimen. The tentacles of a hydra capture food, which is stuffed into the gastrovascular cavity (Fig. 23.6). Observe a living *Hydra* in a small petri dish for a few minutes. What is the current behavior of your hydra? ______________________________

____

Most often a hydra is attached to a hard surface by its basal disk. A hydra can move, however, by turning somersaults:

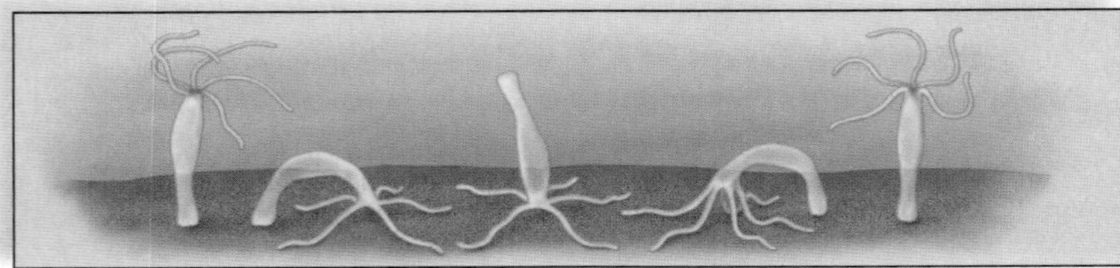

After a few minutes, tap the edge of the petri dish. What is the reaction of your hydra? ______________

____

____

4. Mount a living *Hydra* on a depression glass slide with a coverslip and examine a tentacle. Unique to cnidarians are specialized stinging cells, called **cnidocytes,** which give the phylum its name. Each cnidocyte has a fluid-filled capsule called a **nematocyst** (see Fig. 23.5, far right), which contains a long, spirally coiled hollow thread. The threads trap and/or sting prey. Note the cnidocytes as swellings on the tentacles. Add a drop of vinegar (5% acetic acid) and note what happens to the cnidocytes. Did your hydra discard any nematocysts? __________ Describe. ______________________________

____

____

Of what benefit is it to *Hydra* to have cnidocytes? ______________________________

____

Conclusions: Anatomy of Cnidarians

- The anatomy and behavior of a hydra aid its survival and its ability to reproduce. How does a hydra
  - **a.** acquire and digest food? ______________________________
  - **b.** protect itself from predators? ______________________________
  - **c.** reproduce asexually and sexually? ______________________________

Diversity of Cnidarians

Cnidarians consist of a large number of mainly marine animals (Fig. 23.7). Sea anemones, sometimes called the flowers of the sea, are solitary polyps often found in coral reefs, areas of biological abundance in shallow tropical seas. Stony corals have a calcium carbonate skeleton that contributes greatly to the building of coral reefs. Portuguese man-of-war is a colony of modified polyps and medusae. Jellyfishes are a part of the zooplankton, suspended animals that serve as food for larger animals in the ocean.

a. Sea anemone, *Corynactis*

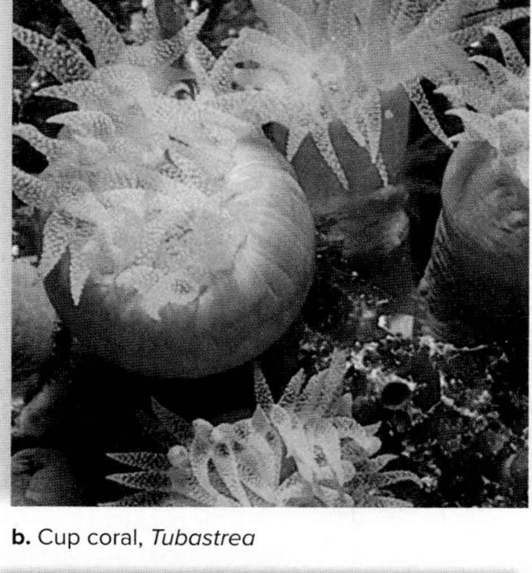

b. Cup coral, *Tubastrea*

c. Portuguese man-of-war, *Physalia*

d. Jellyfish, *Aurelia*

Figure 23.7 Cnidarian diversity.
(a) ©Danita Delimont/Alamy Stock Photo RF; (b) ©Ron Taylor/Bruce Coleman/Photoshot; (c) Image courtesy of Islands in the Sea 2002, NOAA/OER; (d) ©Ermis Koukaris/Shutterstock

23.4 Flatworms

Flatworms (phylum Platyhelminthes) are bilaterally symmetrical animals that can be either free-living or parasitic. Free-living flatworms, called planarians, are more complex than the cnidarians. In addition to the germ layers ectoderm and endoderm, mesoderm is also present. Flatworms are usually **hermaphroditic,** which means they possess both male and female sex organs (Fig. 23.8*c*). Why is it advantageous for an animal to be hermaphroditic? ___

___

Planarians practice cross-fertilization when the penis of one is inserted into the genital pore of the other. The fertilized eggs are enclosed in a cocoon and hatch as tiny worms in 2 or 3 weeks.

Planarians

Planarians such as *Dugesia* live in lakes, ponds, and streams, where they feed on small living or dead organisms. In planarians, the three germ layers give rise to various organs aside from the reproductive organs (Fig. 23.8). The three-part **gastrovascular cavity** ramifies throughout the body. Excretory organs called **flame cells** (because their cilia reminded early investigators of a flickering flame of a candle) collect fluids from inside the body and send them via a tube to an excretory pore. The nervous system contains a brain and lateral nerve cords connected by transverse nerves, which is why it is called **ladderlike.** *Complete the labels in both 23.8*b *and* d. *Label excretory canal, brain, and nerve cord.* Why would you expect an animal that lives in fresh water to have a well-developed excretory system? ___

___

Figure 23.8 Planarian anatomy.
a. When a planarian extends the pharynx, food is sucked up into a gastrovascular cavity that branches throughout the body. **b.** The excretory system has flame cells. **c.** The reproductive system has both male (blue) and female (pink) organs. **d.** The nervous system looks like a ladder.

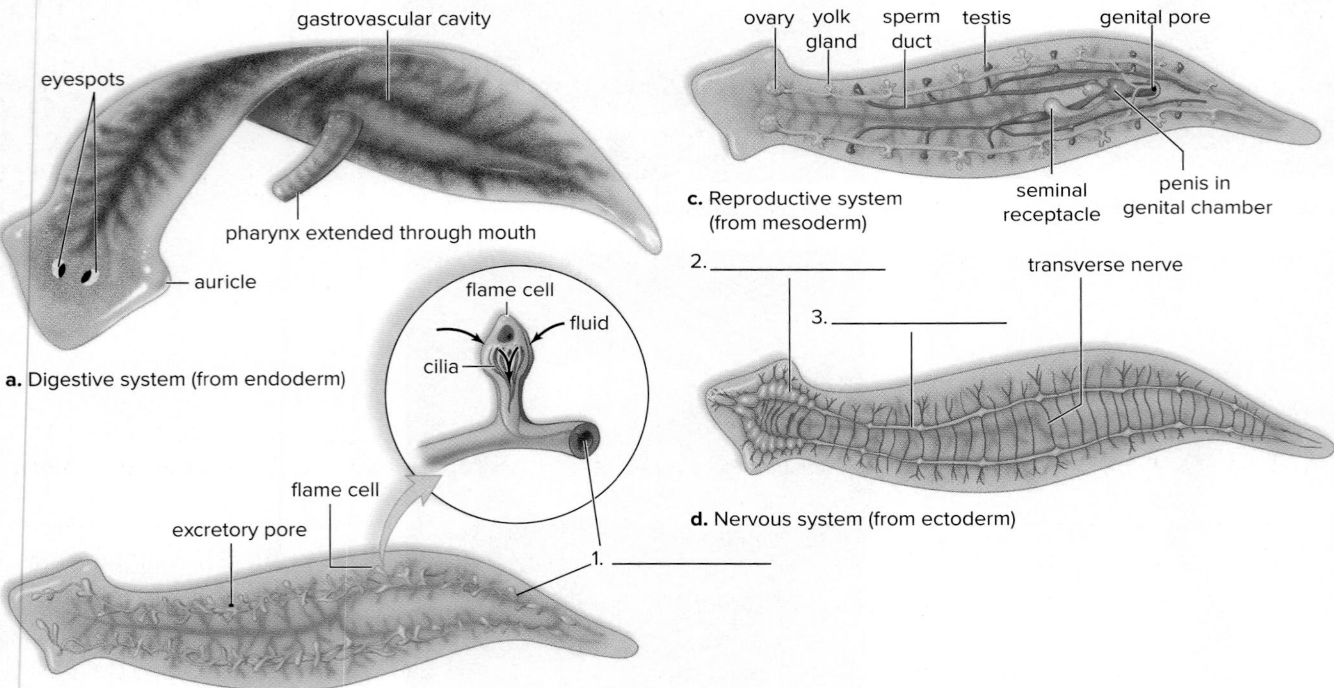

Observation: Planarians

Whole Mount

Examine a whole mount of a planarian that shows the branching gastrovascular cavity (Fig. 23.8*a*). What is the advantage of a gastrovascular cavity that ramifies through the body? ______________________

____

Prepared Slide

Examine a cross section of a planarian under the microscope. Can you locate the structures shown in Figure 23.9? Does a planarian have a body cavity?__________ Explain. ______________________

____

Figure 23.9 Planarian micrograph.
Cross section of a planarian at the pharynx.
©Carolina Biological Supply Company/Phototake

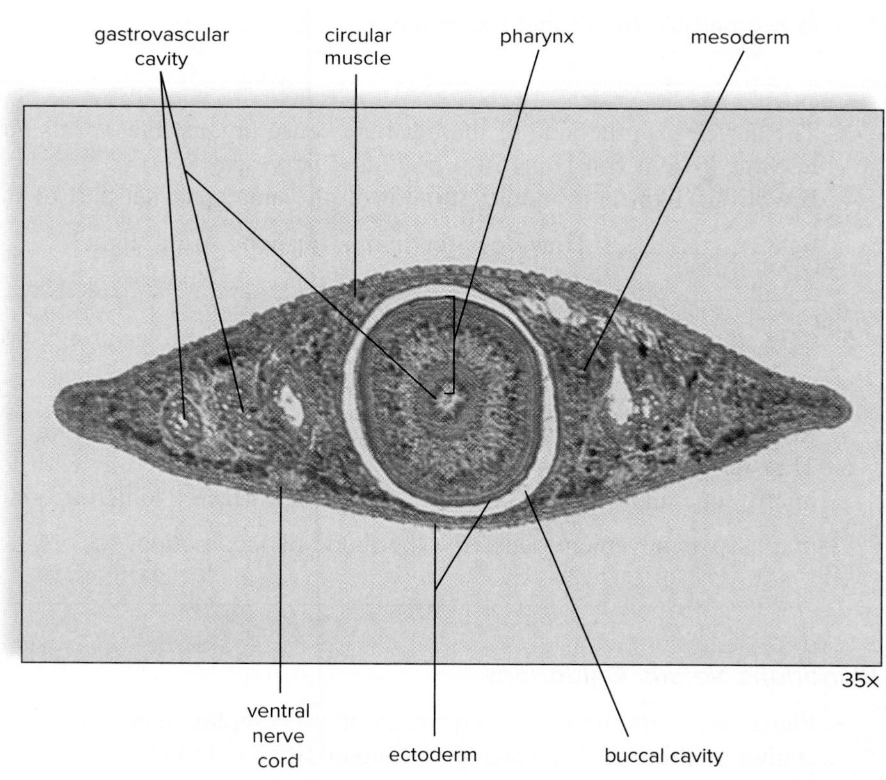

Living Planarian

1. Examine the behavior of a living planarian (Fig. 23.10) in a petri dish. Describe the behavior of the animal.

   ____

2. Does the animal move in a definite direction?__________ Why is it advantageous for a predator such as a planarian to have bilateral symmetry and a definite head region? ______________________

   ____

Figure 23.10 Micrograph of planarian feeding and the gastrovascular cavity. The pharynx protrudes through the mouth and sucks up food when a planarian feeds.

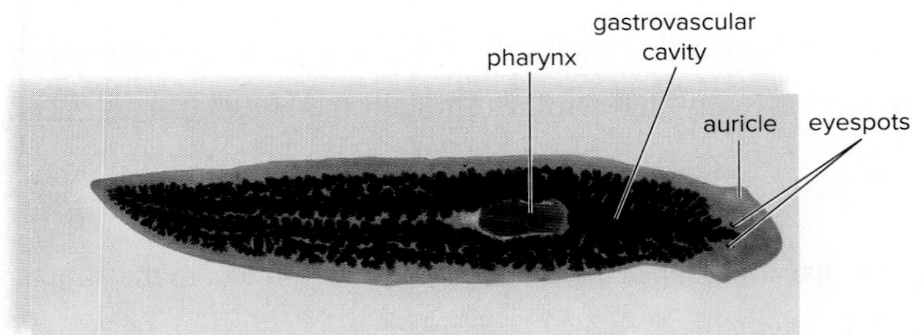

3. Gently touch the animal with a probe. What three types of cells must be present for flatworms to be able to respond to stimuli and move about? ______________________

____

The auricles on the side of the head are sense organs. Flatworms have well-developed muscles and a nervous system consisting of a brain and nerves.

4. If a strong light is available, shine it on the animal. What part of the animal would be able to detect light? __________ How does the animal respond to the light? ______________________

____

5. Offer the worm some food, such as a small piece of liver or egg yolk, and describe its manner of eating.

____

Roll the animal away from its food and note the pharynx extending from the body (Fig. 23.10).

6. Transfer your worm to a concave depression slide and cover with a coverslip. Examine with a microscope and note the cilia on the ventral surface. Numerous gland cells secrete a mucous material that assists movement. Describe the mode of locomotion. ______________________

____

Planarians Versus Cnidarians

- Planarians, with three germ layers, are more complex than cnidarians. Contrast a hydra with a planarian by stating in Table 23.1 the significant organ differences between them.
- Planarians have no respiratory or circulatory system. As with cnidarians, each individual __________ takes care of its own needs for these two life functions.

Table 23.1 Contrasts Between a Hydra and a Planarian

	Digestive System	Excretory System	Nervous Organization
Hydra			
Planarian			

Tapeworms

Tapeworms are parasitic flatworms known as cestodes. They live in the intestines of vertebrate animals, including humans (Fig. 23.11). The worms consist of a **scolex** (head), usually with suckers and hooks, and **proglottids** (segments of the body). Ripe proglottids detach and pass out with the host's feces, scattering fertilized eggs on the ground. If pigs or cattle happen to ingest these, larvae called bladder worms develop and eventually become encysted in muscle, which humans may then eat in poorly cooked or raw meat. A bladder worm that escapes from a cyst develops into a mature tapeworm attached to the intestinal wall.

1. How do humans get infected with the pig tapeworm? ___________________________

2. What is the function of a tapeworm's hooks and suckers? ___________________________

3. Why would you expect a tapeworm to have a reduced digestive system? ___________________________

4. Proglottids mature into "bags of eggs." Given the life cycle of the tapeworm, why might a tapeworm have an expanded reproductive system compared to a planarian? ___________________________

Figure 23.11 Life cycle of the tapeworm *Taenia*.
The secondary host (pig) is the means by which the worm is dispersed to the primary host (humans).

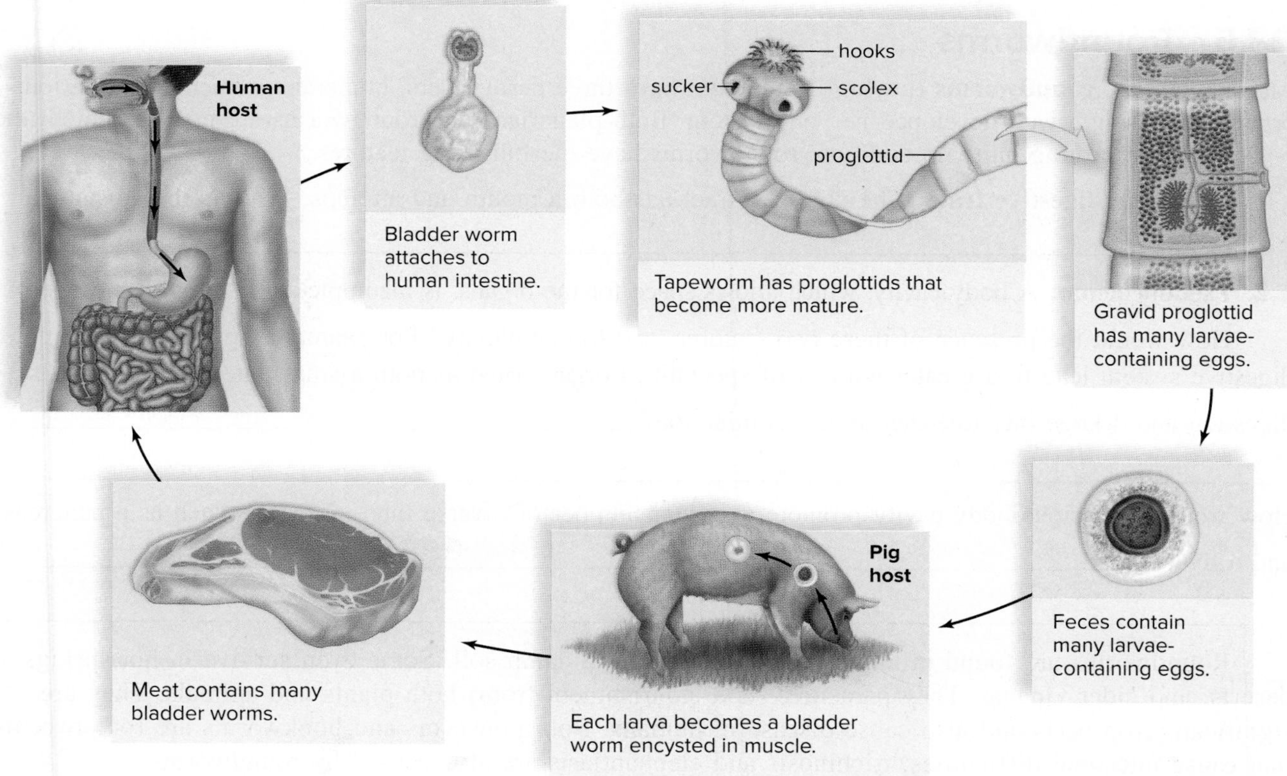

Observation: Tapeworm Anatomy

1. Examine a preserved specimen and/or slide of *Taenia pisiformis,* a tapeworm.
2. With the help of Figure 23.12, identify the scolex, with hooks and suckers, and the proglottids.

Figure 23.12 Anatomy of *Taenia*.
The adult worm is modified for its parasitic way of life. It consists of **a.** the scolex and **b.** many proglottids, which become bags of eggs. (a) ©C. James Webb/Phototake; (b) ©Ed Reschke/Getty Images

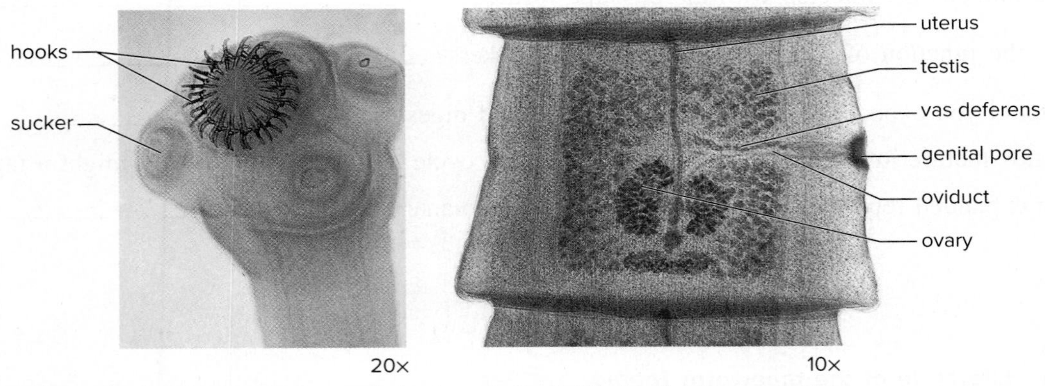

a. The scolex functions to attach worm to host.

b. Proglottids function in reproduction.

23.5 Roundworms

Like planarians, **roundworms** (phylum Nematoda) have three germ layers, bilateral symmetry, and various organs, including a well-developed nervous system. Both planarians and roundworms are nonsegmented; the body has no repeating units. In addition, roundworms have the following features:

1. **Complete digestive tract:** The digestive tract has both a mouth and an anus. What is the advantage of this? ______________________________
2. **Pseudocoelom:** A body cavity, which allows space for the organs, is incompletely lined with mesoderm.

How would the presence of these two features lead to complexity? For example, how would a complete digestive system lead to a greater number of specialized organs such as both a small intestine that assists digestion and a large intestine that assists elimination? ______________________________

____

How would a spacious body cavity promote a greater number of diverse internal organs such as a pancreas and a liver? ______________________________

____

Roundworms are found in all aquatic habitats and in damp soil. Some even survive in hot springs, deserts, and cider vinegar. They parasitize (take nourishment from) both plants and animals. They are significant crop pests and also cause disease in humans. Both pinworms and hookworms are roundworms that cause intestinal difficulties; trichinosis and elephantiasis are also caused by roundworms.

Ascaris

Ascaris, a large, primarily tropical intestinal parasite, is often studied as an example of this phylum.

Observation: Ascaris

Examine preserved specimens of *Ascaris,* both male and female (Fig. 23.13). In roundworms, the sexes are separate. The male is smaller and has a curved posterior end. Be sure to examine specimens of each sex.

Figure 23.13 Roundworm anatomy.
a. Photograph of male *Ascaris.* **b.** Male reproductive system. **c.** Photograph of female *Ascaris.* **d.** Female reproductive system.
(both) ©P&R fotos/age fotostock/Superstock

a. Male *Ascaris*

c. Female *Ascaris*

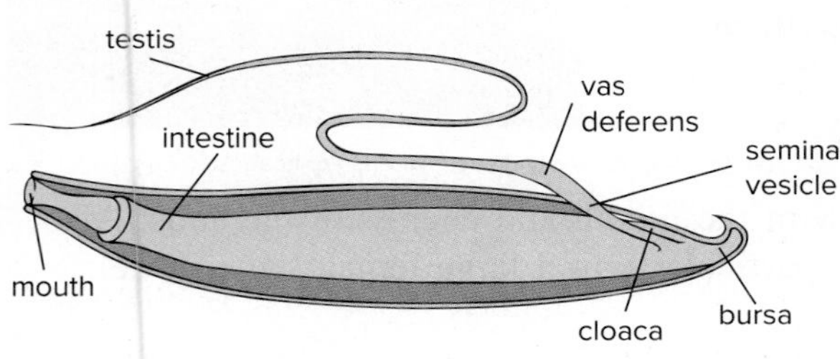

b. Male reproductive system

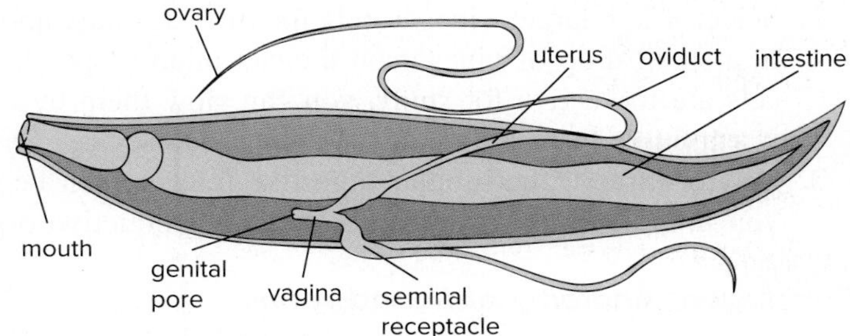

d. Female reproductive system

Trichinella

Trichinella is a parasitic roundworm that causes the disease **trichinosis.** When pigs or humans eat raw or undercooked pork infected with *Trichinella* cysts, juvenile worms are released in the digestive tract, where they penetrate the wall of the small intestine and mature sexually. After male and female worms mate, females produce juvenile worms that migrate and form cysts in various muscles (Fig. 23.14). A human with trichinosis has muscular aches and pains that can lead to death if the respiratory muscles fail.

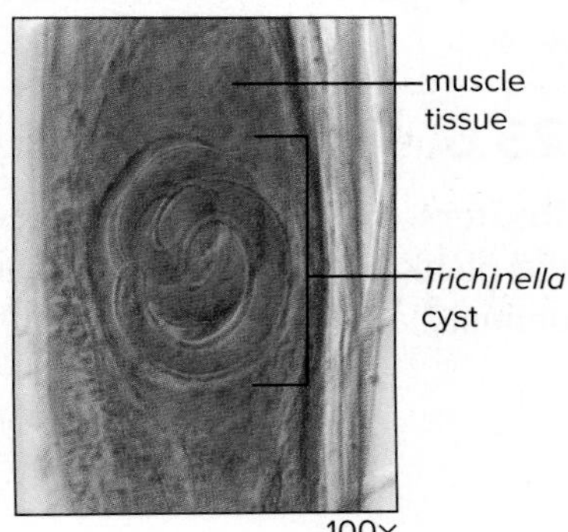

Figure 23.14 Larva of the roundworm *Trichinella* embedded in a muscle.
A larva coils in a spiral and is surrounded by a sheath derived from a muscle fiber.
©Carolina Biological Supply Company/Phototake

Observation: Trichinella

1. Examine preserved, infected muscle or a slide of infected muscle, and locate the *Trichinella* cysts, which contain the juvenile worms.
2. How can trichinosis be prevented in humans? ____

____
3. How can pig farmers help to stamp out trichinosis so that humans are not threatened by the disease? ____

____

Filarial Worm

A roundworm called a **filarial worm** infects lymphatic vessels and blocks the flow of lymph. The condition is called **elephantiasis** because when a leg is affected, it becomes massively swollen.

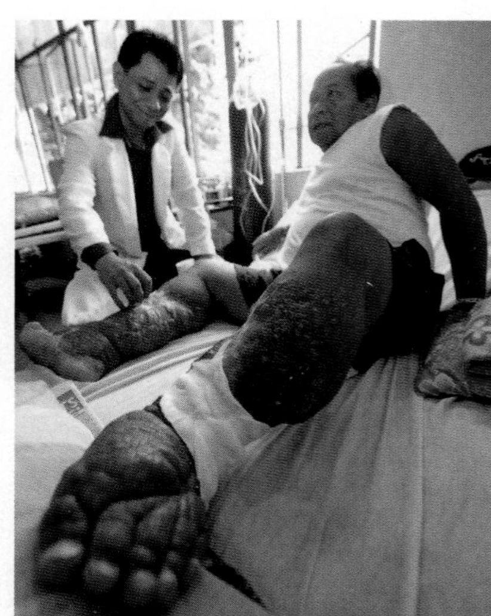

©Jay Directo/AFP/Getty Images

Living Vinegar Eels

Vinegar eels are tiny, free-living nematodes that can live in unpasteurized vinegar.

1. Examine live vinegar eels, and observe their active, whiplike swimming movements. This thrashing motion may be a result of nematodes having longitudinal muscles only; they lack circular muscles.
2. Select a few larger vinegar eels for further study, and place them in a small drop of vinegar on a clean microscope slide. If the eels are too active for study, you can slow them by briefly warming them or by adding methyl cellulose.
3. Try to observe the tubular digestive tract, which begins with the mouth and ends with the anus. Also, you may be able to see some of the reproductive organs, particularly in a large female vinegar eel.

Conclusion: Anatomy of Roundworms

- Nematodes are extremely plentiful, in terms of both their variety and their overall number. From your knowledge of adaptive radiation, explain why there might be so many different types of nematodes. ____

____

____

23.6 Rotifers

Rotifers (phylum Rotifera) are common and abundant freshwater animals. They are important constituents of the plankton of lakes, ponds, and streams, and are a significant food source for many species of fish and other animals.

Like roundworms, rotifers have a pseudocoelom and a complete digestive tract with a mouth and anus. The corona is a crown of cilia around the mouth. To some, the movement of the cilia resembles that of a rapid wheel; and in Latin, *rotifer* means "wheel-bearer." The cilia draw water into the mouth, and from there food is ground up by trophi (jaws) before entering the stomach from which nondigested remains pass through the cloaca and anus (Fig. 23.15).

Figure 23.15 Live *Philodina*, a common rotifer.
a. The contents of the digestive system take on a color in a live common rotifer.
b. Line art showing the complex anatomy of a rotifer.
(a) ©Blickwinkel/Alamy Stock Photo

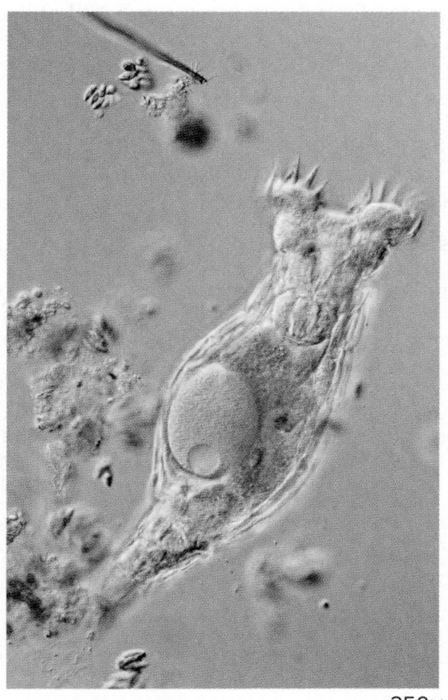

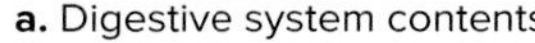

a. Digestive system contents 250×

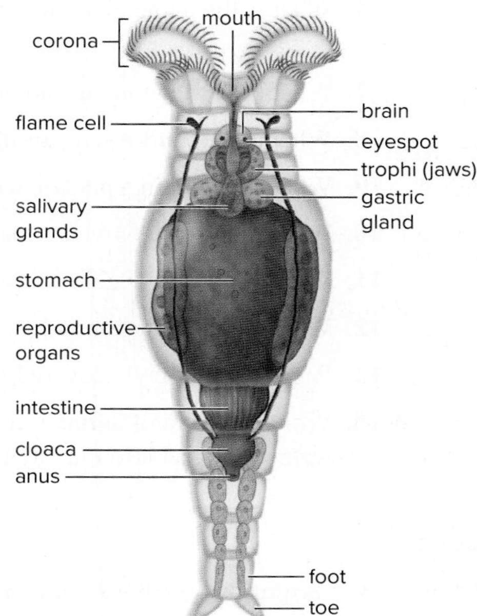

b. Complex anatomy

Observation: Rotifers

Living Rotifers

1. Use a pipet to obtain a living rotifer specimen near a clump of vegetation. Place the liquid and rotifer on a concave depression slide. Do not add a coverslip. Study the animal's behavior and appearance.
2. Describe the rotifer's behavior. ____

   ____
3. Observe the elongated, cylindrical body that can be divided into three general regions: the head, the trunk, and a posterior foot. The rotifers have no true segmentation, but the cuticle covering the body can be divided into a number of superficial segments.
4. Observe the telescoping of the segments when the animal retracts its head.
5. Note also the large, ciliated **corona** at the anterior end. Most rotifers have a conspicuous corona that serves both for locomotion and for feeding. It creates a current that brings smaller microorganisms (e.g., algae, protozoans, bacteria) close enough to be swallowed.

Laboratory Review 23

_______________ **1.** What do invertebrates lack that is present in vertebrates?

_______________ **2.** What is an incompletely lined body cavity called?

_______________ **3.** What type of symmetry is associated with one plane of symmetry and definite right and left halves?

_______________ **4.** What is the name of the main opening in a sponge through which water is expelled?

_______________ **5.** What are the hard and sharp projections of a calcereous sponge called?

_______________ **6.** What is the name given to the body form of a cnidarian that has a mouth directed upward?

_______________ **7.** What kind of symmetry do organisms classified as cnidarians have?

_______________ **8.** What is the fluid-filled capsule in a cnidocyte called?

_______________ **9.** What structure in a planarian digests food and ensures that nutrients get to all the cells?

_______________ **10.** What is the name given to the ciliated cells that perform excretion for a planarian?

_______________ **11.** What structures enable a tapeworm to attach itself to the intestinal wall of its host?

_______________ **12.** What are the anterior and posterior ends of a complete digestive tract called?

_______________ **13.** Which of the phyla covered in this lab exercise have complete digestive tracts?

_______________ **14.** Which phylum of animals covered in this lab exercise relies on a crown of cilia to draw water and food into the mouth?

Thought Questions

15. How does the process of acquiring and digesting food in cnidarians differ from the process in sponges?

16. The anterior end of an animal is often referred to as a head. The anterior end of a tapeworm is referred to as a scolex. Explain why a different word is used for the anterior end of the tapeworm.

17. a. Do the animals with a gastrovascular cavity have a body cavity?

b. How do the animals with a pseudocoelom digest their food?

c. What advantage is there to having a body cavity and a separate mouth and anus?

LABORATORY

24

Invertebrate Coelomates

Learning Outcomes

Introduction

- State the advantages of a coelom.

24.1 Molluscs

- Describe the general characteristics of molluscs and the specific features of selected groups.
- Identify and/or locate external and internal structures of a clam and a squid.
- Compare clam anatomy to squid anatomy and tell how each is adapted to its way of life.

24.2 Annelids

- Describe the general characteristics of annelids and the specific features of the three major groups.
- Identify and/or locate anatomical structures of an earthworm.
- Compare clam anatomy to that of an earthworm.

24.3 Arthropods

- Describe the general characteristics of arthropods.
- Identify and/or locate external and internal structures of the crayfish and grasshopper.
- Contrast the anatomy of the crayfish and the grasshopper, indicating how each is adapted to its way of life.
- Contrast complete and incomplete metamorphosis.

24.4 Echinoderms

- Describe the general characteristics of echinoderms.
- Identify and locate external and internal structures of a sea star.

Introduction

The animals studied in this laboratory are **coelomates** because they have a body cavity completely lined with mesoderm, the last of the germ layers to appear during the evolution of animals. A coelom offers many advantages such as the digestive system and body wall can move independently; internal organs can become more complex; coelomic fluid can assist respiration, circulation, and excretion; and, in some animals, the coelom also serves as a hydrostatic skeleton because muscles can work against a fluid-filled cavity.

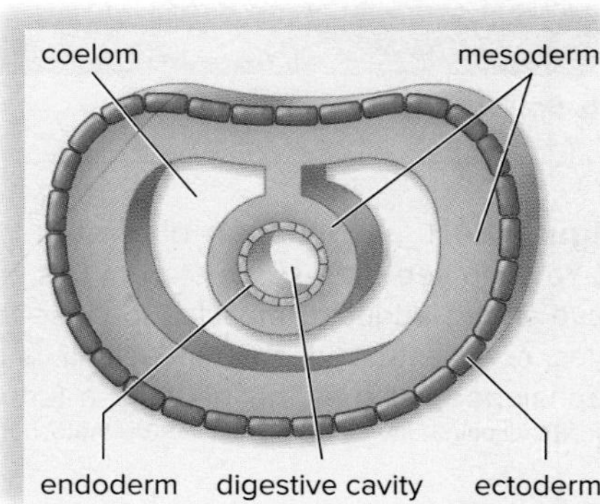

Coelomate (molluscs, annelids, arthropods, echinoderms, chordates)

Among the coelomates, the molluscs, annelids, and arthropods are **protostomes,** animals in which the first (*protos*) embryonic opening becomes the mouth (*stoma*), while the echinoderms and the vertebrates (studied in Laboratory 25) are **deuterostomes.** In the deuterostomes, the first opening becomes the anus, and the second (*deutero*) opening becomes the mouth. The evolutionary relationship between protostomes and deuterostomes is shown in Figure 24.1.

24.1 Molluscs

Most **molluscs** (phylum Mollusca) are marine, but there are also some freshwater and terrestrial molluscs (Fig. 24.1). Among molluscs, the grazing marine herbivores, known as **chitons,** have a body flattened dorsoventrally covered by a shell consisting of eight plates (Fig. 24.1*a*). The **bivalves** contain marine and freshwater sessile filter feeders, such as clams and scallops, with a body enclosed by a shell consisting of two valves (Fig. 24.1*b*). The **gastropods** contain marine, freshwater, and terrestrial species. In snails, the shell, if present, is coiled (Fig. 24.1*c*). The **cephalopods** contain marine active predators, such as squids and nautiluses. Tentacles are about the head (Fig. 24.1*d*).

All molluscs have a three-part body consisting of (1) a muscular **foot** specialized for various means of locomotion; (2) **visceral mass** that includes the internal organs; and (3) a **mantle,** a thin tissue that encloses the visceral mass and may secrete a shell. **Cephalization** is the development of a head region. On the lines provided in Figure 24.1, write *cephalization* or *no cephalization* as appropriate for this mollusc.

a. Chiton, *Tonicella*

____

c. Snail, *Helix* is a gastropod.

____

b. Scallop, *Aequipecten* is a bivalve.

____

d. Nautilus, *Nautilus,* is a cephalopod.

____

Figure 24.1 Molluscan diversity.
a. You can see the exoskeleton of this chiton but not its dorsally flattened foot. **b.** A scallop doesn't have a foot but it does have strong adductor muscles to close the shell. In this specimen, the edge of the mantle bears tentacles and many blue eyes. **c.** A gastropod, such as a snail, is named for the location of its large foot beneath the visceral mass. **d.** In a cephalopod, such as this nautilus, a funnel (its foot) opens in the area of the tentacles and allows it to move by jet propulsion.
(a) ©Randimal/Shutterstock RF; (b) ©NHPA/Photoshot; (c) ©Rosemary Calvert/Getty Images; (d) ©Douglas Faulkner/Science Source

Anatomy of a Clam

Clams are bivalved because they have right and left shells secreted by the mantle. Clams have no head, and they burrow in sand by extending a **muscular foot** between the valves. Clams are **filter feeders** and feed on debris that enters the mantle cavity. In the visceral mass, the blood leaves the heart and enters sinuses (cavities) by way of anterior and posterior aortas. There are many different types of clams. The one examined here is the freshwater clam *Venus*.

Observation: Anatomy of a Clam

External Anatomy

1. Examine the external shell (Fig. 24.2) of a preserved clam (*Venus*). The shell is an **exoskeleton.**
2. Find the posterior and anterior ends. The more pointed end of the **valves** (the halves of the shell) is the posterior end.
3. Determine the clam's dorsal and ventral regions. The valves are hinged together dorsally.
4. What is the function of a heavy shell? ______________________________

Figure 24.2 External view of the clam shell.

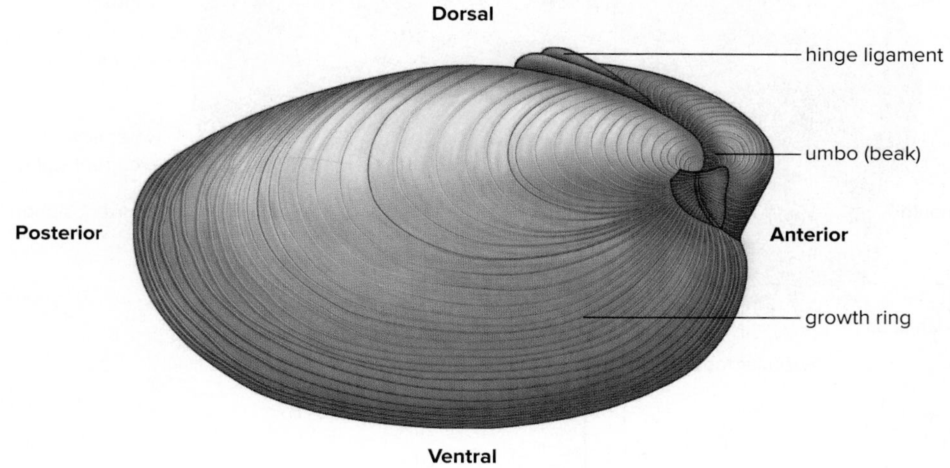

Internal Anatomy

1. Place the clam in the dissecting pan, with the **hinge ligament** and **umbo** (blunt dorsal protrusion) down. Carefully separate the **mantle** from the right valve by inserting a scalpel into the slight opening of the valves. What is a mantle? ______________________________

____
2. Insert the scalpel between the mantle and the valve you just loosened.
3. The **adductor muscles** hold the valves together. Cut the adductor muscles at the anterior and posterior ends by pressing the scalpel toward the dissecting pan. After these muscles are cut, the valve can be carefully lifted away. What is the advantage of powerful adductor muscles? ______________________________

____
4. Examine the inside of the valve you removed. Note the concentric lines of growth on the outside, the hinge teeth that interlock with the other valve, the adductor muscle scars, and the mantle line. The inner layer of the shell is mother-of-pearl.
5. Examine the rest of the clam (Fig. 24.3) attached to the other valve. Notice the adductor muscles and the mantle, which lies over the visceral mass and foot.
6. Bring the two halves of the mantle together. Explain the term *mantle cavity.* ______________________________

____
7. Identify the **incurrent** (more ventral) and **excurrent siphons** at the posterior end (Fig. 24.3). Explain how water enters and exits the mantle cavity. ______________________________

____

Figure 24.3 Anatomy of a bivalve.
The mantle has been removed to reveal the internal organs. **a.** Drawing. **b.** Dissected specimen.
(b) ©Ken Taylor/Wildlife Images

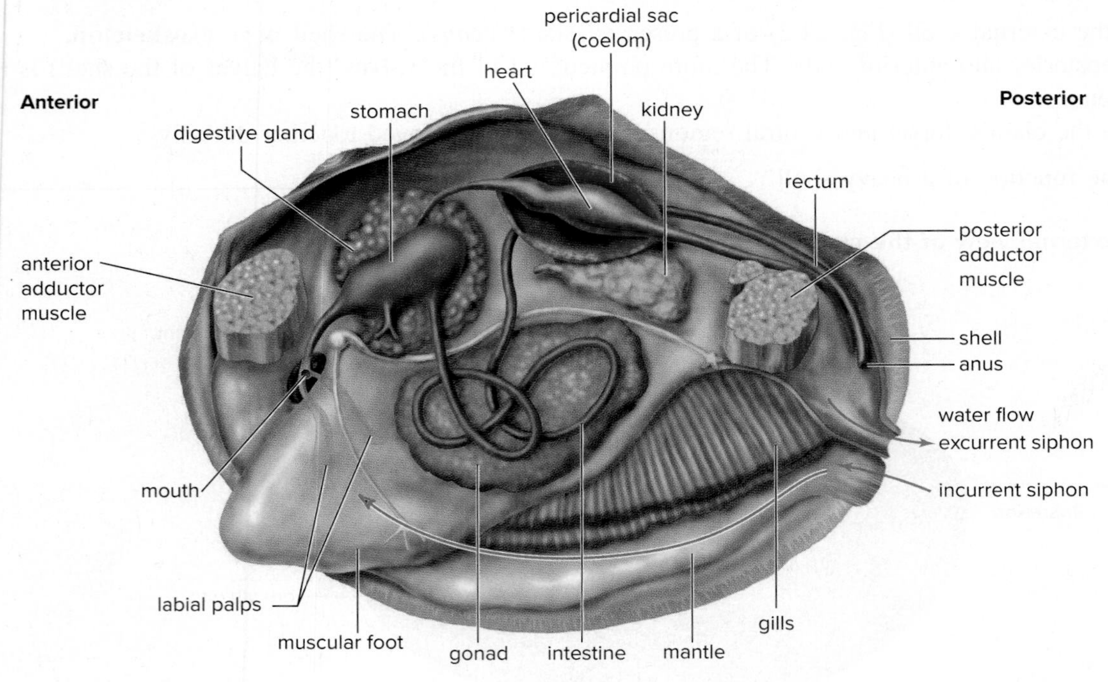

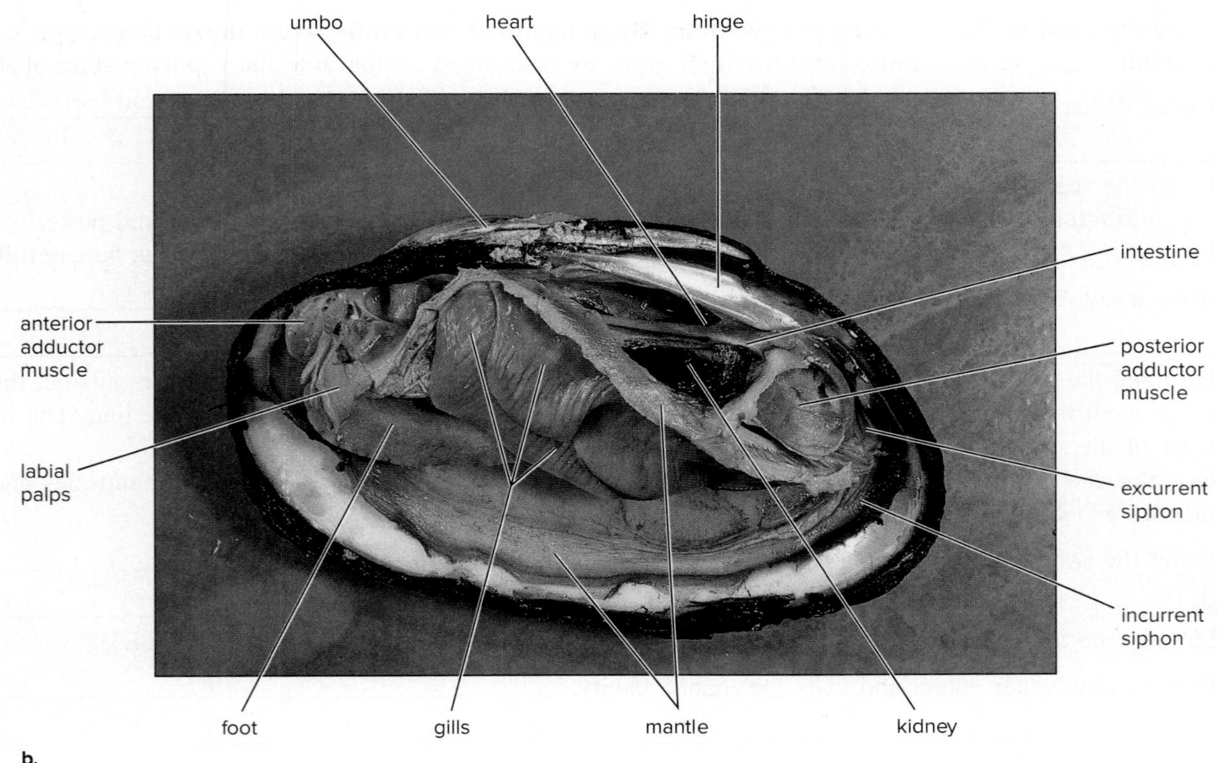

8. Cut away the free-hanging portion of the mantle to expose the **gills.** Does the clam have a respiratory organ? ____________ If so, what type of respiratory organ? ____________________________

9. A mucous layer on the gills entraps food particles brought into the mantle cavity, and the cilia on the gills convey these food particles to the mouth. Why is the clam called a filter feeder? ______________

____

____

10. The nervous system is composed of three pairs of ganglia (located anteriorly, posteriorly, and in the foot), all connected by nerves. The clam does not have a brain. A ganglion contains a limited number of neurons, whereas a brain is a large collection of neurons in a definite head region.
11. Identify the **foot,** a tough, muscular organ for locomotion, and the **visceral mass,** which lies above the foot and is soft and plump. The visceral mass contains the digestive and reproductive organs.
12. Identify the **labial palps** that channel food into the open mouth.
13. Identify the **anus,** which discharges into the excurrent siphon.
14. Find the **intestine** by its dark contents. Trace the intestine forward until it passes into a sac, the clam's only evidence of a coelom.
15. Locate the **pericardial sac (pericardium)** that contains the heart. The intestine passes through the heart. The heart pumps blood into the aortas, which deliver it to blood sinuses (open spaces) in the tissues.

    A clam has an **open circulatory system.** Explain your answer. ______________________________

    ____

16. Cut the visceral mass and the foot into exact left and right halves, and examine the cut surfaces. Identify the digestive glands, greenish-brown; the stomach, embedded in the digestive glands; and the intestine, which winds about in the visceral mass. Reproductive organs (gonads) are also present.

Anatomy of a Squid

Squids are cephalopods because they have a well-defined head; the foot became the funnel surrounded by two arms and the many tentacles about the head. The head contains a brain and bears sense organs. The squid moves quickly by jet propulsion of water, which enters the mantle cavity by way of a space that encircles the head. When the cavity is closed off, water exits by means of the funnel. Then the squid moves rapidly in the opposite direction.

The squid seizes fish with its tentacles; the mouth has a pair of powerful, beaklike jaws and a **radula,** a beltlike organ containing rows of teeth. The squid has a **closed circulatory system** composed of vessels and three hearts, one of which pumps blood to all the internal organs, while the other two pump blood to the gills located in the mantle cavity.

Observation: Anatomy of a Squid

1. Examine a preserved squid.
2. Refer to Figure 24.4 for help in identifying the mouth (defined by beaklike jaws and containing a radula) and the tentacles and arms, which encircle the mouth.
3. Locate the head with its sense organs, notably the large, well-developed eye.
4. Find the funnel, where water exits from the mantle cavity, causing the squid to move backward.
5. If the squid has been dissected, note the heart, gills, and blood vessels.

Figure 24.4 Anatomy of a squid.
The squid is an active predator and lacks the external shell of a clam. It captures fish with its tentacles and bites off pieces with its jaws. A strong contraction of the mantle forces water out the funnel, resulting in "jet propulsion." **a.** Drawing. **b.** Dissected specimen. (b) ©Ken Taylor/Wildlife Images

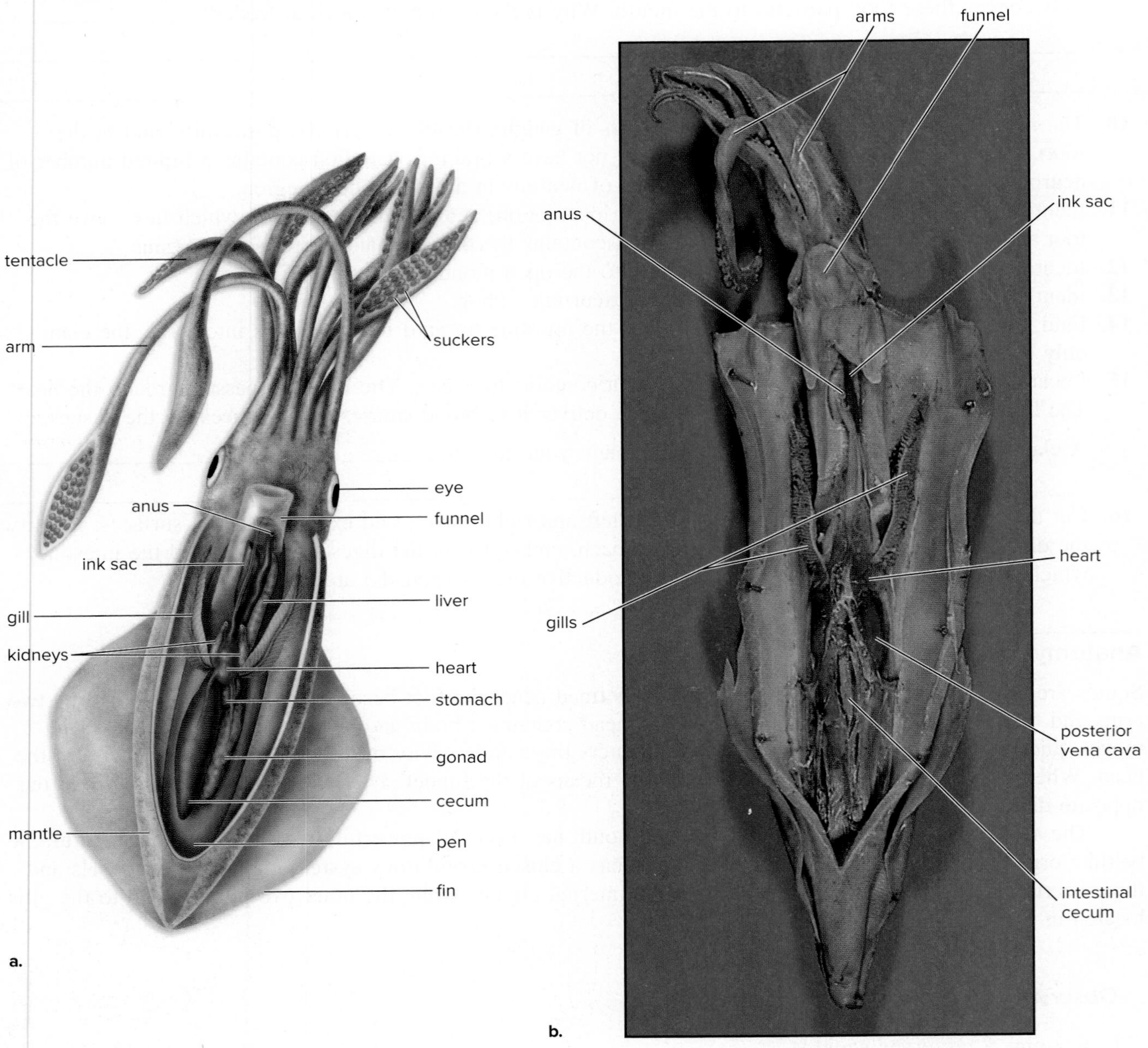

Conclusion: Comparison of Clam to Squid

- Compare clam anatomy to squid anatomy by completing Table 24.1.
- Explain how both clams and squids are adapted to their way of life. ______________________________

____

____

____

Table 24.1 Comparison of Clam to Squid

	Clam	Squid
Feeding mode		
Skeleton		
Circulation		
Cephalization		
Locomotion		

24.2 Annelids

Annelids (phylum Annelida) are the **segmented** worms, so called because the body is divided into a number of segments and has a ringed appearance. The circular and longitudinal muscles work against the fluid-filled coelom to produce changes in width and length (Fig. 24.5). Therefore, annelids are said to have a **hydrostatic skeleton.**

Figure 24.5 Locomotion in the earthworm.
Contraction of first circular muscles and then longitudinal muscles allow the earthworm to move forward.

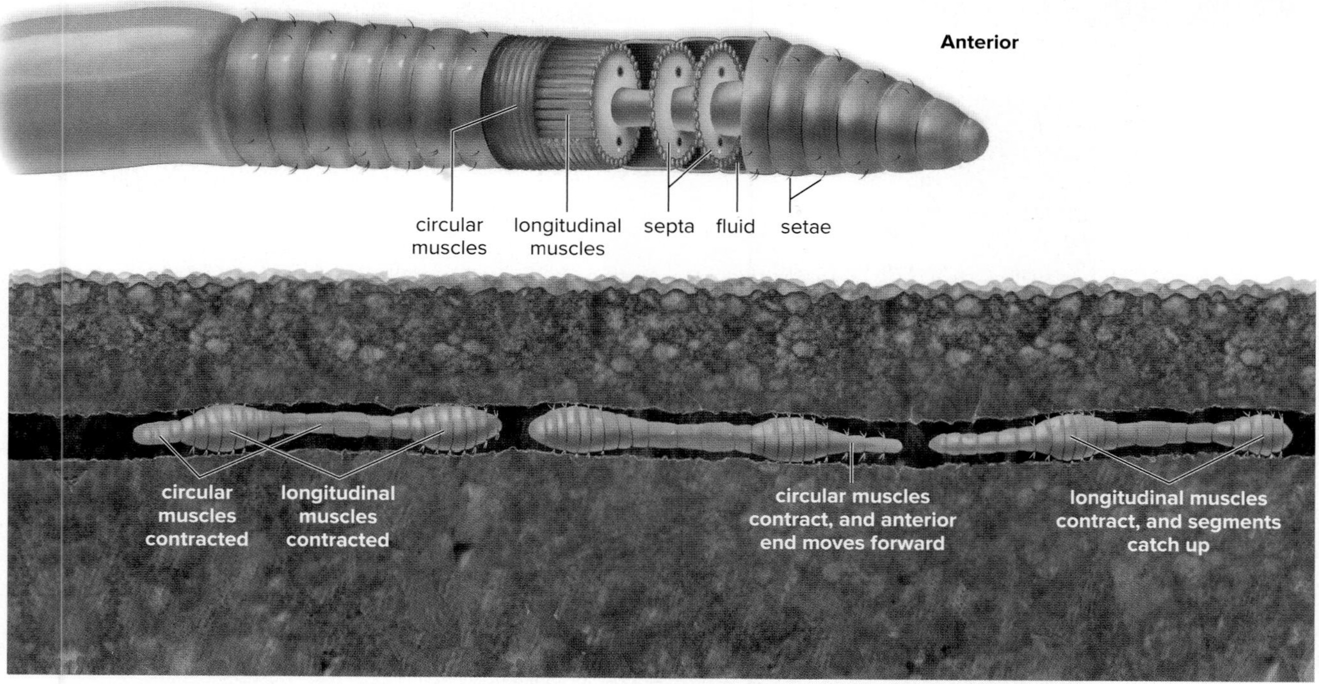

Among annelids (Fig. 24.6), **polychaetes** have many slender bristles called **setae.** The polychaetes, almost all marine, are plentiful from the intertidal zone to the ocean depths. They are quite diverse, ranging from jawed forms that are carnivorous to fanworms that live in tubes and extend feathery filaments when filter feeding. Earthworms are called **oligochaetes** because they have few setae. Earthworms, which have a worldwide distribution in almost any soil, are often present in large numbers, and may reach a length of as much as 3 meters. **Leeches,** annelids without setae, include the medicinal leech, which has been used in the practice of bloodletting for centuries. Most people simply called them bloodsuckers.

Show that the annelids are the segmented worms by *labeling a segment in 24.6*a, b, *and* d. In which group would you expect the animals to be predators based on the type of head region? ______________________

Figure 24.6 Annelid diversity.
Aside from **a.** earthworms, which are oligochaetes, there are **b.** clam worms and **c.** fan worms, both of which are polychaetes, and **d.** leeches. Note the obvious segmentation.
(a) ©MikeLane45/iStock360/Getty Images RF; (b) ©Blickwinkel/Alamy Stock Photo; (c) ©Borut Furlan/Getty Images; (d) ©Bildagentur/age fotostock

a. Earthworms, *Lumbricus*, mating

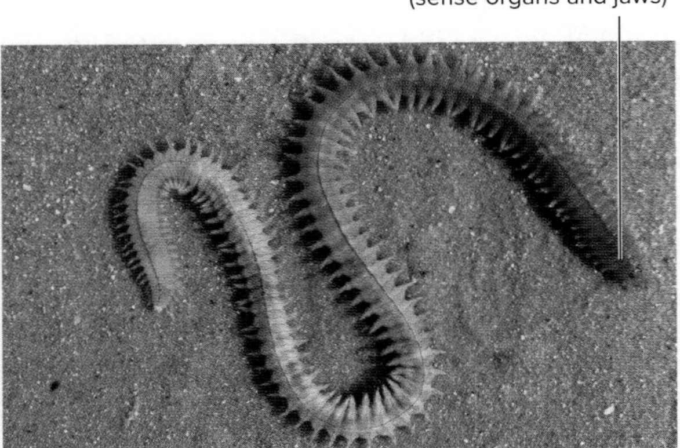

b. Clam worm, *Nereis*

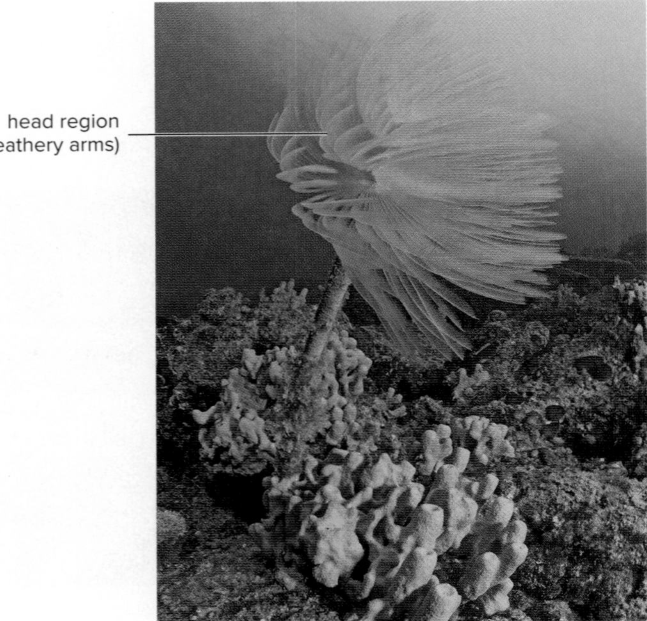

c. European fan worm, *Sabella*

d. Leech, *Hirudo*

Anatomy of the Earthworm

Earthworms are segmented in that the body has a series of ringlike segments. Earthworms have no head, and they burrow in the soil by alternately expanding and contracting segments along the length of the body.

Earthworms are scavengers that feed on decaying organic matter in the soil. They have a well-developed coelomic cavity, providing room for a well-developed digestive tract and both sets of reproductive organs. Earthworms are **hermaphroditic.**

Observation: Anatomy of the Earthworm

External Anatomy

1. Examine a live or preserved specimen of an earthworm. Locate the small projection that sticks out over the mouth. Has cephalization occurred? __________ Explain your answer. ______________________________
____
2. Count the total number of segments, beginning at the anterior end. The sperm duct openings are on segment 15 (somite XV) (Fig. 24.7). The enlarged section around a short length of the body is the **clitellum.** The clitellum secretes mucus that holds the worms together during mating. It also functions as a cocoon, in which fertilized eggs hatch and young worms develop. The anus is located on the worm's terminal segment.
3. Lightly pass your fingers over the earthworm's ventral and lateral sides. Do you feel the setae? ________
____
Earthworms insert these slender bristles into the soil. Setae, along with circular and longitudinal muscles, enable the worm to locomote. Explain the action. ____
____

Figure 24.7 External anatomy of an earthworm.
In this drawing, the segments are numbered.

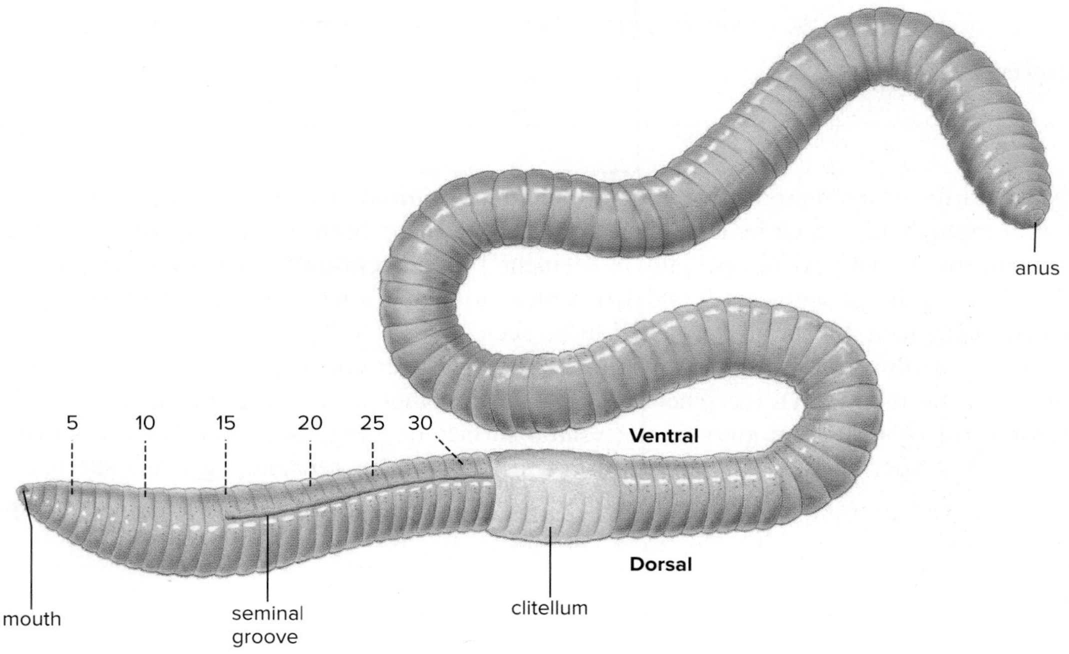

Internal Anatomy

1. Place a preserved earthworm on its ventral side in a dissecting pan. With a scalpel or razor blade, make a shallow incision slightly to the side of the blackish median dorsal blood vessel (Fig. 24.8*a*). Start your incision about 10 segments after the clitellum, and proceed anteriorly to the mouth. If you see black ooze, you have accidentally cut the intestine.
2. Identify the thin partitions, the **septa,** between segments.
3. Lay out the body wall, and pin every 10th segment to the wax in your pan. Add water to prevent drying out. While alive, the earthworm body wall is always moist, and this facilitates gas exchange across the body wall. Notice the earthworm has no respiratory organ.
4. An earthworm feeds on **detritus** (organic matter) in the soil. Identify the digestive tract, which begins at the mouth and extends through body segments 1, 2, and 3 (Fig. 24.8*a* and *b*). It opens into the swollen, muscular, thick-walled **pharynx,** which extends from segment 3 to segment 6. The **esophagus** extends from the pharynx through segment 14 to the **crop.** Next is the **gizzard,** which lies in segments 17 through 28. The intestine extends from the gizzard to the anus. Does the digestive system show specialization of parts?__________ Explain your answer. ____

   ____

   ____
5. Identify the earthworm's circulatory system. The blood is always contained within vessels and never runs free. The **dorsal blood vessel** is readily seen along the dorsal side of the digestive tract. A series of aortic arches or "hearts" encircles the esophagus between segments 7 and 11, connecting the dorsal blood vessel with the ventral blood vessel. Does the earthworm have an open or closed circulatory system? __________

   Explain your answer. ____

   ____
6. Locate the earthworm's nervous system. The two-lobed brain is located on the dorsal surface of the pharynx in segment 3. Two nerves, one on each side of the pharynx, connect the brain to a ganglion that lies below the pharynx in segment 4. The **ventral nerve cord** then extends along the floor of the body cavity to the last segment.
7. Find the earthworm's excretory system, which consists of a pair of minute, coiled, white tubules, the **nephridia,** located in every segment except the first three and the last. Each nephridium opens to the outside by means of an excretory pore. Does the excretory system show that the earthworm is segmented? __________ Explain your answer. ____

   ____

   ____
8. Identify the earthworm's reproductive system, including **seminal vesicles,** light-colored bodies in segments 9 through 12, which house maturing sperm that have been formed in two pairs of testes within them; **sperm ducts** that pass to openings in segment 15; and **seminal receptacles** (four small, white, special bodies that lie in segments 9 and 10), which store sperm received from another worm. **Ovaries** are located in segment 13 but are too small to be seen.
9. During mating, earthworms are arranged so that the sperm duct openings of one worm are just about, but not quite, opposite the seminal receptacle openings of the other worm. After being released, the sperm pass down a pair of seminal grooves on the ventral surface (see Fig. 24.7) and then cross over at the level of the seminal receptacles of the opposite worm. Once the worms separate, eggs and sperm are released into a cocoon secreted by the clitellum. Is the earthworm hermaphroditic? ______________ Explain your answer. ____

   ____

10. Does the earthworm have a respiratory organ or system (i.e., gills or lungs)? __________ How does the worm exchange gases? ____

____

Why would you expect an earthworm to lack an exoskeleton? ______________________________

____

Figure 24.8 Internal anatomy of an earthworm, dorsal view.
a. Drawing shows internal organs and a cross section. The segments are numbered. **b.** Dissected specimen.
(b) ©Ken Taylor/Wildlife Images

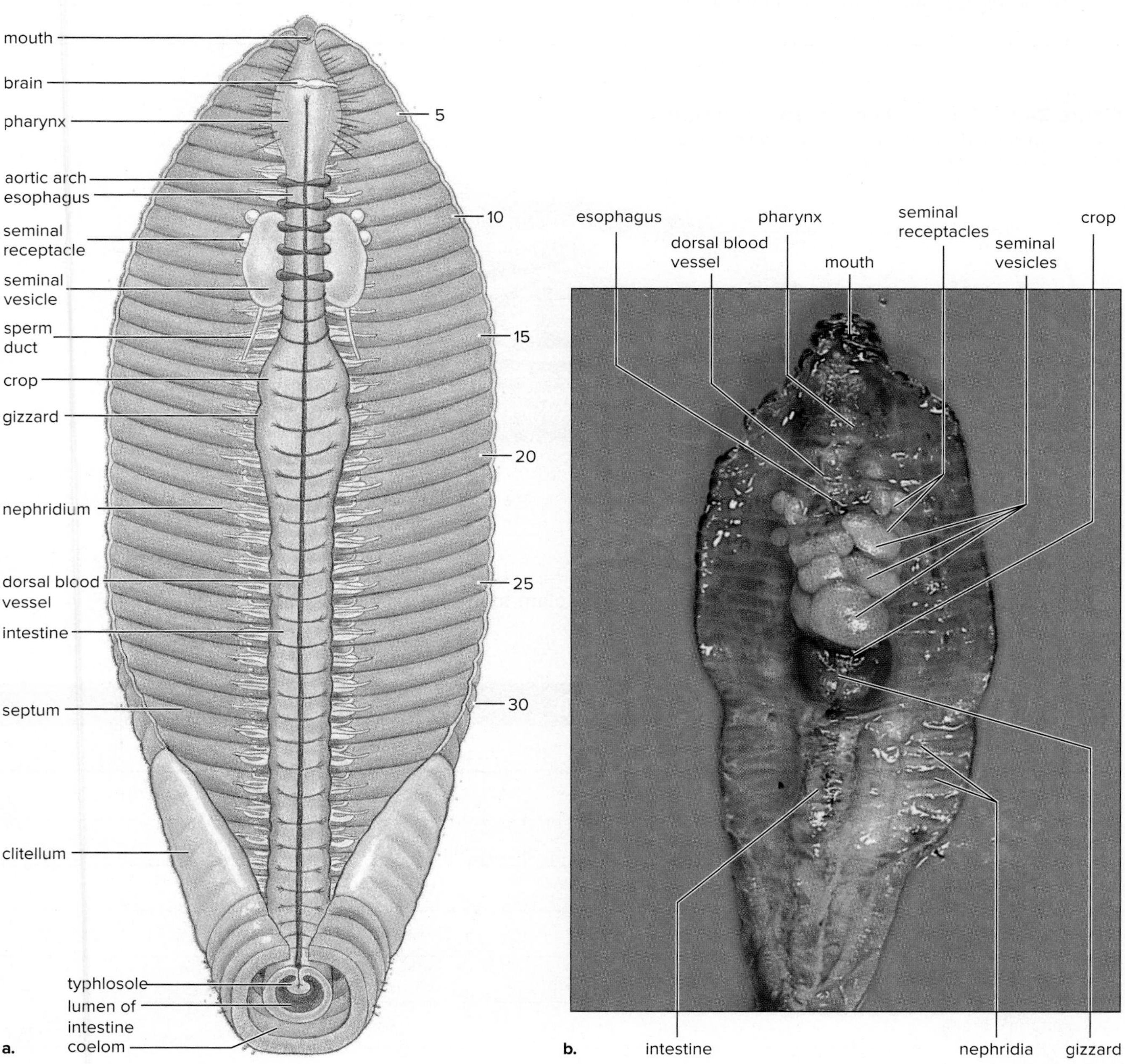

Prepared Slide

1. Obtain a prepared slide of a cross section of an earthworm (Fig. 28.9). Examine the slide under the dissecting microscope and under the light microscope.
2. Identify the following structures.
   a. **Body wall:** A thick outer circle of tissue, consisting of the **cuticle** and the **epidermis.**
   b. **Coelom:** A relatively clear space with scattered fragments of tissue.
   c. **Intestine:** An inner circle with a suspended fold.
   d. **Typhlosole:** A fold that increases the intestine's surface area.
   e. **Ventral nerve cord:** A white, threadlike structure.
3. Does the typhlosole help in nutrient absorption? __________ Explain your answer. ____________________

____

Figure 24.9 Cross section of an earthworm.
Cross-section slide as it would appear under the microscope.
©Carolina Biological Supply Company/Phototake

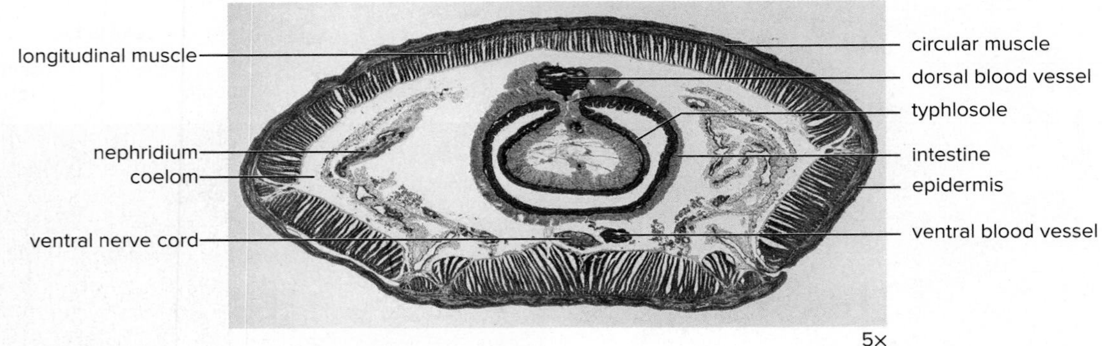

Conclusion: Comparison of Clam to Earthworm

- Complete Table 24.2 to compare the anatomy of a clam to that of an earthworm.

Table 24.2 Comparison of Clam to Earthworm

	Clam	Earthworm
Habitat (where they live)		
Feeding mode		
Skeleton		
Segmentation		
Circulation		
Respiratory organs		
Locomotion		
Reproductive organs		

24.3 Arthropods

Arthropods (phylum Arthropoda) have paired, jointed appendages and a hard exoskeleton that contains chitin. The chitinous exoskeleton consists of hardened plates separated by thin, membranous areas that allow movement of the body segments and appendages.

Figure 24.10*a* features insects and relatives. **Insects** with three pairs of legs, with or without wings, and three distinct body regions comprise 95% of all arthropods. **Millipedes** have two pairs of legs per segment, while **centipedes** have one pair of legs per segment. Figure 24.10*b* features spiders and relatives. Spiders and scorpions have four pairs of legs, no antennae, and a cephalothorax (head and thorax are fused). The horseshoe crab is a living fossil. It has remained unchanged for thousands of years. The **crustaceans** (Fig. 24.10*c*), which include crabs, shrimp, and lobsters, have three to five pairs of legs, and two pairs of antennae. Barnacles are unusual, in that their legs are used to gather food.

For each animal in Figure 24.10, circle the obvious types of appendages.

Figure 24.10 Arthropod diversity.
a. Among arthropods, insects, millipedes, and centipedes are possibly related. **b.** Spiders, scorpions, and horseshoe crabs are related. **c.** Crabs, shrimp, and barnacles, among others, are crustaceans.
(a) (left) Source: USDA-ARS/Stephen Ausmus; (a) (center) ©blickwinkel/Alamy Stock Photo; (a) (right) ©Larry Miller/Science Source; (b) (left) ©Horia Bogdan/Getty Images RF; (b) (center) ©Papilio/Alamy Stock Photo; (b) (right) Source: U.S. Fish & Wildlife Service/Robert Pos; (c) (left) ©Tom McHugh/Science Source; (c) (center) ©imageBroker/Alamy Stock Photo; (c) (right) ©L. Newman & A. Flowers/Science Source

a. Insects and relatives

Honeybee, *Apis mellifera*

Millipede, *Rhapidostreptus virgator*

Centipede, *Scolopendra* sp.

b. Spider and relatives

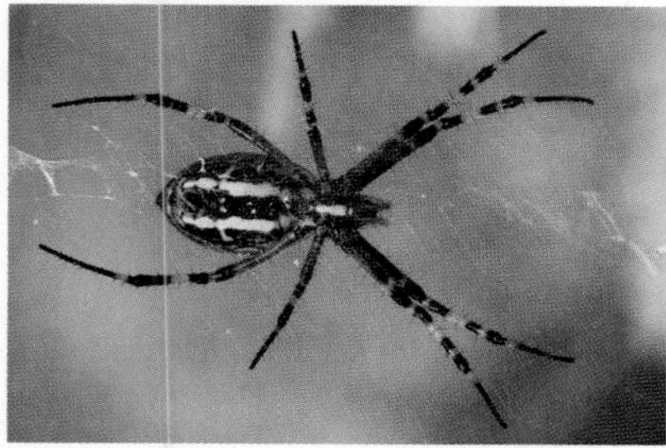
Spider, *Argiope bruennichi*

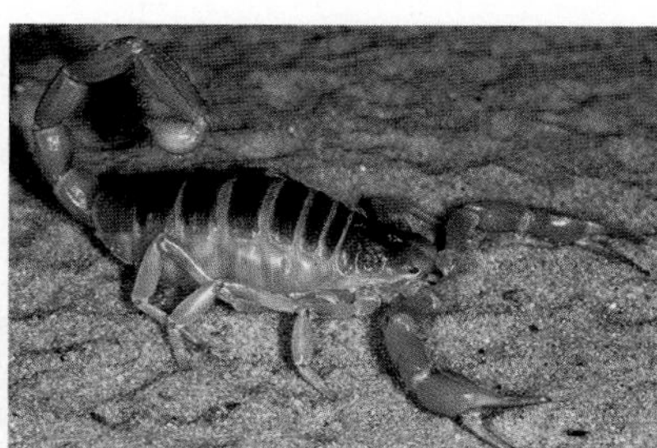
Scorpion, *Hadrurus arizonensis*

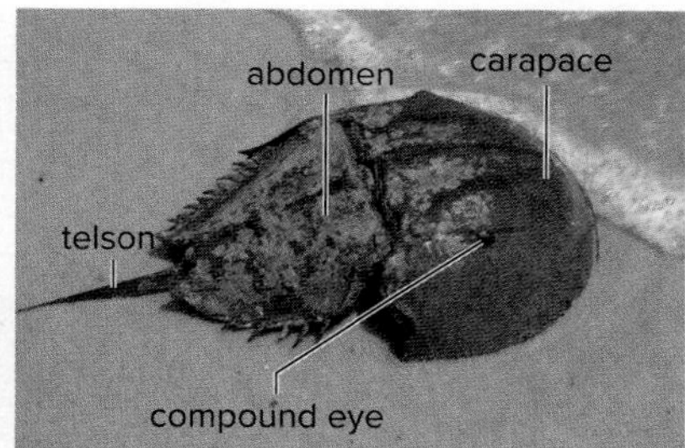

Horseshoe crab, *Limulus polyphemus*

c. Crustaceans

Crab, *Cancer productus*

Shrimp, *Lysmata grabhami*

Barnacles, *Lepas anatifera*

Anatomy of a Crayfish

Crayfish also belong to the group of arthropods called crustaceans. Crayfish are adapted to an aquatic existence. They are known to be scavengers, but they also prey on other invertebrates. The mouth is surrounded by appendages modified for feeding, and there is a well-developed digestive tract. Dorsal, anterior, and posterior arteries carry **hemolymph** (blood plus lymph) to tissue spaces (hemocoel) and sinuses. Therefore, a crayfish has an open circulatory system.

Observation: Anatomy of a Crayfish

External Anatomy

1. In a preserved crayfish, identify the chitinous **exoskeleton.** With the help of Figure 24.11, identify the head, thorax, and abdomen. Together, the head and thorax are called the **cephalothorax;** the

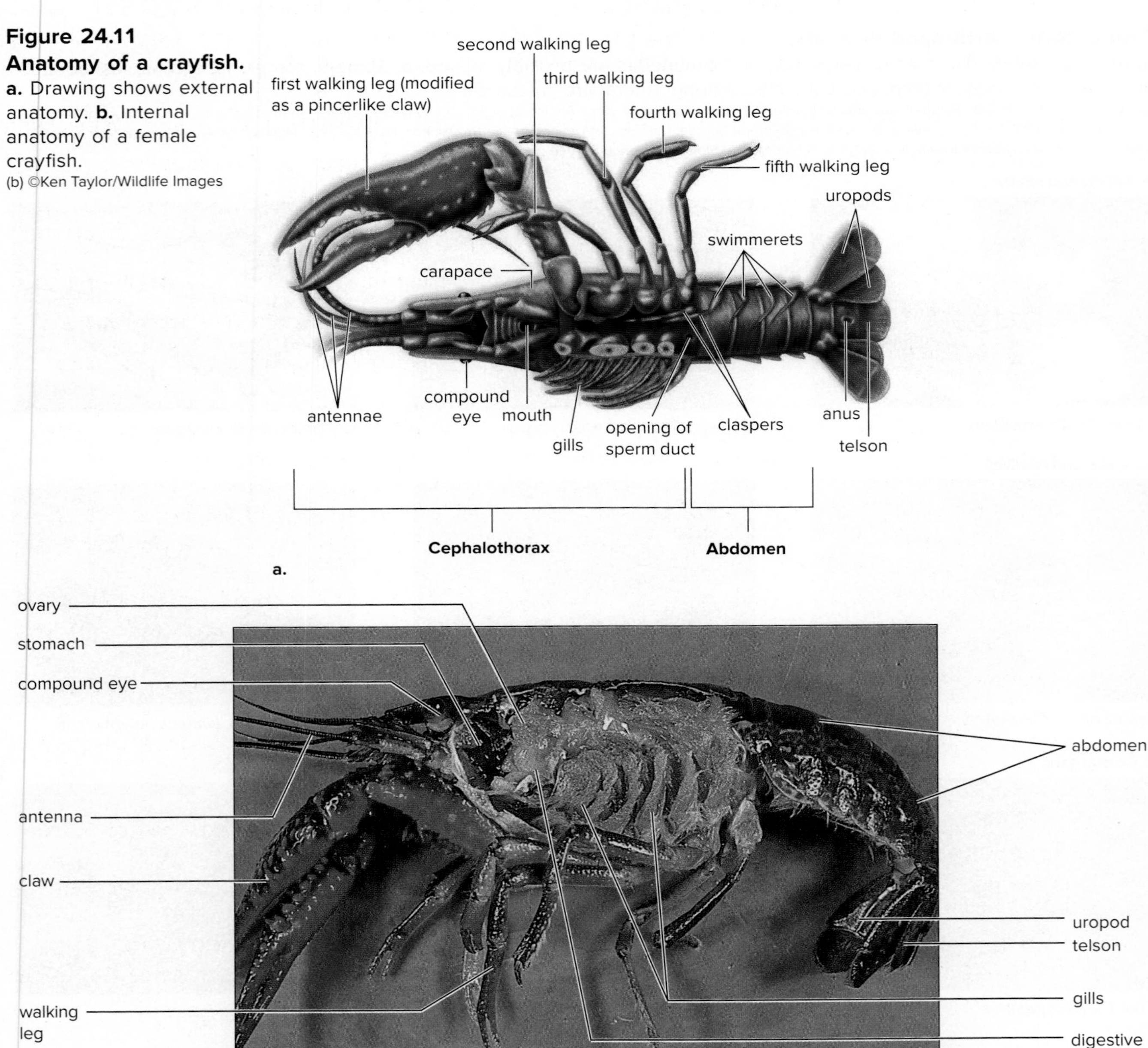

Figure 24.11 Anatomy of a crayfish. **a.** Drawing shows external anatomy. **b.** Internal anatomy of a female crayfish.
(b) ©Ken Taylor/Wildlife Images

cephalothorax is covered by the **carapace.** Has specialization of segments occurred? ___________

Explain your answer. ___________

2. Find the **antennae,** which project from the head. At the base of each antenna, locate a small, raised nipple containing an opening for the **green glands,** the organs of excretion. Crayfish excrete a liquid nitrogenous waste.
3. Locate the **compound eyes,** composed of many individual units for sight. Do crayfish demonstrate cephalization? ___________ Explain your answer. ___________
4. Identify the six pairs of appendages around the mouth for handling food.
5. Find the five pairs of walking legs attached to the cephalothorax. The most anterior pair is modified as pincerlike claws. In the female, identify the **seminal receptacles,** a swelling located between the bases of the third and fourth pairs of walking legs. Sperm from the male are deposited in the seminal receptacles. In the male, identify the opening of the sperm duct located at the base of the fifth walking leg.
6. Locate the five pairs of **swimmerets** on the abdomen. Crayfish use their swimmerets to swim and move water over their gills. In males, the anterior two pairs are stiffened and folded forward. They are claspers that aid in the transfer of sperm during mating. In females, clusters of eggs and larvae attach to the swimmerets.
7. Find the last abdominal segment, which bears a pair of broad, fan-shaped **uropods** that, together with a terminal extension of the body, form a tail. Has specialization of appendages occurred? ___________

   Explain your answer. ___________

Internal Anatomy

1. Cut away the lateral surface of the carapace with scissors to expose the **gills** (Fig. 24.11*b*). Observe that the gills occur in distinct, longitudinal rows. How many rows of gills are there in your specimen? ______

   The outer row of gills is attached to the base of certain appendages. Which ones? ___________

   ____
2. Remove a gill with your scissors by cutting it free near its point of attachment, and place it in a watch glass filled with water. Observe the numerous gill filaments arranged along a central axis.
3. Carefully cut away the dorsal surface of the carapace with scissors and a scalpel. The epidermis that adheres to the exoskeleton secretes the exoskeleton. Remove any epidermis adhering to the internal organs.
4. Identify the diamond-shaped heart lying in the middorsal region. A crayfish has an open circulatory system. Carefully remove the heart.
5. Locate the **gonads** anterior to the heart in both the male and female. The gonads are tubular structures bilaterally arranged in front of the heart and continuing behind it as a single mass. In the male, the testes are highly coiled, white tubes.
6. Find the **mouth;** the short, tubular **esophagus;** and the two-part **stomach,** with the attached **digestive gland,** that precedes the intestine.
7. Identify the **green glands,** two excretory structures just anterior to the stomach, on the ventral segment wall.
8. Remove the thoracic contents previously identified.
9. Identify the **brain** in front of the esophagus. The brain is connected to the ventral nerve cord by a pair of nerves that pass around the esophagus.
10. Remove the animal's entire digestive tract, and float it in water. Observe the various parts, especially the connections of the digestive gland to the stomach.
11. Cut through the stomach, and notice in the anterior region of the stomach wall the heavy, toothlike projections, called the **gastric mill,** which grind up food. Do you see any grinding stones ingested by the crayfish? ___________

    If possible, identify what your specimen had been eating. ___________

Anatomy of a Grasshopper

The grasshopper is an **insect.** All insects have a head, a thorax, and an abdomen. Their appendages always include (1) three pairs of jointed legs and usually (2) two antennae as sensory organs. Grasshoppers are adapted to live on land. Wings and jumping legs are suitable for locomotion on land; **Malpighian tubules** save water by secreting a solid nitrogenous waste; the **tracheae** are tiny tubules that deliver air directly to the muscles; and the male has a penis with attached claspers to deliver sperm to the seminal receptacles of a female so they do not dry out.

Observation: Anatomy of a Grasshopper

External Anatomy

1. Obtain a preserved grasshopper (*Romalea*), and study its external anatomy with the help of Figure 24.12*a*. Identify the head, thorax, and abdomen.
2. Use a hand lens or dissecting microscope to examine the grasshopper's special sense organs of the **head.** Identify the **antennae** (a pair of long, jointed feelers), the **compound eyes,** and the three dotlike **simple eyes.** The labial palps, labeled in Figure 24.12*a,* have sense organs for tasting food.
3. Note the sturdy **mouthparts,** which are used for chewing plant material. A grasshopper's mouthparts are quite different from those of a piercing and sucking insect.
4. Locate the leathery **forewings** and the inner, membranous **hindwings** attached to the **thorax.** Which pair of legs is used for jumping? ________________ How many segments does each leg have? __________
5. Is locomotion in the grasshopper adapted to land? ____
Explain your answer. ____
____
6. In the **abdomen,** identify the **tympana** (sing., **tympanum**), one on each side of the first abdominal segment (Fig. 24.12*a*). The grasshopper detects sound vibrations with these membranes.
7. Locate the **spiracles,** along the sides of the abdominal segments. These openings allow air to enter the tracheae, which constitute the respiratory system.
8. Find the **ovipositors** (Figs. 24.12*a* and 24.13*a*), four curved and pointed processes projecting from the abdomen of the female. These are used to dig a hole in which eggs are laid. The male has a **penis** with **claspers** used during copulation (Fig. 24.13*b*).

Figure 24.12 Female grasshopper.
a. External anatomy. **b.** Internal anatomy.

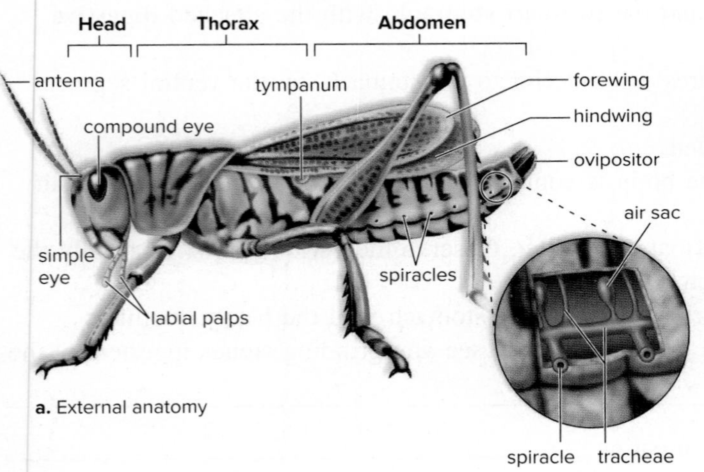

a. External anatomy

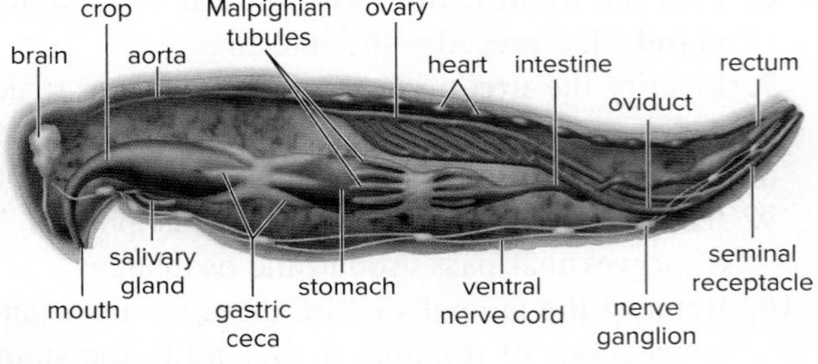

b. Internal anatomy

Figure 24.13 Grasshopper genitalia.
a. Females have an ovipositor, and **b.** males have claspers at the distal end of the penis.

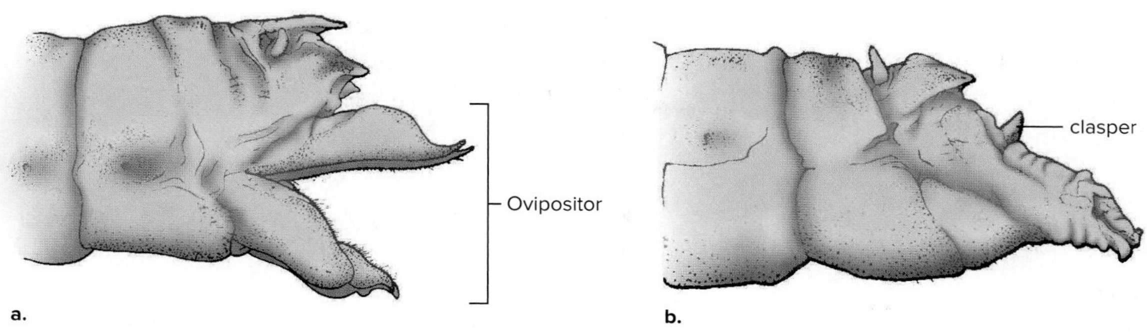

Internal Anatomy

Observe a longitudinal section of a grasshopper if available on demonstration. Try to locate the structures shown in Figure 24.12*b*. What is the function of Malpighian tubules?____________________________________

Conclusion: Comparison of Crayfish to Grasshopper

Compare the adaptations of a crayfish to those of a grasshopper by completing Table 24.3. Put a star beside each item that indicates an adaptation to life in the water (crayfish) and to life on land (grasshopper). Check with your instructor to see if you identified the maximum number of adaptations.

Table 24.3 Comparison of Crayfish to Grasshopper

	Crayfish	Grasshopper
Locomotion		
Respiration		
Sense organs		
Nervous system		
External reproductive features Male Female		

Grasshopper Metamorphosis

Metamorphosis means a change, usually a drastic one, in form and shape. Grasshoppers undergo *incomplete metamorphosis,* a gradual change in form rather than a drastic change. The immature stages of the grasshopper are called **nymphs,** and they are recognizable as grasshoppers even though they differ somewhat in shape and form (Fig. 24.14*a*). Some insects undergo what is called *complete metamorphosis,* in which case they have three stages of development: **larvae, pupa,** and **adult** (Fig. 24.14*b*). Metamorphosis occurs during the pupa stage when the animal is enclosed within a hard covering. The animals best known for complete metamorphosis are the butterfly and the moth, whose larval stage is called a caterpillar and whose pupa stage is the cocoon; the adult is the butterfly or moth.

Figure 24.14 Metamorphosis.
a. During incomplete metamorphosis of a grasshopper, a series of **nymphs** leads to a full-grown grasshopper. **b.** During complete metamorphosis of a moth, a series of larvae lead to pupation. The **adult** hatches out of the pupa.

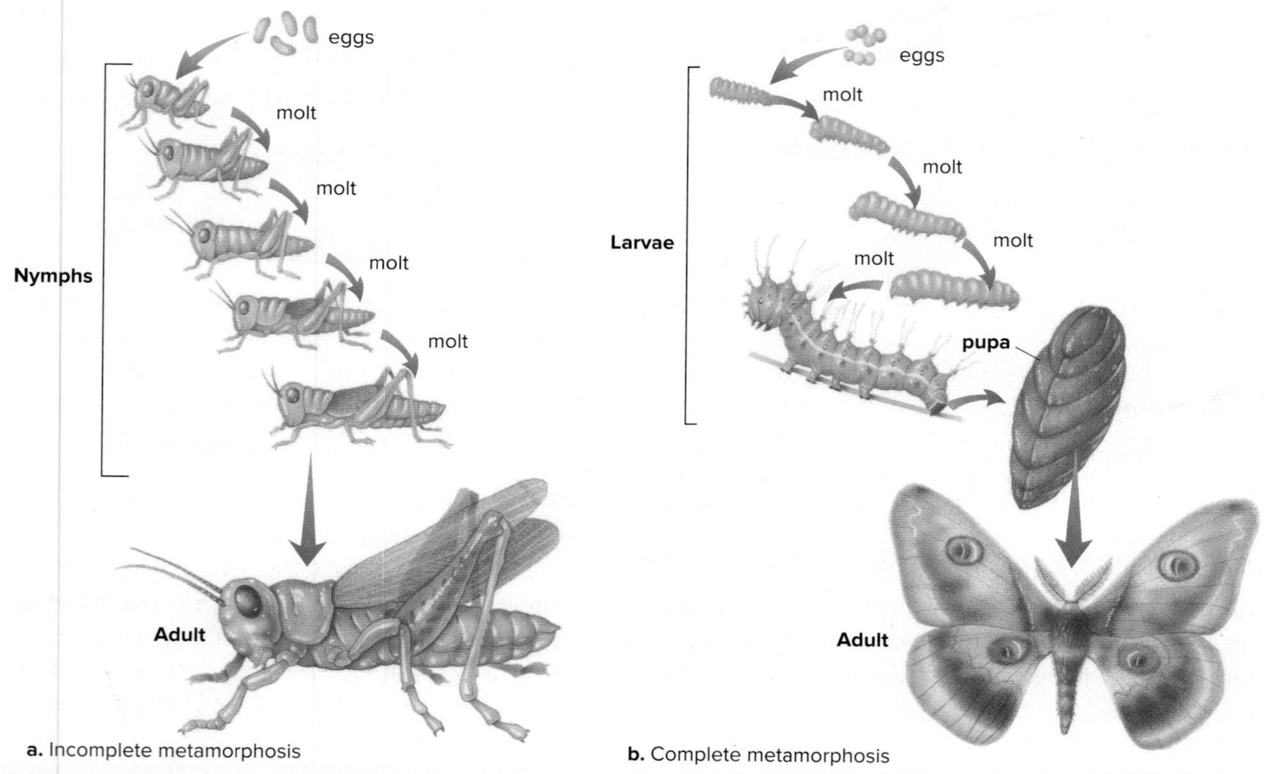

a. Incomplete metamorphosis

b. Complete metamorphosis

Observation: Grasshopper Metamorphosis

1. Use Figure 24.14 to add the grasshopper and the moth to Table 24.4.
2. Examine any specific life cycle displays or plastomounts that illustrate complete and incomplete metamorphosis and add these examples to Table 24.4.

Table 24.4 Insect Metamorphosis

Specimen	Complete or Incomplete Metamorphosis

Conclusions: Insect Metamorphosis

- In insects with incomplete metamorphosis, do the nymphs or the adult have better developed wings? __________ What is the benefit of wings to an insect?______________________________
____
- What stage is missing when an insect has incomplete metamorphosis? ______________ What happens during this stage? ____

- What form, the larvae or the adult, disperses new individuals in flying insects that exhibit complete metamorphosis? ______________________________
  How is this a benefit? ______________________________
  ____

- In insects that undergo complete metamorphosis, the larvae and the adults utilize different food sources and habitats. Why might this be a benefit? ______________________________
  ____

24.4 Echinoderms

The echinoderms (phylum Echinodermata) are the only invertebrate group that shares deuterostome development with the vertebrates (see Fig. 27.1). Unlike the vertebrates, all **echinoderms** are marine, and they dwell on the seabed, either attached to it, like sea lilies, or creeping slowly over it. The name *echinoderm* means "spiny-skinned," and most members of the group have defensive spines on the outside of their bodies. The spines arise from an **endoskeleton** composed of calcium carbonate plates. The endoskeleton supports the body wall and is covered by living tissue that may be soft (as in sea cucumbers) or hard (as in sea urchins).

Especially note that (1) adult echinoderms have radial symmetry, with generally five points of symmetry arranged around the axis of the mouth but the larvae have bilateral symmetry, and (2) the echinoderms' most unique feature is their **water vascular system.** In those echinoderms in which the arms make contact with the substratum, the **tube feet** associated with the water vascular system are used for locomotion. In other echinoderms, the tube feet are used for gas exchange and food gathering.

Echinoderms belong to one of five groups: sea lilies and feather stars; sea stars; brittle stars; sea urchins and sand dollars; and sea cucumbers (Fig. 24.15). *Where appropriate in Figure 24.15, write "ORS" for obvious radial symmetry or "RSNO" for radial symmetry not obvious on the lines provided.*

Figure 24.15 Echinoderm diversity.
(a) ©Borut Furlan/Getty Images; (b) ©Jonathan Bird/Getty Images; (c) ©Robert Dunne/Science Source; (d) ©Neil McDan/Science Source; (e) ©Andrew J. Martinez/Science Source; (f) ©Andrey Nekrasov/Alamy Stock Photo

a. Feather star, *Oxymetra erinacea* ______________

b. Sea star, *Odontaster validus* ______________

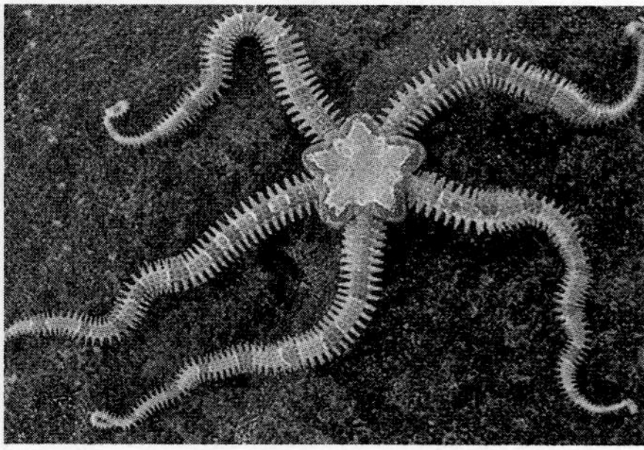
c. Brittle star, *Ophiopholis aculeata* ______________

d. Sea urchin, *Stronglocentrotus pranciscanus* ______________

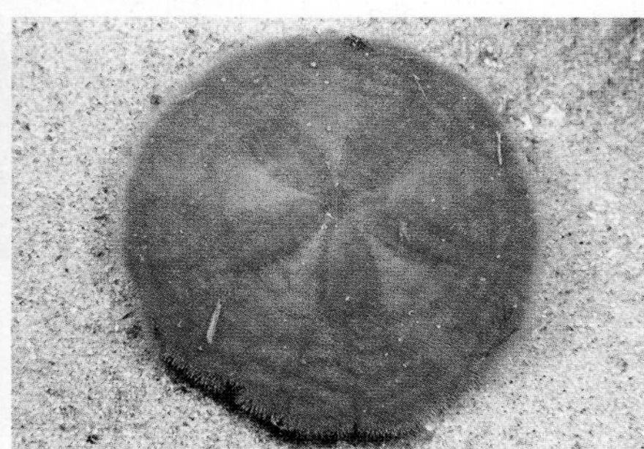
e. Sand dollar, *Echinarachnius parma* ______________

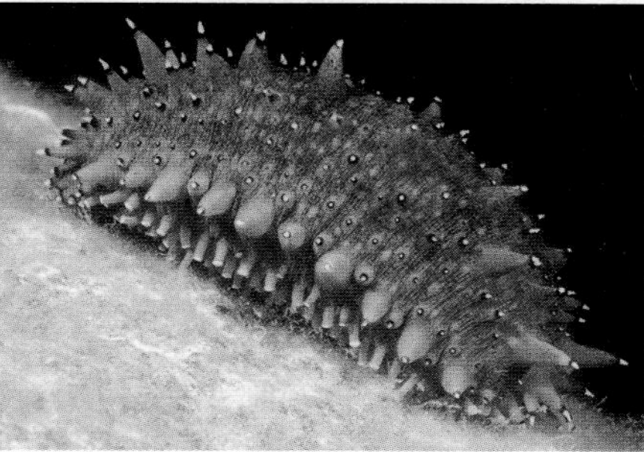
f. Sea cucumber, *Apostichopus japonicus* ______________

Anatomy of a Sea Star

Sea stars (starfish) usually have five arms that radiate from a central disk. The mouth is normally oriented downward, and when sea stars feed on clams, they use the suction of their tube feet to force the shells open a crack. Then they evert the cardiac portion of the stomach, which releases digestive juices into the mantle cavity. Partially digested tissues are taken up by the pyloric portion of the stomach; digestion continues in this portion of the stomach and in the digestive glands found in the arms.

Observation: Anatomy of a Sea Star

External Anatomy

1. Place a preserved sea star in a dissecting pan so that the aboral side is uppermost.
2. With the help of Figure 24.16, identify the **central disk** and five arms. What type of symmetry does an adult sea star have? ____________________

Figure 24.16 Anatomy of a sea star.
a. Diagram and **(b)** image of dissected sea star. Both show the aboral side. **c.** Image of cut arm. **d.** Canals and tube feet of water vascular system. Seen from aboral side. (b–c) ©BiologyImaging.com

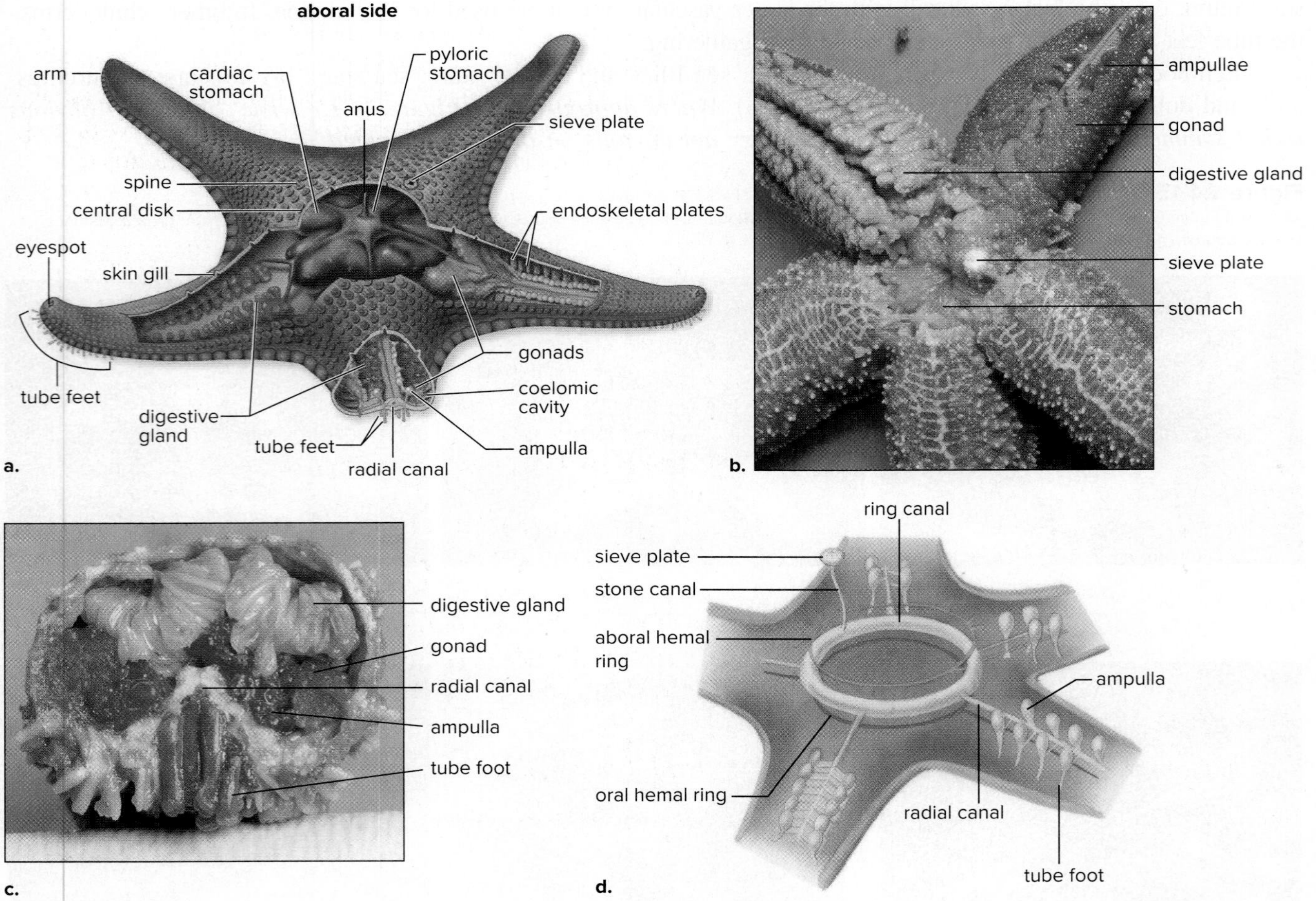

3. With the oral side uppermost, find the mouth, located in the center and protected by spines. Why is this side of the sea star called the oral side? ________________________________

4. Locate the groove that runs along the middle of each arm and the **tube feet** (suctionlike disks) in rows on either side of the groove. Pluck away the tube feet from one area. How many rows of feet are there? __________

5. Turn the sea star over to its aboral side.

6. Locate the anal opening (Fig. 24.16). Why is this side of the sea star called the aboral side? __________
____

7. Lightly run your fingers over the spines extending from calcium carbonate plates that lie buried in the body wall beneath the surface. The plates form an endoskeleton of the animal.

8. Identify the **sieve plate** (**madreporite**), a brownish, circular spot between two arms where water enters the water vascular system.

Internal Anatomy

1. Place the sea star so that the aboral side is uppermost. Refer to Figure 24.16*a* and *b* as you dissect the sea star following these instructions:
2. Cut the tip of one of the arms and, with scissors, carefully cut through the body wall along each side of this arm.
3. Carefully lift up the upper body wall. Separate any internal organs that may be adhering so that all internal organs are left intact.
4. Cut off the body wall near the central disk, but leave the sieve plate (madreporite) in place.
5. Remove the body wall of the central disk, being careful not to injure the internal organs.
6. Identify the digestive system. The mouth leads into a short **esophagus,** which is connected to the saclike **cardiac stomach.** When a sea star eats, the cardiac stomach sticks out through the sea star's mouth and starts digesting the contents of a clam or oyster. Above the cardiac stomach is the **pyloric stomach,** which leads to a short intestine. Each arm contains one pair of **digestive glands.**

   To which stomach do the digestive glands attach? ________________________________

7. Cut off a portion of an arm, and examine the cut edge (Fig. 24.16*c*). Identify the digestive glands, ambulacral groove, radial canal, ampullae, and tube feet.
8. Remove the digestive glands of one arm.
9. Identify the **gonads** extending into the arm. What is the function of gonads? ________________
____

   It is not possible to distinguish male sea stars from females by this observation.

10. Remove both stomachs.
11. In the **water vascular system** (Fig. 24.16*d*), you have already located the sieve plate and tube feet. Now try to identify the following components:
    a. **Stone canal:** Takes water from the sieve plate to the ring canal.
    b. **Ring canal:** Surrounds the mouth and takes water to the radial canals.
    c. **Radial canals:** Send water into the ampulla. When the ampullae contract, water enters the tube feet. Each tube foot has an inner muscular sac called an ampulla. The ampulla contracts and forces water into the tube foot.

    What is the function of the water vascular system? ________________________________
____

Laboratory Review 24

_______________ **1.** What kind of body cavity is characteristic of the animals in the phyla covered in this lab exercise?

_______________ **2.** Is an animal a protostome or a deuterostome if the first opening becomes the mouth?

_______________ **3.** What do all molluscs have in addition to a muscular foot and a mantle?

_______________ **4.** What characteristic does an animal with a definite head region have?

_______________ **5.** What organ is found inside the pericardial sac (pericardium)?

_______________ **6.** What kind of skeleton do annelids have?

_______________ **7.** To what system do the crop and gizzard of an earthworm belong?

_______________ **8.** What is the location of the nerve cord in an earthworm?

_______________ **9.** What phylum of animals have chitinous exoskeletons?

_______________ **10.** What are the excretory structures in a crayfish called?

_______________ **11.** How is air taken directly to the muscles in a grasshopper?

_______________ **12.** What system of the echinoderms enables them to move about?

_______________ **13.** What kind of symmetry does an adult starfish have?

_______________ **14.** Where does water enter the water vascular system in a starfish?

Thought Questions

15. Explain the presence of specialized digestive organs in the animals studied in this laboratory given that they all have a coelom and a complete digestive system with a separate mouth and anus.

16. Explain why an earthworm will die if it dries out based on the type of skeleton annelids have and the absence of a respiratory organ or system in the earthworm.

17. Contrast an arthropod, the crayfish or the grasshopper, with the representative echinoderm, the starfish, with regard to signs of cephalization and how the type of symmetry each has is related.

25

The Vertebrates

Learning Outcomes

Introduction
- State the characteristics that all chordates have in common.

25.1 Evolution of Chordates
- Discuss the evolution of chordates in terms of seven derived characteristics.

25.2 Invertebrate Chordates
- Use the observation of lancelet anatomy to point out the common characteristics of all chordates.

25.3 Vertebrates
- Name the types of vertebrates, and give an example of an animal in each group.
- Identify and locate external and internal structures of a frog.
- Trace the path of air, food, and urine in a frog, and explain the term *urogenital system.*

25.4 Comparative Vertebrate Anatomy
- Compare the organ systems (except musculoskeletal) of a frog, perch, pigeon, and rat.
- Relate aspects of an animal's respiratory system to the animal's environment and to features of the animal's circulatory system.

Introduction

Vertebrates are **chordates.** All chordates have (1) a dorsal tubular nerve cord; (2) a dorsal supporting rod, called a notochord, at some time in their life history; (3) a postanal tail (e.g., tailbone or coccyx); and, (4) pharyngeal pouches. In chordates that breathe by means of gills, these pouches become gill slits. In terrestrial chordates, the pharyngeal pouches are modified for other purposes.

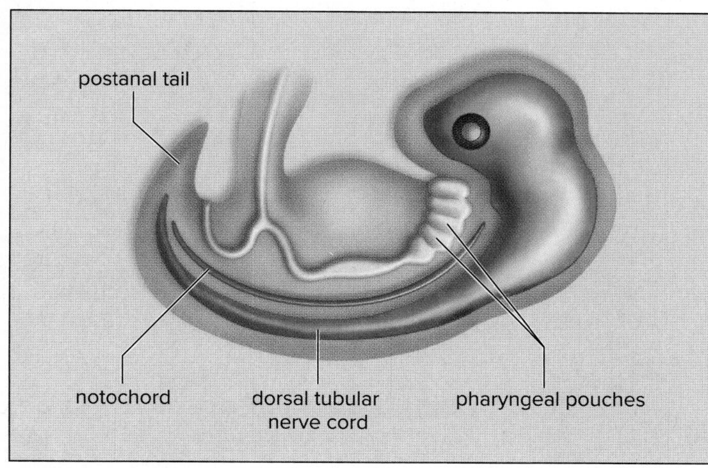

25.1 Evolution of Chordates

The evolutionary tree of the chordates (Fig. 25.1) shows that some chordates, notably the **tunicates** and **lancelets,** are not vertebrates. These chordates retain the notochord and are called the **invertebrate chordates.**

Explain the term *invertebrate chordates.* ______________________________

The other animal groups in Figure 25.1 are **vertebrates** in which the notochord has been replaced by the vertebral column. Fishes include three groups: the **jawless fishes** were the first to evolve, followed by the **cartilaginous fishes** and then the **bony fishes.** The bony fishes include the **ray-finned** fishes (the largest group of vertebrates) and the **lobe-finned fishes.** The first lobe-finned fishes had a bony skeleton, fleshy appendages, and a lung. These lobe-finned fishes lived in shallow pools and gave rise to the amphibians.

What three features called out in Figure 25.1 evolved among fishes? ______________________________

____

The terrestrial vertebrates are all **tetrapods** because they have four limbs. The limbs of tetrapods are ______________ appendages just like those of arthropods. Amphibians still return to the water to reproduce, but **reptiles** are fully adapted to life on land because among other features they produce an **amniotic egg.** The amniotic egg is so named because the embryo is surrounded by an amniotic membrane that encloses amniotic fluid. Therefore, amniotes develop in an aquatic environment of their own making. Do all animals develop in a water environment? __________ Explain. ______________________________

____

In placental mammals, such as humans, the fertilized egg develops inside the female, where the unborn mammal receives nutrients via the placenta. Reptiles (including birds) and mammals have many other adaptations that are suitable to living on land, as we will stress in later sections.

Figure 25.1 Phylogenetic tree of the chordates.
Evolution of chordates is marked by at least seven innovations.

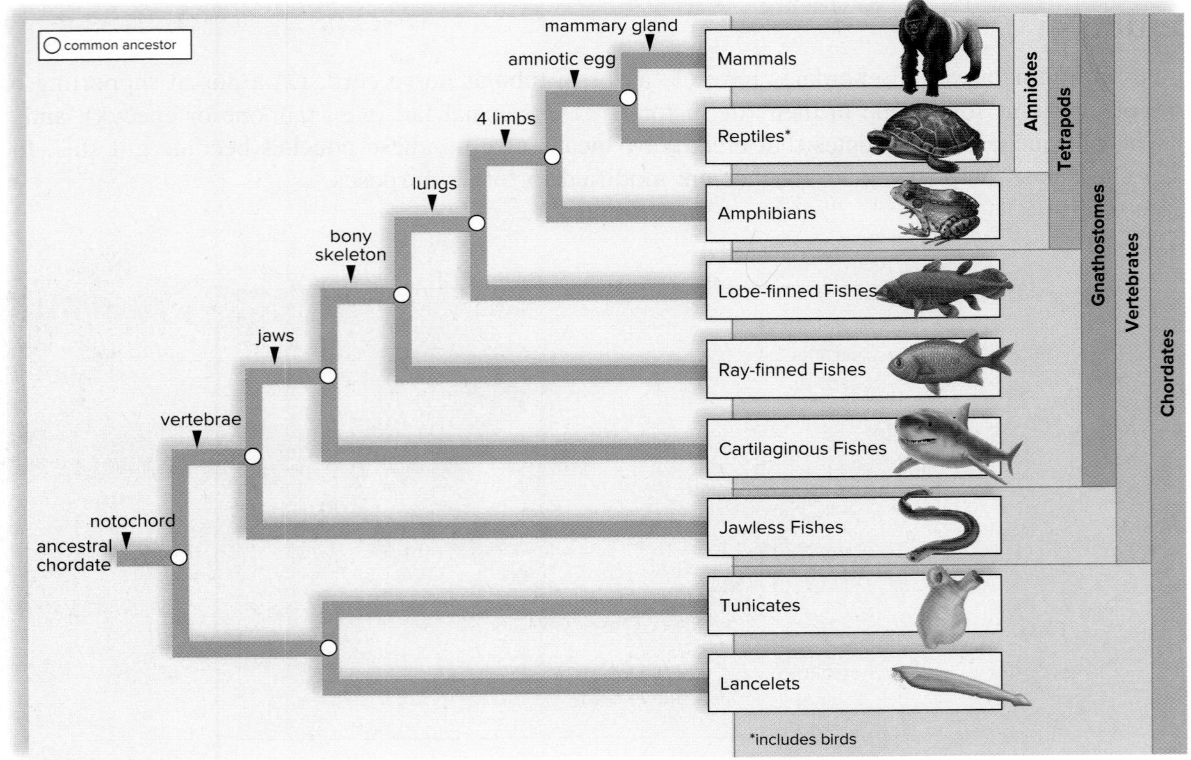

25.2 Invertebrate Chordates

Among the chordates (phylum Chordata), two groups contain invertebrates, and the others contain the vertebrates.

Invertebrate Chordates

The two types of invertebrate chordates are urochordates and cephalochordates.

1. **Urochordates.** The tunicates, or sea squirts (Fig. 25.2), come in varying sizes and shapes, but all have incurrent and excurrent siphons. **Gill slits** are the only remaining chordate characteristic in adult tunicates. Examine any examples of tunicates on display.
2. **Cephalochordates.** Lancelets, also known as amphioxus (*Branchiostoma*), are small, fishlike animals that occur in shallow marine waters in most parts of the world. They spend most of their time buried in the sandy bottom, with only the anterior end projecting.

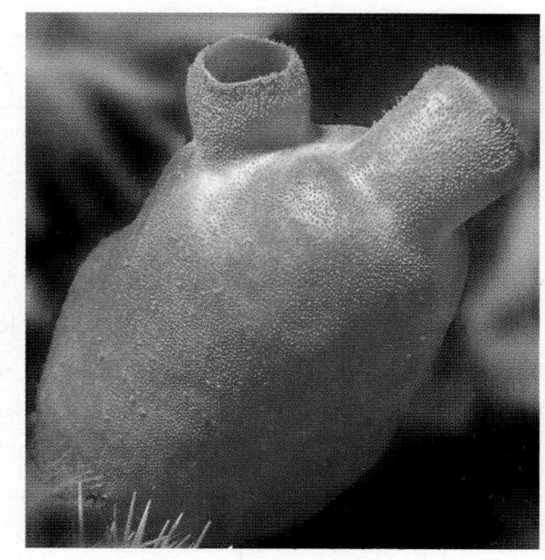

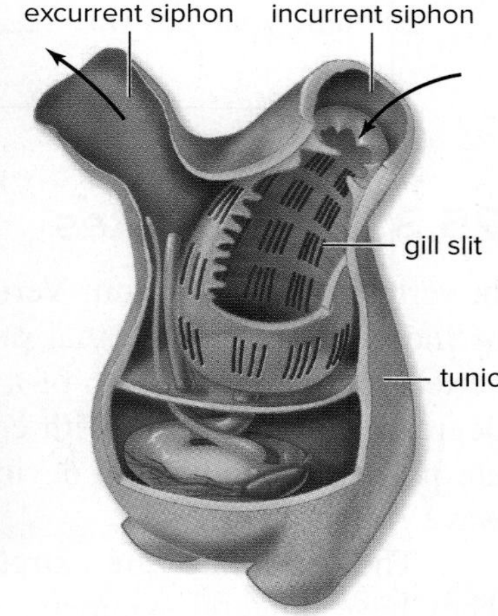

Figure 25.2 Urochordates.
The gill slits of a tunicate are the only chordate characteristic remaining in the adult. (left) ©Andrew J. Martinez/Science Source

Observation: Lancelet Anatomy

Preserved Specimen

1. Examine a preserved lancelet (Fig. 25.3).
2. Identify the **caudal fin** (enlarged tail) used in locomotion, the **dorsal fin,** and the short **ventral fin.**
3. Examine the lancelet's V-shaped muscles.

Figure 25.3 Anatomy of the lancelet, *Branchiostoma*.
Lancelets feed on microscopic particles filtered out of the constant stream of water that enters the mouth and exits through the gill slits into a protective atrium formed by body folds. The water exits at the atriopore.
(right) ©Natural Visions/Alamy Stock Photo

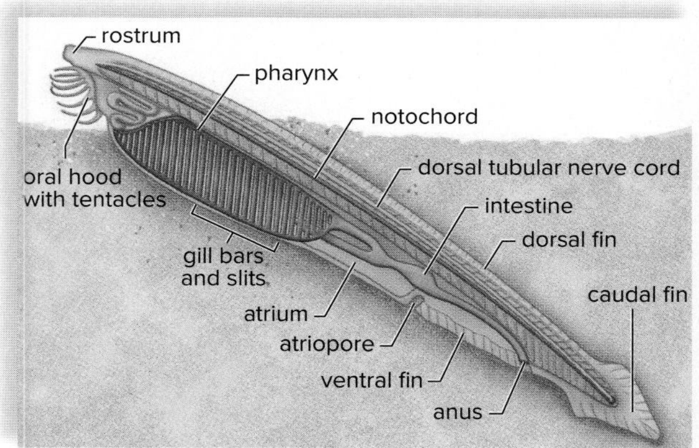

4. Find the tentacled **oral hood,** located anterior to the mouth and covering a vestibule. Water entering the mouth is channeled into the **pharynx,** where food particles are trapped before the water exits at the **atriopore.** Lancelets are filter feeders. Has cephalization occurred? ____________________

   Explain your answer. ____________________

   ____

25.3 Vertebrates

In vertebrates (subphylum Vertebrata) the embryonic notochord is replaced by a vertebral column composed of individual vertebrae that protect the nerve cord. The internal jointed skeleton consists not only of a vertebral column, but also of a skull that encloses and protects the well-developed brain. There is an extreme degree of cephalization with complex sense organs. The eyes develop as outgrowths of the brain. The ears are primarily equilibrium devices in aquatic vertebrates; in land vertebrates, they also function as sound wave receivers.

The vertebrates are extremely motile and have well-developed muscles and usually paired appendages. They have bilateral symmetry and are segmented, as witnessed by the vertebral column. There is a large body cavity, a complete gut with both a mouth and an anus (or instead, a cloacal opening), and the circulatory system consists of a well-developed heart and many blood vessels. They have an efficient means of extracting oxygen from water (gills) or air (lungs) as appropriate. The kidneys are important excretory and water-regulating organs that conserve or rid the body of water as necessary. The sexes are generally separate, and reproduction is usually sexual.

Figure 25.4 Vertebrate groups.
(shark) ©Alastair Pollock Photography/Getty Images RF; (fish) ©Hal Beral/Corbis/Getty Images RF; (frog) ©Rod Planck/Science Source; (turtle) ©Suzanne and Joseph Collins/Science Source; (bird) ©Rick & Nora Bowers/Alamy Stock Photo; (fox) ©Craig Lorenz/Science Source

Tiger shark

Cartilaginous fishes
Lack operculum and swim bladder; tail fin usually asymmetrical (sharks, skates, and rays)

Banded butterflyfish

Bony fishes
Operculum; swim bladder or lungs; tail fin usually symmetrical: lung-fishes, lobe-finned fishes, and ray-finned fishes (herring, salmon, sturgeon, eels, and sea horse)

Northern leopard frog

Amphibians
Tetrapods with nonamniotic egg; nonscaly skin; some show metamorphosis; three-chambered heart (salamanders, frogs, and toads)

Pearl River redbelly turtle

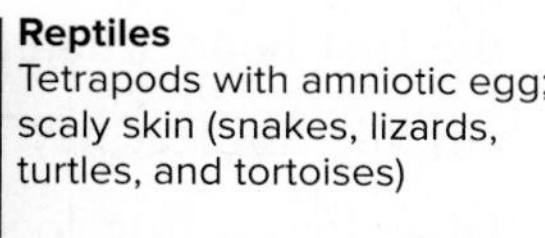

Reptiles
Tetrapods with amniotic egg; scaly skin (snakes, lizards, turtles, and tortoises)

Scissor-tailed flycatcher

Birds
Now grouped with reptiles; tetrapods with feathers; bipedal with wings; double circulation (sparrows, penguins, and ostriches)

Gray fox

Mammals
Tetrapods with hair, mammary glands; double circulation; teeth differentiated: monotremes (spiny anteater and duckbill platypus), marsupials (opossum and kangaroo), and placental mammals (whales, rodents, dogs, cats, elephants, horses, bats, and humans)

Anatomy of the Frog

In this laboratory the anatomy of the frog will be considered typical of vertebrates. Frogs are amphibians, a group of animals in which metamorphosis occurs. Metamorphosis includes a change in structure, as when an aquatic tadpole becomes a frog with lungs and limbs (Fig. 25.5). Amphibians were the first vertebrates to be adapted to living on land; however, they typically return to the water to reproduce. Underline every structure mentioned in the following Observation that represents an adaptation to a land environment.

Figure 25.5 External frog anatomy.
©Rod Planck/Science Source

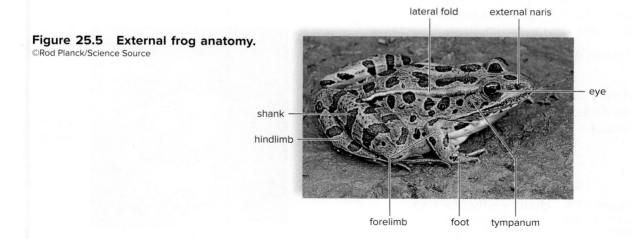

Observation: External Anatomy of the Frog

1. Place a preserved frog (*Rana pipiens*) in a dissecting tray.
2. Identify the bulging eyes, which have a nonmovable upper and lower lid but can be covered by a **nictitating membrane** that serves to moisten the eye.
3. Locate the **tympanum** behind each eye (Fig. 25.5). What is the function of a tympanum? ______________________________
4. Examine the external **nares** (sing., **naris,** or **nostril**). Insert a probe into an external naris, and observe that it protrudes from one of the paired small openings, the internal nares (Fig. 25.6), inside the mouth cavity. What is the function of the nares? ______________________________
5. Identify the paired limbs. The bones of the fore- and hindlimbs are the same as in all tetrapods, in that the first bone articulates with a girdle and the limb ends in phalanges. The hind feet have five phalanges, and the forefeet have only four phalanges. Which pair of limbs is longest? ______________________________

   How does a frog locomote on land? ______________________________

   What is a frog's means of locomotion in the water? ______________________________

Observation: Internal Anatomy of the Frog

Mouth

1. Open your frog's mouth very wide (Fig. 25.6), cutting the angles of the jaws if necessary.
2. Identify the tongue attached to the lower jaw's anterior end.
3. Find the **auditory (eustachian) tube** opening in the angle of the jaws. These tubes lead to the ears. Auditory tubes equalize air pressure in the ears.
4. Examine the **maxillary teeth** located along the rim of the upper jaw. Another set of teeth—**vomerine teeth**—is present just behind the midportion of the upper jaw.
5. Locate the **glottis,** a slit through which air passes into and out of the **trachea,** the short tube from glottis to lungs. What is the function of a glottis? ______________________________
6. Identify the **esophagus,** which lies dorsal and posterior to the glottis and leads to the stomach.

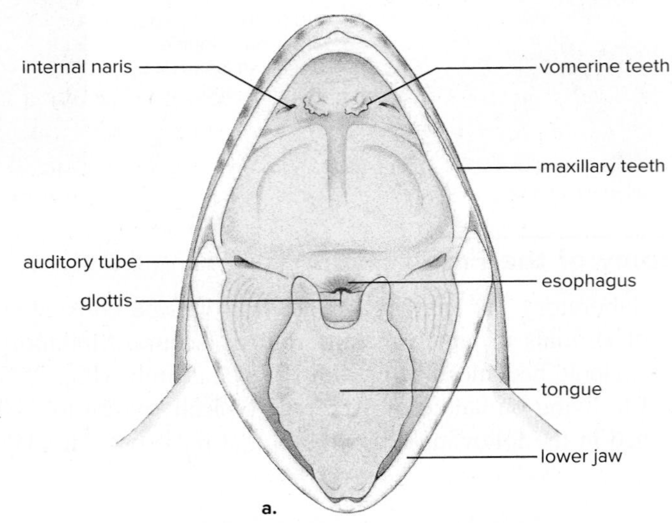

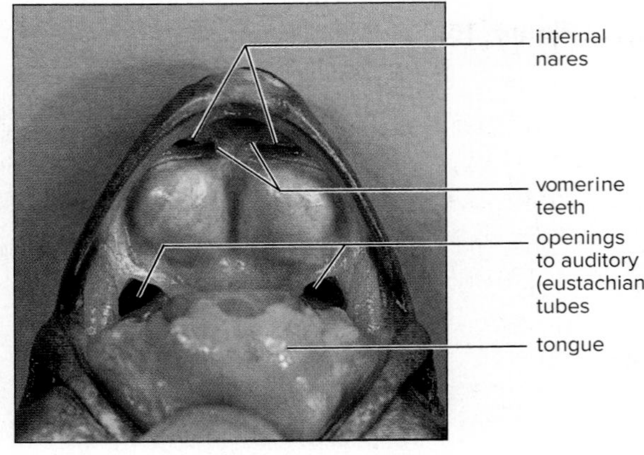

Figure 25.6 Mouth cavity of a frog.
a. Drawing. **b.** Dissected specimen.
(b) ©Carolina Biological Supply Company/Phototake

Opening the Frog

1. Place the frog ventral side up in the dissecting pan. Lift the skin with forceps, and use scissors to make a large, circular cut to remove the skin from the abdominal region as close to the limbs as possible. Cut only skin, not muscle.
2. Now, remove the muscles by cutting through them in the same circular fashion. At the same time, cut through any bones you encounter. A vein, called the abdominal vein, will be slightly attached to the internal side of the muscles.
3. Identify the **coelom,** or body cavity. Recall that vertebrates are deuterostomes in which the first embryonic opening becomes the anus and the second opening becomes the mouth.
4. If your frog is female, the abdominal cavity is likely to be filled by a pair of large, transparent **ovaries,** each containing hundreds of black and white eggs. Gently lift the left ovary with forceps, and find its place of attachment. Cut through the attachment, and remove the ovary in one piece.

Respiratory System and Liver

1. Insert a probe into the glottis, and observe its passage into the trachea. Enlarge the glottis by making short cuts above and below it. When the glottis is spread open, you will see a fold on either side; these are the vocal cords used in croaking.
2. Identify the **lungs,** two small sacs on either side of the midline and partially hidden under the liver (Fig. 25.7). Sequence the organs in the respiratory tract to trace the path of air from the external nares to the lungs. ______________________________

   ____
3. Locate the **liver,** the large, prominent, dark-brown organ in the midventral portion of the trunk (Fig. 25.7). Between the right half and left half of the liver, find the **gallbladder.**

Circulatory System

1. Lift the liver gently. Identify the **heart,** covered by a membranous covering (the **pericardium**). With forceps, lift the covering, and gently slit it open. The heart consists of a single, thick-walled **ventricle** and two (right and left) anterior, thin-walled **atria.**
2. Locate the three large veins that join together beneath the heart to form the **sinus venosus.** (To lift the heart, you may have to snip the slender strand of tissue that connects the atria to the pericardium.) Blood from the sinus venosus enters the right atrium. The left atrium receives blood from the lungs.
3. Find the **conus arteriosus,** a single, wide arterial vessel leaving the ventricle and passing ventrally over the right atrium. Follow the conus arteriosus forward to where it divides into three branches on each side. The middle artery on each side is the **systemic artery,** which fuses behind the heart to become the **dorsal aorta.** The dorsal aorta transports blood through the body cavity and gives off many branches. The **posterior vena cava** begins between the two kidneys and returns blood to the sinus venosus. Which vessel lies above (ventral to) the other? ______________________________

   ____

Digestive Tract

1. Identify the **esophagus,** a very short connection between the mouth and the stomach. Lift the left liver lobe, and identify the stomach, whitish and J-shaped. The **stomach** connects with the esophagus anteriorly and with the small intestine posteriorly.
2. Find the **small intestine** and the **large intestine,** which enters the **cloaca.** The cloaca lies beneath the pubic bone and is a general receptacle for the intestine, the reproductive system, and the urinary system. It opens to the outside by way of the anus. Sequence the organs in the digestive tract to trace the path of food from the mouth to the cloaca. ______________________________

   ____

Accessory Glands

1. You identified the liver and gallbladder previously. Now try to find the **pancreas,** a yellowish tissue near the stomach and intestine.
2. Lift the stomach to see the **spleen,** a small, pea-shaped body.

Figure 25.7 Internal organs of a female frog, ventral view. ©Ken Taylor/Wildlife Images

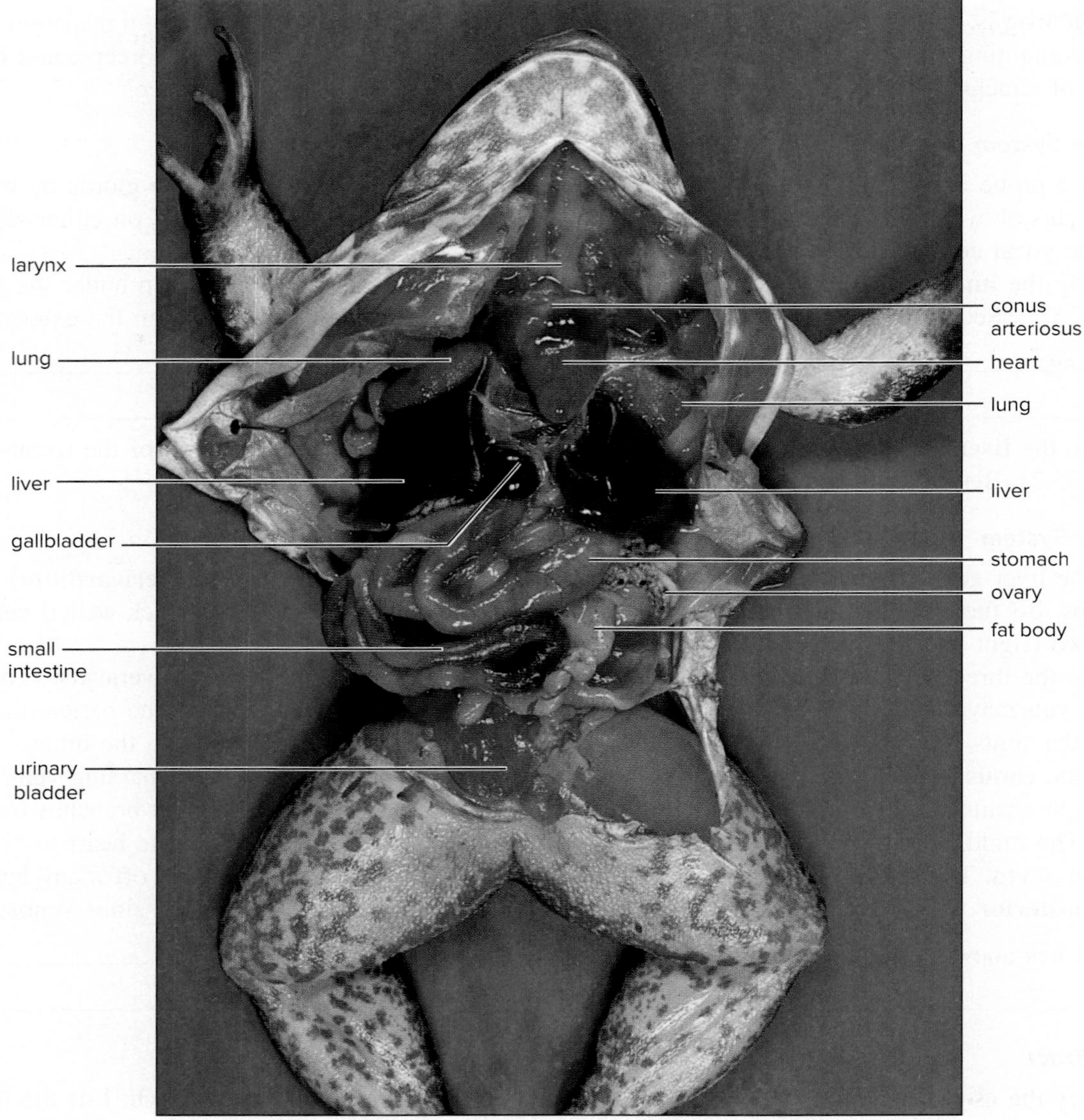

Urogenital System

1. Identify the **kidneys,** long, narrow organs lying against the dorsal body wall (Fig. 25.8).
2. Locate the **testes** in a male frog (Fig. 25.8). Testes are yellow, oval organs attached to the anterior portions of the kidneys. Several small ducts, the **vasa efferentia,** carry sperm into kidney ducts that also carry urine from the kidneys. **Fat bodies,** which store fat, are attached to the testes.

3. Locate the ovaries in a female frog. The ovaries are attached to the dorsal body wall (Fig. 25.9). Fat bodies are also attached to the ovaries. Highly coiled **oviducts** lead to the cloaca. The ostium (opening) of the oviduct is dorsal to the liver.
4. Find the **mesonephric ducts**—thin, white tubes that carry urine from the kidney to the cloaca. In female frogs, you will have to remove the left ovary to see the mesonephric ducts.
5. Locate the **cloaca.** You will need to split through the bones of the pelvic girdle in the midventral line and carefully separate the bones and muscles to find the cloaca.
6. Identify the urinary bladder attached to the ventral wall of the cloaca. In frogs, urine backs up into the bladder from the cloaca.
7. Explain the term *urogenital system.* ______________________________
____
8. The cloaca receives material from (1) ______________________________,
(2) ______________________________, and (3) ______________________________.
9. Compare the frog's urogenital system to the human urinary system, which in females has no connection to the genital system. *Beside each organ listed on the right, tell how the comparable frog organ differs from that of a human.*

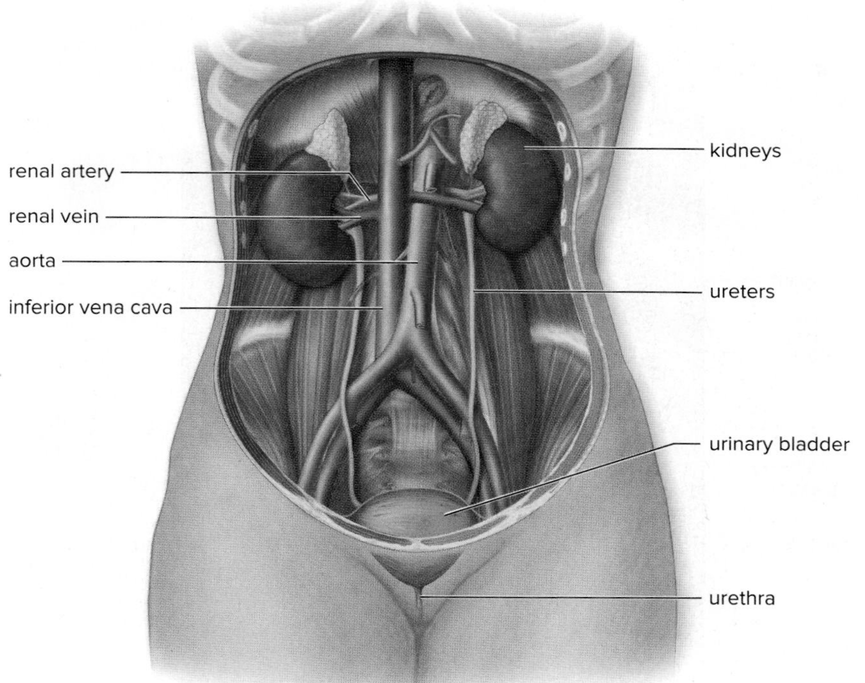

Figure 25.8 Urogenital system of a male frog.
a. Drawing. **b.** Dissected specimen. (b) ©Ken Taylor/Wildlife Images

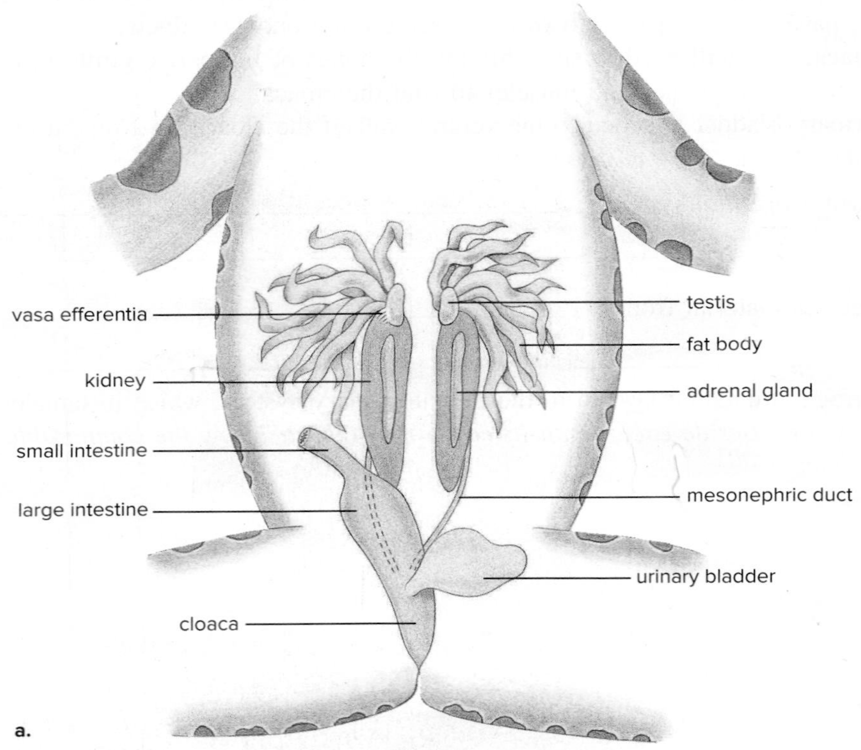

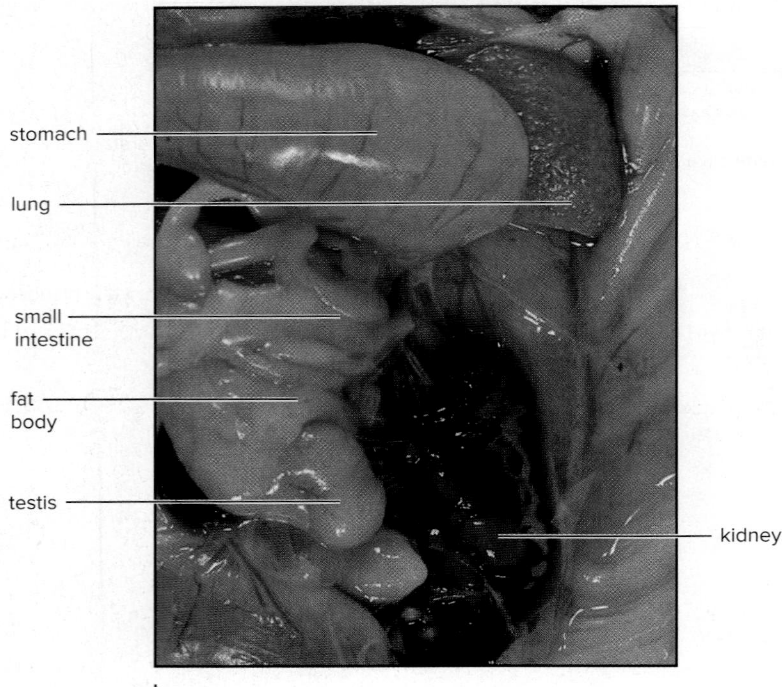

Figure 25.9 Urogenital system of a female frog.
a. Drawing. **b.** Dissected specimen.
(b) ©Ken Taylor/Wildlife Images

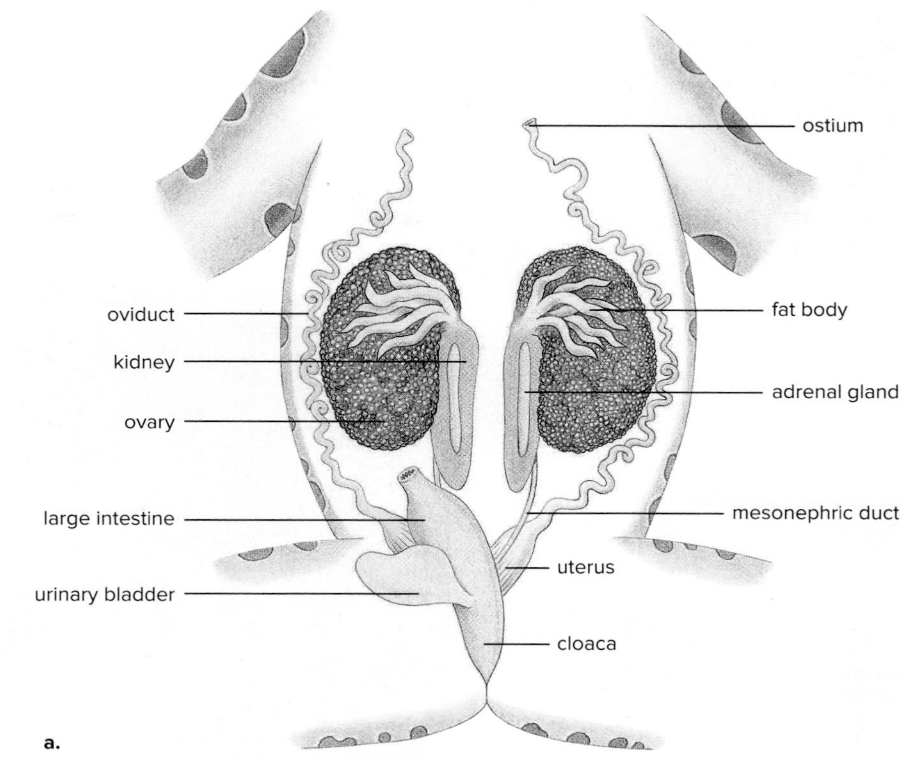

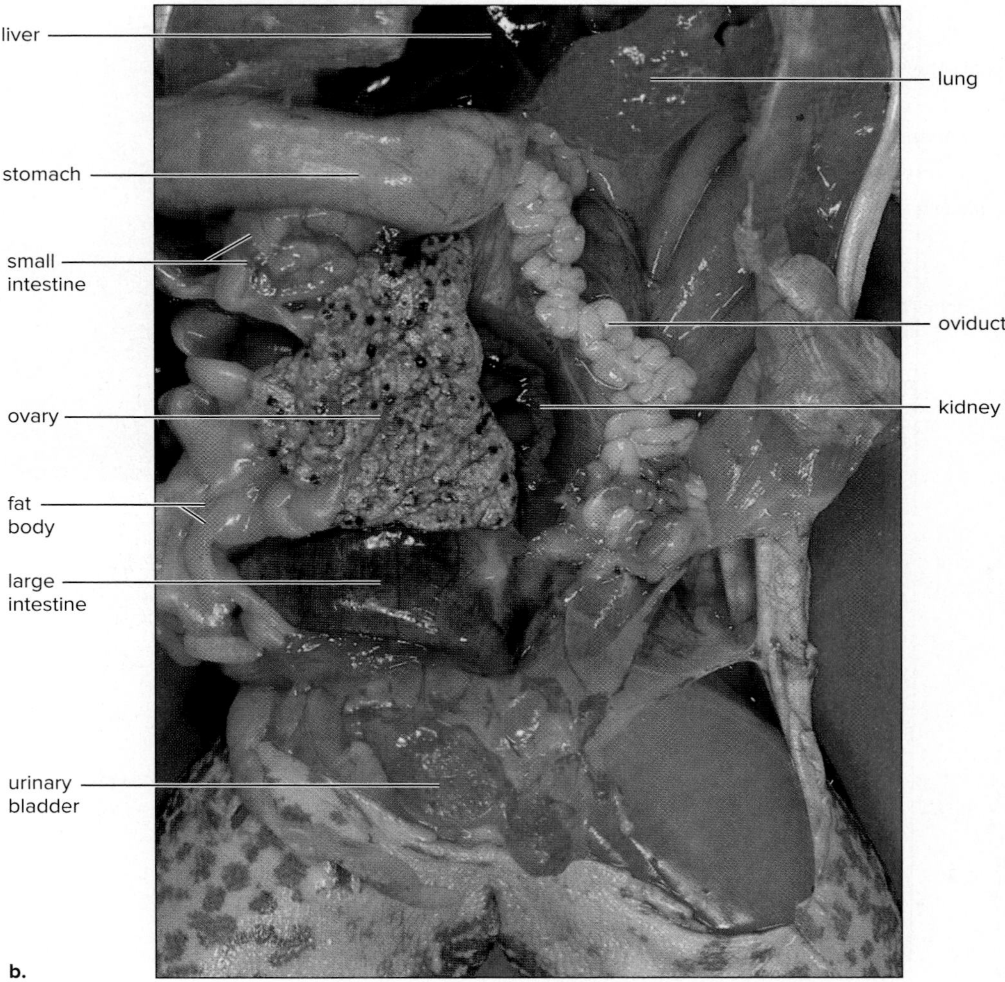

Nervous System

In the frog demonstration dissection, identify the **brain,** lying exposed within the skull. With the help of Figure 25.10, find the major parts of the brain.

Figure 25.10 Frog brain, dorsal view.
a. Drawing. **b.** Dissected specimen. (b) ©Ken Taylor/Wildlife Images

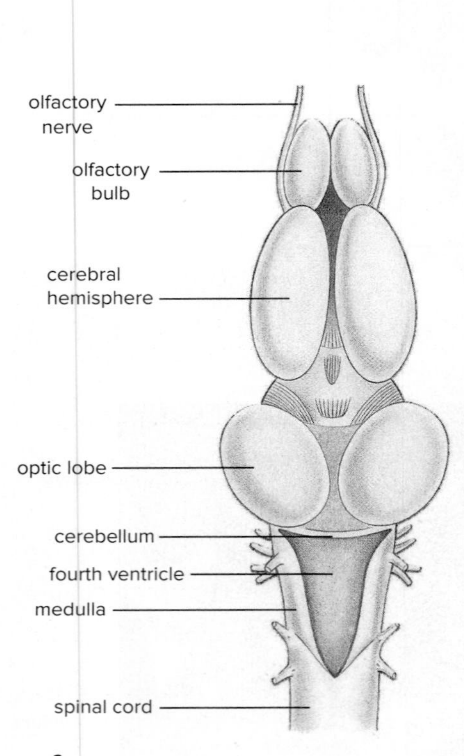

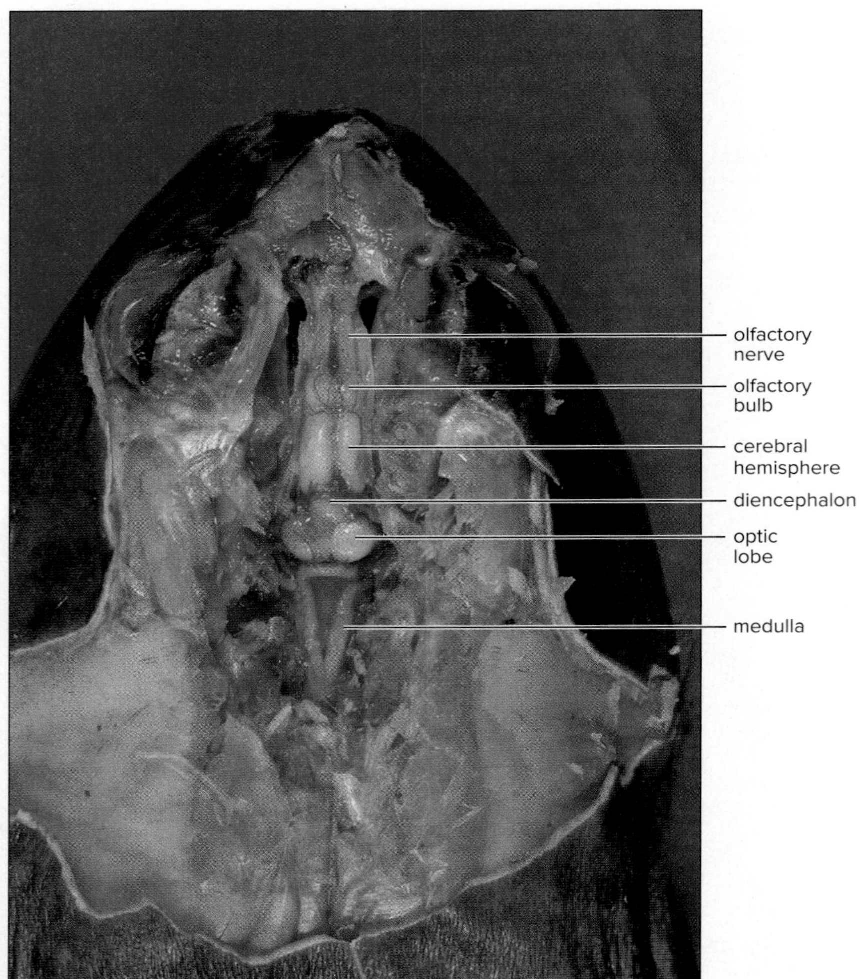

25.4 Comparative Vertebrate Anatomy

In addition to the frog (amphibian), examine the perch (fish), pigeon (reptile), and rat (mammal) on display.

Observation: External Anatomy of Vertebrates

1. Compare the external features of the perch, frog, pigeon, and rat by examining specimens in the laboratory. Answer the following questions and record your observations in Table 25.1.

- **a.** Is the skin smooth, scaly, hairy, or feathery?
- **b.** Is there any external evidence of segmentation?
- **c.** Are all forms bilaterally symmetrical?
- **d.** Is the body differentiated into regions?
- **e.** Is there a well-defined neck?
- **f.** Is there a postanal tail?
- **g.** Are there nares (nostrils)?
- **h.** Is there a cloaca opening, or are there urogenital and anal openings?
- **i.** Are eyelids present? How many?
- **j.** How many appendages are there? (Fins are considered appendages.) (Fig. 25.11)

Table 25.1 Comparison of External Features

	Perch	Frog	Pigeon	Rat
a. Skin				
b. Segmentation				
c. Symmetry				
d. Regions				
e. Neck				
f. Postanal tail				
g. Nares				
h. Cloaca				
i. Eyelids				
j. Appendages				

2. Evidence that birds are reptiles: Birds
   a. have feathers, which are modified scales.
   b. have scales on their feet.
   c. and reptiles both lay eggs.
   d. and reptiles have similar internal organs.
   e. and reptiles also show some skeletal (skull) similarities.

   Which of these can you substantiate by external examination? ______________________

   ____

3. The perch, which lives in fresh water, and the pigeon and rat, which live on land, have a nearly impenetrable covering. Why is this an advantage in each case? ______________________

   ____

4. A frog uses its skin for breathing. Is the skin of a frog thick and dry or thin and moist? Explain. ________

   ____

Figure 25.11 Perch anatomy.
All fins are shown, except the pectoral fin.

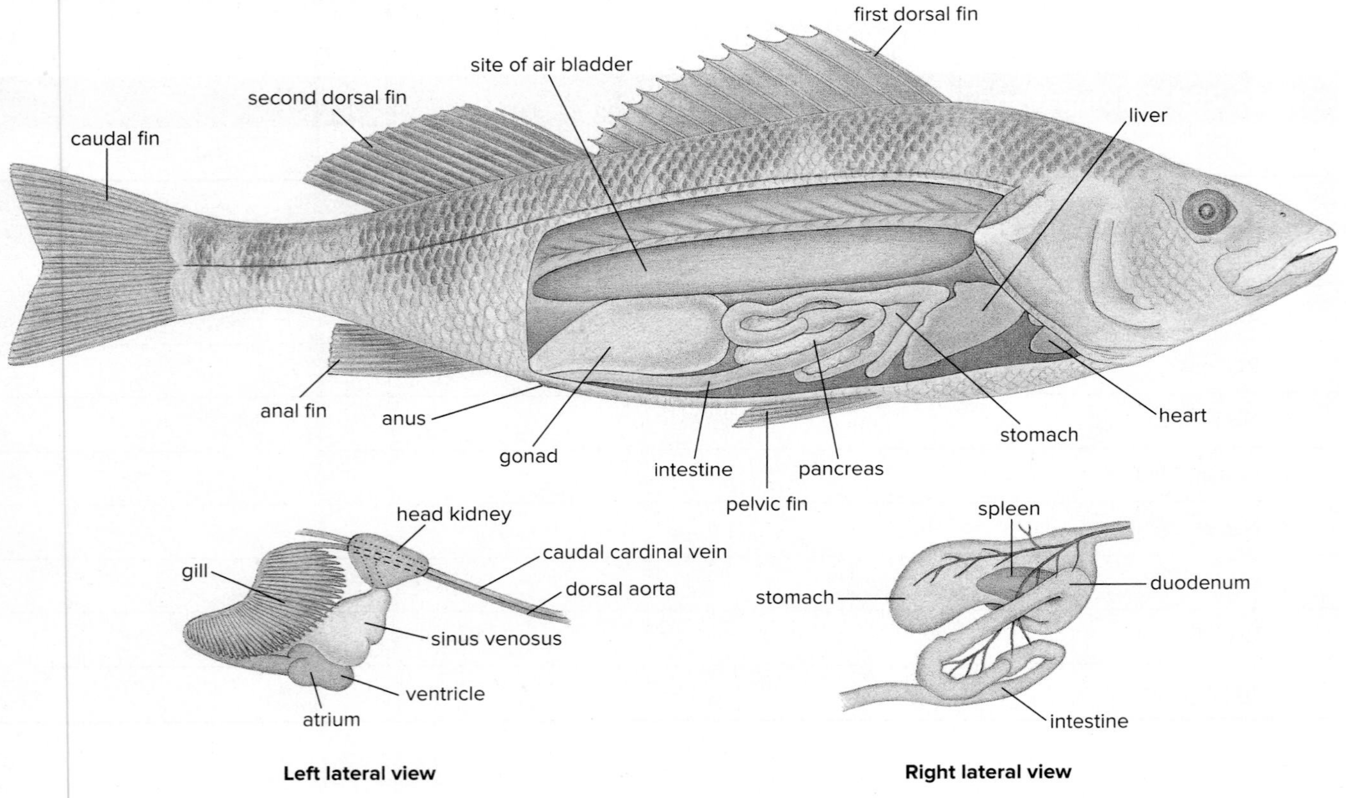

Observation: Internal Anatomy of Vertebrates

1. Examine the internal organs of the perch, frog, pigeon (Fig. 25.12) and rat (Fig. 25.13).
2. If necessary, make a median longitudinal incision in the ventral body wall, from the jaws to the anus. The body cavity is called a coelom because it is completely lined by mesoderm.
3. Which of these animals has a **diaphragm** dividing the body cavity into **thorax** and **abdomen?** ________

   ____

Figure 25.12 Pigeon anatomy.

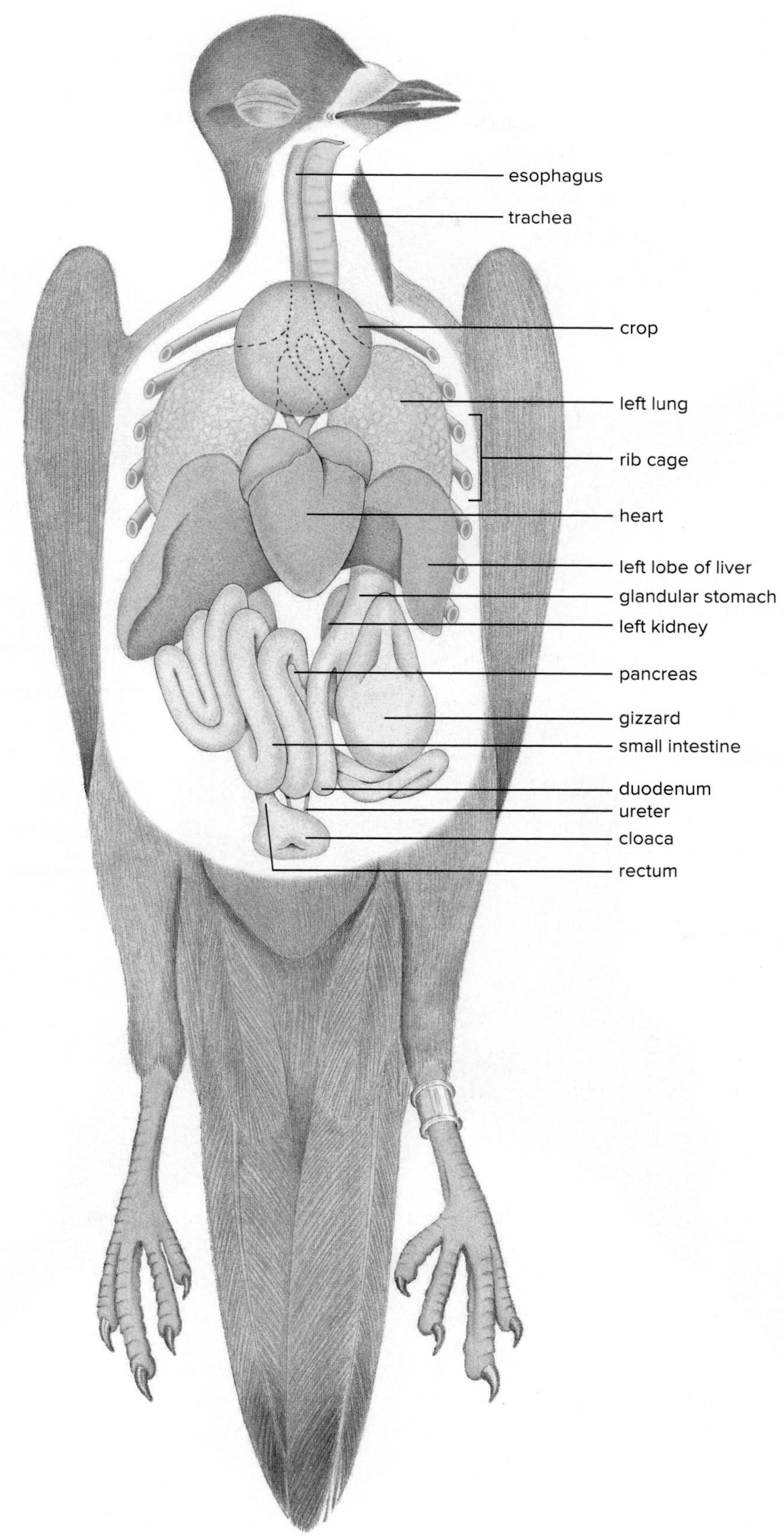

Figure 25.13 Rat anatomy.

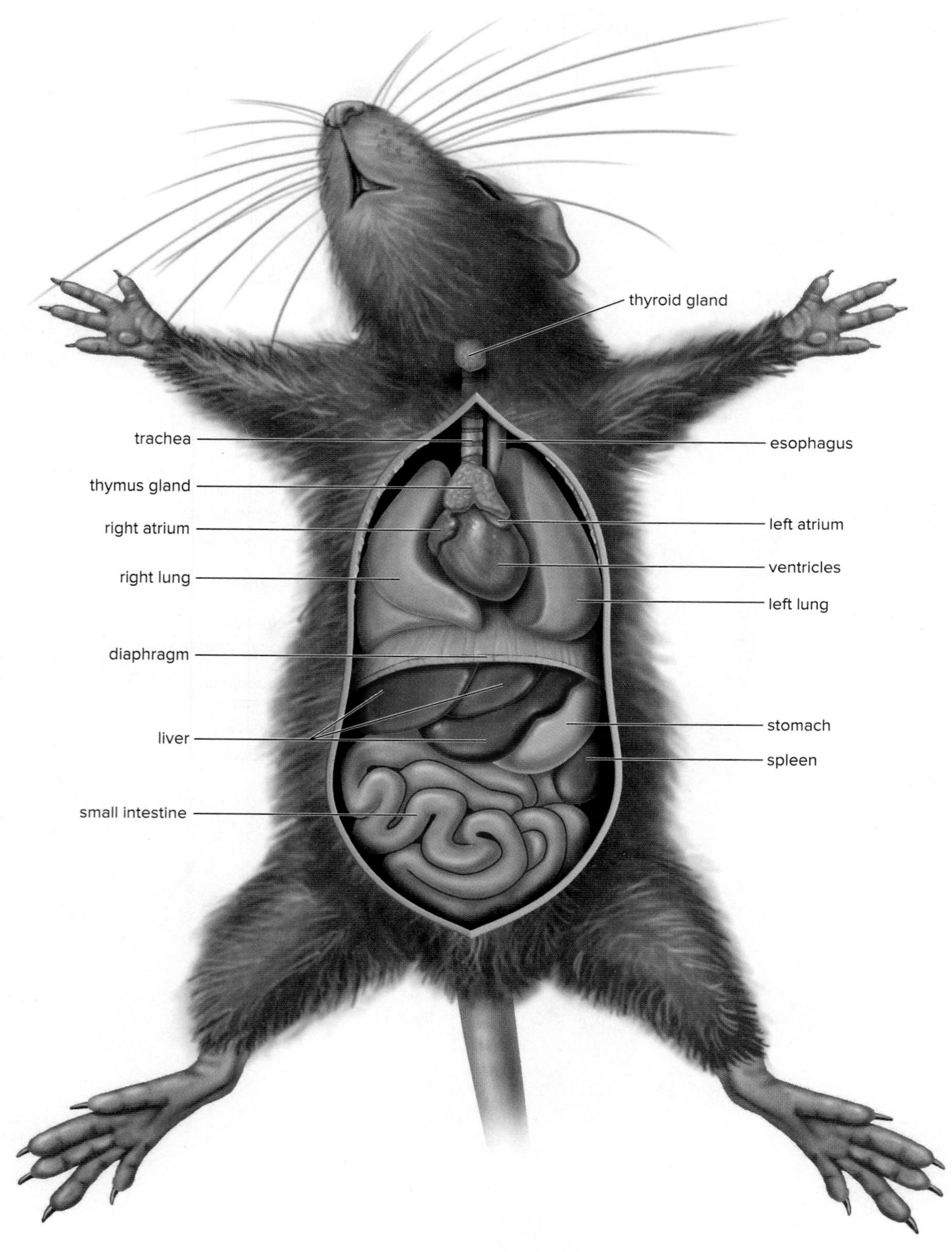

Digestive Systems and Urogenital Systems

1. All the vertebrates have a stomach, small intestine, and large intestine where food is processed. They also have a liver and pancreas. Which is the larger, more prominent organ (liver or pancreas) in the pigeon and rat? ____________________

2. All vertebrates have an anus. As noted in Table 25.1, which vertebrates studied have a cloaca (receptacle for the urogenital and digestive systems)? ____________________
   In the other vertebrates studied, the urogenital and digestive systems are separate.

3. **Urogenital systems.** All the animals have gonads and kidneys. Which of these organs is involved in urine production? ____________________
   The kidneys . . . the kidneys help maintain the proper balance of fluid and salts in the blood. Which of these organs is involved in reproduction? ____________________
   The sexes are separate in vertebrates: Females have ovaries and males have testes. In reptiles and mammals, males usually have a penis to pass sperm to the female. What is the chief biological benefit of the penis in terrestrial animals? ____________________

Circulatory Systems

Study heart models for a fish, amphibian, bird, and mammal. Trace the path of the vessel that leaves the ventricle(s), and determine whether the animals have a **pulmonary system** (Fig. 25.14). The word *pulmonary* comes from the Latin *pulmonarius,* meaning "of the lungs."

Figure 25.14 Cardiovascular systems in vertebrates.
a. In a fish, the blood moves in a single loop. The heart has a single atrium and ventricle, which pumps the blood into the gill region, where gas exchange takes place. **b.** Amphibians have a double-loop system in which the heart pumps blood to both the lungs and the body. **c.** In birds and mammals, the right side pumps blood to the lungs, and the left side pumps blood to the body.

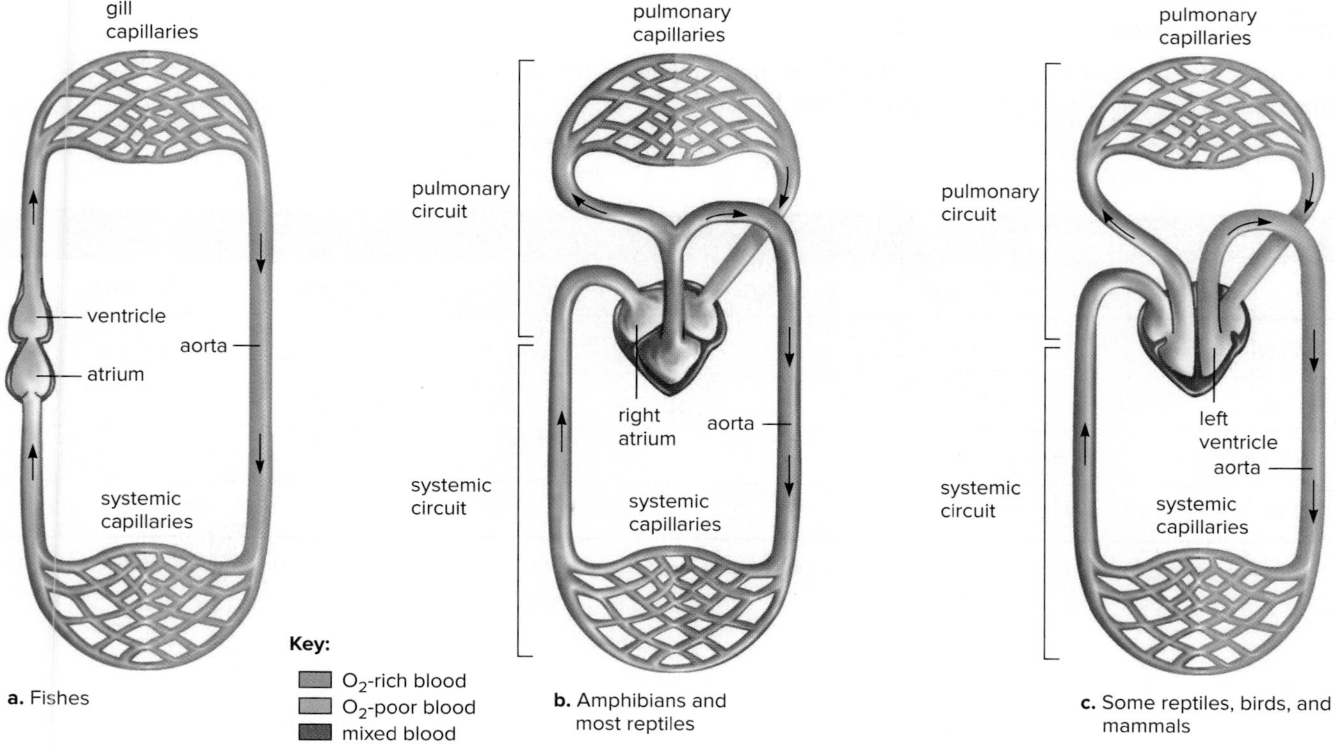

Complete Table 25.2.

Table 25.2 Comparative Circulatory Systems

Animal	Number of Heart Chambers	Pulmonary Circuit (Yes or No)
Perch		
Frog		
Pigeon		
Rat		

1. Do fish have a blood vessel that returns blood from the gills to the heart? ________ Would you expect blood pressure to be high or low after blood has moved through the gills? ________
2. What animals studied have pulmonary vessels that take blood from the heart to the respiratory organ and back to the heart? ________ What is the advantage of a pulmonary circuit? ________
3. Which of these animals has a four-chambered heart? ________
   What is the advantage of having separate ventricles? ________
4. The circulatory system distributes the heat of muscle contraction in birds and mammals. Is the anatomy of birds and mammals conducive to maintaining a warm internal temperature? ________
   Explain your answer. ________

Respiratory Systems

Compare the respiratory systems of the perch, frog, pigeon, and rat, and complete Table 25.3 by checking the anatomical features that appear in each animal.

Table 25.3 Respiratory Systems

	Gills	Trachea	Lungs	Rib Cage*	Diaphragm	Air Sacs
Perch						
Frog						
Pigeon						
Rat						

*A rib cage consists of ribs plus a sternum. Some ribs are connected to the sternum, which lies at the midline in the anterior portion of the rib cage.

1. Among the animals studied, only a perch breathes by ____________. Can the particular respiratory organ be related to the environment of the animals? ____________ Explain your answer. ____________

____

2. Knowing that gills are attached to the pharynx (throat in humans), explain why fish have no trachea. ____

____

____

3. A rib cage is present in the rat and pigeon but missing in the frog. Can this difference be related to the fact that frogs breathe by positive pressure, while birds and mammals breathe by negative pressure? ____________ (A frog swallows air and then pushes the air into its lungs; in birds and mammals, the thorax expands first, and then the air is drawn in.) Explain. ____________

____

4. A diaphragm is present only in mammals (e.g., rat). Of what benefit is this feature to the expansion of lungs in mammals? ____________

____

____

5. Air sacs (not shown in Fig. 25.12) are present only in birds. This feature allows air to pass one way through the lungs of a bird and greatly increases the bird's ability to extract oxygen from the air.

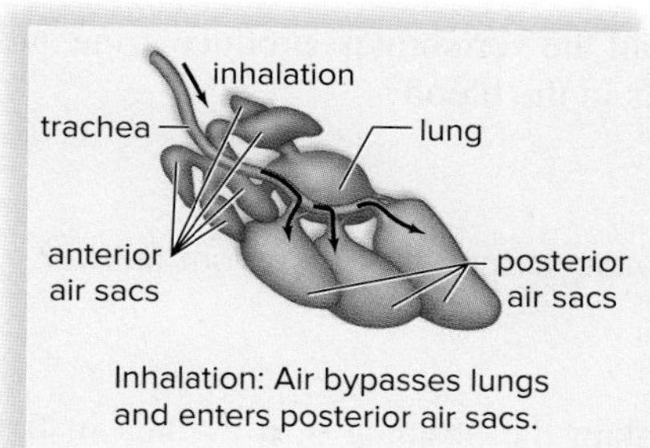

Inhalation: Air bypasses lungs and enters posterior air sacs.

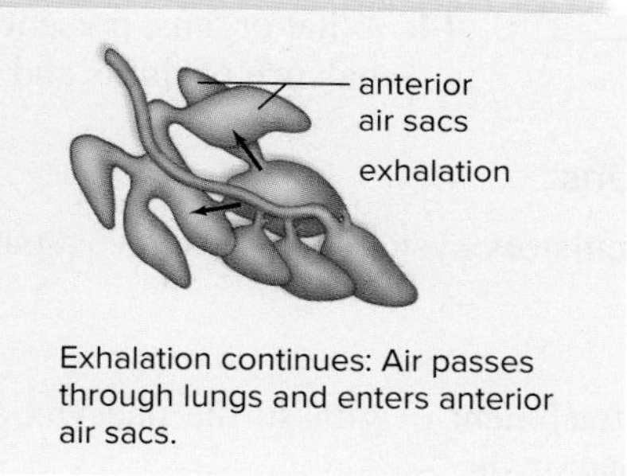

Exhalation continues: Air passes through lungs and enters anterior air sacs.

Laboratory Review 25

_______________ 1. What kind of development, protostome or deuterostome, is common to chordates and echinoderms?

_______________ 2. Is the nerve cord dorsal or ventral in the invertebrate chordates, tunicates and lancelets?

_______________ 3. What structure is replaced by the vertebral column in vertebrate chordates?

_______________ 4. Do cartilaginous fishes or ray-finned fishes have a bony skeleton?

_______________ 5. What term refers to terrestrial vertebrates that have four limbs?

_______________ 6. What do reptiles produce that ensures their embryos won't dry out on land?

_______________ 7. What term refers to the changes that occur when an aquatic tadpole becomes a frog with limbs and lungs?

_______________ 8. What is the name of the membranous covering associated with the heart?

_______________ 9. What structure in the frog receives materials from the large intestine, mesonephritic ducts, and the reproductive organs?

_______________ 10. Which animals have modified scales covering their bodies, have scales on their feet, lay shelled eggs, and have internal organs and skulls that provide evidence to classify them as reptiles?

_______________ 11. What organs are present if a vertebrate has a pulmonary system of circulation with pulmonary capillaries?

_______________ 12. How many ventricles are present in the heart of an amphibian and most reptiles?

_______________ 13. Which vertebrates have air sacs attached to their lungs that make the lungs more efficient?

_______________ 14. What organs, present in all the vertebrates, produce urine and maintain the proper balance of fluids and salts in the blood?

Thought Questions

15. a. Does the circulatory system of fishes have a pulmonary circuit? Explain your answer.

b. Given the attachment of gills to the pharynx, explain the absence of a trachea in fishes and the need for a trachea in mammals.

16. What is the role of the four-chambered hearts of birds and mammals in maintaining a warm body temperature?

17. a. Which of the animals studied in this laboratory exercise have a rib cage? Explain the importance of a rib cage to respiration.

b. Which of the animals studied in this laboratory exercise have a diaphragm? Explain the importance of a diaphragm to respiration.

26

Animal Organization

Learning Outcomes

26.1 Epithelial Tissue
- Identify slides and models of various types of epithelium.
- Explain where particular types of epithelium are located in the body, and describe their function.

26.2 Connective Tissue
- Identify slides and models of various types of connective tissue.
- Explain where particular connective tissues are located in the body, and describe their function.

26.3 Muscular Tissue
- Identify slides and models of three types of muscular tissue.
- Explain where each type of muscular tissue is located in the body, and describe their function.

26.4 Nervous Tissue
- Identify a slide and model of a neuron.
- Explain where nervous tissue is located in the body, and describe its function.

26.5 Organ Level of Organization
- Identify a slide of the intestinal wall and the layers in the wall. Describe the function of each tissue.
- Identify a slide of skin and the two regions of skin. Describe the function for each region of skin.

Introduction

Humans, as well as all other organisms, are made up of **cells.** Groups of cells that have the same structural characteristics and perform the same functions are called **tissues.** Figure 26.1 shows the four categories of tissues in the human body. An **organ** is composed of different types of tissues, and various organs form **organ systems.** Humans thus have the following levels of biological organization:
cells ⟶ tissues ⟶ organs ⟶ organ systems.

The photomicrographs of tissues in this laboratory were obtained by viewing prepared slides with a light microscope. Preparation required the following sequential steps:

1. **Fixation:** The tissue is immersed in a preservative solution to maintain the tissue's existing structure.
2. **Embedding:** Water is removed with alcohol, and the tissue is impregnated with paraffin wax.
3. **Sectioning:** The tissue is cut into extremely thin slices by an instrument called a microtome. When the section runs the length of the tissue, it is called a longitudinal section (l.s.); when the section runs across the tissue, it is called a cross section (c.s.).
4. **Staining:** The tissue is immersed in dyes that stain different structures. The most common dyes are hematoxylin and eosin stains (H & E). They give a differential blue and red color to the basic and acidic structures within the tissue. Other dyes are available for staining specific structures.

Figure 26.1 The major tissues in the human body.
The many kinds of tissues in the human body are grouped into four types: epithelial tissue, muscular tissue, nervous tissue, and connective tissue.

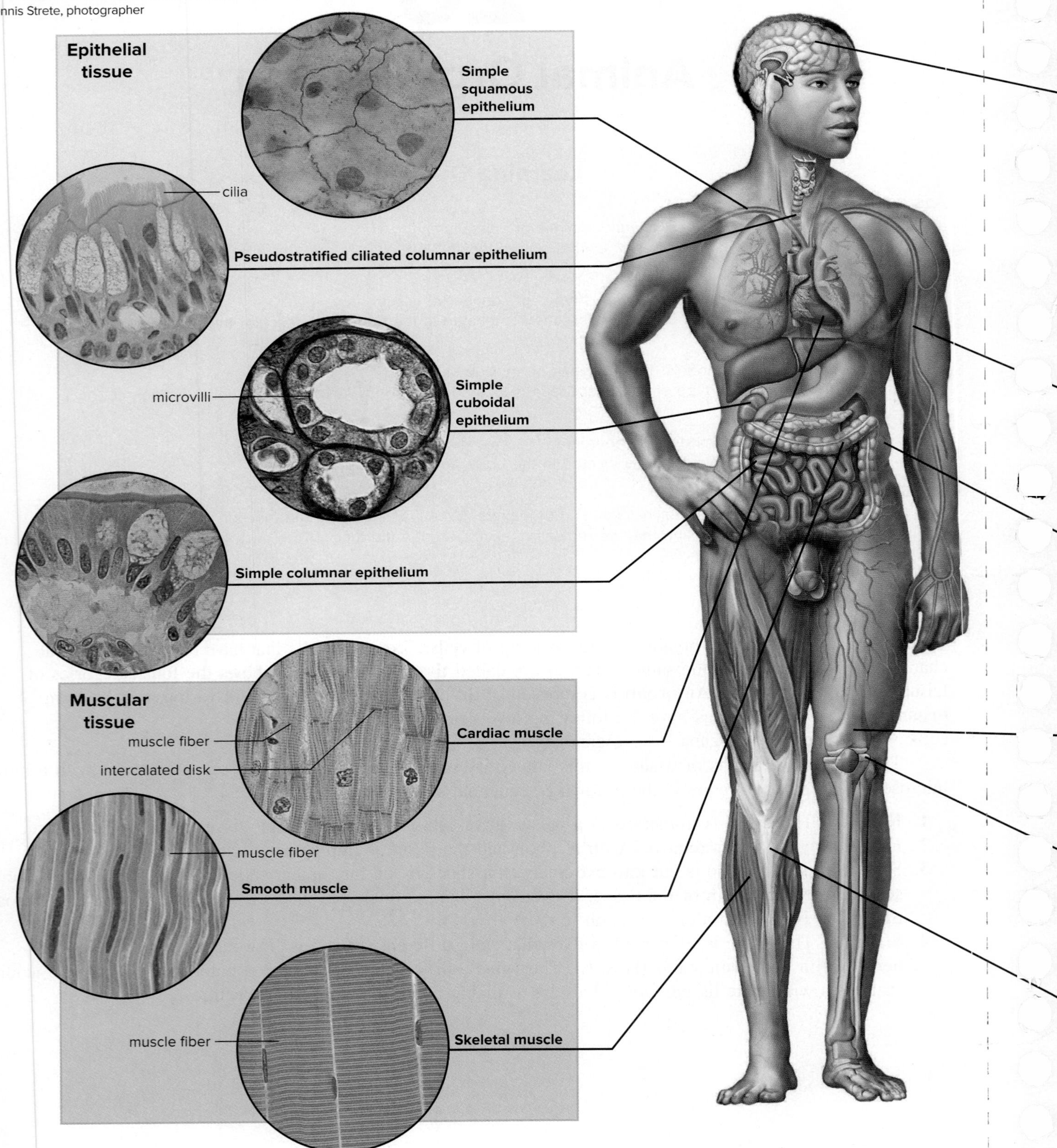

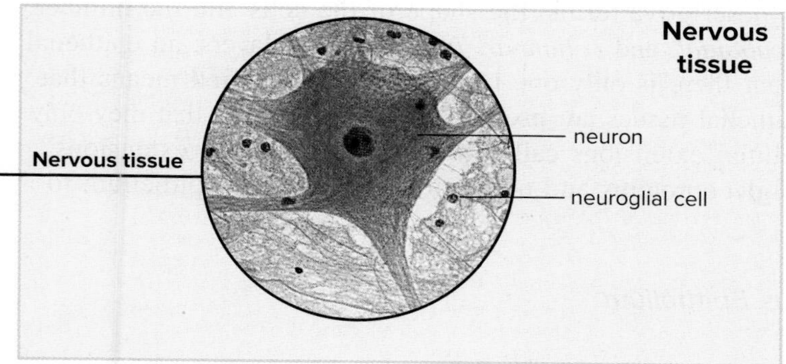
Nervous tissue
Nervous tissue
neuron
neuroglial cell

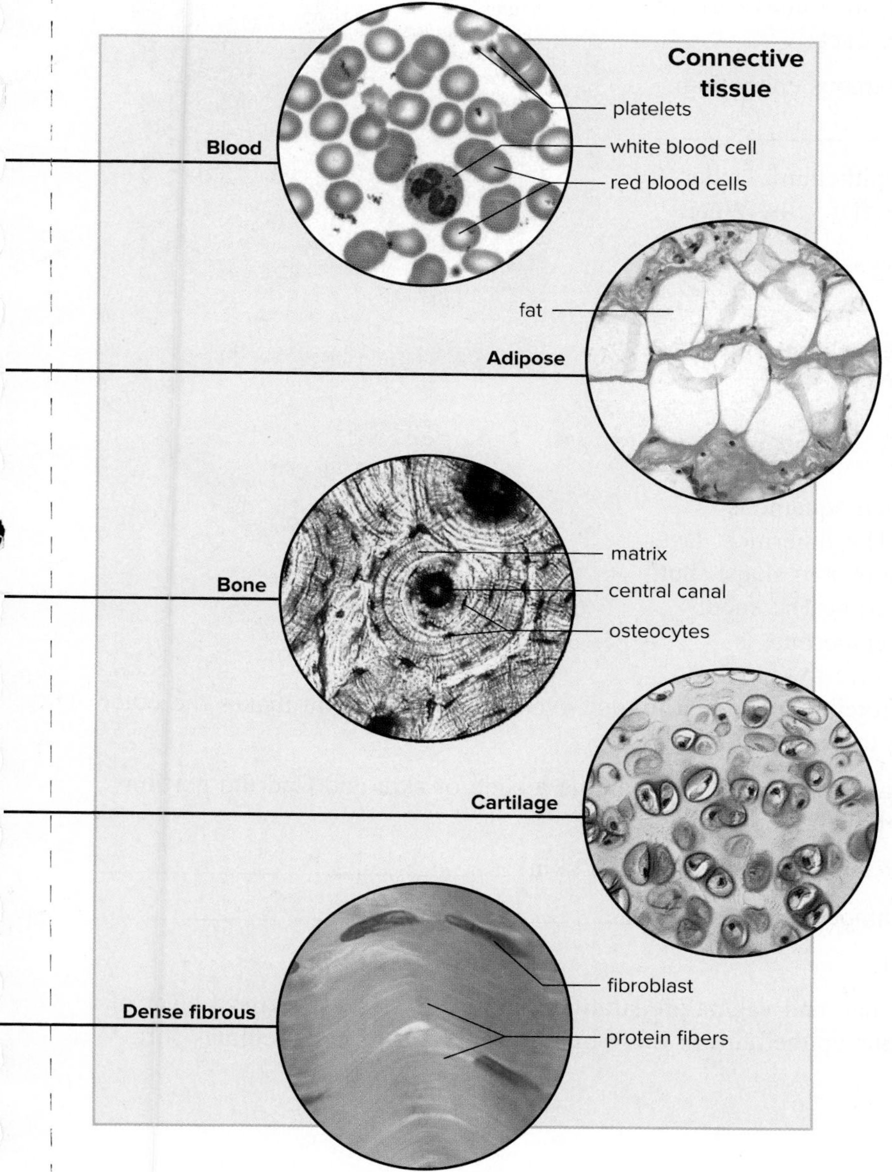
Connective tissue
Blood
platelets
white blood cell
red blood cells
fat
Adipose
Bone
matrix
central canal
osteocytes
Cartilage
fibroblast
Dense fibrous
protein fibers

26.1 Epithelial Tissue

Epithelial tissue (epithelium) forms a continuous layer, or sheet, over the entire body surface and most of the body's inner cavities. Externally, it forms a covering that protects the animal from infection, injury, and drying out. Some epithelial tissues produce and release secretions. Others absorb nutrients.

The name of an epithelial tissue includes two descriptive terms: the shape of the cells and the number of layers. The three possible shapes are *squamous, cuboidal,* and *columnar.* With regard to layers, an epithelial tissue may be simple or stratified. **Simple** means that there is only one layer of cells; **stratified** means that cell layers are placed on top of each other. Some epithelial tissues are **pseudostratified,** meaning that they only appear to be layered. Epithelium may also have cellular extensions called **microvilli** or hairlike extensions called **cilia.** A **basement membrane** consisting of glycoproteins and collagen fibers joins an epithelium to underlying connective tissue.

Observation: Simple and Stratified Squamous Epithelium

Simple Squamous Epithelium

Simple squamous epithelium is a single layer of thin, flat, many-sided cells, each with a central nucleus. It lines internal cavities, the heart, and all the blood vessels. It also lines parts of the urinary, respiratory, and male reproductive tracts.

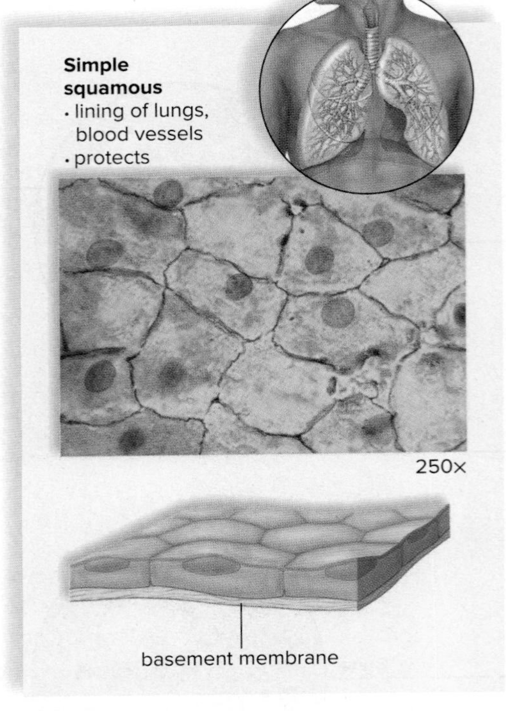

©Ed Reschke

1. Study a model or diagram of simple squamous epithelium.

   What does squamous mean? ______________________
2. Examine a prepared slide of squamous epithelium. Under low power, note the close packing of the flat cells. What

   shapes are the cells? ______________________

   ____
3. Under high power, examine an individual cell, and identify the plasma membrane, cytoplasm, and nucleus.
4. Add a sketch of this tissue to Table 26.1.

Stratified Squamous Epithelium

As would be expected from its name, stratified squamous epithelium consists of many layers of cells. The innermost layer produces cells that are first cuboidal or columnar in shape, but as the cells push toward the surface, they become flattened.

The outer region of the skin, called the epidermis, is stratified squamous epithelium. As the cells move toward the surface, they flatten, begin to accumulate a protein called **keratin,** and eventually die. Keratin makes the outer layer of epidermis tough, protective, and able to repel water.

1. Either now or when you are studying skin in section 26.2, examine a slide of skin and find the portion of the slide that is stratified squamous epithelium.
2. Approximately how many layers of cells make up this portion of skin? ______________________
3. Which layers of cells best represent squamous epithelium? ______________________
4. Add a sketch of this tissue to Table 26.1.

The linings of the mouth, throat, anal canal, and vagina are stratified epithelium. The outermost layer of cells surrounding the cavity is simple squamous epithelium. In these organs, this layer of cells remains soft, moist, and alive.

Observation: Simple Cuboidal Epithelium

Simple cuboidal epithelium is a single layer of cube-shaped cells, each with a central nucleus. It is found in tubules of the kidney and in the ducts of many glands, where it has a protective function. It also occurs in the secretory portions of some glands—that is, where the tissue produces and releases secretions.

1. Study a model or diagram of simple cuboidal epithelium.
2. Examine a prepared slide of simple cuboidal epithelium. Move the slide until you locate cube-shaped cells that line a lumen (cavity). Are these cells ciliated? ________________
3. Add a sketch of this tissue to Table 26.1.

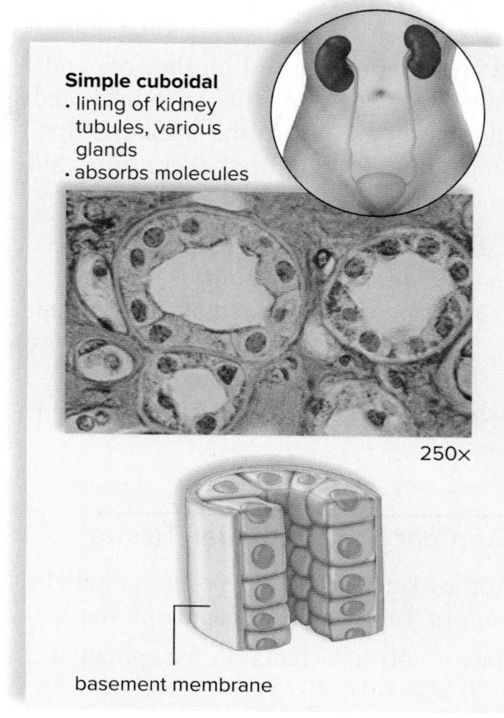

©Ed Reschke

Observation: Simple Columnar Epithelium

Simple columnar epithelium is a single layer of tall, cylindrical cells, each with a nucleus near the base. This tissue, which lines the digestive tract from the stomach to the anus, protects, secretes, and allows absorption of nutrients.

1. Study a model or diagram of simple columnar epithelium.
2. Examine a prepared slide of simple columnar epithelium. Find tall and narrow cells that line a lumen. Under high power, focus on an individual cell. Identify the plasma membrane, the cytoplasm, and the nucleus. Epithelial tissues are attached to underlying tissues by a basement membrane composed of extracellular material containing protein fibers.
3. The tissue you are observing contains mucus-secreting cells. Search among the columnar cells until you find a **goblet cell,** so named because of its goblet-shaped, clear interior. This region contains mucus, which may be stained a light blue. In the living animal, the mucus is discharged into the gut cavity and protects the lining from digestive enzymes.
4. Add a sketch of this tissue to Table 26.1.

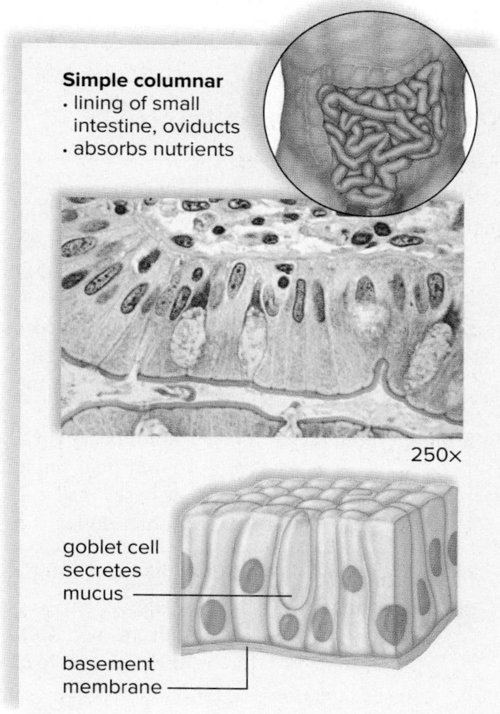

©Ed Reschke

Observation: Pseudostratified Ciliated Columnar Epithelium

Pseudostratified ciliated columnar epithelium appears to be layered, while actually all cells touch the basement membrane. Many cilia are located on the free end of each cell. In the human trachea, the cilia wave back and forth, moving mucus and debris up toward the throat so that they cannot enter the lungs. Smoking destroys these cilia, but they will grow back if smoking is discontinued.

1. Study a model or diagram of pseudostratified ciliated columnar epithelium.
2. Examine a prepared slide of pseudostratified ciliated columnar epithelium. Concentrate on the part of the slide that resembles the model. Identify the cilia.
3. Add a sketch of this tissue to Table 26.1.

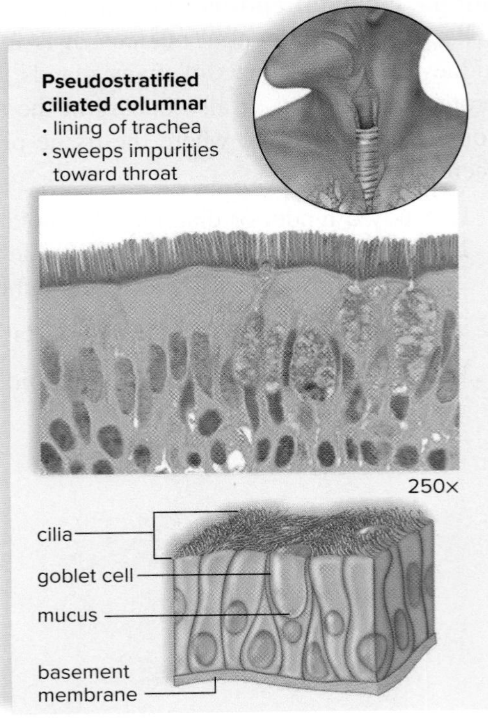

©Ed Reschke

Summary of Epithelial Tissue

Add a sketch of each type of epithelial tissue in the third column of Table 26.1. Recognizing that structure suits function, state a universal function for epithelial tissue. ________________

____

Table 26.1 Epithelial Tissue

Sketch	Structure	Sketch	Location
Simple squamous	Tightly packed thin flat cells		Walls of capillaries, lining of blood vessels, air sacs of lungs, lining of internal cavities
Stratified squamous	Innermost layers are cuboidal or columnar; outermost layers are flattened dead cells.		Skin, linings of mouth, throat, anal canal, vagina
Simple cuboidal	Tightly packed cuboidal cells that often have microvilli at one end		Surface of ovaries, linings of ducts and glands, lining of kidney tubules
Simple columnar	Columnlike—tall, cylindrical nucleus at base; may contain goblet cells.		Lining of uterus, tubes of digestive tract
Pseudostratified ciliated columnar	Tightly packed columnar cells appear to be layered but are not. Cells are ciliated and may contain goblet cells.		Linings of respiratory passages

26.2 Connective Tissue

Connective tissue joins different parts of the body together. There are four general classes of connective tissue: connective tissue proper, bone, cartilage, and blood. All types of connective tissue consist of cells surrounded by a matrix that usually contains fibers. Elastic fibers are composed of a protein called elastin. Collagenous fibers contain the protein collagen.

Observation: Connective Tissue

There are several different types of connective tissue. The accompanying illustrations will help you understand the name of each tissue, where it occurs, and its functions. We will study loose fibrous connective tissue, dense fibrous connective tissue, adipose tissue, bone, cartilage, and blood. **Loose fibrous connective tissue,** so named because it has space between components, supports epithelium and also many internal organs, such as muscles, blood vessels, and nerves. Its loose construction allows organs to freely move. **Dense fibrous connective tissue** contains many collagenous fibers packed closely together, as in tendons, which connect muscles to bones, and in ligaments, which connect bones to other bones at joints.

1. Examine a slide of loose fibrous connective tissue, and compare it to the figure below (*left*). What is the function of loose fibrous connective tissue? ______________________
2. Examine a slide of dense fibrous connective tissue, and compare it to the figure below (*right*). What two kinds of structures in the body contain dense fibrous connective tissue? ______________________

   ____
3. Add sketches of these tissues to Table 26.2.

Loose fibrous connective tissue
• has space between components.
• occurs beneath skin and most epithelial layers.
• functions in support and binds organs.
fibroblast
elastic fiber
collagen fiber
250×

Dense fibrous connective tissue
• has collagenous fibers closely packed.
• is in dermis of skin, tendons, ligaments.
• functions in support.

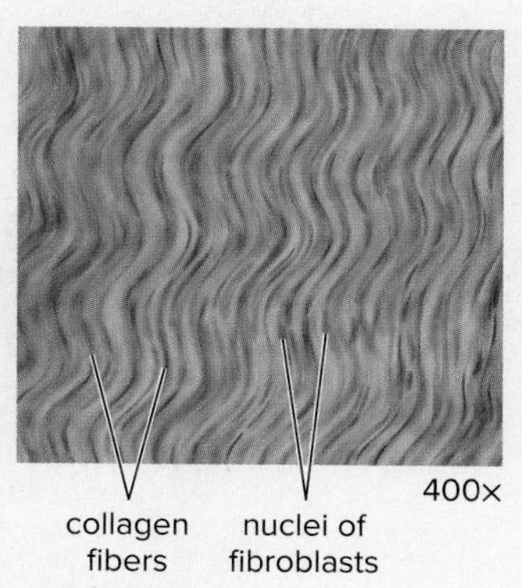

Observation: Adipose Tissue

In **adipose tissue,** the cells have a large, central, fat-filled vacuole that causes the nucleus and cytoplasm to be at the perimeter of the cell. Adipose tissue occurs beneath the skin, where it insulates the body, and around internal organs, such as the kidneys and heart. It cushions and helps protect these organs.

1. Examine a prepared slide of adipose tissue. Why is the nucleus pushed to one side? ______________________________

   ____

2. State a location for adipose tissue in the body. ______________________________

   ____

   ____

   What are two functions of adipose tissue at this location? ______________________________

   ____

   ____

3. Add a sketch of this tissue to Table 26.2.

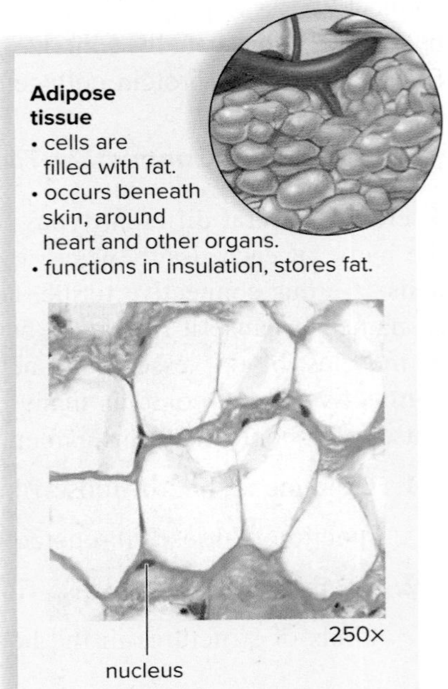

Adipose tissue
• cells are filled with fat.
• occurs beneath skin, around heart and other organs.
• functions in insulation, stores fat.
250×
nucleus

©McGraw-Hill Education/Al Telser, photographer

Observation: Compact Bone

Compact bone is found in the bones that make up the skeleton. It consists of **osteons** (Haversian system) with a **central canal,** and concentric rings of spaces called **lacunae,** connected by tiny crevices called **canaliculi.** The central canal contains a nerve and blood vessels, which service bone. The lacunae contain bone cells called **osteocytes,** whose processes extend into the canaliculi. Separating the lacunae is a matrix that is hard because it contains minerals, notably calcium salts. The matrix also contains collagenous fibers.

1. Study a model or diagram of compact bone. Then look at a prepared slide and identify the central canal, lacunae, and canaliculi.
2. What is the function of the central canal and canaliculi? ______________________________

   ____

   ____

3. Add a sketch of this tissue to Table 26.2.

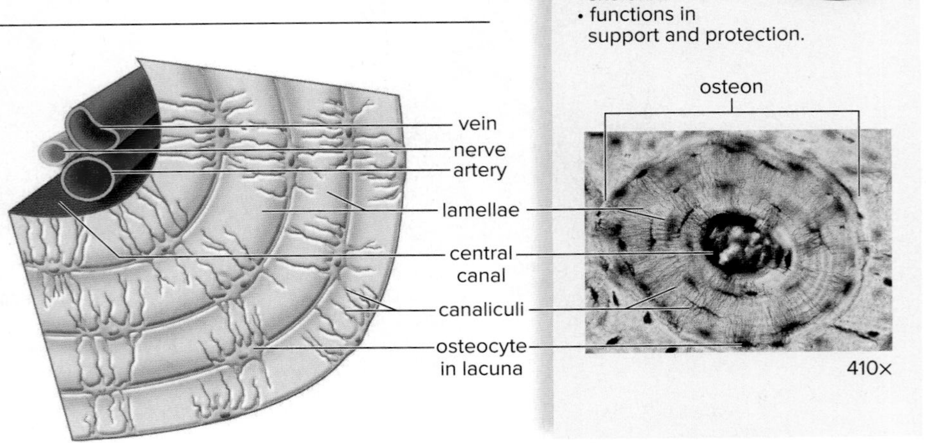

©Ed Reschke/Getty Images

Observation: Hyaline Cartilage

In **hyaline cartilage,** cells called **chondrocytes** are found in twos or threes in lacunae. The lacunae are separated by a flexible matrix containing weak collagenous fibers.

1. Study the diagram and photomicrograph of hyaline cartilage in the figure at the right. Then study a prepared slide of hyaline cartilage, and identify the matrix, lacunae, and chondrocytes.
2. Compare compact bone and hyaline cartilage. Which of these types of connective tissue is more organized? ____________________

   Why? ____________________
3. Which of these two types of connective tissue lends more support to body parts? Why? ____________________

   ____
4. Add a sketch of this tissue to Table 26.2.

Hyaline cartilage
- has cells in lacunae.
- occurs in nose and walls of respiratory passages; at ends of bones, including ribs.
- functions in support and protection.

chondrocyte within lacunae

matrix

250×

©Ed Reschke

Observation: Blood

Blood is a connective tissue in which the matrix is an intercellular fluid called **plasma. Red blood cells** (erythrocytes) have a biconcave appearance and lack a nucleus. These cells carry oxygen combined with the respiratory pigment hemoglobin. **White blood cells** (leukocytes) have a nucleus and are typically larger than the more numerous red blood cells. These cells fight infection.

1. Study a prepared slide of human blood. With the help of Figure 26.2, identify the red blood cells and the white blood cells, which appear faint because of the stain.
2. Try to identify a neutrophil, which has a multilobed nucleus, and a lymphocyte, which is the smallest of the white blood cells, with a spherical or slightly indented nucleus.
3. Add a sketch of this tissue to Table 26.2.

Figure 26.2 Blood cells.
Red blood cells are more numerous than white blood cells. White blood cells can be separated into five distinct types. If you have blood work done that includes a complete blood count (CBC), the doctor is getting a count of each of these types of white blood cells. (a-e: Magnification 1,050×)

(a–d) ©Ed Reschke; (e) ©Ed Reschke/Getty Images

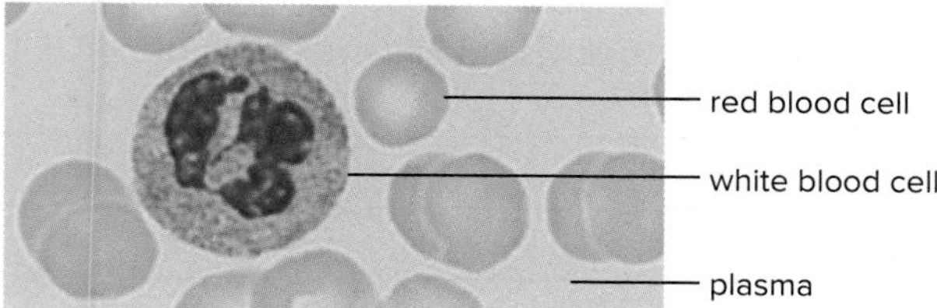

a. Neutrophil

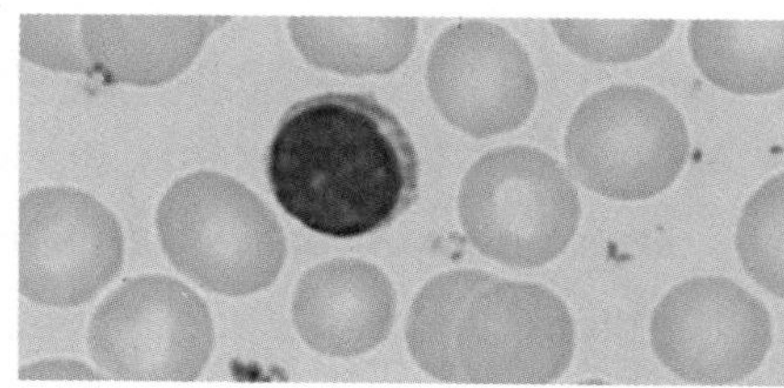

b. Lymphocyte

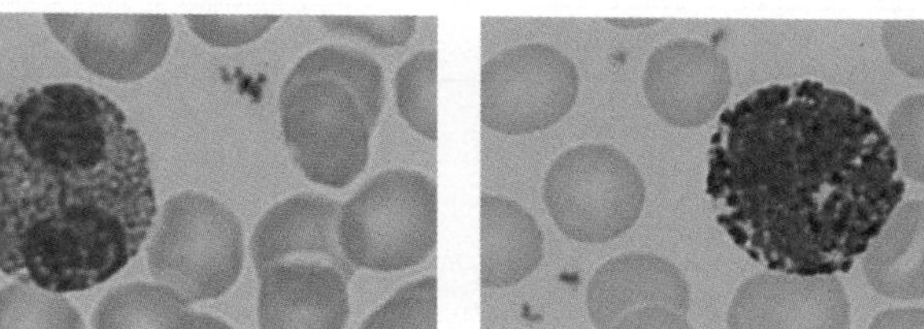

c. Eosinophil

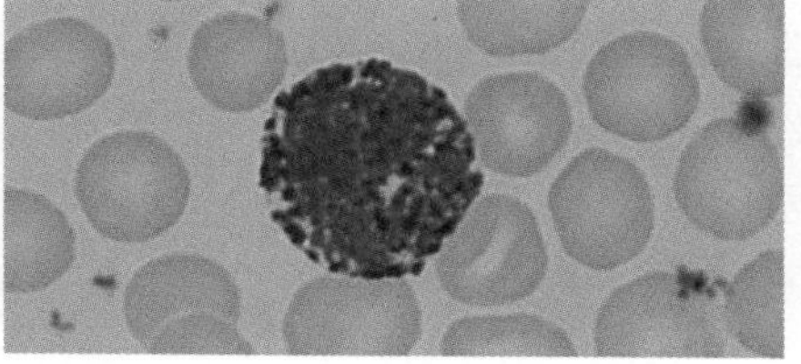

d. Basophil

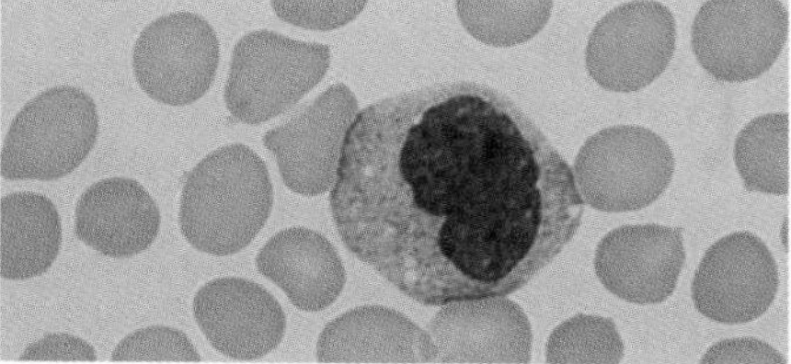

e. Monocyte

Summary of Connective Tissue

1. Add a sketch of each type of connective tissue in the third column of Table 26.2. Recognizing that structure suits function, state a universal function for connective tissue. ____________________

____

Table 26.2 Connective Tissue

Type	Structure	Sketch	Location
Loose fibrous connective tissue	Fibers are widely separated.		Between the muscles; beneath the skin; beneath most epithelial layers
Dense fibrous connective tissue	Fibers are closely packed.		Tendons, ligaments
Adipose	Large cells with fat-filled vacuoles; nuclei are pushed to one side.		Beneath the skin; around the kidney and heart; in the breast
Compact bone	Concentric circles of cells separated by a hard matrix		Bones of skeleton
Hyaline cartilage	Cells in lacunae within a flexible matrix		Nose; ends of bones; rings in walls of respiratory passages; between ribs and sternum
Blood	Red and white cells floating in plasma		Blood vessels

2. Working with others in a group, decide how the structure of each connective tissue suits its function.

Loose fibrous connective tissue ____________________

Dense fibrous connective tissue ____________________

Adipose tissue ____________________

Compact bone ____________________

Hyaline cartilage ____________________

Blood ____________________

26.3 Muscular Tissue

Muscular (contractile) tissue is composed of cells called muscle fibers. Muscular tissue has the ability to contract, and contraction usually results in movement. The body contains skeletal, cardiac, and smooth muscle.

Observation: Skeletal Muscle

Skeletal muscle occurs in the muscles attached to the bones of the skeleton. The contraction of skeletal muscle is said to be **voluntary** because it is under conscious control. Skeletal muscle is striated; it contains light and dark bands. The striations are caused by the arrangement of contractile filaments (actin and myosin filaments) in muscle cells often called fibers. Each fiber contains many nuclei, all peripherally located.

1. Study a model or diagram of skeletal muscle, and note that striations are present. You should see several muscle fibers, each marked with striations.
2. Examine a prepared slide of skeletal muscle. The striations may be difficult to make out, but bringing the slide in and out of focus may also help.
3. Each skeletal muscle cell is not only multinucleated, but also cylindrical and elongated. Counting partial cells, how many muscle cells are in the adjoining micrograph?________________

____

Skeletal muscle
- has striated cells with multiple nuclei.
- occurs in muscles attached to skeleton.
- functions in voluntary movement of body.

striation nucleus 250×

©Ed Reschke

Observation: Cardiac Muscle

Cardiac muscle is found only in the heart. It is called **involuntary** because its contraction does not require conscious effort. Cardiac muscle is striated in the same way as skeletal muscle. However, the fibers are branched and bound together at **intercalated disks,** where their folded plasma membranes touch. This arrangement aids communication between fibers.

1. Study a model or diagram of cardiac muscle, and note that striations are present.
2. Examine a prepared slide of cardiac muscle. Find an intercalated disk. What is the function of cardiac muscle? ____

____

____

3. Aside from being striated, cardiac muscle cells are rectangular and branched. What is the benefit of this arrangement? ______

____

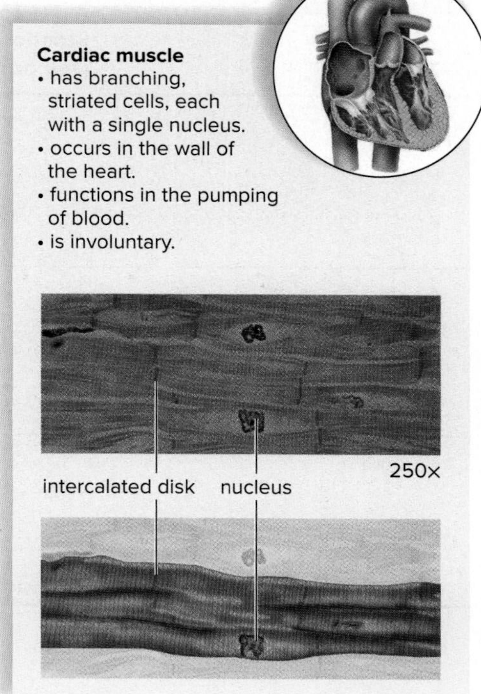

Cardiac muscle
• has branching, striated cells, each with a single nucleus.
• occurs in the wall of the heart.
• functions in the pumping of blood.
• is involuntary.
250×
intercalated disk
nucleus

©Ed Reschke

Observation: Smooth Muscle

Smooth muscle is sometimes called **visceral muscle** because it makes up the walls of the internal organs, such as the intestines and the blood vessels. Smooth muscle is involuntary because its contraction does not require conscious effort.

1. Study a model or diagram of smooth muscle, and note the shape of the cells and the centrally placed nucleus. Smooth muscle has spindle-shaped cells. What does *spindle-shaped* mean? ____
____
____

2. Examine a prepared slide of smooth muscle. Distinguishing the boundaries between the different cells may require you to take the slide in and out of focus.

Smooth muscle
- has spindle-shaped cells, each with a single nucleus.
- cells have no striations.
- functions in movement of substances in lumens of body.
- is involuntary.
- is found in blood vessel walls and walls of the digestive tract.

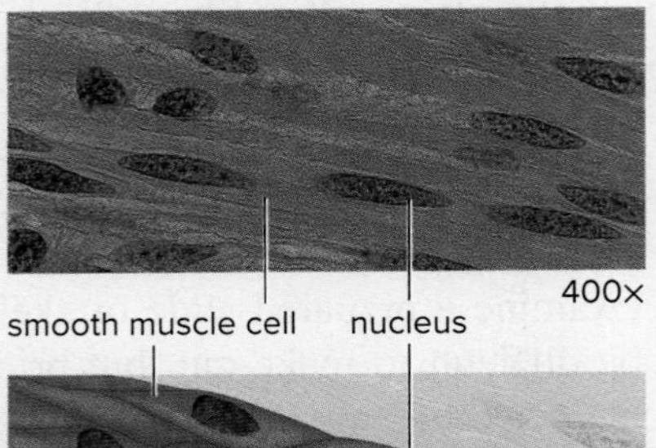

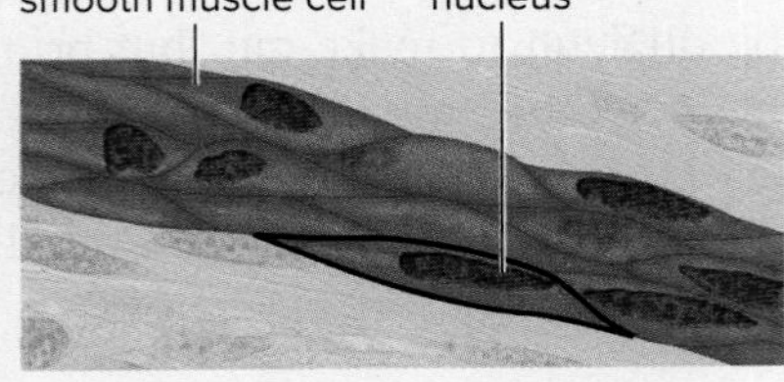

Summary of Muscular Tissue

1. Complete Table 26.3 to summarize your study of muscular tissue.

2. How does it benefit an animal that skeletal muscle is voluntary while cardiac and smooth muscle are involuntary?______________
____
____

Table 26.3 Muscular Tissue

Type	Striations (Yes or No)	Branching (Yes or No)	Conscious Control (Yes or No)
Skeletal			
Cardiac			
Smooth			

26.4 Nervous Tissue

Nervous tissue is found in the brain, spinal cord, and nerves. Nervous tissue receives and integrates incoming stimuli before conducting nerve impulses, which control the glands and muscles of the body. Nervous tissue is composed of two types of cells: **neurons** that transmit messages and **neuroglia** that support and nourish the neurons. Motor neurons, which take messages from the spinal cord to the muscles, are often used to exemplify typical neurons. Motor neurons have several **dendrites,** processes that take signals to a **cell body,** where the nucleus is located, and an **axon** that takes nerve impulses away from the cell body.

Observation: Nervous Tissue

1. Study a model or diagram of a neuron, and then examine a prepared slide. Most likely, you will not be able to see neuroglia because they are much smaller than neurons.
2. Identify a cell body, the nucleus, a dendrite, and the axon in Figure 26.3*a* and *label the micrograph. Also label the neuroglia surrounding the neuron.*
3. Explain the appearance and function of the parts of a motor neuron:
   a. Dendrites ____
   ____
   b. Cell body ____
   ____
   c. Axon ____
   ____

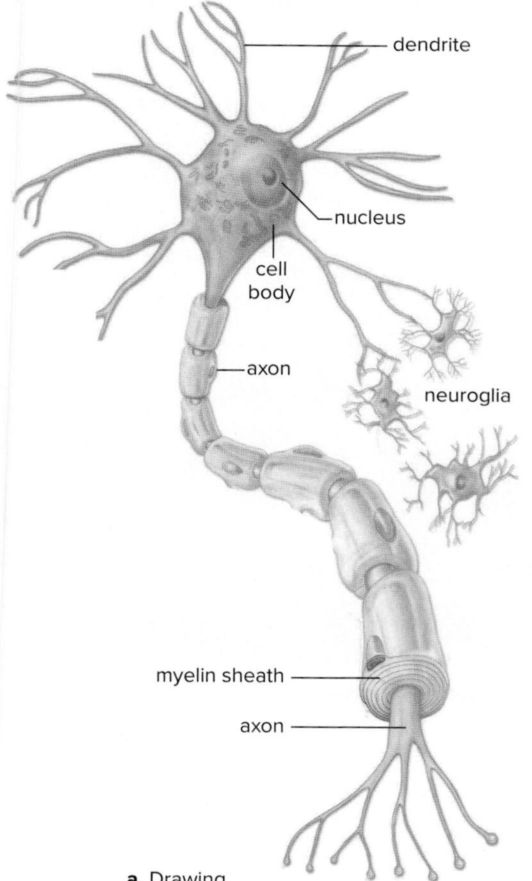

a. Drawing

Figure 26.3 Motor neuron anatomy.
(b) ©Ed Reschke

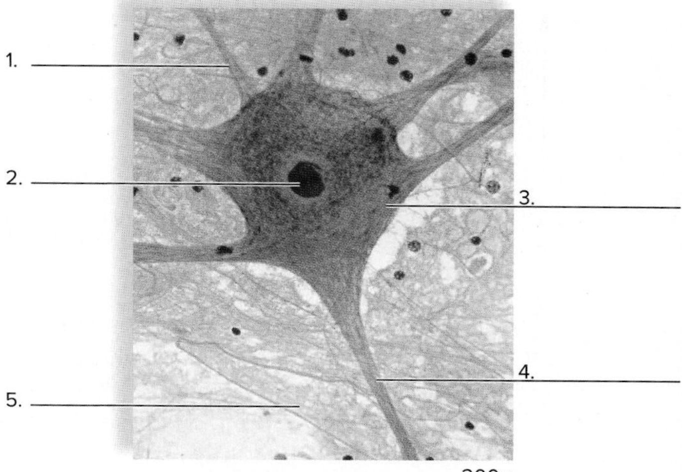

b. Photomicrograph of a neuron

26.5 Organ Level of Organization

Organs are structures composed of two or more types of tissue that work together to perform particular functions. You may tend to think that a particular organ contains only one type of tissue. For example, muscular tissue is usually associated with muscles and nervous tissue with the brain. However, muscles and the brain also contain other types of tissue—for example, loose connective tissue and blood. Here we will study the compositions of two organs—the intestine and the skin.

Intestine

The **intestine,** a part of the digestive system, processes food and absorbs nutrient molecules.

Observation: Intestinal Wall

Study a slide of a cross section of intestinal wall. With the help of Figure 26.4, identify the following layers:

1. **Mucosa** (mucous membrane layer): This layer, which lines the central lumen (cavity), is made up of columnar epithelium overlying the submucosa. This epithelium is glandular—that is, it secretes mucus from goblet cells and digestive enzymes from the rest of the epithelium. The membrane is arranged in deep folds (fingerlike projections) called **villi,** which increase the small intestine's absorptive surface.
2. **Submucosa** (submucosal layer): This loose fibrous connective tissue layer contains nerve fibers, blood vessels, and lymphatic vessels. The products of digestion are absorbed into these blood and lymphatic vessels.
3. **Muscularis** (smooth muscle layer): Circular muscular tissue and then longitudinal muscular tissue are found in this layer. Rhythmic contraction of these muscles causes **peristalsis,** a wavelike motion that moves food along the intestine.
4. **Serosa** (serous membrane layer): In this layer, a thin sheet of loose fibrous connective tissue underlies a thin, outermost sheet of squamous epithelium. This membrane is part of the **peritoneum,** which lines the entire abdominal cavity.

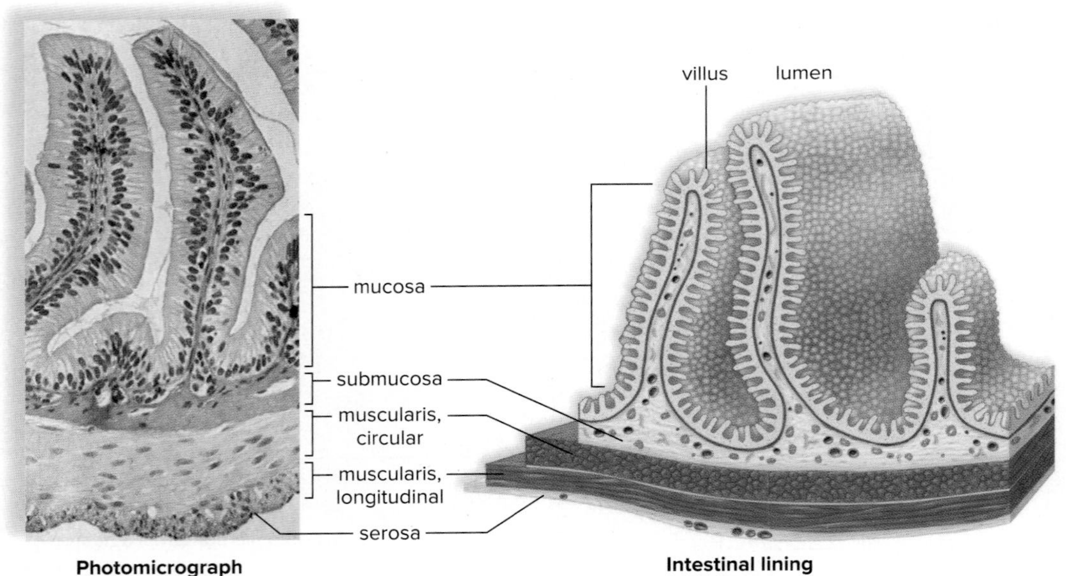

Figure 26.4 Tissues of the intestinal wall.

In Table 26.4, list the types of tissue found in the layers of the intestinal wall.

Table 26.4 Tissues of the Intestinal Wall

	Mucosa	Submucosa	Muscularis	Serosa
Tissue(s)				

Skin

The skin covers the entire exterior of the human body. Skin functions include protection, water retention, sensory reception, body temperature regulation, and vitamin D synthesis.

Observation: Skin

Study a model or diagram and also a prepared slide of the skin. Identify these two skin regions and the subcutaneous layer.

1. **Epidermis:** This region is composed of stratified squamous epithelial cells. The outer cells of the epidermis are nonliving and create a waterproof covering that prevents excessive water loss. These cells are always being replaced because an inner layer of the epidermis is composed of living cells that constantly produce new cells.
2. **Dermis:** This region is a connective tissue containing blood vessels, nerves, sense organs, and the expanded portions of oil (sebaceous) and sweat glands and hair follicles.

   List the structures you can identify on your slide: ______________________________

   ____

   ____

3. **Subcutaneous layer:** This is a layer of loose connective tissue and adipose tissue that lies beneath the skin proper and serves to insulate and protect inner body parts.

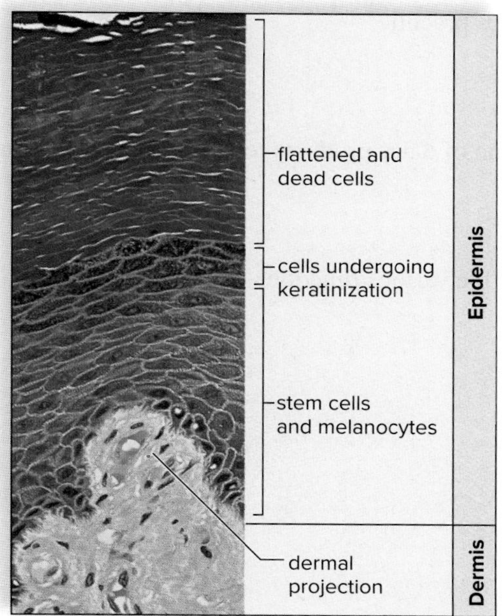

Photomicrograph of skin

Laboratory Review 26

_______________ **1.** What are cells with similar structural features and a common function called?

_______________ **2.** If organ systems are present in an animal, what else must be present?

_______________ **3.** What kind of tissue covers the entire body surface and most inner cavities?

_______________ **4.** What term is used to describe cell tissues that only appear to be layered?

_______________ **5.** What is the name of the protein that makes the outer layer of epidermis tough, protective, and able to repel water?

_______________ **6.** What kinds of cells produce mucus that protects the simple columnar epithelium lining in the stomach from digestive enzymes?

_______________ **7.** What type of connective tissue has a hard matrix that contains calcium salts?

_______________ **8.** What type of connective tissue has a fluid matrix called plasma?

_______________ **9.** Which type of muscle is controlled voluntarily?

_______________ **10.** What feature aids communication between cardiac muscle fibers?

_______________ **11.** Which type of muscle is found in the bladder?

_______________ **12.** What is the name of the cells in nervous tissue that transmit messages?

_______________ **13.** When two or more tissues work together to perform a specific function, what is the structure formed by the multiple tissues called?

_______________ **14.** In what region of the skin are blood vessels, nerves, sense organs, glands, and hair follicles located?

Thought Questions

15. Why is a muscle, like the biceps brachii in your arm, considered an organ and not just a tissue? What kind of muscle tissue is part of the biceps brachii?

16. Why is an increased consumption of dairy products recommended after someone breaks a bone?

17. Based on your understanding of the various functions of epithelium, explain why long-time smokers tend to cough a lot.

LABORATORY

27

Basic Mammalian Anatomy I

Learning Outcomes

27.1 External Anatomy
- Compare the limbs of a pig to the limbs of a human.
- Identify the sex of a fetal pig.

27.2 Oral Cavity and Pharynx
- Find and identify the teeth, tongue, and hard and soft palates.
- Identify and state a function for the epiglottis, glottis, and esophagus.
- Name the two pathways that cross in the pharynx.

27.3 Thoracic and Abdominal Incisions
- Identify the thoracic cavity and the abdominal cavity.
- Find and identify the diaphragm.

27.4 Neck Region
- Find, identify, and state a function for the thymus gland, the larynx, and the thyroid gland.

27.5 Thoracic Cavity
- Identify the three compartments and the organs of the thoracic cavity.

27.6 Abdominal Cavity
- Find, identify, and state a function for the liver, stomach, spleen, small intestine, gallbladder, pancreas, and large intestine. Describe where these organs are positioned in relation to one another.

27.7 Human Anatomy
- Using a human torso model, find, identify, and state a function for the organs studied in this laboratory.
- Associate each organ with a particular system of the body.

Introduction

In this laboratory, you will dissect a fetal pig. Alternately, your instructor may choose to have you observe a pig that has already been dissected. Both pigs and humans are mammals; therefore, you will be studying mammalian anatomy. The period of pregnancy, or gestation, in pigs is approximately 17 weeks (compared with an average of 40 weeks in humans). The piglets used in class will usually be within 1 to 2 weeks of birth.

The pigs may have a slash in the right neck region, indicating the site of blood drainage. A red latex solution may have been injected into the **arterial system,** and a blue latex solution may have been injected into the **venous system** of the pigs. If so, when a vessel appears red, it is an artery, and when a vessel appears blue, it is a vein.

As a result of this laboratory, you should gain an appreciation of which organs work together. For example, the liver and the pancreas help to digest fat in the small intestine.

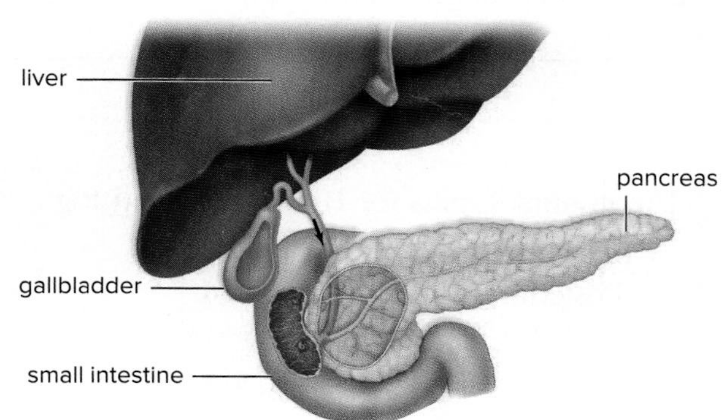

27.1 External Anatomy

Mammals are characterized by the presence of mammary glands and hair. Mammals also occur in two distinct sexes, males and females, often distinguishable by their external **genitals,** the reproductive organs.

Both pigs and humans are placental mammals, which means that development occurs within the uterus of the mother. An **umbilical cord** stretches externally between the fetal animal and the **placenta,** where carbon dioxide and organic wastes are exchanged for oxygen and organic nutrients.

Pigs and humans are tetrapods—that is, they have four limbs. Pigs walk on all four of their limbs; in fact, they walk on their toes, and their toenails have evolved into hooves. In contrast, humans walk only on the feet of their legs.

Observation: External Anatomy

Body Regions and Limbs

> **Latex gloves:** Wear protective latex gloves when handling preserved animal organs. Use protective eyewear and exercise caution when using sharp instruments during this experiment. Wash hands thoroughly upon completion of this experiment.

1. Place your animal in a dissecting pan, and observe the following body regions: the rather large head; the short, thick neck; the cylindrical trunk with two pairs of appendages (forelimbs and hindlimbs); and the short tail (Fig. 27.1*a*). The tail is an extension of the vertebral column.
2. Examine the four limbs, and feel for the joints of the digits, wrist, elbow, shoulder, hip, knee, and ankle.
3. Determine which parts of the forelimb correspond to your arm, elbow, forearm, wrist, and hand.
4. Do the same for the hindlimb, comparing it with your leg.
5. The pig walks on its toenails, which would be like a ballet dancer on "tiptoe." Notice how your heel touches the ground when you walk. Where is the heel of the pig? ______________________

Umbilical Cord

1. Locate the umbilical cord arising from the ventral (toward the belly) portion of the abdomen.
2. Note the cut ends of the umbilical blood vessels. If they are not easily seen, cut the umbilical cord near the end and observe this new surface.
3. What is the function of the umbilical cord? ______________________

   ____

Nipples and Hair

1. Locate the small **nipples,** the external openings of the **mammary glands.** The nipples are *not* an indication of sex, since both males and females possess them. How many nipples does a pig have? ________ When is it advantageous for a pig to have so many nipples? ______________________

   ____
2. Can you find hair on the pig? __________ Where? ______________________

   ____

Directional Terms for Dissecting Fetal Pig

Anterior: toward the head end	Ventral: toward the belly
Posterior: toward the hind end	Dorsal: toward the back

Figure 27.1 External anatomy of the fetal pig.
a. Body regions and limbs. **b, c.** The sexes can be distinguished by the external genitals.

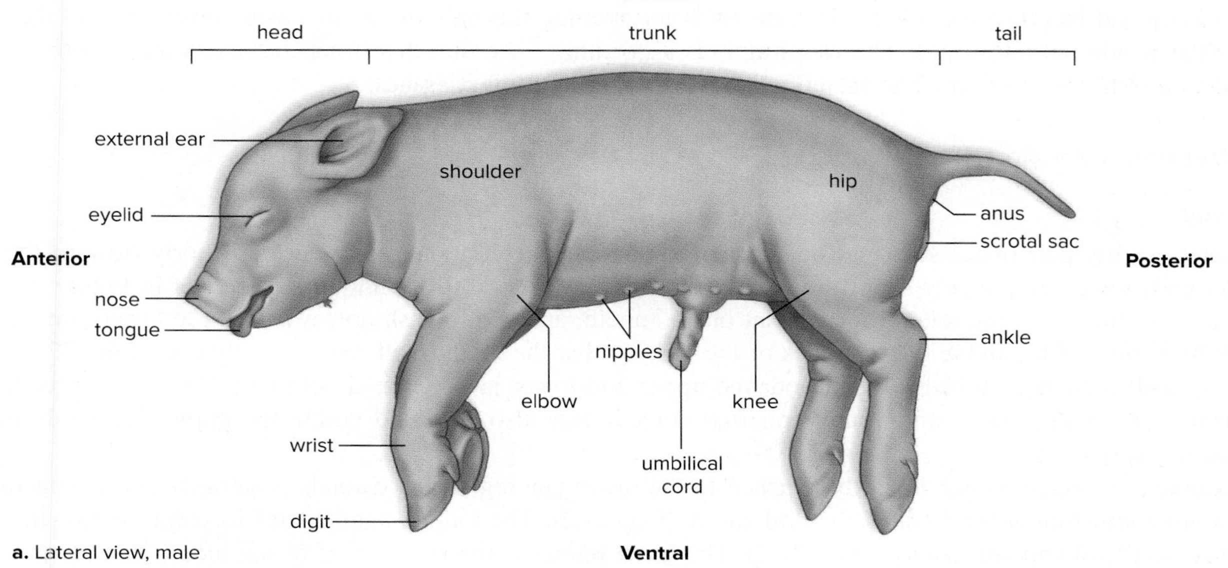

a. Lateral view, male

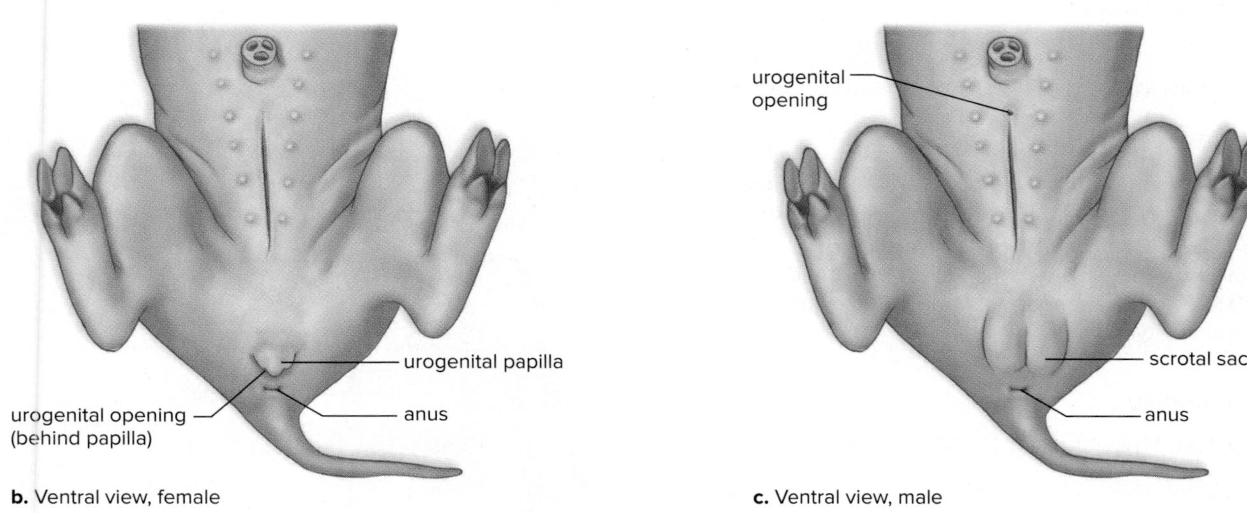

b. Ventral view, female

c. Ventral view, male

Anus and External Genitals

1. Locate the **anus** under the tail. Name the organ system that ends in the opening called the anus. ___________
2. In females, locate the **urogenital opening,** just anterior to the anus, and a small, fleshy **urogenital papilla** projecting from the urogenital opening (Fig. 27.1*b*).
3. In males, locate the urogenital opening just posterior to the umbilical cord (Fig. 27.1*c*). The duct leading to it runs forward from between the legs in a long, thick tube, the **penis,** which can be felt under the skin. In males, the urinary system and the genital system are always joined.
4. You are responsible for identifying pigs of both sexes. What sex is your pig? ___________________
Be sure to look at a pig of the opposite sex that another group of students is dissecting.

27.2 Oral Cavity and Pharynx

The **oral cavity** is the space in the mouth that contains the tongue and the teeth. The **pharynx** is dorsal to the oral cavity and has three openings: The **glottis** is an opening through which air passes on its way to the **trachea** (the windpipe) and lungs. The **esophagus** is a portion of the digestive tract that leads through the neck and thorax to the stomach. The **nasopharynx** leads to the nasal passages.

Observation: Oral Cavity and Pharynx

Oral Cavity

1. Insert a sturdy pair of scissors into one corner of the specimen's mouth, and cut posteriorly (toward the hind end) for approximately 4 cm. Repeat on the opposite side until the mouth is open as in Figure 27.2.
2. Place your thumb on the tongue at the front of the mouth, and gently push downward on the lower jaw. This will tear some of the tissue in the angles of the jaws so that the mouth will remain partly open (Fig. 27.2).
3. Note small, underdeveloped teeth in both the upper and lower jaws. Care should be taken because teeth can be very sharp. Other embryonic, nonerupted teeth may also be found within the gums. The teeth are used to chew food.
4. Examine the tongue, which is partly attached to the lower jaw region but extends posteriorly and is attached to a bony structure at the back of the oral cavity (Fig. 27.2). The tongue manipulates food for swallowing.
5. Locate the hard and soft palates (Fig. 27.2). The **hard palate** is the ridged roof of the mouth that separates the oral cavity from the nasal passages. The **soft palate** is a smooth region posterior to the hard palate. An extension of the soft palate—the **uvula**—hangs down into the throat in humans. (A pig does not have a uvula.)

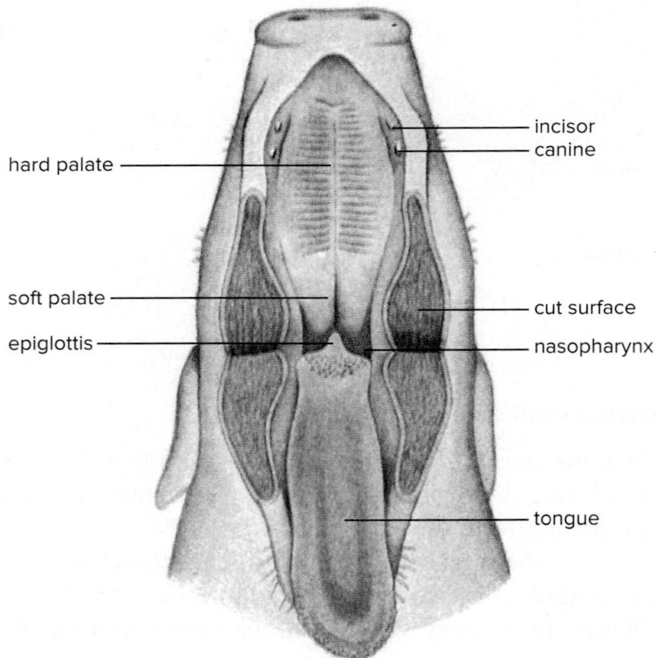

Figure 27.2 Oral cavity of the fetal pig. The roof of the oral cavity contains the hard and soft palates, and the tongue lies above the floor of the oral cavity.

Pharynx

1. Push down on the tongue until you open the jaws far enough to see a slightly pointed flap of tissue pointing dorsally (toward the back) (Fig. 27.2). This flap is the **epiglottis,** which covers the glottis. The **glottis** leads to the trachea (Fig. 27.3*a*).
2. Posterior and dorsal to the glottis, find the opening into the **esophagus,** a tube that takes food to the stomach. Note the proximity of the glottis and the opening to the esophagus. Each time the pig—or a human—swallows, the epiglottis instantly closes to keep food and fluids from going into the lungs via the trachea.
3. Insert a blunt probe into the glottis, and note that it enters the trachea. Remove the probe, insert it into the esophagus, and note the position of the esophagus beneath (dorsal to) the trachea.
4. Make a midline cut in the soft palate from the epiglottis to the hard palate. Then make two lateral cuts at the edge of the hard palate.
5. Posterior to the soft palate, locate the openings to the nasal passages.
6. Explain why it is correct to say that the air and food passages cross in the pharynx. ______________

____

____

____

Figure 27.3 Air and food passages in the fetal pig.
The air and food passages cross in the pharynx. **a.** Drawing. **b.** Dissection of specimen. (b) ©Ken Taylor/Wildlife Images

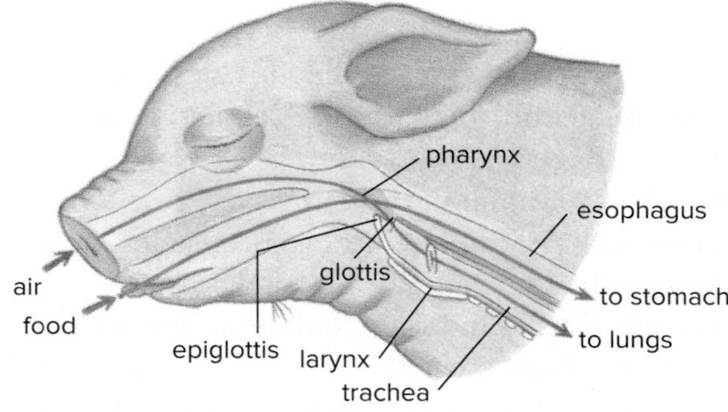

a.

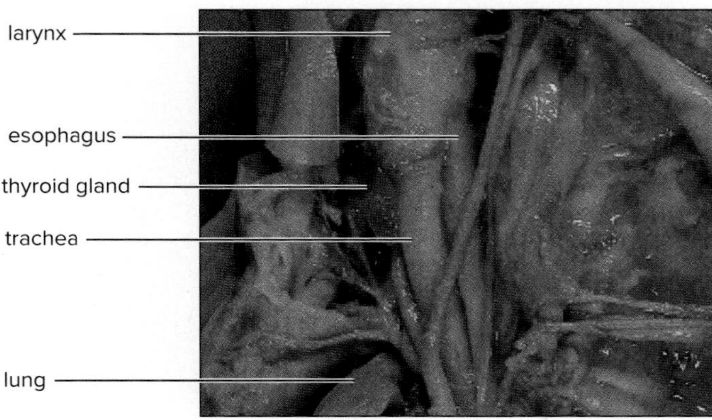

b.

27.3 Thoracic and Abdominal Incisions

First, prepare your pig according to the following directions, and then make thoracic and abdominal incisions so that you will be able to study the internal anatomy of your pig.

Preparation of Pig for Dissection

1. Place the fetal pig on its back in the dissecting pan.
2. Tie a cord around one forelimb, and then bring the cord around underneath the pan to fasten back the other forelimb.
3. Spread the hindlimbs in the same way.
4. With scissors always pointing up (never down), make the following incisions to expose the thoracic and abdominal cavities. The incisions are numbered on Figure 27.4 to correspond with the following steps.

Thoracic Incisions

1. Starting at the diaphragm, a structure that separates the thoracic cavity from the abdominal cavity, cut anteriorly until you reach the hairs in the throat region.
2. Make two lateral cuts, one on each side of the midline incision anterior to the forelimbs, taking extra care not to damage the blood vessels around the heart.
3. Make two lateral cuts, one on each side of the midline just posterior to the forelimbs and anterior to the diaphragm, following the ends of the ribs. Pull back the flaps created by these cuts (do not remove them) to expose the **thoracic cavity.** List the organs you find in the thoracic cavity. ____________________

   ____

Abdominal Incisions

4. With scissors pointing up, cut posteriorly from the diaphragm to the umbilical cord.
5. Make a flap containing the umbilical cord by cutting a semicircle around the cord and by cutting posteriorly to the left and right of the cord.
6. Make two cuts, one on each side of the midline incision posterior to the diaphragm. Examine the diaphragm, attached to the chest wall by radially arranged muscles. The central region of the diaphragm, called the **central tendon,** is a membranous area.
7. Make two more cuts, one on each side of the flap containing the umbilical cord and just anterior to the hindlimbs. Pull back the side flaps created by these cuts to expose the **abdominal cavity.**
8. Lifting the flap with the umbilical cord requires cutting the **umbilical vein.** Before cutting the umbilical vein, tie a thread on each side of where you will cut to mark the vein for future reference.
9. Rinse out your pig as soon as you have opened the abdominal cavity. If you have a problem with excess fluid, obtain a disposable plastic pipet to suction off the liquid.
10. Name the two cavities separated by the diaphragm. ____________________
11. List the organs located in the abdominal cavity. ____________________

    ____

Figure 27.4 Ventral view of the fetal pig indicating incisions.
These incisions are to be made preparatory to dissecting the internal organs. They are numbered here in the order they should be done.

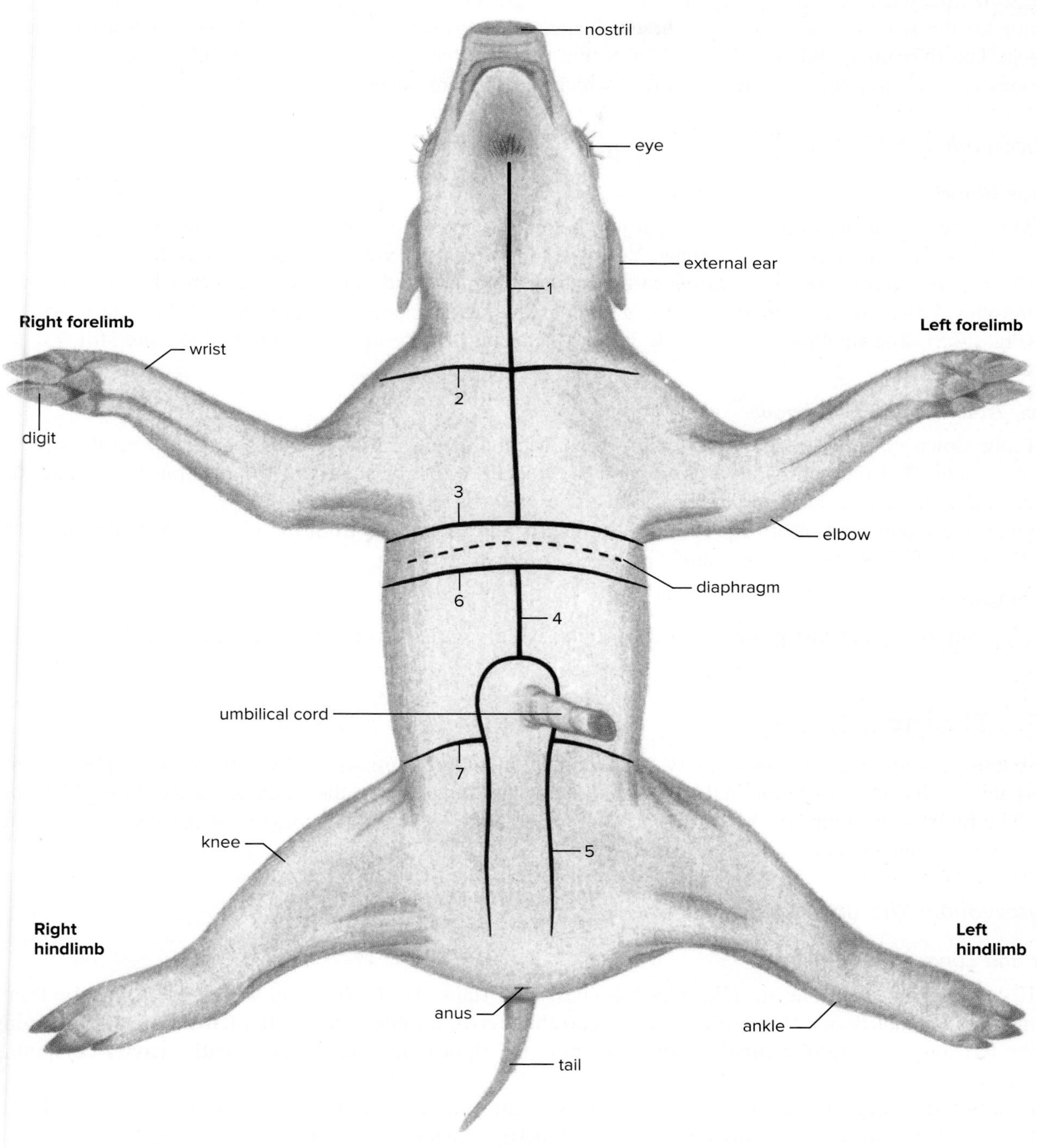

27.4 Neck Region

You will locate several organs in the neck region. Use Figures 27.3*b* and 27.5 as a guide, but *keep all the flaps in* order to close the thoracic and abdominal cavities at the end of the laboratory session.

The **thymus gland** is a part of the lymphatic system. Certain white blood cells called T (for thymus) lymphocytes mature in the thymus gland and help us fight disease. The **larynx,** or voice box, sits atop (anterior to) the **trachea,** or windpipe. The esophagus is a portion of the digestive tract that leads to the stomach. The **thyroid gland** secretes hormones that travel in the blood and act upon other body cells. These hormones (e.g., thyroxine) regulate the rate at which metabolism occurs in cells.

Observation: Neck Region

Thymus Gland

1. Move the skin apart in the neck region just below the hairs mentioned earlier. If necessary, cut the body wall laterally to make flaps. You will most likely be viewing exposed muscles.
2. *Cut through and clear away muscle* to expose the thymus gland, a diffuse gland that lies among the muscles. Later you will notice that the glandular thymus flanks the thyroid and overlies the heart (Fig. 27.5). The thymus is particularly large in fetal pigs, since their immune systems are still developing.

Larynx, Trachea, and Esophagus

1. Probe down into the deeper layers of the neck. Medially (toward the center), beneath several strips of muscle, find the hard-walled larynx and the trachea, which are parts of the respiratory passage to be examined later. Dorsal to the trachea, find the esophagus.
2. Open the mouth and insert a probe into the glottis and esophagus from the pharynx to better understand the orientation of these two organs.

Thyroid Gland

Locate the thyroid gland just posterior to the larynx, lying ventral to (on top of) the trachea.

27.5 Thoracic Cavity

As previously mentioned, the body cavity of mammals, including humans, is divided by the diaphragm into the thoracic cavity and the abdominal cavity. The heart and lungs are in the thoracic cavity (Figs. 27.5 and 27.6). The **heart** is a pump for the cardiovascular system, and the **lungs** are organs of the respiratory system where gas exchange occurs.

Observation: Thoracic Cavity

Heart and Lungs

1. If you have not yet done so, fold back the chest wall flaps. To do this, you will need to tear the thin membranes that divide the thoracic cavity into three compartments: the **left pleural cavity** containing the left lung, the **right pleural cavity** containing the right lung, and the **pericardial cavity** containing the heart.
2. Examine the lungs. Locate the four lobes of the right lung and the three lobes of the left lung. The trachea, dorsal to the heart, divides into the **bronchi,** which enter the lungs. Later, when the heart is removed, you will be able to see the trachea and bronchi.
3. Sequence the organs of the respiratory tract to trace the path of air from the nasal passages to the lungs.

____

____

Figure 27.5 Internal anatomy of the fetal pig.
The major organs are featured in this drawing. In the fetal pig, a red color tells you a vessel is an artery, and a blue color tells you it is a vein. (It does not tell you whether this vessel carries O_2-rich or O_2-poor blood.) Contrary to this drawing, *keep all the flaps on your pig* so you can close the thoracic and abdominal cavities at the end of the laboratory session.

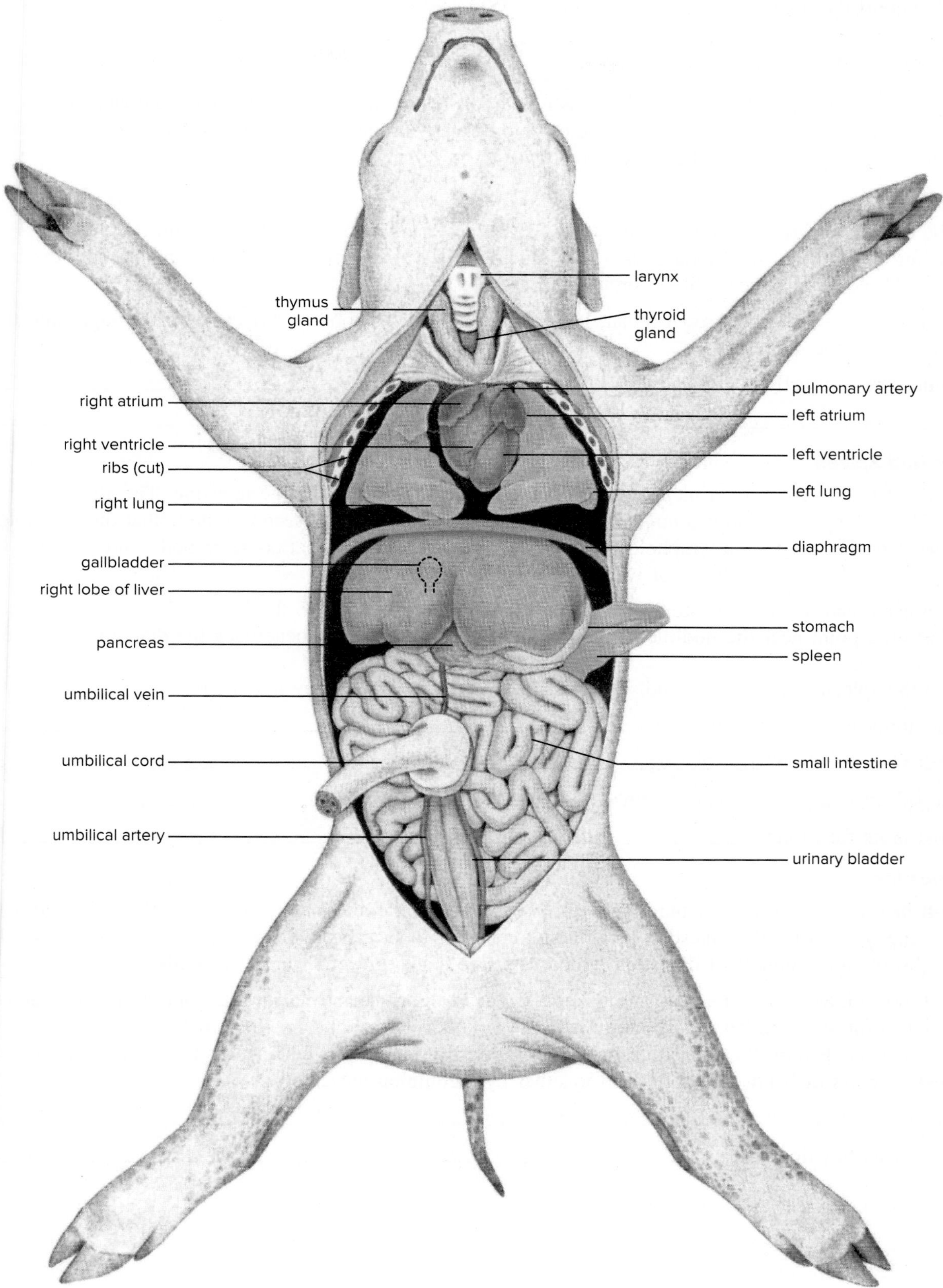

27.6 Abdominal Cavity

The abdominal wall and organs are lined by a membrane called **peritoneum,** consisting of epithelium supported by connective tissue. Double-layered sheets of peritoneum, called **mesenteries,** project from the body wall and support the organs.

Observation: Abdominal Cavity

If your pig is partially filled with dark, brownish material, take your animal to the sink and rinse it out. This material is clotted blood. Consult your instructor before removing any red or blue latex masses, since they may enclose organs you will need to study.

Liver

The **liver,** the largest organ in the abdomen (Fig. 27.6), performs numerous vital functions, including (1) disposing of worn-out red blood cells, (2) producing bile, (3) storing glycogen, (4) maintaining the blood glucose level, and (5) producing blood proteins.

1. Locate the liver, a large, brown organ. Its anterior surface is smoothly convex and fits snugly into the concavity of the diaphragm.
2. Name several functions of the liver. ____

   ____

Stomach and Spleen

The organs of the digestive tract include the stomach, small intestine, and large intestine. The **stomach** (see Fig. 27.5) stores food and has numerous gastric glands. These glands secrete a juice that digests protein. The **spleen** (see Fig. 27.5) is a lymphoid organ in the lymphatic system that contains both white and red blood cells. It purifies blood and disposes of worn-out red blood cells.

1. Push aside and identify the stomach, a large sac dorsal to the liver on the left side.
2. Locate the point near the midline of the body where the **esophagus** penetrates the diaphragm and joins the stomach.
3. Find the spleen, a long, flat, reddish organ attached to the stomach by mesentery.
4. The stomach is a part of what system? ____________________________________

   What is its function? ____
5. The spleen is a part of what system? ____________________________________

   What is its function? ____

Small Intestine

The **small intestine** is the part of the digestive tract that receives secretions from the pancreas and gallbladder. Besides being an area for the digestion of all components of food, carbohydrate, protein, and fat, the small intestine absorbs the products of digestion: glucose, amino acids, glycerol, and fatty acids.

1. Look posteriorly where the stomach makes a curve to the right and narrows to join the anterior end of the small intestine, called the **duodenum.**
2. From the duodenum, the small intestine runs posteriorly for a short distance and is then thrown into an irregular mass of bends and coils held together by a common mesentery.
3. The small intestine is a part of what system? ______________________________

   What is its function? ____

Figure 27.6 Internal anatomy of the fetal pig.
Most of the major organs are shown in this photograph. The stomach has been removed. The spleen, gallbladder, and pancreas are not visible. *Do not* remove any organs or flaps from your pig. ©Ken Taylor/Wildlife Images

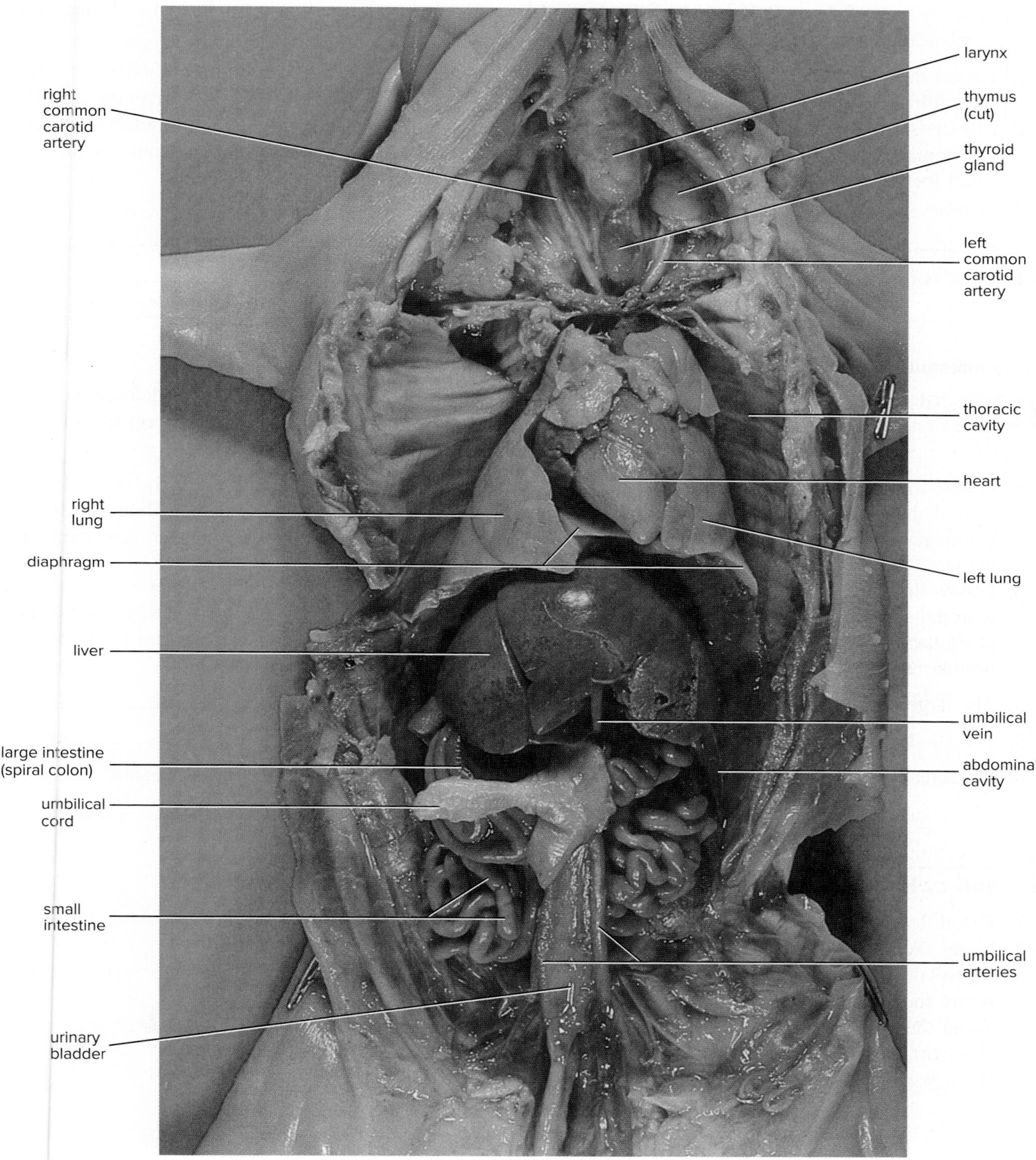

Gallbladder and Pancreas

The **gallbladder** stores and releases bile, which aids in the digestion of fat. The **pancreas** (see Fig. 27.5) is both an exocrine and an endocrine gland. As an exocrine gland, it produces and secretes pancreatic juice, which digests all the components of food in the small intestine. Both bile and pancreatic juice enter the duodenum by way of ducts. As an endocrine gland, the pancreas secretes the hormones insulin and glucagon into the bloodstream. Insulin and glucagon regulate blood glucose levels.

1. Locate the **bile duct,** which runs in the mesentery stretching between the liver and the duodenum. Find the gallbladder, embedded in the liver on the underside of the right lobe. It is a small, greenish sac.
2. Lift the stomach and locate the pancreas, the light-colored, diffuse gland lying in the mesentery between the stomach and the small intestine. The pancreas has a duct that empties into the duodenum of the small intestine.
3. What is the function of the gallbladder? ______________________________

   ____
4. What is the function of the pancreas? ______________________________

   ____

Large Intestine

The **large intestine** is the part of the digestive tract that absorbs water and prepares feces for defecation at the anus. The first part of the large intestine, called the **cecum,** has a projection called the vermiform (meaning wormlike) appendix.

1. Locate the distal (far) end of the small intestine, which joins the large intestine posteriorly, in the left side of the abdominal cavity (right side in humans). At this junction, note the cecum, a blind pouch.
2. Compare the large intestine of your pig to Figure 27.7. The organ does not have the same appearance in humans.
3. Follow the main portion of the large intestine, known as the **colon,** as it runs from the point of juncture with the small intestine into a tight coil (spiral colon), then out of the coil anteriorly, then posteriorly again along the midline of the dorsal wall of the abdominal cavity. In the pelvic region, the **rectum** is the last portion of the large intestine. The rectum leads to the **anus.**
4. The large intestine is a part of what system? ______________________________
5. What is the function of the large intestine? ______________________________
6. Sequence the organs of the digestive system to trace the path of food from the mouth to the anus. __________

   ____

Storage of Pigs

1. Before leaving the laboratory, place your pig in the plastic bag provided.
2. Expel excess air from the bag, and tie it shut.
3. Write your *name* and *section* on the tag provided, and attach it to the bag. Your instructor will indicate where the bags are to be stored until the next laboratory period.
4. Clean the dissecting tray and tools, and return them to their proper location.
5. Wipe off your goggles.
6. Wash your hands.

27.7 Human Anatomy

Humans and pigs are both mammals, and their organs are similar. A human torso model shows the exact location of the organs in humans (Fig. 27.7). You should learn to associate each human organ with its particular system. Six systems are color-coded in Figure 27.7.

Figure 27.7 Human internal organs.

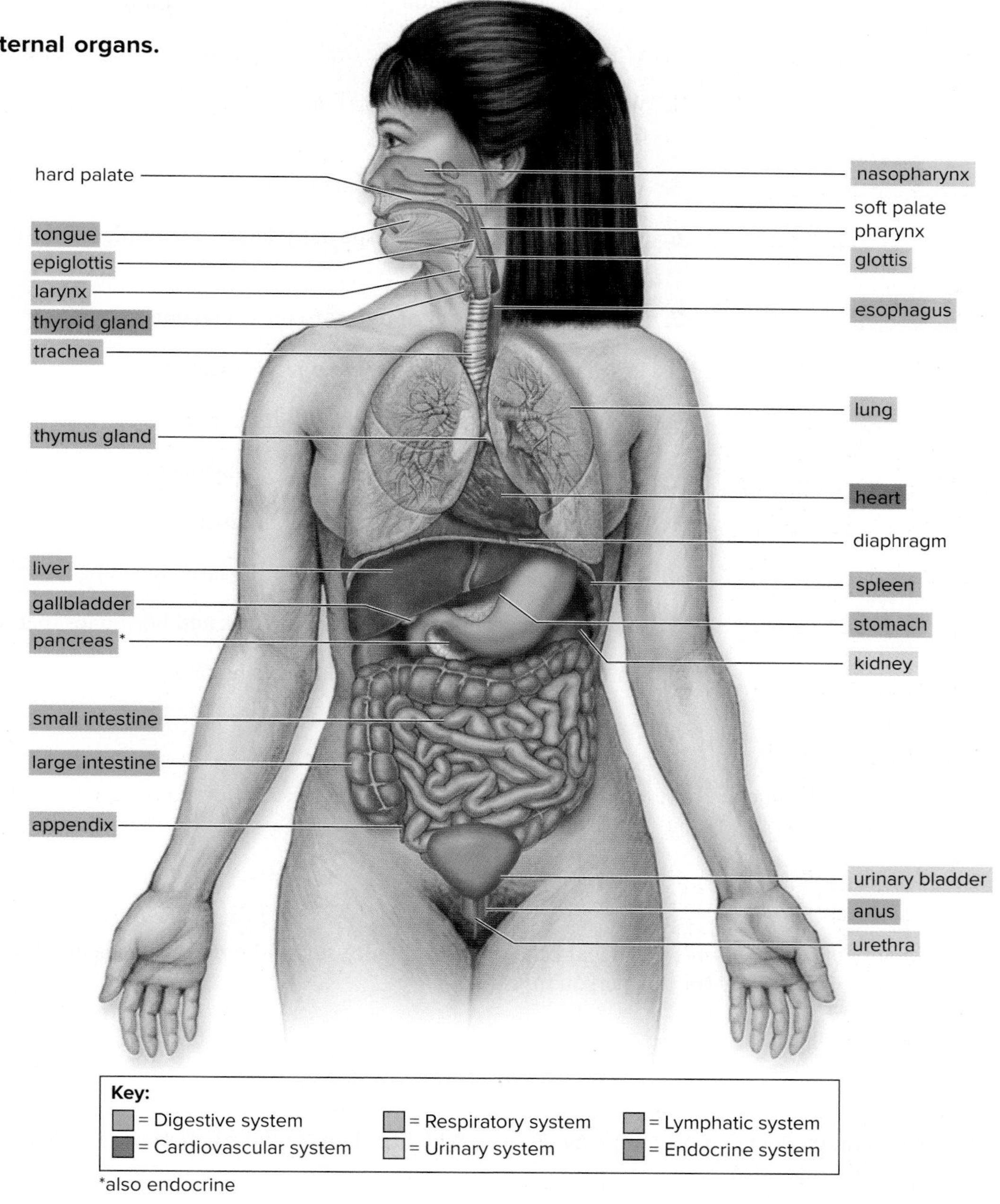

*also endocrine

Observation: Human Torso

1. Examine a human torso model, and using Figure 27.7 as a guide, locate the same organs just dissected in the fetal pig.
2. In your studies so far, have you seen any major differences between pig internal anatomy and human internal anatomy? ______________________________

____

Laboratory Review 27

_________________ **1.** What directional term refers to the head end of the fetal pig?

_________________ **2.** If the urogenital opening is just anterior to the anus, is the fetal pig male or female?

_________________ **3.** What space is dorsal to the oral cavity and has three openings?

_________________ **4.** What is the name of the flap that closes during swallowing to prevent food and fluids from entering the trachea?

_________________ **5.** What is the structure that separates the thoracic cavity from the abdominal cavity?

_________________ **6.** What glandular organ is especially large in fetal pigs since their immune systems are still developing?

_________________ **7.** Is the trachea ventral or dorsal to the esophagus?

_________________ **8.** In what thoracic cavities are the lungs located?

_________________ **9.** What is the largest organ in the abdominal cavity?

_________________ **10.** What organ do secretions from the gallbladder and pancreas enter?

_________________ **11.** What is the name of the tissue that holds the small intestine together?

_________________ **12.** What is the name of the first part of the large intestine?

_________________ **13.** What organ produces digestive enzymes and hormones that regulate blood glucose?

_________________ **14.** To what organ system do the kidneys belong?

Thought Questions

15. Why would someone born without a thymus gland be likely to suffer from a greater number of infections?

16. Treatment for a person's laryngeal cancer might require surgical removal of the epiglottis. What would postsurgical therapy for this person involve?

17. Based on the location of the spleen, explain why it is damaged so often in car accidents, especially when the driver's side is struck.

LABORATORY

28

Chemical Aspects of Digestion

Learning Outcomes

Introduction

- Sequence the organs of the digestive tract from the mouth to the anus.
- State the contribution of each organ, if any, to the process of chemical digestion.

28.1 Protein Digestion by Pepsin

- Associate the enzyme pepsin with the ability of the stomach to digest protein.
- Explain why stomach contents are acidic and how a warm body temperature aids digestion.

28.2 Fat Digestion by Pancreatic Lipase

- Associate the enzyme lipase with the ability of the small intestine to digest fat.
- Explain why the emulsification process assists the action of lipase.
- Explain why a change in pH indicates that fat digestion has occurred.
- Explain the relationship between time and enzyme activity.

28.3 Starch Digestion by Pancreatic Amylase

- Associate the enzyme pancreatic amylase with the ability of the small intestine to digest starch.

28.4 Requirements for Digestion

- Assuming a specific enzyme, list four factors that can affect the activity of all enzymes.
- Explain why the operative procedure that reduces the size of the stomach causes an individual to lose weight.

Introduction

In this lab, we will examine the process of digestion by learning the organs associated with digestion and studying the action of digestive enzymes.

Enzymes are molecules (typically proteins) that catalyze chemical reactions. They are very specific and usually participate in only one type of reaction. The active site of an enzyme has a shape that accommodates its substrate, and if an environmental factor such as a boiling temperature or a wrong pH alters this shape, the enzyme loses its ability to function well, if at all. We will have an opportunity to make these observations with controlled experiments.

Planning Ahead Be advised that protein digestion requires 1½ hours and fat digestion requires 1 hour. Also a boiling water bath is required for starch digestion.

Figure 28.1 Organs of the digestive tract (right) and accessory organs (left).

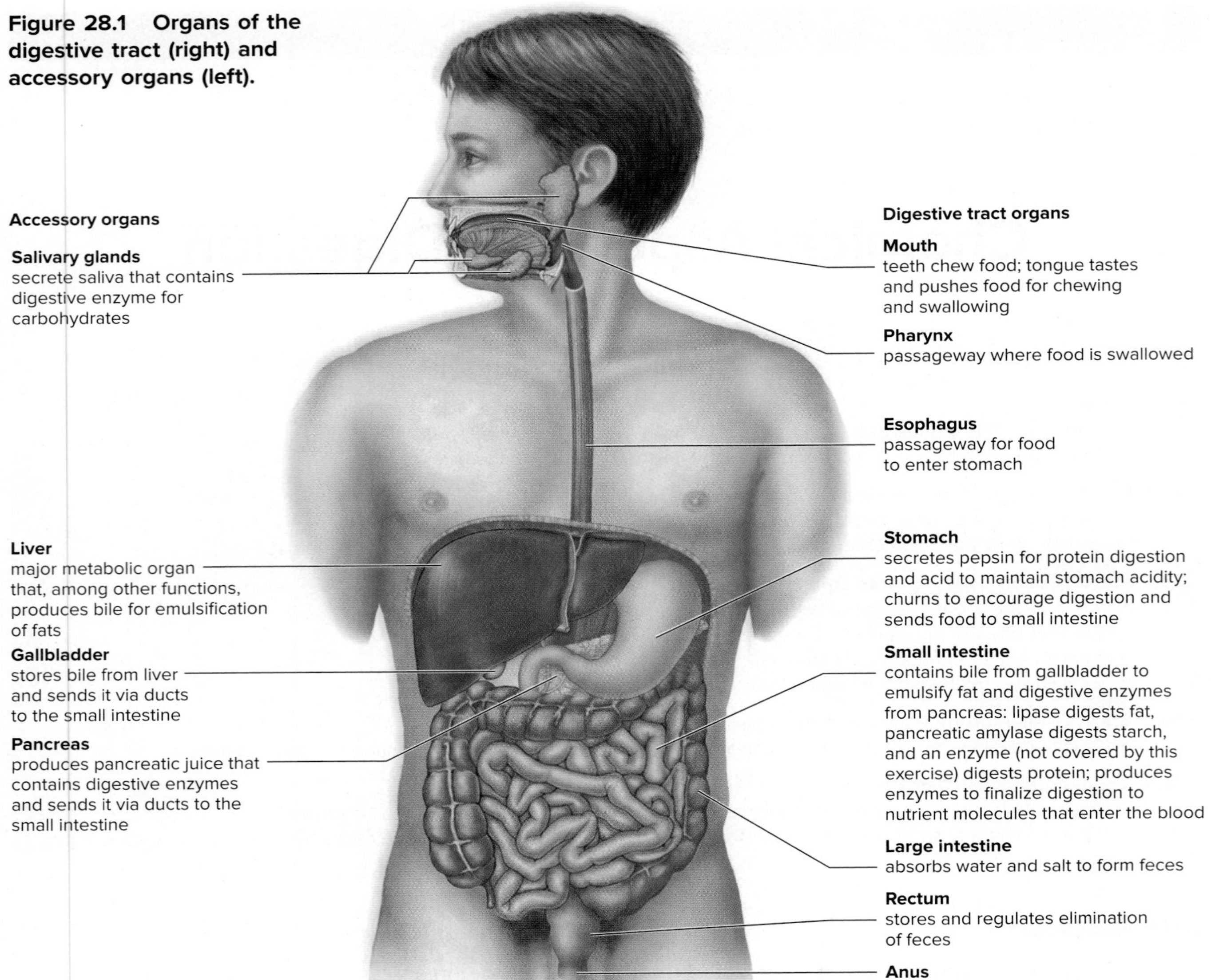

Accessory organs
Salivary glands
secrete saliva that contains digestive enzyme for carbohydrates
Liver
major metabolic organ that, among other functions, produces bile for emulsification of fats
Gallbladder
stores bile from liver and sends it via ducts to the small intestine
Pancreas
produces pancreatic juice that contains digestive enzymes and sends it via ducts to the small intestine
Digestive tract organs
Mouth
teeth chew food; tongue tastes and pushes food for chewing and swallowing
Pharynx
passageway where food is swallowed
Esophagus
passageway for food to enter stomach
Stomach
secretes pepsin for protein digestion and acid to maintain stomach acidity; churns to encourage digestion and sends food to small intestine
Small intestine
contains bile from gallbladder to emulsify fat and digestive enzymes from pancreas: lipase digests fat, pancreatic amylase digests starch, and an enzyme (not covered by this exercise) digests protein; produces enzymes to finalize digestion to nutrient molecules that enter the blood
Large intestine
absorbs water and salt to form feces
Rectum
stores and regulates elimination of feces
Anus

What Is a Control?

The experiments in today's laboratory have both a positive control and a negative control, *which should be saved for comparison purposes until the experiment is complete.* The **positive control** goes through all the steps of the experiment and does contain the substance being tested. Therefore, positive results are expected. The **negative control** goes through all the steps of the experiment, except it does not contain the substance being tested. Therefore, negative results are expected.

For example, if a test tube contains glucose (the substance being tested) and Benedict's reagent (blue) is added, a red color develops upon heating. This test tube is the positive control; it tests positive for glucose. If a test tube does not contain glucose and Benedict's reagent is added, Benedict's is expected to remain blue. This test tube is the negative control; it tests negative for glucose.

What benefit is a positive control? Positive controls give you a standard by which to tell if the substance being tested is present (or acting properly) in an unknown sample. Negative controls ensure that the experiment is giving reliable results; after all, if a negative control should happen to give a positive result, then the entire experiment may be faulty and unreliable.

28.1 Protein Digestion by Pepsin

Certain foods, such as meat and egg whites, are rich in protein. Egg whites contain albumin, which is the protein used in this Experimental Procedure. Protein is digested by **pepsin** in the stomach (Fig. 28.2), a process described by the following reaction:

$$\text{protein} + \text{water} \xrightarrow{\text{pepsin (enzyme)}} \text{peptides}$$

The stomach has a very low pH. Does this indicate that pepsin works effectively in an acidic or a basic environment? ______________________ This is the pH that allows the enzyme to maintain its normal shape so that it will combine with the substrate. A warm temperature causes molecules to move about more rapidly and increases the encounters between enzyme and substrate. Therefore, you would hypothesize that the yield from this enzymatic reaction will be higher if the pH is __________________ and the temperature is __________________ (body temperature 37°C).

Test for Protein Digestion

Biuret reagent is used to test for protein digestion. If digestion has not occurred, biuret reagent turns purple, indicating that protein is present. If digestion has occurred, biuret reagent turns pinkish-purple, indicating that peptides are present.

> ⚠ **Biuret reagent** is highly corrosive. Exercise care in using this chemical. If any should spill on your skin, wash the area with mild soap and water. Follow your instructor's directions for its disposal.

Experimental Procedure: Protein Digestion

1. Label four clean test tubes (1 to 4). Using the designated graduated pipet, add 2 ml of the albumin solution to all tubes. Albumin is a protein.
2. Add 2 ml of the pepsin solution to tubes 1 to 3, as listed in Table 28.1.
3. Add 2 ml of 0.2% HCl to tubes 1 and 2. HCl simulates the acidic conditions of the stomach.
4. Add 2 ml of water to tube 3 and 4 ml of water to tube 4, as listed in Table 28.1.
5. Swirl to mix the tubes. Tube 2 remains at room temperature, but the other three are incubated for 1½ hours. Record the temperature for each tube in Table 28.1.
6. Remove the tubes from the incubator and place all four tubes in a tube rack. Add 2 ml of biuret reagent to all tubes and observe. Record your results in Table 28.1 as + or – to indicate digestion or no digestion.

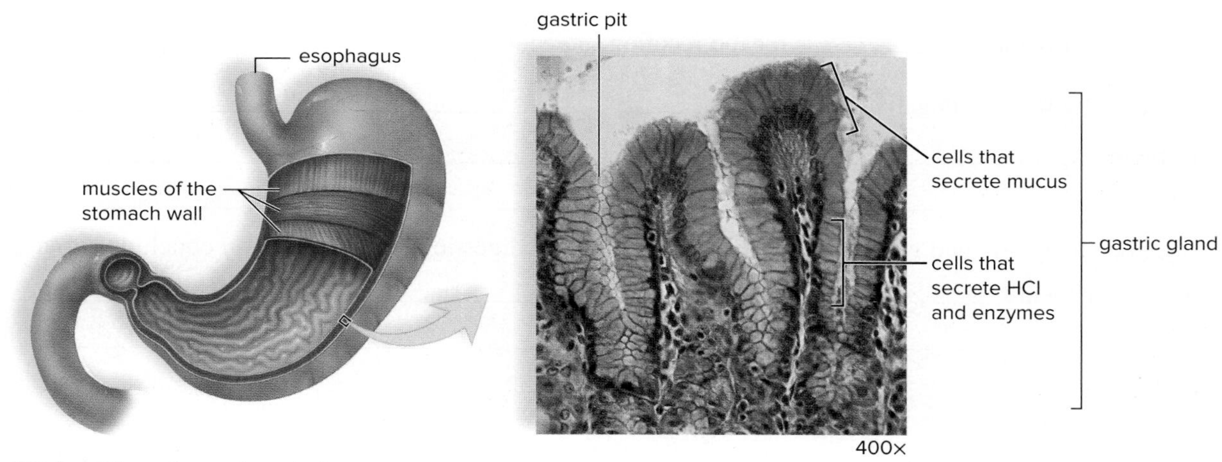

Figure 28.2 Digestion of protein.
Pepsin, produced by the gastric glands of the stomach, helps digest protein. (b) ©Ed Reschke

Table 28.1 Protein Digestion by Pepsin				
Tube	**Contents**	**Temperature**	**Digestion (+ or −)**	**Explanation**
1	Albumin Pepsin HCl Biuret reagent			
2	Albumin Pepsin HCl Biuret reagent			
3	Albumin Pepsin Water Biuret reagent			
4	Albumin Water Biuret reagent			

Conclusions: Protein Digestion

- Explain your results in Table 28.1 by giving an explanation why digestion did or did not occur. To be complete, consider all the requirements for an enzymatic reaction as listed in Table 28.4. Now show here that tube 1 met all the requirements for digestion:

  Pepsin is the correct ______________.

  Albumin is the correct ______________.

  37°C is the optimum ______________.

  HCL provides the optimum ______________.

  1½ hours provides ______________ for the reaction to occur.

- Which tube was the negative control? ______________________________

  Explain. ______________________________

  ____

- If this control tube had given a positive result for protein digestion, what could you conclude about this experiment? ______________________________

  ____

28.2 Fat Digestion by Pancreatic Lipase

Lipids include fats (e.g., butterfat) and oils (e.g., sunflower, corn, olive, and canola). Lipids are digested by **pancreatic lipase** in the small intestine (Fig. 28.3).

Figure 28.3 Emulsification and digestion of fat.
Bile from the liver (stored in the gallbladder) enters the small intestine, where lipase in pancreatic juice from the pancreas digests fat.

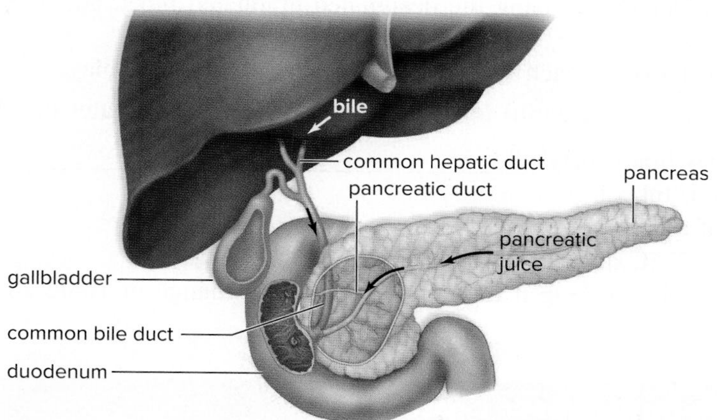

The following two reactions describe fat digestion:

1. $$\text{fat} \xrightarrow{\text{bile (emulsifier)}} \text{fat droplets}$$

2. $$\text{fat droplets} + \text{water} \xrightarrow{\text{lipase (enzyme)}} \text{glycerol} + \text{fatty acids}$$

With regard to the first step, consider that fat is not soluble in water; yet, lipase makes use of water when it digests fat. Therefore, bile is needed to emulsify fat—cause it to break up into fat droplets that disperse in water. The reason for dispersal is that bile contains molecules with two ends. One end (the nonpolar end) is soluble in fat, and the other end (the polar end) is soluble in water. Bile can emulsify fat because of this.

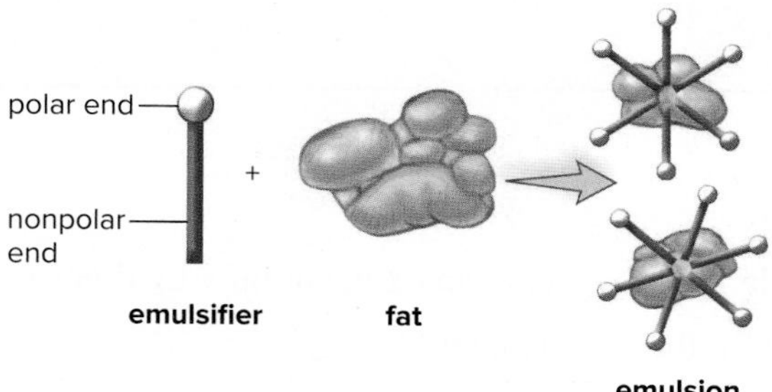

With regard to the second step, would the pH of the solution be lower before or after the enzymatic reaction? (*Hint:* Remember that an acid decreases pH and a base increases pH.) ______________________

____

Test for Fat Digestion

In the test for fat digestion, you will be using a pH indicator, which changes color as the solution in the test tube goes from basic conditions to acidic conditions. Phenol red is a pH indicator that is red in basic solutions and yellow in acidic solutions.

Experimental Procedure: Fat Digestion

1. Label three clean test tubes (1 to 3). Using the designated graduated pipet, add 1 ml of vegetable oil to all tubes.
2. Add 2 ml of phenol red solution to each tube. What role does phenol red play? ____________________
3. Add 2 ml of pancreatic lipase (pancreatin) to tubes 1 and 2 and 2 ml of water to tube 3, as listed in Table 28.2. What role does lipase play? ____________________
4. Add a pinch of bile salts to tube 1.
5. Record the initial color of all tubes in Table 28.2.
6. Incubate all three tubes at 37°C and check every 20 minutes.
7. Record any color change and how long it took to see this color change in Table 28.2.

Table 28.2 Fat Digestion by Pancreatic Lipase

Tube	Contents	Color		Time Taken	Explanation
		Initial	*Final*		
1	Vegetable oil Phenol red Pancreatin Bile salts				
2	Vegetable oil Phenol red Pancreatin				
3	Vegetable oil Phenol red Water				

Conclusions: Fat Digestion

- Explain your results in Table 28.2 by giving an explanation why digestion did or did not occur.
- What role did bile salts play in this experiment? ____________________
- What role did phenol red play in this experiment? ____________________
- Which test tube in this experiment could be considered a negative control? ____________________

28.3 Starch Digestion by Pancreatic Amylase

Starch is present in bakery products and in potatoes, rice, and corn. Starch is digested by **pancreatic amylase** in the small intestine, a process described by the following reaction:

$$\text{starch} + \text{water} \xrightarrow{\text{amylase (enzyme)}} \text{maltose}$$

1. If digestion *does not* occur, which will be present—starch or maltose? ______________________
2. If digestion *does* occur, which will be present—starch or maltose? ______________________

Tests for Starch Digestion

You will be using two tests for starch digestion:

1. If digestion has not taken place, the iodine test for starch will be positive (+) and a blue-black color will be observed. If digestion has occurred, the iodine test for starch will be negative (–) and the iodine will remain yellowish-brown.
2. If digestion has taken place, the Benedict's test for sugar (maltose) will be positive (+) and a color change ranging from green to red will be observed. If digestion has not taken place, the Benedict's test for sugar will be negative (–) and the solution will remain blue.

   To test for sugar, add five drops of Benedict's reagent to each test tube. Place the tube in a boiling water bath for a few minutes, and note any color changes. Boiling the test tube is necessary for the Benedict's reagent to react.

> ⚠ **Benedict's reagent** is highly corrosive. Use protective eyewear when performing this experiment. Exercise care in using this chemical. If any should spill on your skin, wash the area with mild soap and water. Follow your instructor's directions for disposal of this chemical.

Experimental Procedure: Starch Digestion

1. Label six clean test tubes (1 to 6).
2. Using the designated graduated transfer pipet, add 1 ml of pancreatic amylase solution to tubes 1 to 4 and 1 ml of water to tubes 5 and 6.
3. Test tubes 1 and 2 immediately.

   Tube 1 Shake the starch solution and add 1 ml of starch solution. Immediately add five drops of iodine to test for starch. Put this tube in a test tube rack and record your results in Table 28.3.

   Tube 2 Shake the starch solution and add 1 ml of starch solution. Immediately test for sugar with the Benedict's test following the preceding directions. Put this tube in a test tube rack and record your results in Table 28.3.

4. Shake the starch solution and add 1 ml of starch solution to tubes 3 to 6. Allow the tubes to stand for 30 minutes.

   Tubes 3 and 5 After the 30 minutes have passed, test for starch using the iodine test. Place these tubes in the test tube rack and record your results in Table 28.3.

   Tubes 4 and 6 After the 30 minutes have passed, test for sugar with the Benedict's test following the preceding directions. Place these tubes in the test tube rack and record your results in Table 28.3.

5. Examine all your tubes in the test tube rack and decide whether digestion occurred (+) or did not occur (–). Complete Table 28.3.

Table 28.3 Starch Digestion by Amylase					
Tube	**Contents**	**Time***	**Type of Test**	**Results**	**Explanation**
1	Pancreatic amylase Starch	0	Iodine	+	
2	Pancreatic amylase Starch				
3	Pancreatic amylase Starch				
4	Pancreatic amylase Starch				
5	Water Starch				
6	Water Starch				

* Enter either 0 for immediately or T for after 30 minutes.

Conclusions: Starch Digestion

- Considering tubes 1 and 2, this experimental procedure showed that __________ must pass for digestion to occur.
- Considering tubes 5 and 6, this experimental procedure showed that an active __________ must be present for digestion to occur.
- Why would you not recommend doing the test for starch and the test for sugar on the same tube? __________

  ____

  ____
- Which test tubes served as a negative control in this experiment? ____________________

  Explain your answer. ______________________________

  ____

Absorption of Sugars and Other Nutrients

Figure 28.4 shows that the folded lining of the small intestine has many fingerlike projections called villi. The small intestine not only digests food; it also absorbs the products of digestion, such as sugars from carbohydrate digestion, amino acids from protein digestion, and glycerol and fatty acids from fat digestion at the villi.

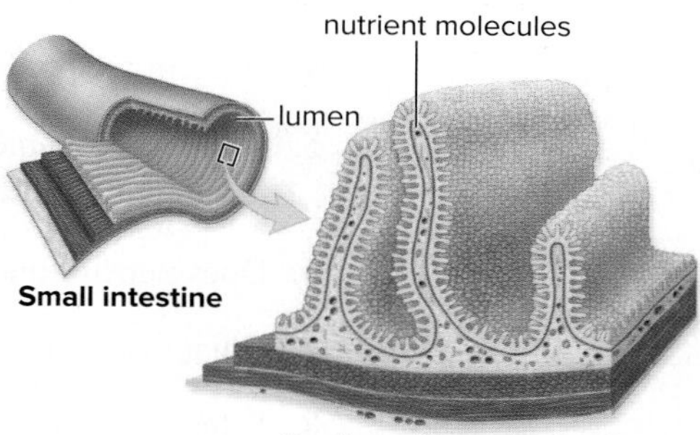

Figure 28.4 Anatomy of the small intestine. Nutrients enter the bloodstream across the much-convoluted walls of the small intestine.

28.4 Requirements for Digestion

Explain in Table 28.4 how each of the requirements listed influences effective digestion.

Table 28.4 Requirements for Digestion

Requirement	Explanation
Specific enzyme	
Specific substrate	
Warm temperature	
Specific pH	
Time	
Fat emulsifier	

To lose weight, some obese individuals undergo an operation in which (1) the stomach is reduced to the size of a golf ball, and (2) food bypasses the duodenum (first 2 feet) of the intestine. Answer these questions to explain how this operation would affect the requirements for digestion.

1. How is the amount of substrate reduced? ______
2. How is the amount of digestive enzymes reduced? ______
3. How is time reduced? ______
4. What makes the pH of the small intestine higher than before? ______
5. How is fat emulsification reduced? ______
6. How does surgery to reduce obesity sometimes result in malnutrition? ______

Laboratory Review 28

_______________ 1. What part of an enzyme is specific for and accommodates its substrate?

_______________ 2. What two environmental conditions affect the ability of an enzyme to function properly?

_______________ 3. Where are carbohydrates first digested by an enzyme?

_______________ 4. What organ produces the lipase and amylase that perform digestion in the small intestine?

_______________ 5. What is the name given to a sample that contains the factor being tested and that goes through all the steps of the experiment?

_______________ 6. Does pepsin function in an acidic or a basic pH?

_______________ 7. What results from the digestion of proteins by pepsin?

_______________ 8. What reagent was used to test for the digestion of proteins?

_______________ 9. What molecule emulsifies fats in the small intestine?

_______________ 10. What organ stores bile?

_______________ 11. What component of foods is digested by pancreatic amylase?

_______________ 12. What molecule is present after pancreatic amylase digests starch?

_______________ 13. What two tests were done to test for starch digestion?

_______________ 14. What structures absorb the products of digestion in the small intestine?

Thought Questions

15. Salivary amylase digests starch to maltose in the mouth while chewing takes place. Why is it beneficial to have the enzyme that digests starch, amylase, in two different locations?

16. How would the digestion of fat be impacted by the removal of the gallbladder? Explain.

17. Explain why consuming excess antacids might affect protein digestion. How could you investigate the impact of consuming antacids on protein digestion?

LABORATORY

29

Basic Mammalian Anatomy II

Learning Outcomes

29.1 Urinary System

- Locate and identify the organs of the urinary system.
- State a function for the organs of the urinary system.

29.2 Male Reproductive System

- Locate and identify the organs of the male reproductive system.
- State a function for the organs of the male reproductive system.
- Compare the male pig reproductive system with that of the human male.

29.3 Female Reproductive System

- Locate and identify the organs of the female reproductive system.
- State a function for the organs of the female reproductive system.
- Compare the female pig reproductive system with that of the human female.

29.4 Review of the Respiratory, Digestive, and Cardiovascular Systems

- Using preserved specimens, images, or charts, locate and identify the individual organs of the respiratory, digestive, and cardiovascular systems.
- Using preserved specimens, images, or charts, locate and identify the hepatic portal system. State a function for this system.

Introduction

The **urinary system** and the **reproductive system** are so closely associated in mammals that they are often considered together as the **urogenital system.** They are particularly associated in males, where certain structures function in both systems. In this laboratory, we will focus first on dissecting the urinary and reproductive systems in the fetal pig.

The kidneys of the urinary system produce urine, which is stored in the bladder before being released to the exterior. As the kidneys produce urine, they also regulate the volume and the composition of the blood so that the water and salt balance and the acid-base balance of the blood stays within normal limits.

In mammalian reproductive systems, the testes are the male gonads, and the ovaries are the female gonads. The testes produce sperm, and the ovaries produce oocytes that become eggs. We will compare the anatomy of the reproductive systems in pigs with those in humans.

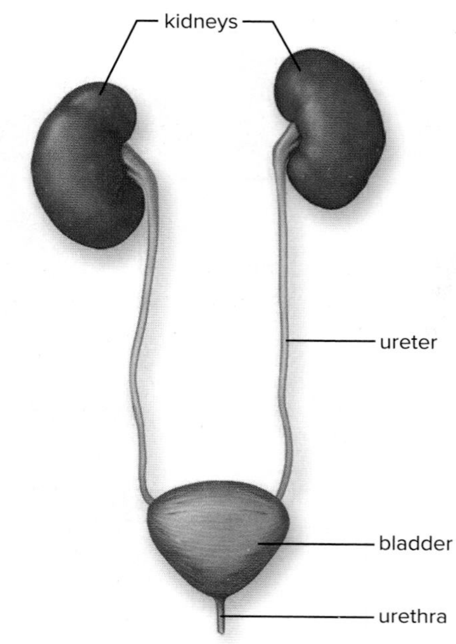

Urinary system

Figure 29.1 Urinary system of the fetal pig.
In **(a)** females and **(b)** males, urine is made by the kidneys, transported to the bladder by the ureters, stored in the bladder, and then excreted from the body through the urethra.

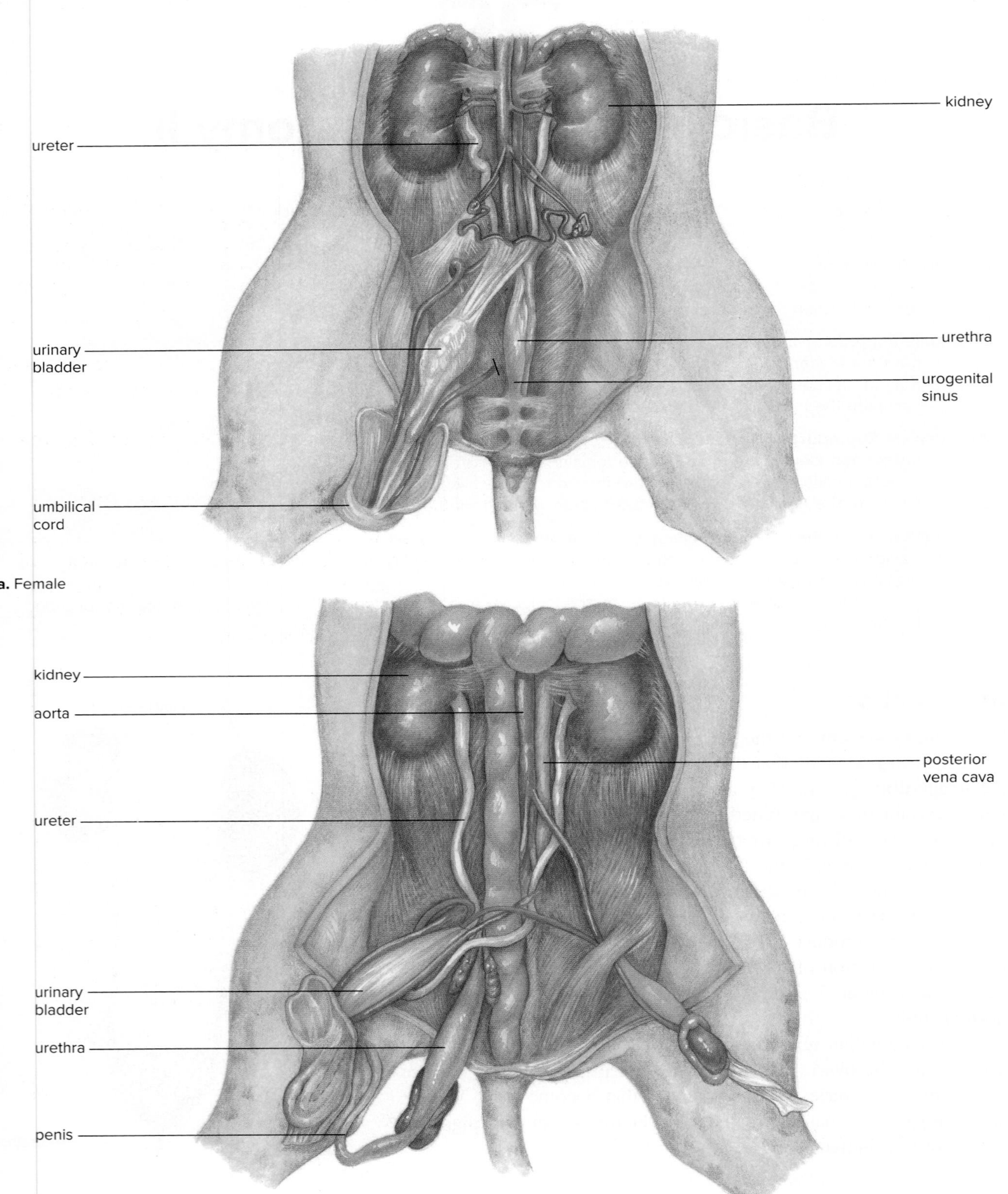

a. Female

b. Male

29.1 Urinary System

The urinary system consists of the **kidneys,** which produce urine; the **ureters,** which transport urine to the **urinary bladder,** where urine is stored; and the **urethra,** which transports urine to the outside. In males, the urethra also transports sperm during ejaculation.

Observation: Urinary System in Pigs

During this dissection, compare the urinary system structures of male and female fetal pigs. Later in this laboratory period, exchange specimens with a neighboring team for a more thorough inspection.

> ⚠ **Wear protective latex gloves and eyewear** when handling preserved animal organs. Exercise caution when using sharp instruments during this laboratory. Wash hands thoroughly upon completion of this laboratory.

1. The large, paired kidneys (Fig. 29.1) are reddish organs covered by **peritoneum,** a membrane that anchors them to the dorsal wall of the abdominal cavity, sometimes called the **peritoneal cavity.** Clean the peritoneum away from one of the kidneys, and study it more closely.
2. Locate the ureters, which leave the kidneys and run posteriorly under the peritoneum (Fig. 29.1).
3. Clean the peritoneum away, and follow a ureter to the urinary bladder, which normally lies in the ventral portion of the abdominal cavity. The urinary bladder is on the inner surface of the flap of tissue to which the umbilical cord was attached.
4. The urethra arises from the bladder posteriorly and joins the urogenital sinus. Follow the urethra until it passes from view into the ring formed by the pelvic girdle.
5. Sequence the organs in the urinary system to trace the path of urine from its production to its exit. ________

____

6. Using a scalpel, section one of the kidneys in place, cutting it lengthwise (Fig. 29.2). At the center of the medial portion of the kidney is an irregular, cavity-like reservoir, the **renal pelvis.** The outermost portion of the kidney (the **renal cortex**) shows many small striations perpendicular to the outer surface. This region and the more even-textured **renal medulla** contain **nephrons** (excretory tubules), microscopic organs that produce urine.

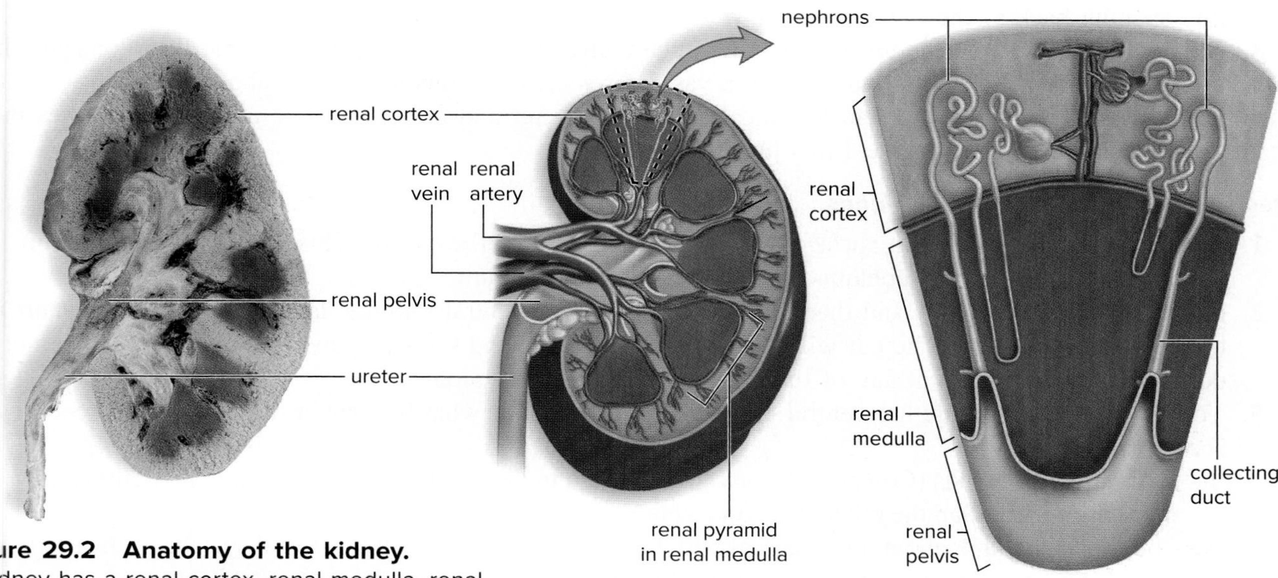

Figure 29.2 Anatomy of the kidney. A kidney has a renal cortex, renal medulla, renal pelvis, and microscopic tubules called nephrons.

29.2 Male Reproductive System

The **male reproductive system** consists of the **testes** (sing., testis), which produce sperm, and the **epididymides** (sing., epididymis), which store sperm before they enter the **vasa deferentia** (sing., vas deferens). Just prior to ejaculation, sperm leave the vasa deferentia and enter the urethra, which eventually passes into the penis. The **penis** is the male organ of sexual intercourse. **Seminal vesicles,** the **prostate gland,** and the **bulbourethral glands** (Cowper's glands) add fluid to semen after sperm reach the urethra. Table 29.1 summarizes the male reproductive organs.

Table 29.1 Male Reproductive Organs and Functions

Organ	Function
Testis	Produces sperm and sex hormones
Epididymis	Stores sperm as they mature
Vas deferens	Conducts and stores sperm
Seminal vesicle	Contributes secretions to semen
Prostate gland	Contributes secretions to semen
Urethra	Conducts sperm
Bulbourethral glands	Contribute secretions to semen
Penis	Organ of copulation

The testes begin their development in the abdominal cavity, just anterior and dorsal to the kidneys. Before birth, however, they gradually descend into paired **scrotal sacs** within the scrotum, suspended anterior to the anus. Each scrotal sac is connected to the body cavity by an **inguinal canal,** the opening of which can be found in the pig. The passage of the testes from the body cavity into the scrotal sacs is called the descent of the testes. The testes in most of the male fetal pigs being dissected will probably be partially or fully descended.

Observation: Male Reproductive System in Pigs

While doing this dissection, consult Table 29.1 for the function of the male reproductive organs.

Inguinal Canal, Testis, Epididymis, and Vas Deferens

1. Locate the opening of the left inguinal canal, which leads to the left scrotal sac (Fig. 29.3).
2. Expose the canal and sac by making an incision through the skin and muscle layers from a point over this opening back to the left scrotal sac.
3. Open the sac, and find the testis. Note the much-coiled tubule—the epididymis—that lies alongside the testis. An epididymis is continuous with a vas deferens, which runs toward the abdominal cavity.
4. Each vas deferens loops over an umbilical artery and ureter and unites with the urethra as it leaves the urinary bladder. We will dissect this juncture below.

Penis, Urethra, and Accessory Glands

1. Cut through the ventral skin surface just posterior to the umbilical cord. This will expose the rather undeveloped penis, which contains a long portion of the urethra.
2. Lay the penis to one side, and then cut down through the ventral midline, laying the legs wide apart in the process (Fig. 29.4). The cut will pass between muscles and through pelvic cartilage (bone has not developed yet). Do not cut any of the ducts or tracts in the region.
3. You will now see the urethra ventral to the rectum. It is somewhat heavier in the male due to certain accessory glands:
   - **a.** Bulbourethral glands (Cowper's glands), about 1 cm in diameter, are further along the urethra and are more prominent than the other accessory glands.
   - **b.** The prostate gland, about 4 mm across and 3 mm thick, is located on the dorsal surface of the urethra, just posterior to the juncture of the urinary bladder with the urethra. It is often difficult to locate and is not shown in Figures 29.3 and 29.4.
   - **c.** Small, paired seminal vesicles may be seen on either side of the prostate gland.

Figure 29.3 Male reproductive system of the fetal pig.
In males, the urinary system and the reproductive system are joined. The vasa deferentia (sing., vas deferens) enter the urethra, which also carries urine.

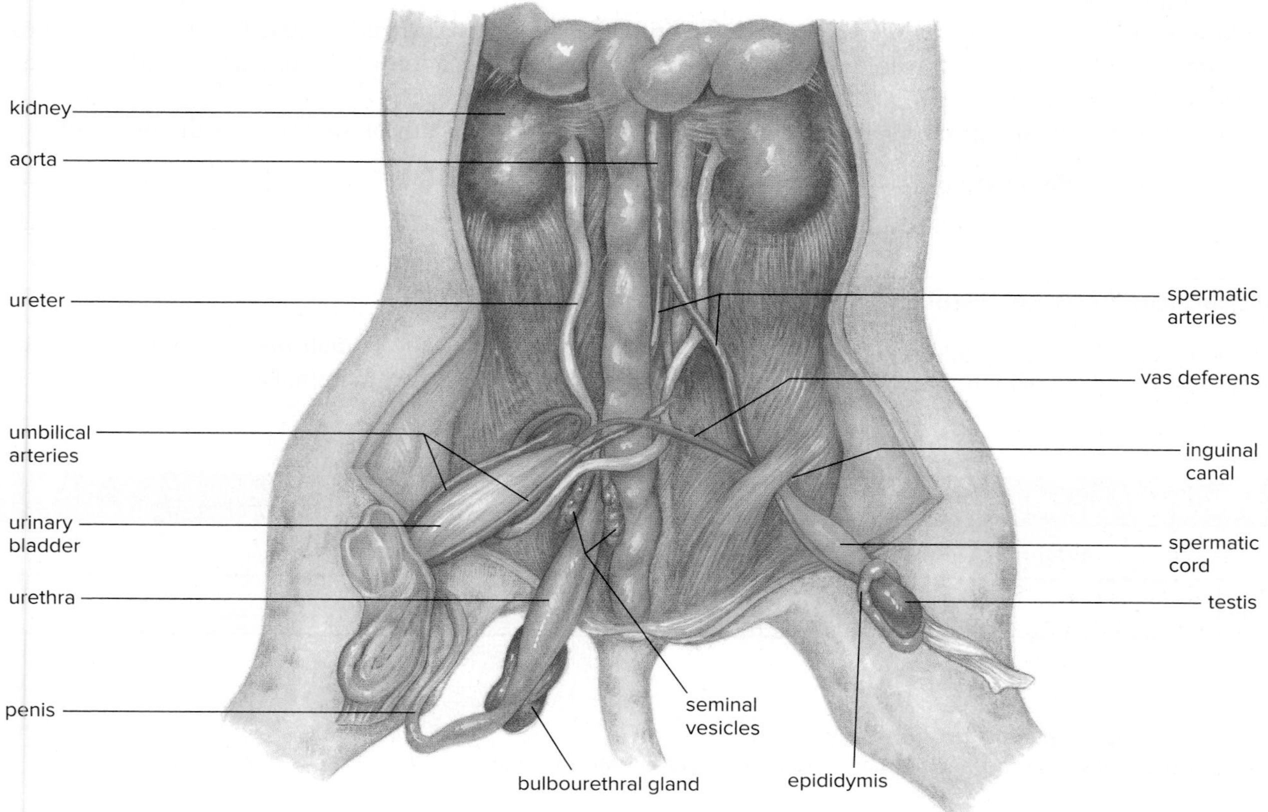

Figure 29.4 Photograph of the male reproductive system of the fetal pig.
Compare Figure 29.3 to this photograph to help identify the structures of the male urogenital system.

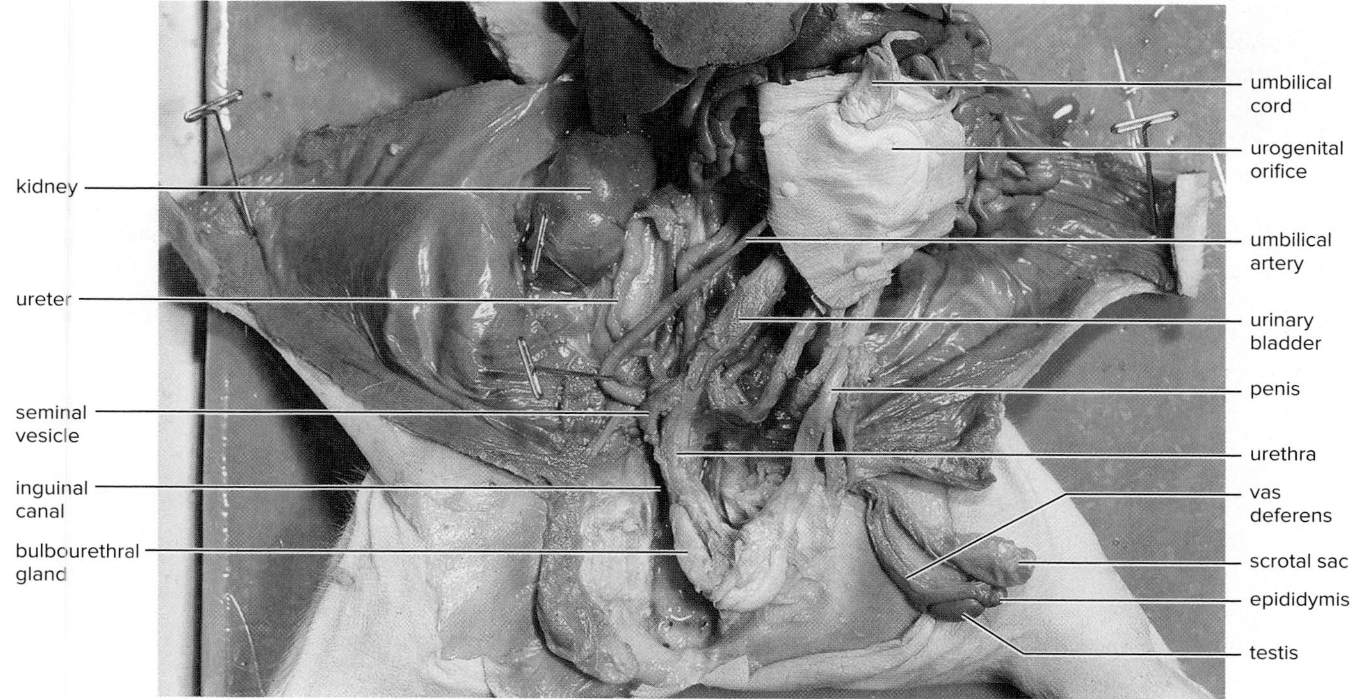

4. Trace the urethra as it leaves the bladder. When it nears the end of the abdominal cavity, it turns rather abruptly and runs anteriorly just under the skin where you have just dissected it. This portion of the urethra is within the penis.
5. Now you should be able to see the vasa deferentia enter the urethra. If necessary, free these structures from surrounding tissue to see them enter the urethra near the location of the prostate gland. In males, the urethra transports sperm and urine to the urogenital opening.
6. Sequence the organs in the male reproductive system to trace the path of sperm from the organ of production to the penis. ____

____

Comparison of Male Fetal Pig and Human Male

Use Figure 29.5 to help compare the male pig reproductive system with the human male reproductive system. Complete Table 29.2, which compares the location of the penis in these two mammals.

Table 29.2 Location of Penis in Male Fetal Pig and Human Male

	Fetal Pig	Human
Penis		

Figure 29.5 Human male urogenital system.
In the fetal pig, but not in the human male, the penis lies beneath the skin and exits at the urogenital opening.

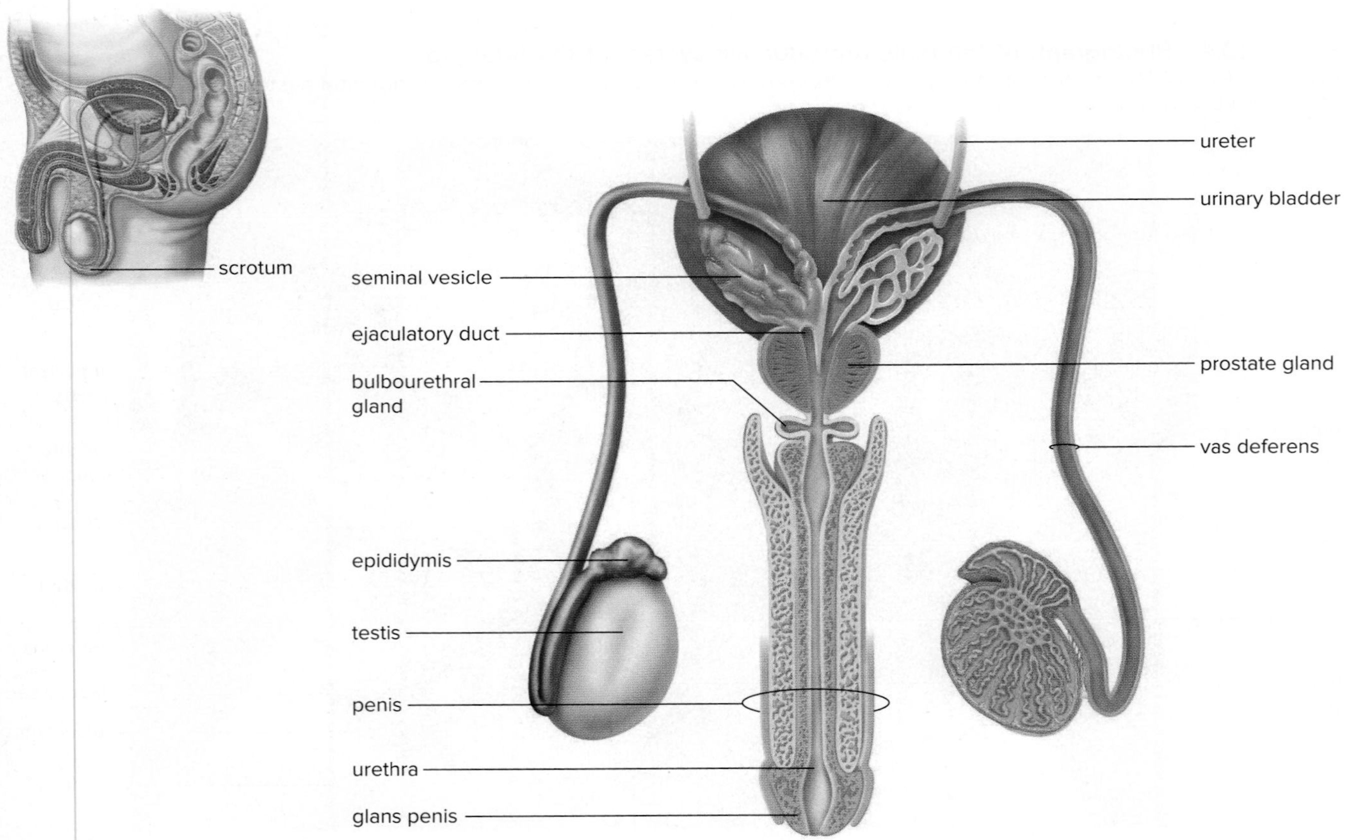

29.3 Female Reproductive System

Organ	Function
Ovary	Produces egg and sex hormones
Oviduct (fallopian tube)	Conducts egg toward uterus
Uterus	Houses developing fetus
Vagina	Receives penis during copulation and serves as birth canal

Table 29.3 Female Reproductive Organs and Functions

The **female reproductive system** (Table 29.3) consists of the **ovaries** (sing., ovary), which produce eggs, and the **oviducts,** which transport eggs to the **uterus,** where development occurs. In the fetal pig, the uterus does not form a single organ, as in humans, but is partially divided into structures called **uterine horns,** which connect with the oviduct. The **vagina** is the birth canal and the female organ of sexual intercourse.

Observation: Female Reproductive System in Pigs

While doing this dissection, consult Table 29.3 for the function of the female reproductive organs.

Ovaries and Oviducts

1. Locate the paired ovaries, small bodies suspended from the peritoneal wall in mesenteries, posterior to the kidneys (Figs. 29.6 and 29.7).
2. Closely examine one ovary. Note the small, short, coiled oviduct, sometimes called the **fallopian tube.** The oviduct does not attach directly to the ovary but ends in a funnel-shaped structure with fingerlike processes (fimbriae) that partially encloses the ovary. The egg produced by an ovary enters an oviduct, where it is fertilized by a sperm, if reproduction will occur. Any resulting embryo passes to the uterus.

Uterine Horns

1. Locate the **uterine horns.** (Do not confuse the uterine horns with the oviducts; the latter are much smaller and are found very close to the ovaries.)
2. Find the body of the uterus, located where the uterine horns join.

Vagina

1. Separate the hindlimbs of your specimen, and cut down along the midventral line. The cut will pass through muscle and the cartilaginous pelvic girdle. With your fingers, spread the cut edges apart, and use blunt dissecting instruments to separate connective tissue.
2. Now find the vagina, which passes from the uterus to the urogenital sinus. The vagina is dorsal to the urethra, which also enters the urogenital sinus, and ventral to the rectum, which exits at the anus. The urogenital sinus opens at the urogenital papilla (Figs. 29.6 and 29.7).
3. The vagina is the organ of copulation and is the birth canal. The receptacle for the vagina in a pig, the urogenital sinus, is absent in adult humans and several other adult female mammals in which both the urethra and vagina have their own openings.
4. The vagina plays a critical role in reproduction, even though development of the offspring occurs in the uterus. Explain. ____

____

Figure 29.6 Female reproductive system of the fetal pig.
In the fetal pig, both the vagina and the urethra enter the urogenital sinus, which opens at the urogenital papilla.

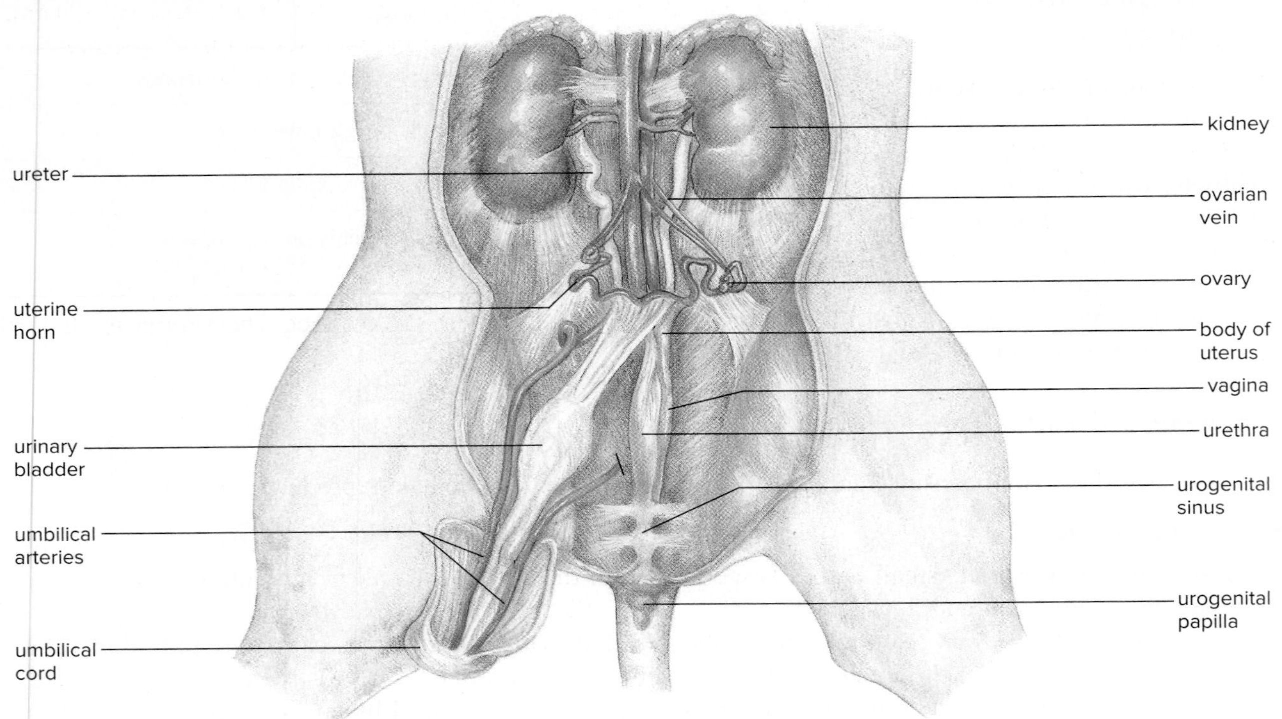

Figure 29.7 Photograph of the female reproductive system of the fetal pig.
Compare Figure 29.6 with this photograph to help identify the structures of the female urinary and reproductive systems.

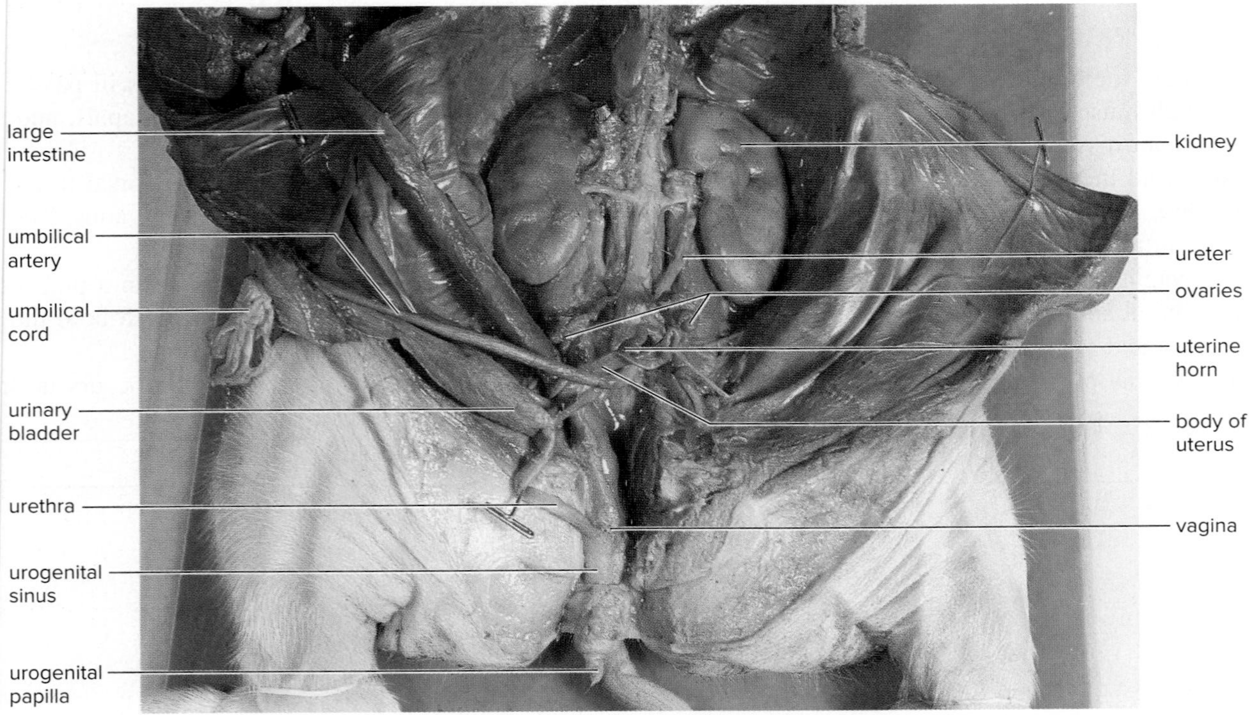

Comparison of Female Fetal Pig with Human Female

Use Figure 29.8 to compare the female pig reproductive system with the human female reproductive system. Complete Table 29.4, which compares the appearance of the oviducts and the uterus, as well as the presence or absence of a urogenital sinus, in these two mammals.

Figure 29.8 Human female reproductive system.
Especially compare the anatomy of the oviducts in humans with that of the uterine horns in a pig. In a pig, the fetuses develop in the uterine horns; in a human female, the fetus develops in the body of the uterus.

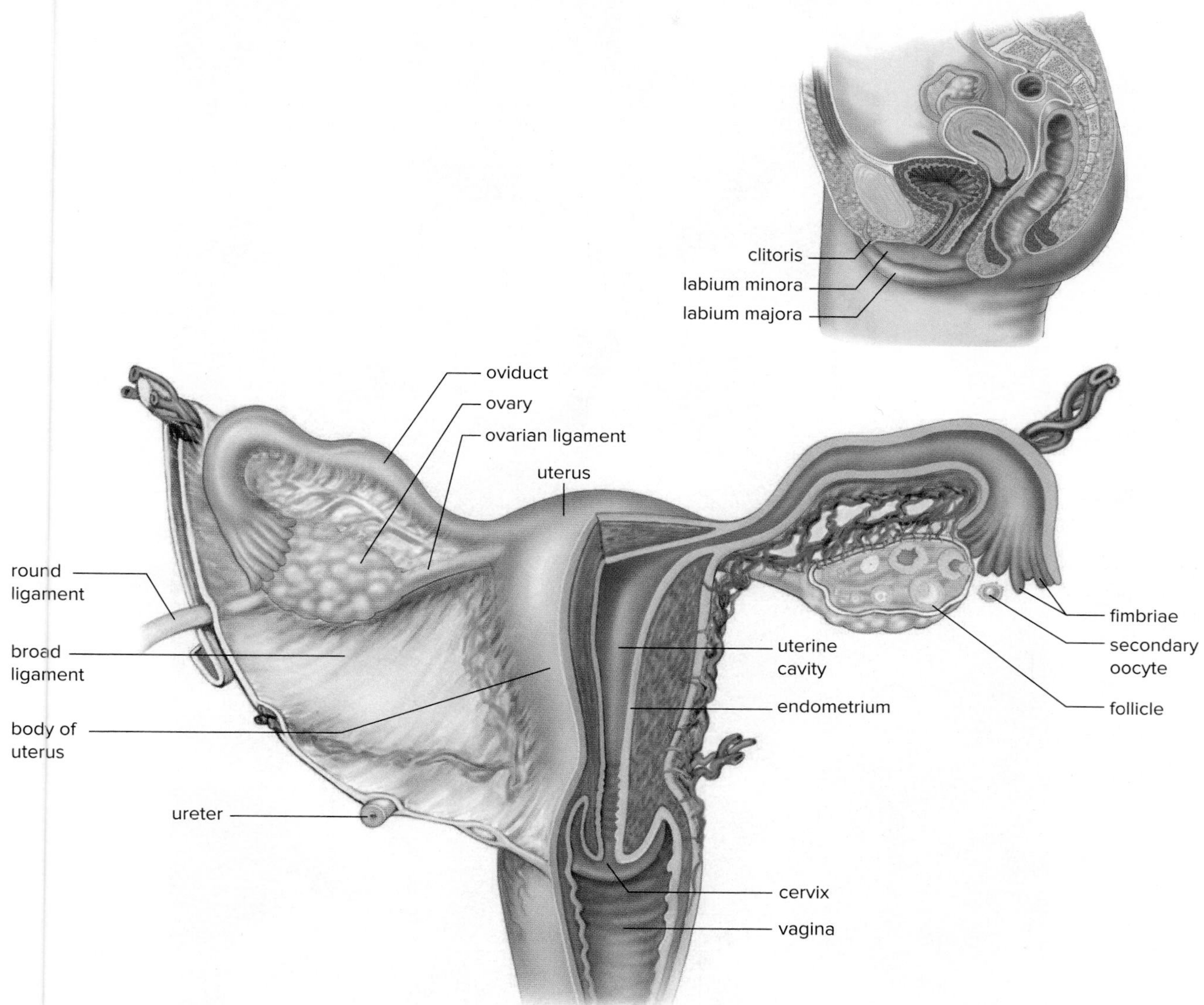

Table 29.4 Comparison of Female Fetal Pig with Human Female

	Fetal Pig	Human
Oviducts		
Uterus		
Urogenital sinus		

Figure 29.9 Internal anatomy of the fetal pig.
Most of the major organs are shown in this photograph. The stomach has been removed. The spleen, gallbladder, and pancreas are not visible. ©Ken Taylor/Wildlife Images

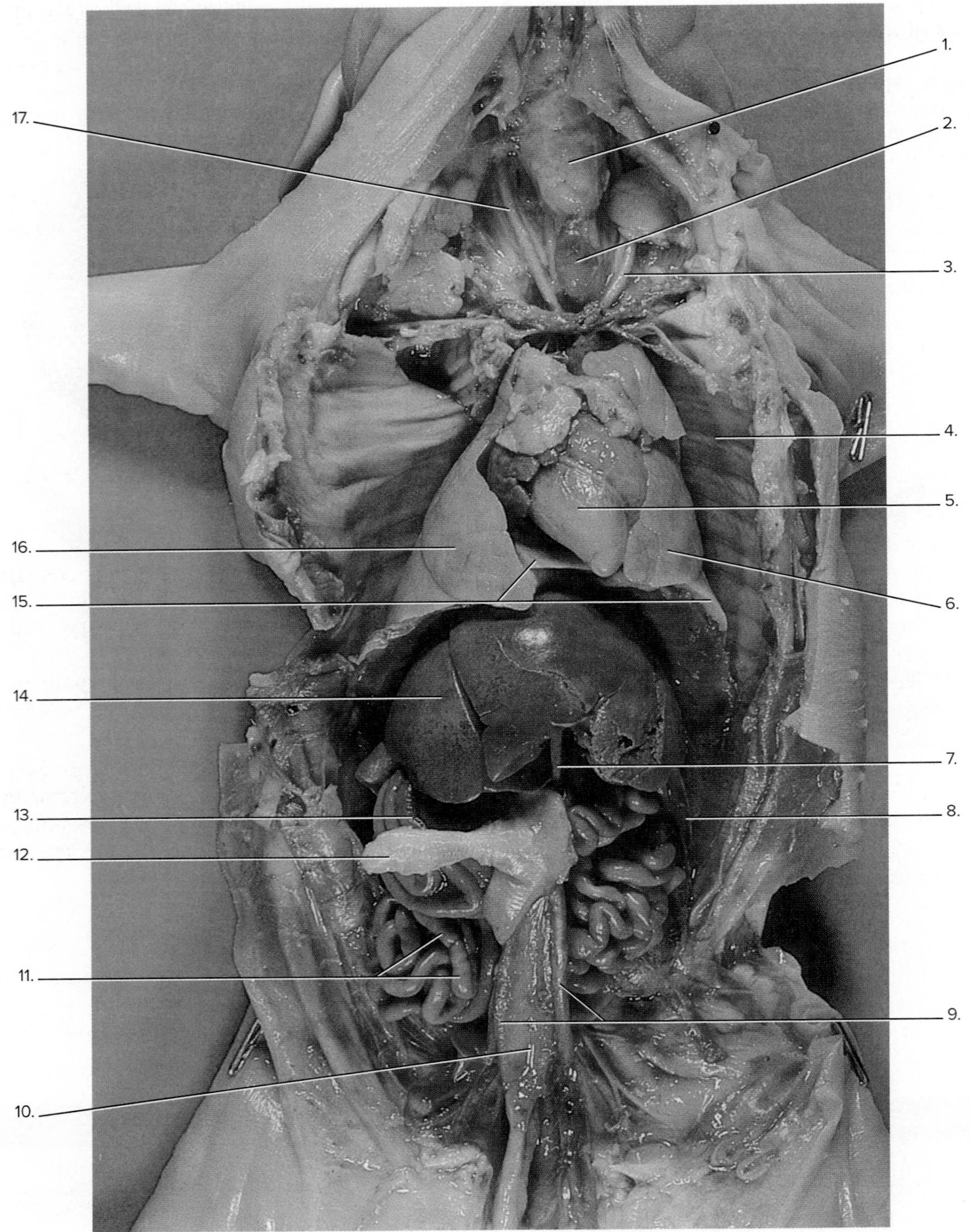

29.4 Review of the Respiratory, Digestive, and Cardiovascular Systems

You previously dissected the respiratory, digestive, and cardiovascular systems of the fetal pig. Review your knowledge of these systems by reexamining your dissection of the fetal pig and *label Figure 29.9.* In this portion of today's lab, you will review each system and examine some organs in more detail. **Do not remove any organs** unless told to do so by your instructor.

Observation: Respiratory System in Pigs

1. Using these terms (bronchiole, bronchus, glottis, larynx, pharynx, trachea), trace the path of air from the nasal passages to the lungs. List the first three organs in the left column and the last three organs in the right column.

nasal passages	______________
______________	______________
______________	______________
______________	lungs

2. Make sure you have cut the corners of the mouth. In the **pharynx,** you should be able to locate the **glottis,** an opening to the ______________________.
3. If necessary, make a midventral incision in the neck to expose the **larynx.**
4. Clear away the "straplike" muscles covering the **trachea.** Now you should be able to feel the cartilaginous rings that hold the trachea open. Locate the esophagus, which lies below (dorsal to) the trachea.
5. If available, observe a slide on display showing a section through the trachea and esophagus. Notice in Figure 29.10 that the air and food pathways cross in the pharynx.
6. Open the pig's mouth, insert a blunt probe into the glottis, and carefully work the probe down through the larynx to the level of the **bronchi.**
7. Observe the **lungs,** and if available, observe a prepared slide of lung tissue.
8. If so directed by your instructor, remove a portion of the trachea, the bronchi, and the lungs, keeping them all in one piece. Place this specimen in a small container of water. Holding the trachea with your forceps, gently but firmly stroke the lung repeatedly with the blunt wooden base of one of your probes. If you work carefully, the alveolar tissue will be fragmented and rubbed away, leaving the branching system of air tubes and blood vessels.

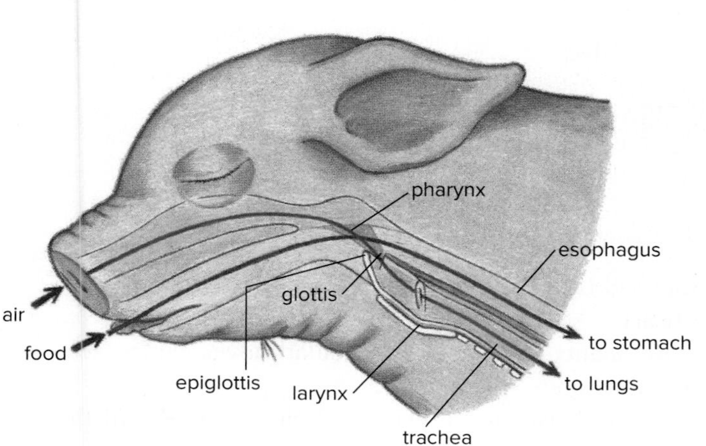

Figure 29.10 Air and food passages in the fetal pig.
A probe can pass from the mouth to the larynx and to the esophagus.

Observation: Digestive System in Pigs

1. Using these terms (esophagus, large intestine, small intestine, stomach), trace the path of food from the mouth to the anus:

   mouth ____________ ____________

   ____________ ____________

   ____________ anus

2. Open the **mouth,** and insert a blunt probe into the esophagus (Fig. 29.10). Then trace the **esophagus** to the stomach.
3. Open one side of the **stomach,** and examine its interior surface. Does it appear smooth or rough? ________
4. Find the pyloric sphincter, the muscle that surrounds the entrance to the duodenum, the first part of the **small intestine.** Record the length of the small intestine. ____________

   If you have not done so before, find the bile duct that empties into the duodenum. The bile duct comes from the ____________.
5. Find the **cecum,** a projection where the small intestine enters the large intestine. How does the appearance of the pig's large intestine differ from that of a human? ____________

   ____
6. Carefully cut the mesenteries holding the colon of the **large intestine** in place, and uncoil the large intestine. Record the length of the large intestine. ________ How does the length of the large intestine compare with that of the small intestine? ____________
7. Locate again the liver, pancreas, and gallbladder, three accessory organs of digestion.

Observation: Cardiovascular System in Pigs

Heart

1. Trace the path of blood through the heart, starting with the vena cava and ending with the aorta. Mention all the chambers of the heart and the valves.

To the heart:	From the lungs:
vena cava	____________
____________	____________
____________ valve	____________ valve
____________	____________
____________ valve	____________ valve
____________	aorta
____________	

2. Keeping the heart inside the pig, cut the pericardial sac (the tissue that surrounds the heart).
3. Look for and identify the vessels attached to the heart.
4. Section the heart, and look for its four chambers. Remnants of the atrioventricular valves can be seen as thin sheets of whitish tissue attached to fine, white, tendinous strands.
5. With your blunt probe, find the oval opening in the wall between the two atrial chambers. Recall that this is a shunt that allows blood to bypass lung circulation prior to birth.

Blood Vessels

1. In general, arteries take blood ____________ the heart, and veins take blood __________ the heart.
2. Locate the following blood vessels in your pig. State their origin and destination.
   a. **coronary artery** ____________________
   b. **cardiac vein** ____________________
   c. **carotid artery** ____________________
   d. **jugular vein** ____________________
   e. **subclavian artery** ____________________
   f. **subclavian vein** ____________________
   g. **renal artery** ____________________
   h. **renal vein** ____________________
   i. **iliac artery** ____________________
   j. **iliac vein** ____________________

Hepatic Portal System and Associated Vessels

A **portal system** goes from one capillary bed to another without passing through the heart. For example, the hepatic portal vein takes blood from the intestinal capillaries to capillaries in the liver. The liver plays an important role in processing and storing materials absorbed from the intestine. The hepatic veins take blood from the liver to the inferior (posterior) vena cava.

1. In your pig, the hepatic portal vein is dorsal to the bile duct and will not be blue if the latex did not enter it. To find the hepatic portal vein, break the mesenteries in the region of the bile duct (Fig. 29.11).
2. The **hepatic veins** consist of three or four vessels from the liver to the inferior vena cava. To see them, scrape away the liver with the blunt side of the scalpel until liver material has been removed and only a mass of cords remains.
3. Identify the **umbilical vein** leading into the liver and, on its posterior surface, the **venous duct,** which is the main channel through the liver in the fetus.

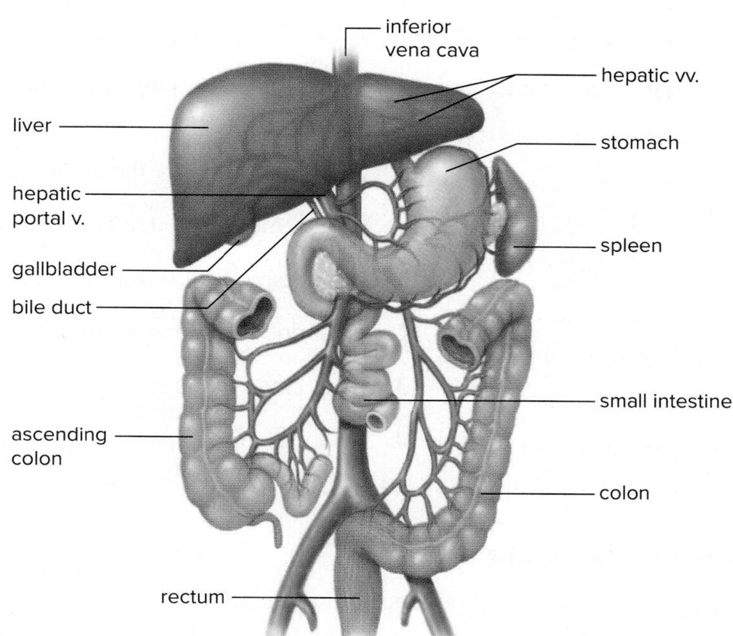

Figure 29.11 Hepatic portal system. The hepatic portal vein lies between the digestive tract and the liver. The hepatic veins enter the inferior vena cava. (v. = vein; vv. = veins)

Storage of Pigs

1. Before leaving the laboratory, place your pig in the plastic bag provided.
2. Expel excess air from the bag, and tie it shut.
3. Write your name and section on the tag provided, and attach it to the bag. Your instructor will indicate where the bags are to be stored until the next laboratory period.
4. Clean the dissecting tray and tools, and return them to their proper location.
5. Wipe off your goggles and wash your hands.

Laboratory Review 29

_______________ **1.** What two individual systems make up the urogenital system?

_______________ **2.** What is the cavity-like space in a kidney called?

_______________ **3.** What organ stores urine before it is transported to the outside?

_______________ **4.** What structure associated with the urinary system transports urine and sperm?

_______________ **5.** Where are sperm stored while they mature?

_______________ **6.** When descending, what do the testes travel through to get to the scrotal sacs?

_______________ **7.** What accessory gland for reproduction in human males is inferior (posterior) to the urinary bladder?

_______________ **8.** What female reproductive structures transport eggs to the location where embryos/fetuses develop?

_______________ **9.** Where, in a pig, are the embryos/fetuses located during development?

_______________ **10.** What female reproductive structure produces sex hormones?

_______________ **11.** What respiratory structure has cartilaginous rings that keep the airway open?

_______________ **12.** What does food travel through to enter the duodenum when leaving the stomach?

_______________ **13.** What is the tissue surrounding the heart called?

_______________ **14.** What does blood pass through when it goes from one capillary bed to another without passing through the heart?

Thought Questions

15. Why do women experience more urinary tract infections than men? It may help to observe Figures 29.5 and 29.8 (the lateral view at top right) and think about where the bladder and urethra are in the drawings. You may also refer to figures of the human female reproductive system in your textbook.

16. What male reproductive structure is severed during a vasectomy? Do the testes still produce sperm following a vasectomy? Why does sterility result from a vasectomy?

17. The hormone insulin lowers blood glucose. How will insulin affect what happens to glucose once blood travels through the hepatic portal system?

LABORATORY

30

Homeostasis

Learning Outcomes

Introduction

- Define homeostasis and the internal environment of vertebrates.

30.1 Heartbeat and Blood Flow

- Describe the cardiovascular system and relate the heartbeat cycle to blood pressure and blood flow.
- Measure blood pressure, and explain the relationship between blood pressure and heart rate.

30.2 Blood Flow and Systemic Capillary Exchange

- Describe the exchange of molecules across a capillary wall, and describe how this exchange relates to blood pressure and osmotic pressure.

30.3 Lung Structure and Human Respiratory Volumes

- Describe the mechanics of breathing and the role of the alveoli in gas exchange.
- Measure respiratory volumes (e.g., tidal volume) and explain their relationship to homeostasis.

30.4 Kidneys

- Understand kidney and nephron structure and blood supply.
- State the three steps in urine formation and how they relate to the parts of a nephron.
- Perform a urinalysis and explain how the results are related to kidney functions.

Introduction

Homeostasis refers to the dynamic equilibrium of the body's internal environment. The **internal environment** consists of blood and tissue fluid. The body's cells take nutrients from tissue fluid and return their waste molecules to it. Tissue fluid, in turn, exchanges molecules with the blood. This is called capillary exchange. All internal organs contribute to homeostasis, but this laboratory specifically examines the contributions of the blood, lungs, and kidneys (Fig. 30.1).

Figure 30.1 Contributions of organs to homeostasis. The lungs exchange gases with blood; the kidneys remove nitrogenous wastes from blood; and the intestinal tract adds nutrients as regulated by the liver to blood.

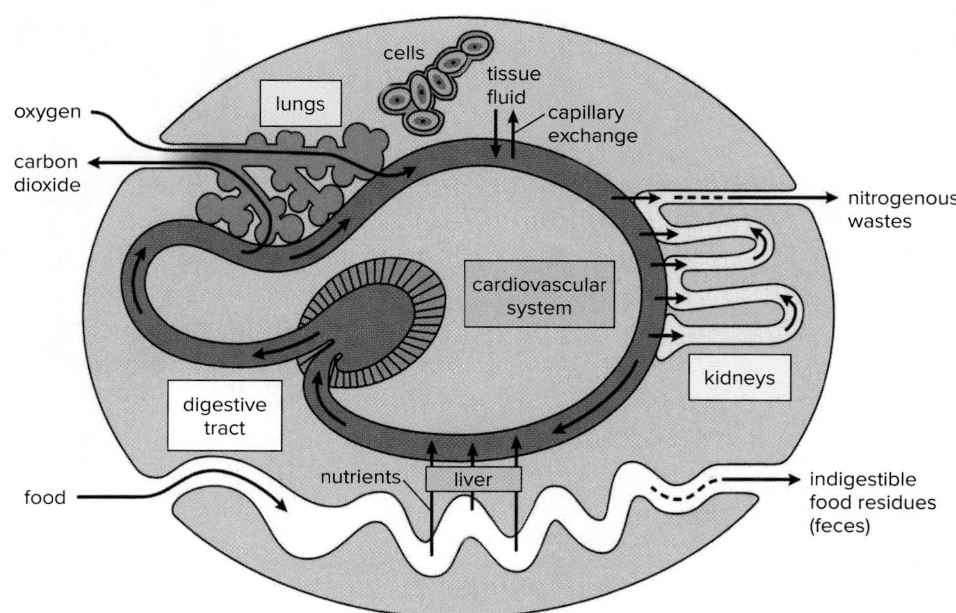

30.1 Heartbeat and Blood Flow

Recall that the cardiovascular system consists of the heart, blood vessels, and blood (Fig. 30.2). Arteries carry blood away from the heart while veins transport blood toward the heart. Arteries branch into smaller vessels called arterioles that enter capillary beds. Capillary beds are present throughout the organs and tissues of the body. An exchange of gases takes place across the thin walls of **pulmonary capillaries.** In the lungs, CO_2 leaves the blood and O_2 enters the blood. An exchange of gases and nutrients for metabolic wastes takes place across the thin walls of **systemic capillaries.** In the body tissues, O_2 and nutrients exit the blood, while CO_2 and metabolic wastes enter the blood.

We will see that the heart is vital to homeostasis because its contraction (called the **heartbeat**) keeps the blood moving in the arteries and arterioles, which take blood to the capillaries. The exchanges that take place across capillaries help maintain homeostasis.

Liver

The hepatic portal vein lies between the ________________________ and the ________________________.

This placement allows the liver to regulate what molecules enter the blood from the digestive tract. For example, if the hormone insulin is present, the liver removes excess glucose and stores it as glycogen. Later, the liver breaks down glycogen to glucose to keep the blood glucose concentration constant.

Figure 30.2 The circulatory system.
The heart provides the pumping action that transports the blood through the arteries, capillary beds, and veins.

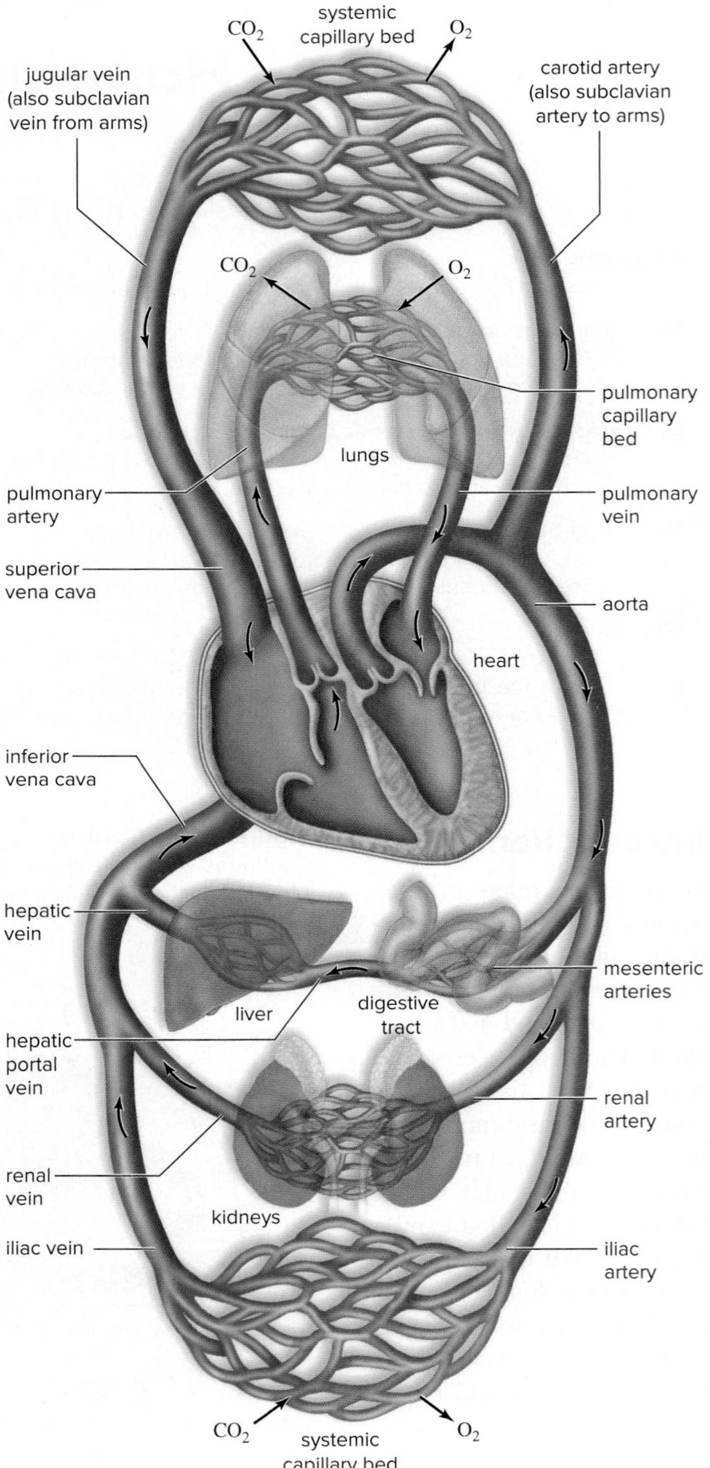

Heartbeat

During a heartbeat, first the atria contract and then the ventricles contract. When a chamber contracts, it is called **systole,** and when a chamber relaxes, it is called **diastole.** The atria and ventricles take turns being in systole.

Time	*Atria*	*Ventricles*
0.15 sec	Systole	Diastole
0.30 sec	Diastole	Systole
0.40 sec	Diastole	Diastole

Usually, there are two heart sounds with each heartbeat (Fig. 30.3). The first sound (*lub*) is low and dull and lasts longer than the second sound. It is caused by the closure of valves following atrial systole. The second sound (*dub*) follows the first sound after a brief pause. The sound has a snapping quality of higher pitch and shorter duration. The *dub* sound is caused by the closure of valves following ventricle systole.

Figure 30.3 The heartbeat sounds.
a. When the atria contract (are in systole), the ventricles fill with blood. **b.** Closure of the valves between atria and ventricles results in a **lub** sound. When the ventricles contract, blood enters the attached arteries (aorta and pulmonary trunk). **c.** Closure of valves between ventricles and arteries results in a **dub** sound. Blood enters the heart from the attached veins (venae cavae and pulmonary veins) once more.

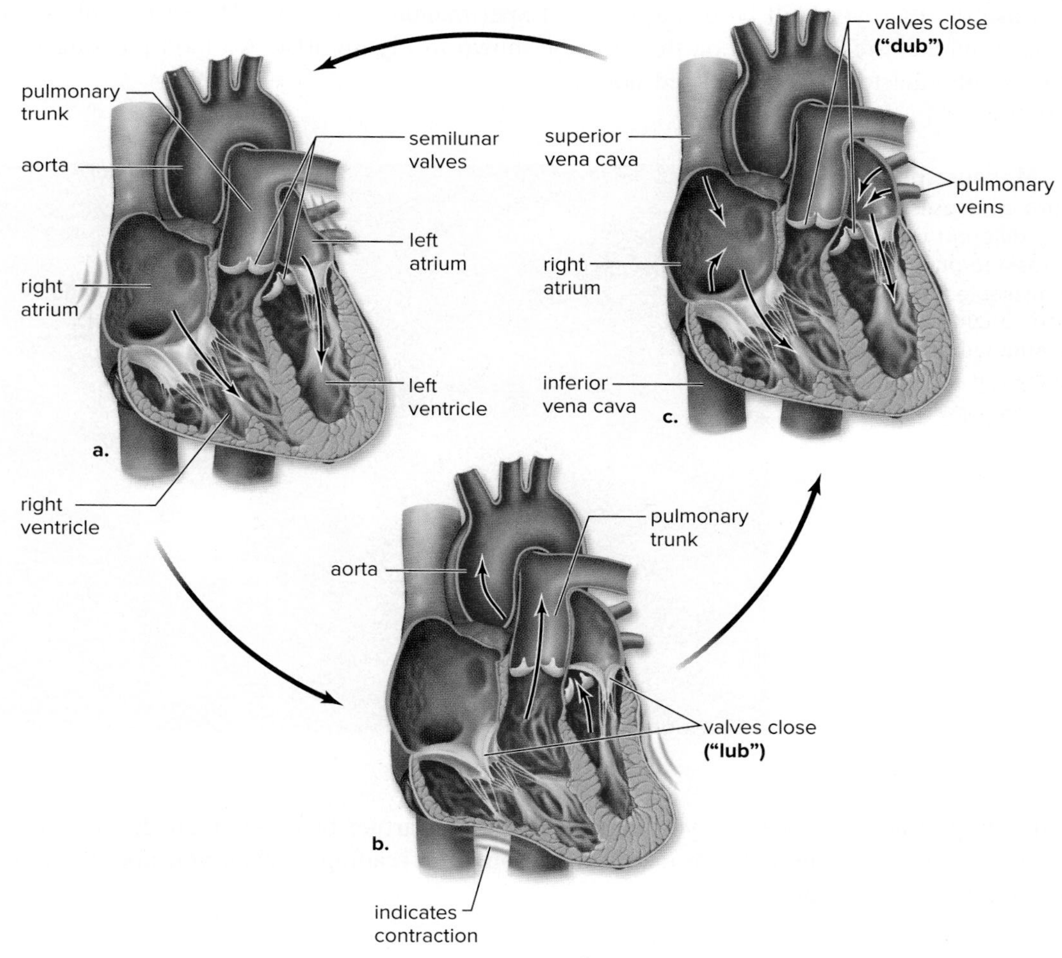

Experimental Procedure: Heartbeat at Rest

In the following procedure, you will work with a partner and use a stethoscope to listen to the heartbeat. It will not be necessary for you to count the number of beats per minute.

1. Obtain a stethoscope, and properly position the earpieces. They should point forward. Place the bell of the stethoscope on the left side of your partner's chest between the fourth and fifth ribs. This is where the apex (tip) of the heart is closest to the body wall.
2. Which of the two sounds (**lub** or **dub**) is louder? ____________________
3. Now switch, and your partner will determine your heartbeat.

Blood Pressure

Blood pressure is highest just after ventricular systole (contraction), and it is lowest during ventricular diastole (relaxation). Why? ____________________

We would expect a person to have lower blood pressure readings at rest than after exercise. Why?

____

Experimental Procedure: Blood Pressure and Pulse at Rest and After Exercise

A number of different types of digital blood pressure monitors are available, and your instructor will instruct you on how to use the type you will be using for this Experimental Procedure. The resting blood pressure readings for an individual are displayed on the monitor shown in Figure 30.4. A blood pressure reading at or below 120/80 (systolic/diastolic) is considered normal.

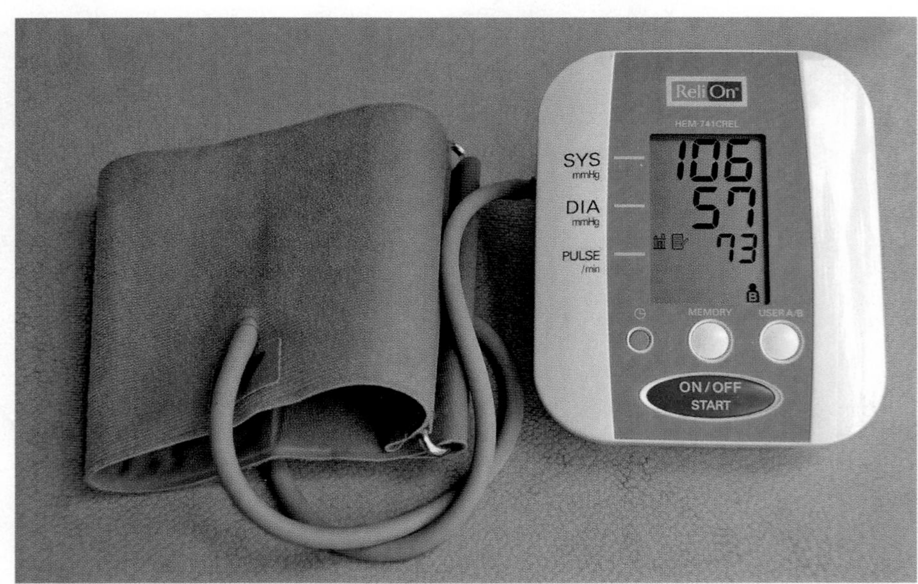

Figure 30.4 Measurement of blood pressure and pulse. There are many different types of digital blood pressure/pulse monitors now available. The one shown here uses a cuff to be placed on the arm. Others use a cuff for the wrist.
©Ilene MacDonald/Alamy Stock Photo RF

During this Experimental Procedure you may work with a partner or by yourself. If working with a partner, each of you will assist the other in taking blood pressure readings. When you note the blood pressure readings, also note the pulse reading.

Blood Pressure and Pulse at Rest

1. Reduce your activity as much as possible.
2. Use the blood pressure monitor to obtain several blood pressure readings, average them, and record your results in Table 30.1. Also note the pulse rates and average. Record in Table 30.1.

Blood Pressure and Pulse After Exercise

1. Run in place for 1 minute.
2. Immediately use the blood pressure monitor to obtain a blood pressure reading, and record it in Table 30.1. Also note the pulse rate and record in Table 30.1.

Table 30.1 Blood Pressure

	Rates at Rest		Rates After Exercise	
	Blood Pressure	**Pulse**	**Blood Pressure**	**Pulse**
Partner				
Yourself				

Conclusions: Blood Pressure

- Knowing that exercise increases the heart rate, offer an explanation for your results. ______________________

  ____
- Under what conditions in everyday life would you expect the heart rate and the blood pressure to increase, even though you were not exercising? ______________________

  ____

  When might this be an advantage? ______________________

  A disadvantage? ______________________

30.2 Blood Flow and Systemic Capillary Exchange

We associate death with lack of a heartbeat, but the real problem is lack of blood flow to the capillaries.

Blood Flow

The beat of the heart moves blood into the aorta, from which other arteries branch off to specific locations or organs. These arteries divide into arterioles and then arterioles divide into capillaries. Venules, which receive blood from capillaries, combine to form veins, which take blood back to the heart.

Experimental Procedure: Blood Flow

1. Observe blood flow through arterioles, capillaries, and venules, either in the tail of a goldfish or in the webbed skin between the toes of a frog, as prepared by your instructor.
2. Examine under low and high power of the microscope.
3. Watch the pulse and the swiftly moving blood in the arterioles.
4. Contrast this with the more slowly moving blood that circulates in the opposite direction in the venules. Many criss-crossing capillaries are visible.
5. Look for blood cells floating in the bloodstream. Don't confuse blood cells with chromatophores, irregular black patches of pigment that may be visible in the skin.

Systemic Capillary Exchange

The beat of the heart creates blood pressure, and blood pressure is necessary to capillary exchange. Blood pressure acts to move water out of a capillary, while osmotic pressure (created by the presence of proteins in the blood) acts to move water into a capillary (Fig. 30.5). Blood pressure is higher than osmotic pressure at the arteriole end of a capillary, and water moves out of a capillary. But blood pressure lessens as the blood moves through a capillary bed. This means that osmotic pressure is higher than blood pressure at the venule end of a capillary, and water moves back into the capillary.

Figure 30.5 Systemic capillary exchange.
At a systemic capillary, an exchange takes place across the capillary wall. Between the arterial end and the venule end, molecules follow their concentration gradient. Oxygen and nutrients move out of a capillary while carbon dioxide and wastes move into the capillary.

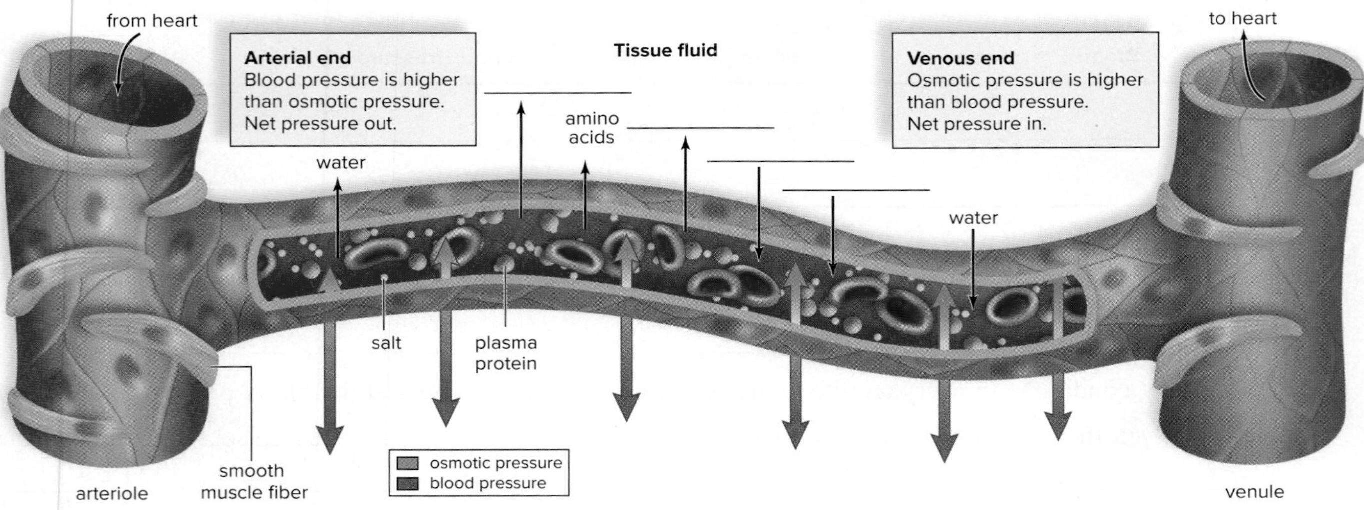

Conclusions: Systemic Capillary Exchange

- What generates blood pressure? ______________________________
- Why are tissue cells always in need of glucose and oxygen? ______________________________

  Add glucose at the end of an appropriate arrow in Figure 30.5. Do the same for oxygen.
- Why are tissue cells always producing carbon dioxide? ______________________________

  Add carbon dioxide at the start of an appropriate arrow in Figure 30.5. Do the same for metabolic wastes.

30.3 Lung Structure and Human Respiratory Volumes

The right and left lungs lie in the thoracic cavity on either side of the heart. Air moves from the nasal passages to the trachea, bronchi, bronchioles, and finally, lungs.

Lung Structure

A **lung** is a spongy organ consisting of irregularly shaped air spaces called **alveoli** (sing., alveolus) (Fig. 30.6*a*). The alveoli are surrounded by a rich network of tiny blood vessels called pulmonary capillaries. *In Figure 30.6*a, *use a labeled arrow to show oxygen entering blood from an alveolus and another labeled arrow to show carbon dioxide entering an alveolus from the blood.*

Observation: Lung Structure

1. Observe a prepared slide of a stained section of a lung (Fig. 30.6*b*). In stained slides, the nuclei of the cells forming the thin alveolar walls appear purple or dark blue.
2. Look for areas that show red or orange disc-shaped **erythrocytes.** These are the red blood cells that contain hemoglobin, which takes up oxygen and transports it to the tissues. When these appear in strings, you are looking at capillary vessels in side view.
3. In some part of the slide, you may even observe an artery. Thicker circular or oval structures with a lumen (cavity) are cross sections of **bronchioles,** tubular pathways through which air reaches the air spaces.
4. In Figure 30.6*c,* note that in emphysema, alveoli have burst. In smokers, small bronchioles collapse, and trapped air in alveoli causes them to burst. Now gas exchange is minimal.

Figure 30.6 Healthy lung tissue versus emphysema.
a. The lungs normally contain many air sacs called alveoli where gas exchange occurs. **b.** Micrograph of normal lung tissue. **c.** In smokers, emphysema can occur; the alveoli burst and gas exchange is inadequate.
(b) ©Dr. Keith Wheeler/Science Source; (c) ©Jim Zuckerman/Corbis Documentary/Getty Images

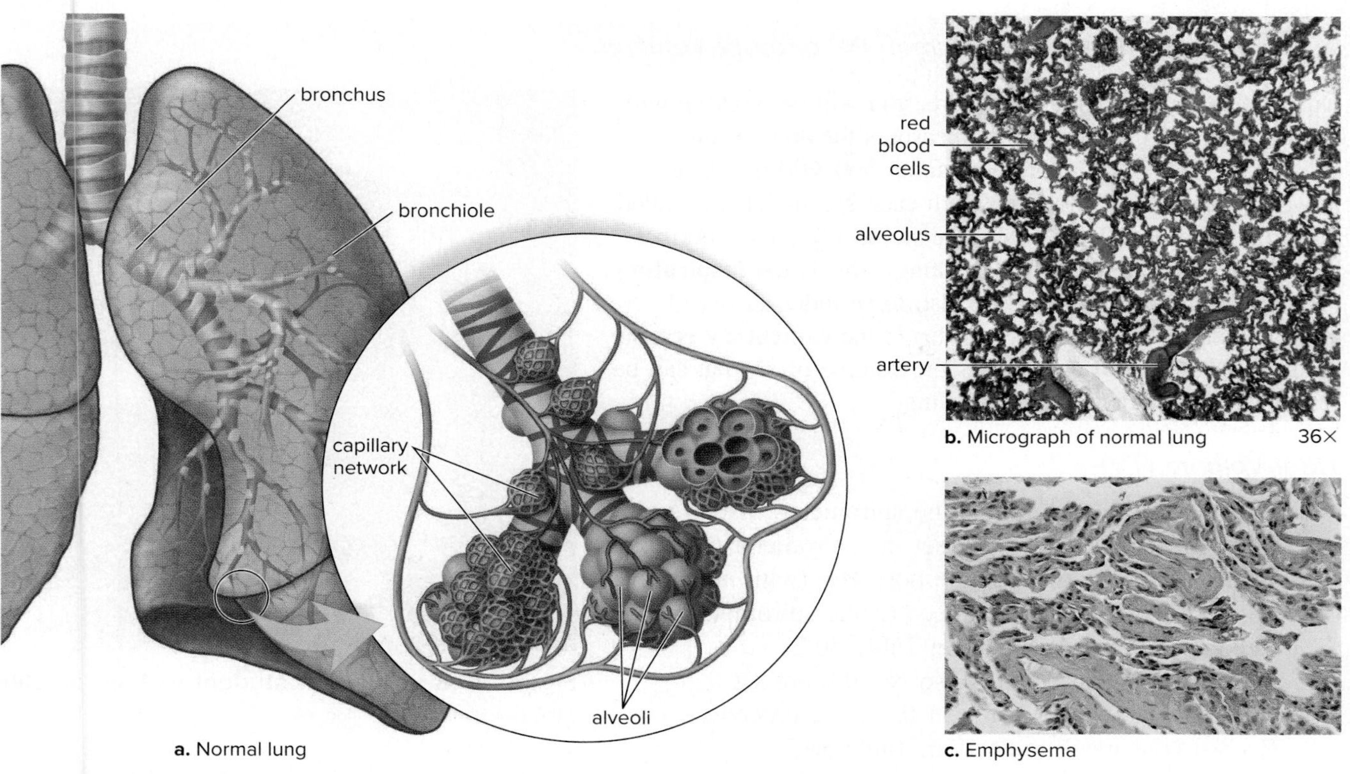

a. Normal lung

b. Micrograph of normal lung

c. Emphysema

Human Respiratory Volumes

Breathing in, called **inspiration** or inhalation, is the active part of breathing because that's when contraction of rib cage muscles causes the rib cage to move up and out, and contraction of the diaphragm causes the diaphragm to lower. Due to an enlarged thoracic cavity, the lungs expand and air is drawn into them. Breathing out, called **expiration** or exhalation, occurs when relaxation of these same muscles causes the thoracic cavity to resume its original capacity. Now air is pushed out of the lungs (Fig. 30.7).

Figure 30.7 Inspiration and expiration.
a. Inspiration occurs after the rib cage moves up and out and the diaphragm moves down. Air rushes into lungs because they expand as the thoracic cavity expands. **b.** Expiration occurs as the rib cage moves down and in and the diaphragm moves up. As the thoracic cavity and lungs get smaller, air is pushed out.

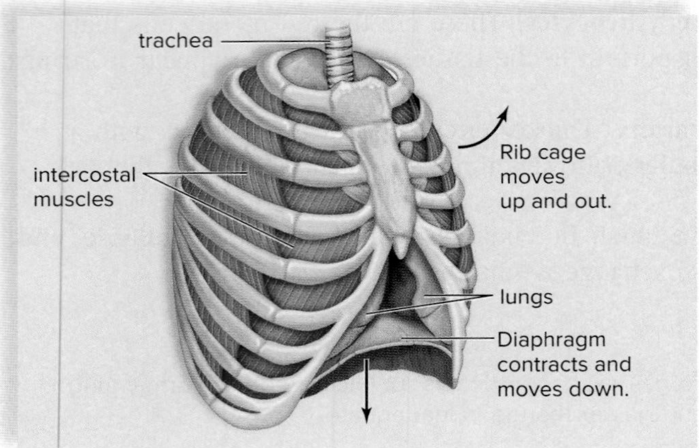

a. Inspiration

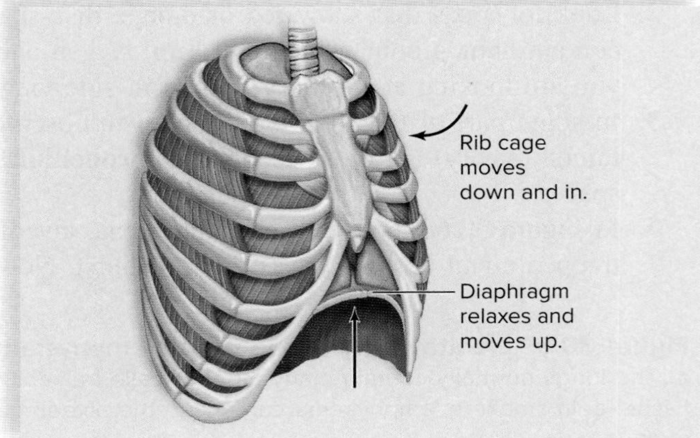

b. Expiration

Experimental Procedure: Human Respiratory Volumes

During this Experimental Procedure, you will be working with a spirometer, an instrument that measures the amount of exhaled air (Fig. 30.8). Normally, about 500–600 ml of air move into and out of the lungs with each breath. This is called the **tidal volume** (**TV**). You can inhale deeply after a normal breath and more air will enter the lungs; this is the **inspiratory reserve volume** (**IRV**). You can also force more air out of your lungs after a normal breath; this is the **expiratory reserve volume** (**ERV**). **Vital capacity** is the volume of air that can be forcibly exhaled after forcibly inhaling.

Figure 30.8 Nine-liter student wet spirometer.
©Phipps & Bird, Inc., Richmond, VA

Tidal Volume (TV)

1. When it's your turn to use the spirometer, install a new disposable mouthpiece and set the spirometer to zero.
2. Inhale normally, then exhale normally (with *no* extra effort) through the mouthpiece of the spirometer. Record your measurement in Table 30.2.
3. Three readings are needed, so twice more set the spirometer to zero and repeat the same procedure. Record your measurements in Table 30.2.
4. If the spirometer being used indicates measurements in liters (l), change your readings to milliliters (ml) by multiplying your values by 1,000. Calculate your average TV in ml by adding your three readings together and dividing by 3.

   In your own words, what is tidal volume? ____

   ____

Expiratory Reserve Volume (ERV)

1. Make sure the spirometer is set to zero.
2. Inhale and exhale normally and then force as much air out as possible into the spirometer. Record your measurement in Table 30.2.

3. Three readings are needed, so twice more, set the spirometer to zero and repeat the same procedure. Record your measurements in Table 30.2.
4. Later, if necessary, change your readings to ml, and calculate your average ERV.

   In your own words, what is expiratory reserve volume? ___

   ___

Vital Capacity (VC)

1. Make sure the spirometer is set to zero.
2. Inhale as much as possible and then exhale as much as possible into the spirometer.
3. Three readings are needed, so twice more, set the spirometer to zero and repeat the same procedure. Record your measurements in Table 30.2.
4. Later, if necessary, change your readings to ml, and calculate your average VC.

   In your own words, what is vital capacity? ___

   ___

Inspiratory Reserve Volume (IRV)

It will be necessary for us to calculate IRV because a spirometer measures only exhaled air, not inhaled air.

Explain. ___

From having measured vital capacity (VC) you can see that VC = TV + IRV + ERV. To calculate IRV, simply subtract the average TV + the average ERV from the value you recorded for the average VC:

IRV = VC − (TV + ERV) = ___ ml. Record your IRV in Table 30.2.

Table 30.2 Measurements of Lung Volumes

Tidal Volume (TV)	Expiratory Reserve Volume (ERV)	Vital Capacity (VC)	Inspiratory Reserve Volume (IRV)
1st	1st	1st	———
2nd	2nd	2nd	———
3rd	3rd	3rd	———
Average ml	Average ml	Average ml	Calculated value = ml

Conclusions: Human Respiratory Volumes

- Vital capacity varies with age, gender, and height; however, typically for men, vital capacity is about 5,200 ml and for women, it is about 4,000 ml. How does your vital capacity compare to the typical values for your gender? ___ If smaller than normal, are you a smoker or is there any health reason why it would be smaller? If larger than normal, do you participate in endurance sports of some kind or do you play a musical instrument that involves inhaling and exhaling deeply? ___

- Diffusion alone accounts for pulmonary gas exchange. Therefore, how does good lung ventilation assist gas exchange? ___

  ___

30.4 Kidneys

The **kidneys** are bean-shaped organs that lie along the dorsal wall of the abdominal cavity.

Kidney Structure

Figure 30.9 shows the structure of a kidney, macroscopic and microscopic. The macroscopic structure of a kidney is due to the placement of over 1 million **nephrons.** Nephrons are tubules that do the work of producing urine.

Figure 30.9 Longitudinal section of a kidney.
a. The kidneys are served by the renal artery and renal vein. **b.** Macroscopically, a kidney has three parts: renal cortex, renal medulla, and renal pelvis. **c.** Microscopically, each kidney contains over a million nephrons.

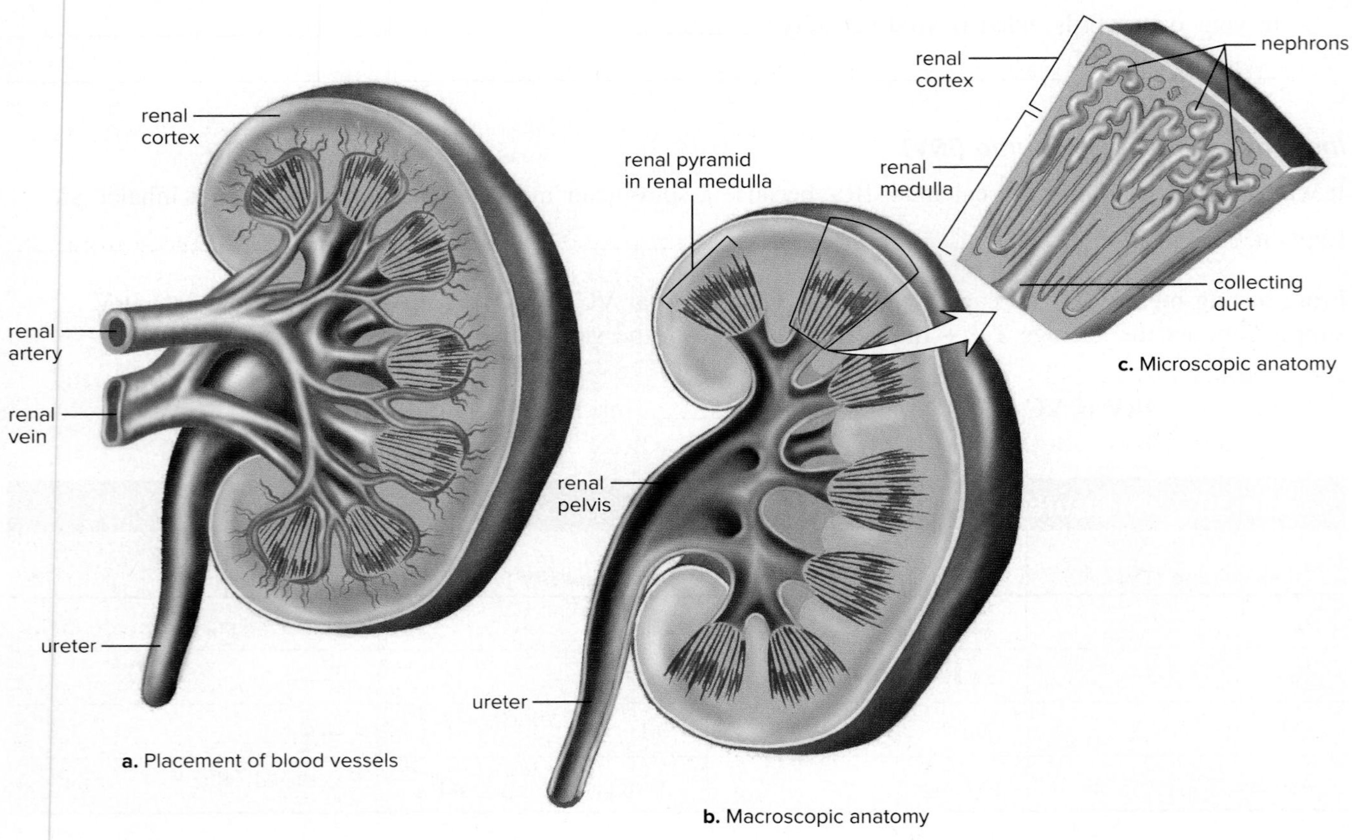

Observation: Kidney Model

Study a model of a kidney, and with the help of Figure 30.9, locate the following:

1. **Renal cortex:** a granular region
2. **Renal medulla:** contains the renal pyramids
3. **Renal pelvis:** where urine collects

Observation: Nephron Structure

Study a nephron model and, with the help of Figure 30.10, identify the following parts of a nephron:

1. **Glomerular capsule** (Bowman's capsule): closed end of the nephron pushed in on itself to form a cuplike structure; the inner layer has pores that allow **glomerular filtration** to occur; substances move from the blood to inside the nephron.
2. **Proximal convoluted tubule:** the inner layer of this region has many microvilli that allow **tubular reabsorption** to occur; substances move from inside the nephron to the blood.
3. **Loop of the nephron:** nephron narrows to form a U-shaped portion that functions in water reabsorption.
4. **Distal convoluted tubule:** second convoluted section that lacks microvilli and functions in **tubular secretion;** substances move from blood to inside nephron.

Several nephrons enter one collecting duct. The **collecting ducts** also function in water reabsorption, and they conduct urine to the pelvis of a kidney.

Observation: Circulation About a Nephron

Study a nephron model and, with the help of Figure 30.10 and Table 30.3, trace the path of blood from the renal artery to the renal vein:

1. **Afferent arteriole:** small vessel that conducts blood from the renal artery to a nephron.
2. **Glomerulus:** capillary network that exists inside the glomerular capsule; small molecules move from inside the capillary to the inside of the glomerulus during glomerular filtration.
3. **Efferent arteriole:** small vessel that conducts blood from the glomerulus to the peritubular capillary network.
4. **Peritubular capillary network:** surrounds the proximal convoluted tubule, the loop of the nephron, and the distal convoluted tubule.
5. **Venule:** takes blood from the peritubular capillary network to the renal vein.

Table 30.3 Blood Vessels Serving the Nephron

Name of Structure	Significance
Afferent arteriole	Brings arteriolar blood to the glomerulus
Glomerulus	Capillary tuft enveloped by glomerular capsule
Efferent arteriole	Takes arteriolar blood away from the glomerulus
Peritubular capillary network	Capillary bed that envelops the rest of the nephron
Venule	Takes venous blood away from the peritubular capillary network

Kidney Function

The kidneys produce urine and in doing so help maintain homeostasis in several ways. Urine formation requires three steps: **glomerular filtration, tubular reabsorption,** and **tubular secretion** (Fig. 30.10).

Figure 30.10 Nephron structure and blood supply.
The three main processes in urine formation are described in boxes and color coded to arrows that show the movement of molecules out of or into the nephron at specific locations. In the end, urine is composed of the substances within the collecting duct (see brown arrow).

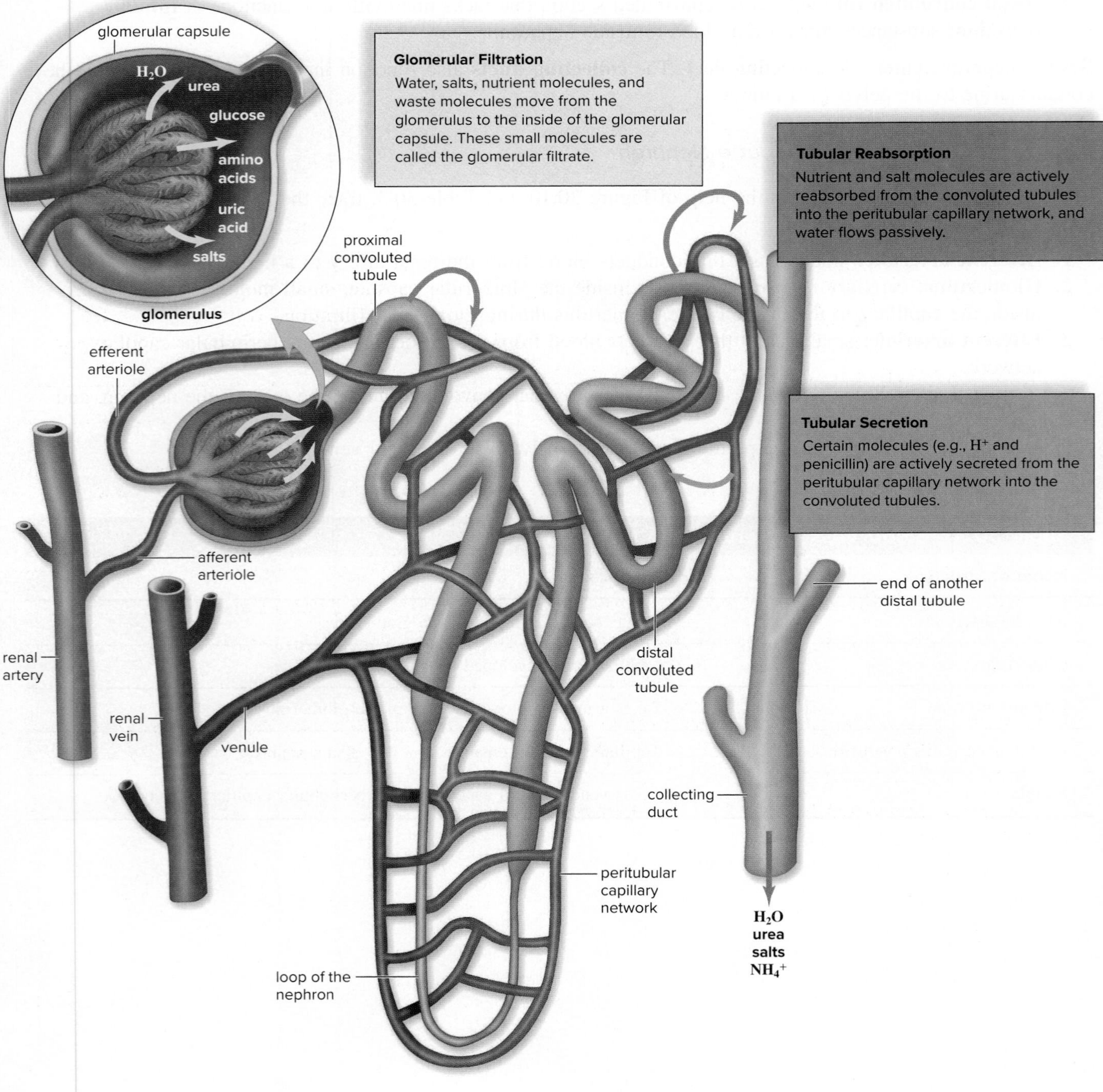

glomerular capsule
H_2O
urea
glucose
amino acids
uric acid
salts
glomerulus
Glomerular Filtration
Water, salts, nutrient molecules, and waste molecules move from the glomerulus to the inside of the glomerular capsule. These small molecules are called the glomerular filtrate.
Tubular Reabsorption
Nutrient and salt molecules are actively reabsorbed from the convoluted tubules into the peritubular capillary network, and water flows passively.
Tubular Secretion
Certain molecules (e.g., H^+ and penicillin) are actively secreted from the peritubular capillary network into the convoluted tubules.
proximal convoluted tubule
efferent arteriole
afferent arteriole
renal artery
renal vein
venule
distal convoluted tubule
end of another distal tubule
collecting duct
peritubular capillary network
loop of the nephron
H_2O
urea
salts
NH_4^+

Glomerular Filtration

1. Blood entering the glomerulus contains blood cells, proteins, glucose, amino acids, salts, urea, and water. Blood cells and proteins are too large to pass through the glomerular wall and enter the filtrate.
2. Blood pressure causes small molecules of glucose, amino acids, salts, urea, and water to exit the blood and enter the glomerular capsule. The fluid in the glomerular capsule is called the **filtrate.**
3. In the list that follows, draw an arrow from left to right for the small molecules that leave the glomerulus and become part of the filtrate.

   Glomerulus **Glomerular (Filtrate)**

   Cells
   Proteins
   Glucose
   Amino acids
   Urea
   Water and salts
4. Complete the second column in Table 30.4. Use an X to indicate that the substance is at the locations noted.

Tubular Reabsorption

1. When the filtrate enters the proximal convoluted tubule, it contains glucose, amino acids, urea, water, and salts. Some water and salts remain in the nephron, but enough are *passively* reabsorbed into the peritubular capillary to maintain blood volume and blood pressure. Use this information to state a way kidneys help maintain homeostasis. ____

   ____
2. The cells that line the proximal convoluted tubule are also engaged in active transport and usually completely reabsorb nutrients (glucose and amino acids) into the peritubular capillary. What would happen to cells if the body lost all its nutrients by way of the kidneys? ____________________

   ____
3. Which of the filtrate substances is reabsorbed the least and will become a part of urine? ____________

   Urea is a nitrogenous waste. State here another way kidneys contribute to homeostasis. ____________

   ____
4. In the list that follows, draw an arrow from left to right for all those molecules passively reabsorbed into the blood of the peritubular capillary. Use darker arrows for those that are normally reabsorbed completely by active transport.

   Proximal Convoluted Tubule **Peritubular Capillary**

   Water and salts
   Glucose
   Amino acids
   Urea

Tubular Secretion

1. During tubular secretion, certain substances—for example, penicillin and histamine—are actively secreted from the peritubular capillary into the fluid of the tubule. Also, hydrogen ions (H^+) and ammonia (NH_3) are secreted as NH_4^+ as necessary. Complete the last column of Table 30.4. Check your entries against Figure 30.10.
2. The blood is buffered, but only the kidneys can excrete H^+. The excretion of H^+ by the kidneys raises the pH of the blood. Use this information to state a third way the kidneys contribute to homeostasis. ________

____

Table 30.4 Urine Constituents

In Glomerulus	In Filtrate	In Urine
Blood cells		
Proteins		
Glucose		
Amino acids		
Urea		
Water and salts		
NH_4^+		

Urinalysis: A Diagnostic Tool

Urinalysis can indicate whether the kidneys are functioning properly or whether an illness such as diabetes mellitus is present. The procedure is easily performed with a Chemstrip test strip, which has indicator spots that produce specific color reactions when certain substances are present in urine.

Experimental Procedure: Urinalysis

A urinalysis has been ordered, and you are to test the urine for a possible illness. (In this laboratory, you will be testing simulated urine.)

Assemble Supplies

1. Obtain three Chemstrip urine test strips each of which tests for leukocytes, pH, protein, glucose, ketones, and blood, as noted in Figure 30.11.
2. The color key on the diagnostic color chart or on the Chemstrip vial label will explain what any color changes mean in terms of the pH level and amount of each substance present in the urine sample. You will use these color blocks to read the results of your test.
3. Obtain three "specimen containers of urine" marked 1 through 3. Among them are a normal specimen and two that indicate the patient has an illness. Have a piece of absorbent paper ready to use.

Test the Specimen

1. Briefly (no longer than 1 second) dip a test strip into the first specimen of urine. Be sure the chemically treated patches on the test strip are totally immersed.
2. Draw the edge of the strip along the rim of the specimen container to remove excess urine.
3. Turn the test strip on its side, and tap once on a piece of absorbent paper to remove any remaining urine and to prevent the possible mixing of chemicals.
4. After 60 seconds, read the results as follows: Hold the strip close to the color blocks on the diagnostic color chart (Fig. 30.11) or vial label, and match carefully, ensuring that the strip is properly oriented to the color chart. Enter the test results in Figure 30.11. Use a negative symbol (−) for items that are not present in the urine, a plus symbol (+) for those that are present, and a number for the pH.
5. Test the other two specimens and complete Figure 30.11.

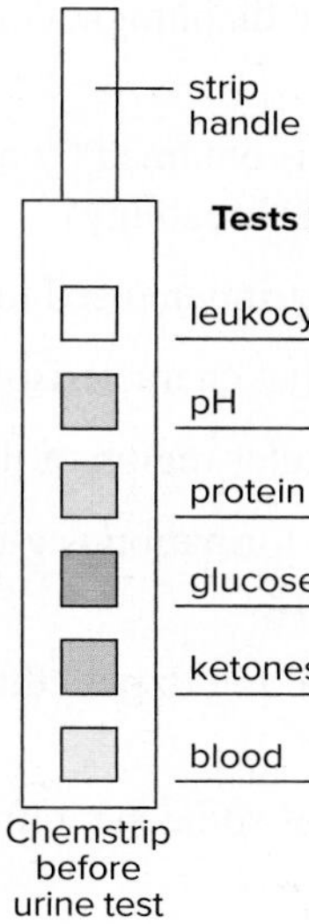

Tests For:	Normal	Results: Test 1	Test 2	Test 3
leukocytes	−			
pH	pH 5			
protein	−			
glucose	−			
ketones	−			
blood	−			

Figure 30.11 Urinalysis test. A Chemstrip test strip can help determine illness in a patient by detecting substances in the urine. If leukocytes (white blood cells), protein, or blood are in the urine, the kidneys are not functioning properly. If glucose and ketones are in the urine, the patient has diabetes mellitus (type 1 or type 2).

Conclusion: Urinalysis

- State below if the urinalysis is normal or indicates a urinary tract infection (leukocytes, blood, and possibly protein in the urine) or that the patient has diabetes mellitus.

  Test strip 1 ____________________

  Test strip 2 ____________________

  Test strip 3 ____________________

- The hormone insulin promotes the uptake of glucose by cells. When glucose is in the urine, either the pancreas is not producing insulin (diabetes mellitus type 1) or cells are resistant to insulin (diabetes mellitus type 2). Ketones (acids) are also in the urine because the cells are metabolizing fat instead of glucose. Explain why. ____________________

  Why is the pH of urine lower than normal? ____________________

- If urinalysis shows that proteins are excreted instead of retained in the blood, would capillary exchange in the tissues (see Fig. 30.5) be normal? ________ Why or why not? ____________________

Laboratory Review 30

_______________ **1.** What organs contribute to homeostasis by removing nitrogenous waste from the blood?

_______________ **2.** What blood vessels are involved in the exchange of gases that takes place in the lungs?

_______________ **3.** Does the liver store or release glucose if the hormone insulin is present?

_______________ **4.** When a heart chamber contracts, is it called systole or diastole?

_______________ **5.** If blood pressure is measured to be 112/72, which number corresponds to the relaxation of the ventricles?

_______________ **6.** What lung structures have burst in someone who has emphysema?

_______________ **7.** During inspiration, does the diaphragm contract or relax, and does that make it move up or down?

_______________ **8.** Which respiratory volume is obtained by measuring the amount of air moved in and out of the lungs with normal breathing?

_______________ **9.** What is the name of the instrument used to measure lung volumes?

_______________ **10.** Besides gender and age, what characteristic affects lung volumes?

_______________ **11.** What is the outermost granular region of the kidney called?

_______________ **12.** During what stage of urine formation is glucose transferred from the tubule to the peritubular capillary network?

_______________ **13.** What blood contents are too big to pass through the glomerular wall and enter the filtrate?

_______________ **14.** What appears in the urine of someone with diabetes mellitus besides glucose?

Thought Questions

15. If someone has high blood pressure, they might be prescribed a beta blocker medication to lower the heart rate or a medication that dilates blood vessels. Explain why lower blood pressure results from treatment with either kind of medication.

16. Why would someone with emphysema complain of shortness of breath and have difficulty climbing stairs? Why would regular exercise be recommended for someone with emphysema?

17. Predict the results (pH and substances present) of a urinalysis on an individual following a high protein, low carbohydrate diet. Explain.

LABORATORY

31

The Nervous System and Senses

Learning Outcomes

31.1 Central Nervous System

- Identify the parts of the sheep brain, and state the functions of each part.
- Label a diagram of the human brain, and give examples to show that the parts of the brain work together.
- Describe the anatomy of the spinal cord, and tell how the cord functions as a relay station.

31.2 Peripheral Nervous System

- Distinguish between cranial nerves and spinal nerves on the basis of location and function.
- Describe the anatomy and physiology of a spinal reflex arc.

31.3 Animal Eyes

- Compare the eyes of various invertebrates, particularly those of an arthropod and a squid.
- Identify the parts of the human eye, and state a function for each part.
- Explain how to test for accommodation and the blind spot.

31.4 Animal Ears

- Compare the tympanum of grasshoppers and the lateral line of fishes to the human ear.
- Identify the parts of the ear, and state a function for each part.

31.5 Sensory Receptors in Human Skin

- Describe the anatomy of the human skin, and explain the distribution and function of sensory receptors.

31.6 Human Chemoreceptors

- Relate the ability to distinguish foods to the senses of smell and taste.

Introduction

The vertebrate nervous system has two major divisions: the central nervous system (CNS), consisting of the brain and spinal cord, and the peripheral nervous system (PNS), which contains cranial nerves and spinal nerves (Fig. 31.1). Sensory receptors detect changes in environmental stimuli, and nerve impulses move along sensory nerve fibers to the brain and the spinal cord. The brain and spinal cord sum up the data before sending impulses via motor nerve fibers to effectors (muscles and glands) so a response to stimuli is possible. Nervous tissue consists of neurons; whereas the brain and spinal cord contain all parts of neurons, nerves contain only axons.

Figure 31.1 The nervous system. The central nervous system (CNS) is in the midline of the body, and the peripheral nervous system (PNS) is outside the CNS.

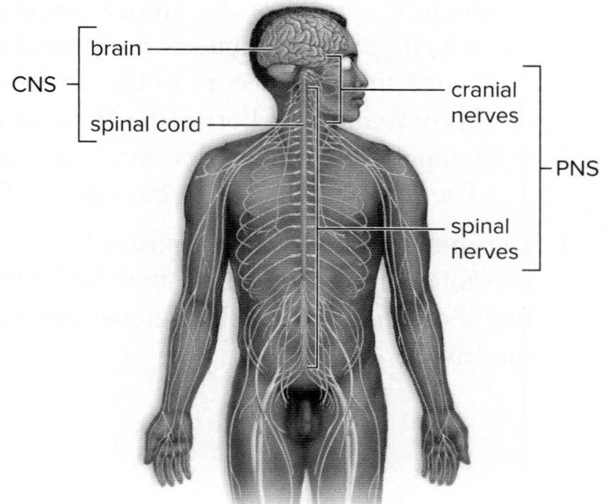

31.1 Central Nervous System

> **Latex gloves** Wear protective latex gloves when handling preserved animal organs. Use protective eyewear and exercise caution when using sharp instruments during this laboratory. Wash hands thoroughly upon completion of this laboratory.

The brain is the enlarged anterior end of the spinal cord; it contains centers that receive input from and can command other regions of the nervous system.

Preserved Sheep Brain

The sheep brain (Fig. 31.2) is often used to study the brain. It is easily available and large enough that individual parts can be identified.

Observation: Preserved Sheep Brain

Examine the exterior and a midsagittal (longitudinal) section of a preserved sheep brain or a model of the human brain, and with the help of Figure 31.1, identify the following:

1. **Ventricles:** Interconnecting spaces that produce and serve as a reservoir for cerebrospinal fluid, which cushions the brain. Toward the anterior, note the lateral ventricle (on one longitudinal section) and similarly a lateral ventricle (on the other longitudinal section). Trace the second ventricle to the third and then the fourth ventricles.
2. **Cerebrum:** Most developed area of the brain; responsible for higher mental capabilities. The cerebrum is divided into the right and left **cerebral hemispheres,** joined by the **corpus callosum,** a broad sheet of white matter. The outer portion of the cerebrum is highly convoluted and divided into the following surface lobes (see Fig. 31.2):
    a. **Frontal lobe:** Controls motor functions and permits voluntary muscle control; it also is responsible for abilities to think, problem solve, speak, and smell.
    b. **Parietal lobe:** Receives information from sensory receptors located in the skin and also the taste receptors in the mouth. A groove called the **central sulcus** separates the frontal lobe from the parietal lobe.
    c. **Occipital lobe:** Interprets visual input and combines visual images with other sensory experiences. The optic nerves split and enter opposite sides of the brain at the optic chiasma, located in the diencephalon.
    d. **Temporal lobe:** Has sensory areas for hearing and smelling. The olfactory bulb contains nerve fibers that communicate with the olfactory cells in the nasal passages and take nerve impulses to the temporal lobe.
3. **Diencephalon:** Portion of the brain where the third ventricle is located. The hypothalamus and thalamus are also located here.
    a. **Thalamus:** Two connected lobes located in the roof of the third ventricle. The thalamus is the highest portion of the brain to receive sensory impulses before the cerebrum. It is believed to control which received impulses are passed on to the cerebrum. For this reason, the thalamus sometimes is called the "gatekeeper to the cerebrum."
    b. **Hypothalamus:** Forms the floor of the third ventricle and contains control centers for appetite, body temperature, blood pressure, and water balance. Its primary function is homeostasis. The hypothalamus also has centers for pleasure, reproductive behavior, hostility, and pain.
4. **Cerebellum:** Located just posterior to the cerebrum as you observe the brain dorsally, the cerebellum's two lobes make it appear rather like a butterfly. In cross section, the cerebellum has an internal pattern that looks like a tree. The cerebellum coordinates equilibrium and motor activity to produce smooth movements.

Figure 31.2 The sheep brain.

(a–c) ©Dr. J. Timothy Cannon

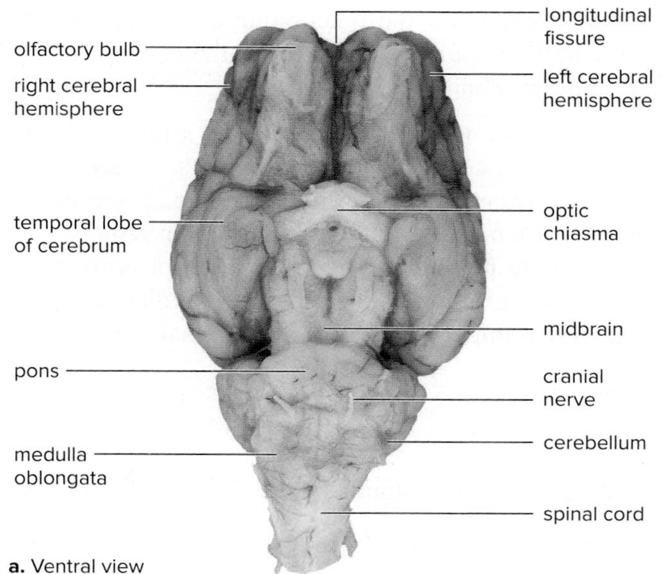

a. Ventral view

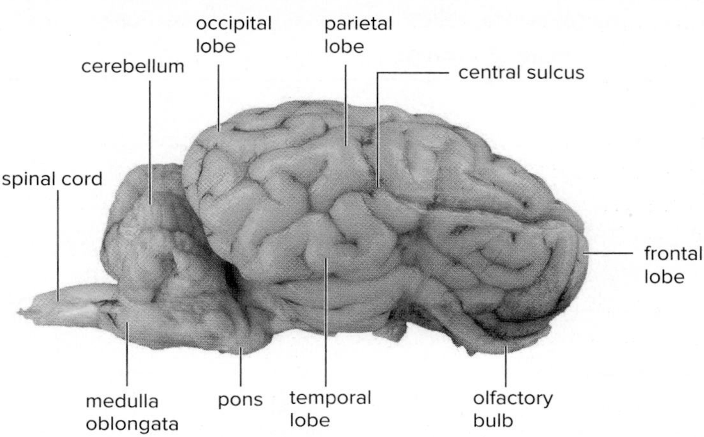

b. Lateral view

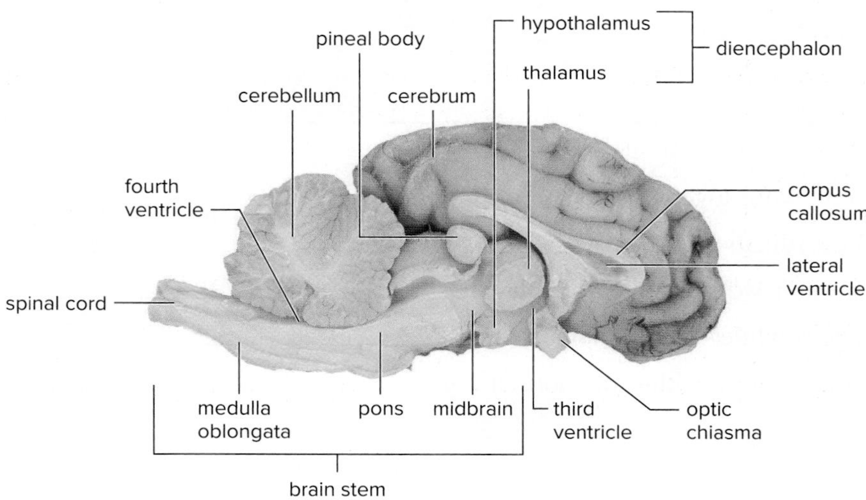

c. Longitudinal section

5. **Brain stem:** Part of the brain that connects with the spinal cord. Because it includes the pons and medulla oblongata, it contains centers for the functioning of internal organs; because of its location, it serves as a relay station for nerve impulses passing from the cord to the brain. Therefore, it helps keep the rest of the brain alert and functioning.
   a. **Midbrain:** Anterior to the pons, the midbrain serves as a relay station for sensory input and motor output. It also contains a reflex center for eye muscles.
   b. **Pons:** The ventral, bulblike enlargement on the brain stem. It serves as a passageway for nerve impulses running between the medulla and the higher brain regions.
   c. **Medulla oblongata** (or simply **medulla**): The most posterior portion of the brain stem. It controls internal organs; for example, blood pressure, cardiac, and breathing control centers are present in the medulla. Nerve impulses pass from the spinal cord through the medulla to and from higher brain regions.

The Human Brain

Based on your knowledge of the sheep brain, complete Table 31.1 by stating the major functions of each part of the brain listed. *Also label Figure 31.3.*

Table 31.1 Summary of Brain Functions

Part	Major Functions
Cerebrum	
Cerebellum	
Diencephalon Thalamus	
Hypothalamus	
Brain stem Midbrain	
Pons	
Medulla oblongata	

Which parts of the brain work together to achieve the following?

1. Good eye–hand coordination ____________________
2. Concentrating on homework when TV is playing ____________________
3. Avoiding dark alleys while walking home at night ____________________
4. Keeping the blood pressure within the normal range ____________________

Figure 31.3 The human brain (longitudinal section).
The cerebrum is larger in humans than in sheep. Label where indicated.

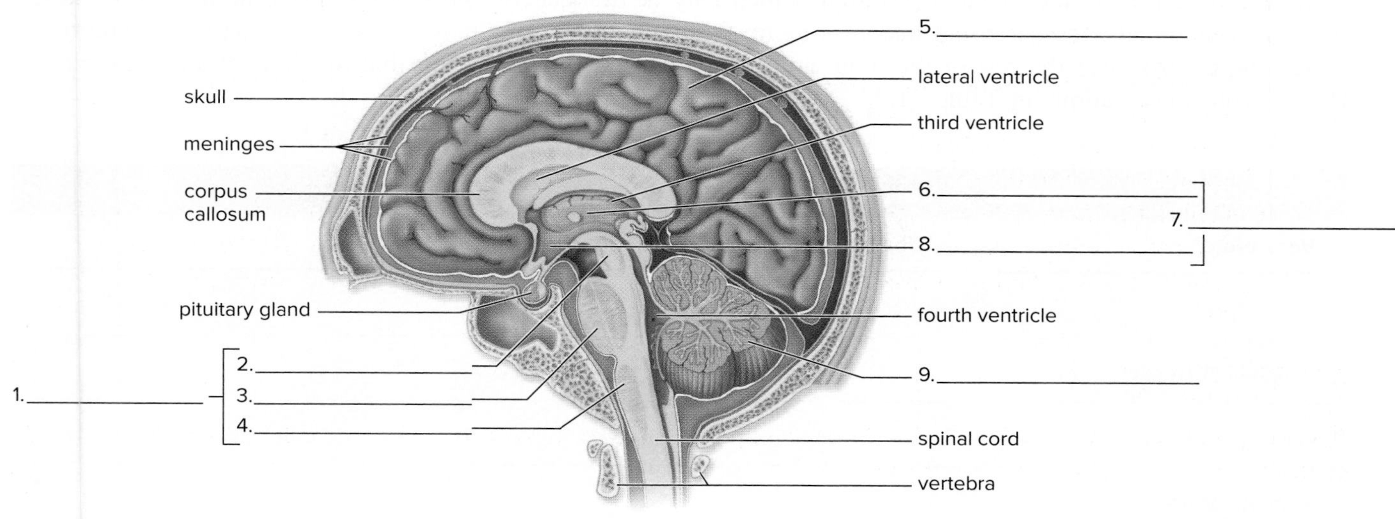

Comparison of Vertebrate Brains

The vertebrate brain has a forebrain, midbrain, and hindbrain. In the earliest vertebrates, the forebrain was largely a center for sense of smell, the midbrain was a center for the sense of vision, and the hindbrain was a center for the sense of hearing and balance. How the functions of these parts changed to accommodate the lifestyles of different vertebrates can be traced.

Observation: Comparison of Vertebrate Brains

1. Examine the brain of a fish, amphibian, bird, and mammal (Fig. 31.4).
2. Compare the sizes of the following three areas:
   a. **Forebrain:** Contains olfactory bulb and cerebrum.
   b. **Midbrain:** Contains optic lobe (and other structures).
   c. **Hindbrain:** Contains cerebellum (and other structures).

Figure 31.4 Comparative vertebrate brains.
Vertebrate brains differ in particular by the comparative sizes of the cerebrum, optic lobe, and cerebellum.

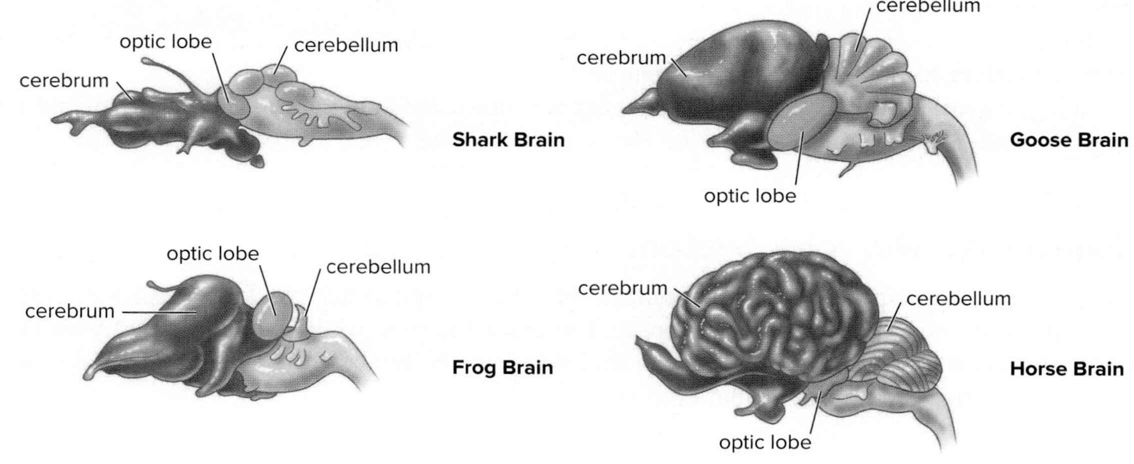

In fishes and amphibians, the midbrain is often the most prominent region of the brain because it contains higher control centers. In these animals, the cerebrum largely has an olfactory function. Fishes, birds, and mammals have a well-developed cerebellum, which may be associated with these animals' agility. In reptiles, birds, and mammals, the cerebrum becomes increasingly complex. This may be associated with the cerebrum's increasing control over the rest of the brain and the evolution of areas responsible for thought and reasoning. Record your observations in Table 31.2.

Table 31.2 Comparison of Vertebrate Brains

Vertebrate	Observations
Fish (shark)	
Amphibian (frog)	
Bird (goose)	
Mammal (horse)	

The Spinal Cord

The spinal cord is a part of the central nervous system. It lies in the middorsal region of the body and is protected by the vertebral column.

Observation: The Spinal Cord

1. Examine a prepared slide of a cross section of the spinal cord under the lowest magnification possible. For example, some microscopes are equipped with a short scanning objective that enlarges about 3.5×, with a total magnification of 35×. If a scanning objective is not available, observe the slide against a white background with the naked eye.
2. Identify the following with the help of Figure 31.5:
   a. **Gray matter:** A central, butterfly-shaped area composed of masses of short nerve fibers, interneurons, and motor neuron cell bodies.
   b. **White matter:** Masses of long fibers that lie outside the gray matter and carry impulses up and down the spinal cord. In animals, white matter appears white because an insulating myelin sheath surrounds long fibers.

Figure 31.5 The spinal cord. Photomicrograph of spinal cord cross section.

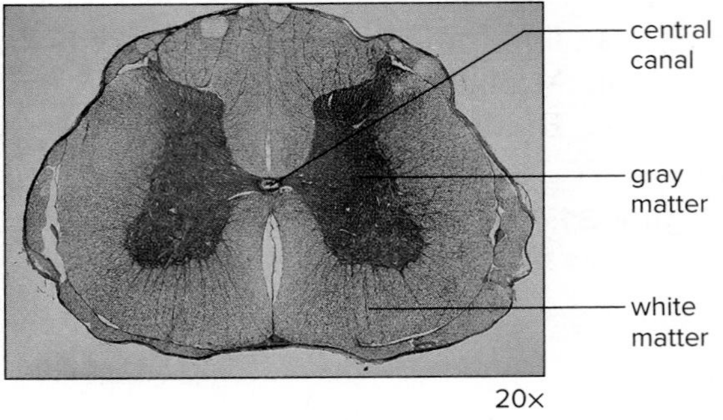

31.2 Peripheral Nervous System

The peripheral nervous system contains the cranial nerves and the spinal nerves. Twelve pairs of cranial nerves project from the inferior surface of the brain. The cranial nerves are largely concerned with nervous communication between the head, neck, and facial regions of the body and the brain. The 31 pairs of spinal nerves emerge from either side of the spinal cord (Fig. 31.6).

Spinal Nerves

Each spinal nerve contains long fibers of sensory neurons and long fibers of motor neurons. In Figure 31.6, identify the following:

1. **Sensory neuron:** Takes nerve impulses from a sensory receptor to the spinal cord. The cell body of a sensory neuron is in the dorsal root ganglion.
2. **Interneuron:** Lies completely within the spinal cord. Some interneurons have long fibers and take nerve impulses to and from the brain. The interneuron in Figure 31.6 transmits nerve impulses from the sensory neuron to the motor neuron.
3. **Motor neuron:** Takes nerve impulses from the spinal cord to an effector—in this case, a muscle. Muscle contraction is one type of response to stimuli.

Suppose you were walking barefoot and stepped on a prickly sandbur. Describe the pathway of information, starting with the pain receptor in your foot, that would allow you to both feel and respond to this unwelcome stimulus. ______________________________

____

____

Spinal Reflexes

A **reflex** is an involuntary and predictable response to a given stimulus that allows a quick response to environmental stimuli without communicating with the brain. When you touch a sharp tack, you immediately withdraw your hand (Fig. 31.6). When a spinal reflex occurs, a sensory receptor is stimulated and generates

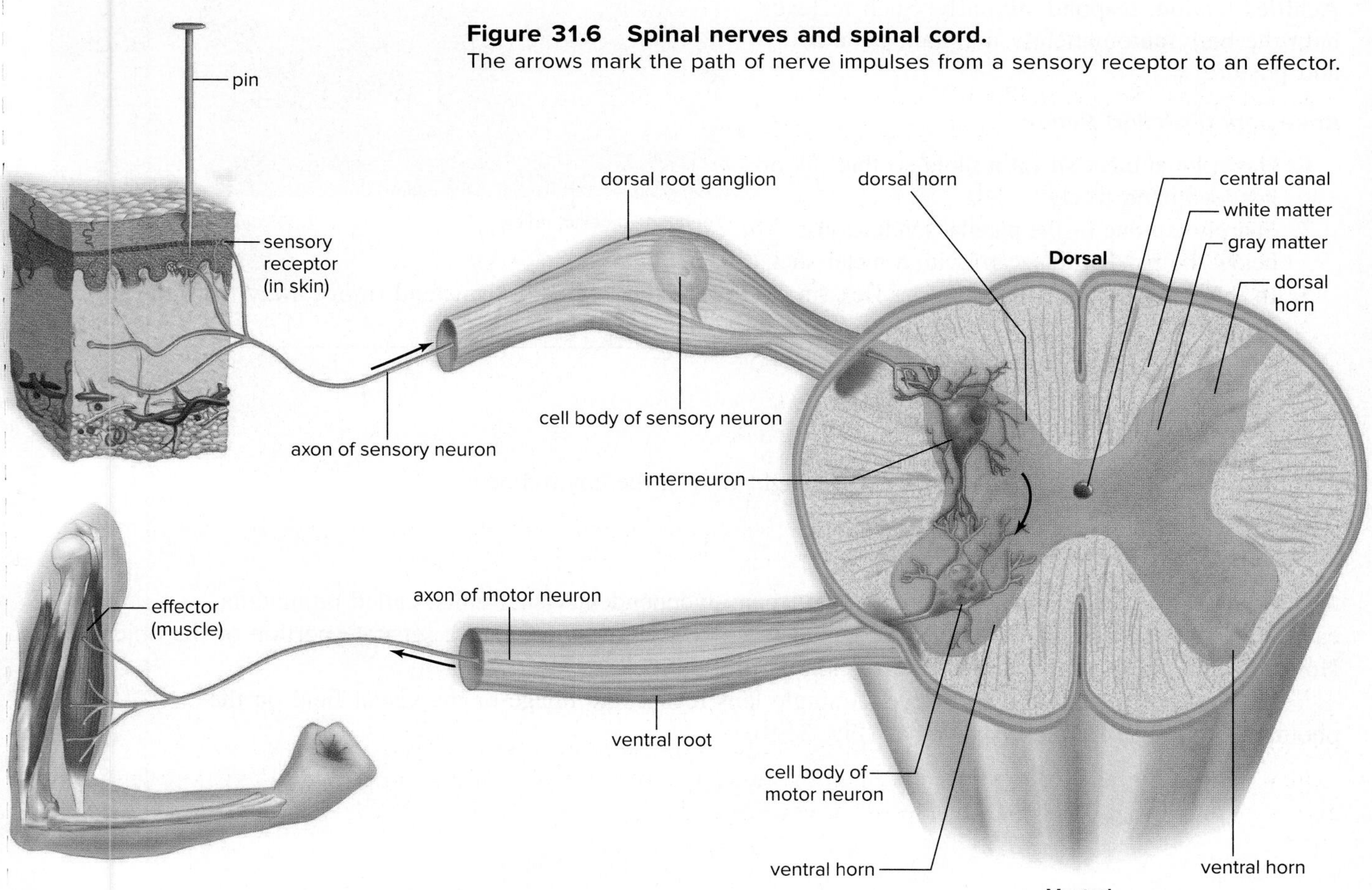

Figure 31.6 Spinal nerves and spinal cord. The arrows mark the path of nerve impulses from a sensory receptor to an effector.

nerve impulses that pass along the three neurons mentioned earlier—the sensory neuron, interneuron, and motor neuron—until the effector responds. In the spinal reflexes that follow, a receptor detects the tap, and sensory neurons conduct nerve impulses to interneurons in the spinal cord. The interneurons send a message via motor neurons to the effectors, muscles in the leg or foot. These reflexes are involuntary because the brain is not involved in formulating the response. *Consciousness* of the stimulus lags behind the response because information must be sent up the spinal cord to the brain before you can become aware of the tap.

Experimental Procedure: Spinal Reflex

Although many reflexes occur in the body, only a tendon reflex is investigated in this Experimental Procedure. One easily tested tendon reflex involves the **patellar tendon.** When this tendon is tapped with a reflex hammer (Fig. 31.7) or, in this experiment, with a meter stick, the attached muscle is stretched. This causes a receptor to generate nerve impulses, which are transmitted along sensory neurons to the spinal cord. Nerve impulses from the cord then pass along motor neurons and stimulate the muscle, causing it to contract. As the muscle contracts, it tugs on the tendon, causing movement of a bone opposite the joint. Receptors in other tendons, such as the Achilles tendon, respond similarly. Such reflexes help the body automatically maintain balance and posture.

Figure 31.7 Knee-jerk reflex. The quick response when the patellar tendon is stimulated by tapping with a reflex hammer indicates that a reflex has occurred.

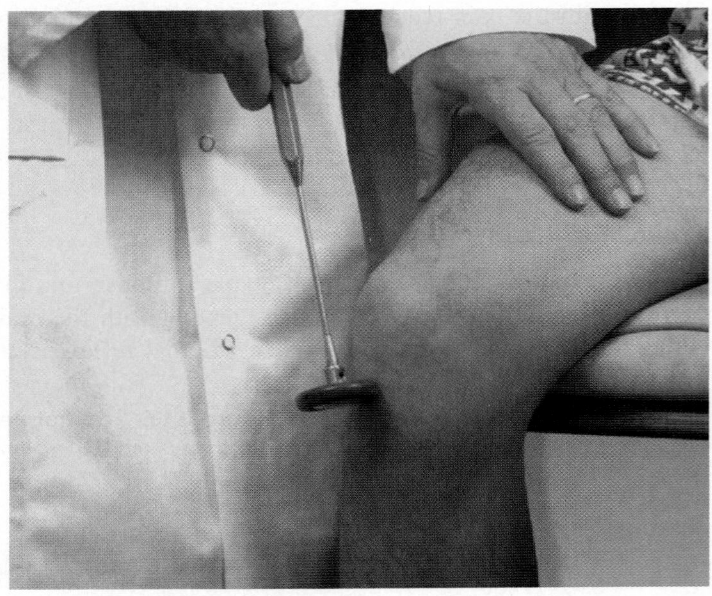

Knee-jerk (patellar) reflex

Knee-Jerk (Patellar) Reflex

1. Have the subject sit on a table so that his or her legs hang freely.
2. Sharply tap one of the patellar tendons just below the patella (kneecap) with a meter stick.
3. In this relaxed state, does the leg flex (move toward the buttocks) or extend (move away from the buttocks)? ______________________________

31.3 Animal Eyes

The eye is a special sense organ for detecting light rays in the environment.

Anatomy of Invertebrate Eyes

Arthropods have **compound eyes** composed of many independent visual units, called ommatidia, each of which has its own photoreceptor cells (Fig. 31.8). Each unit "sees" a separate portion of the object. How well the brain combines this information is not known.

A squid has a camera type of eye. A single lens focuses an image of the visual field on the photoreceptors, packed closely together (Fig. 31.9).

Figure 31.8 Compound eye of a fly.
Each visual unit of a compound eye has a cornea and lens that focus light onto photoreceptor cells. These cells generate nerve impulses transmitted to the brain, where interpretation produces a mosaic image. (right) ©Steve Gschmeissner/Getty Images RF

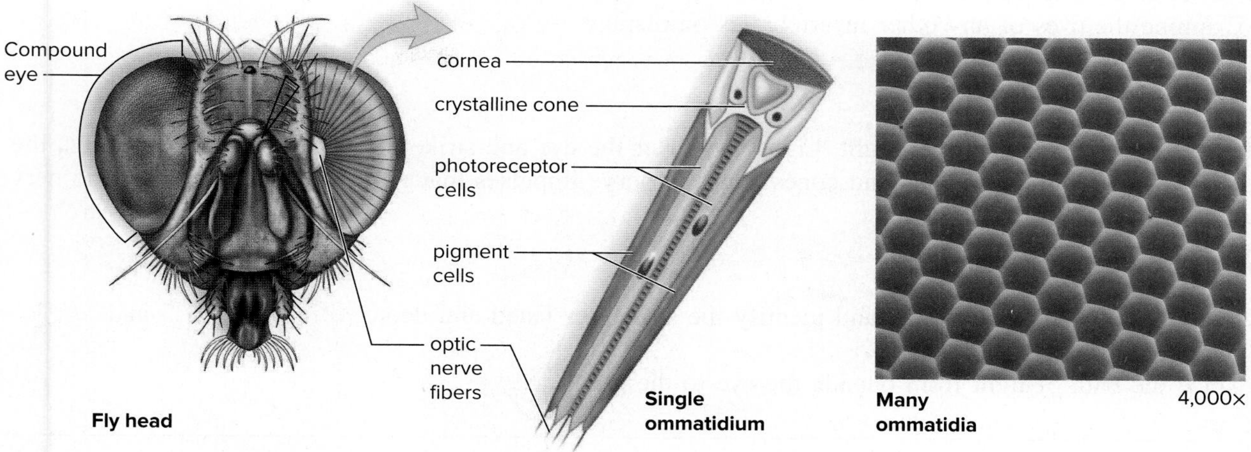

Figure 31.9 Eye of a squid.
The squid has a camera-type eye similar to those of vertebrates.

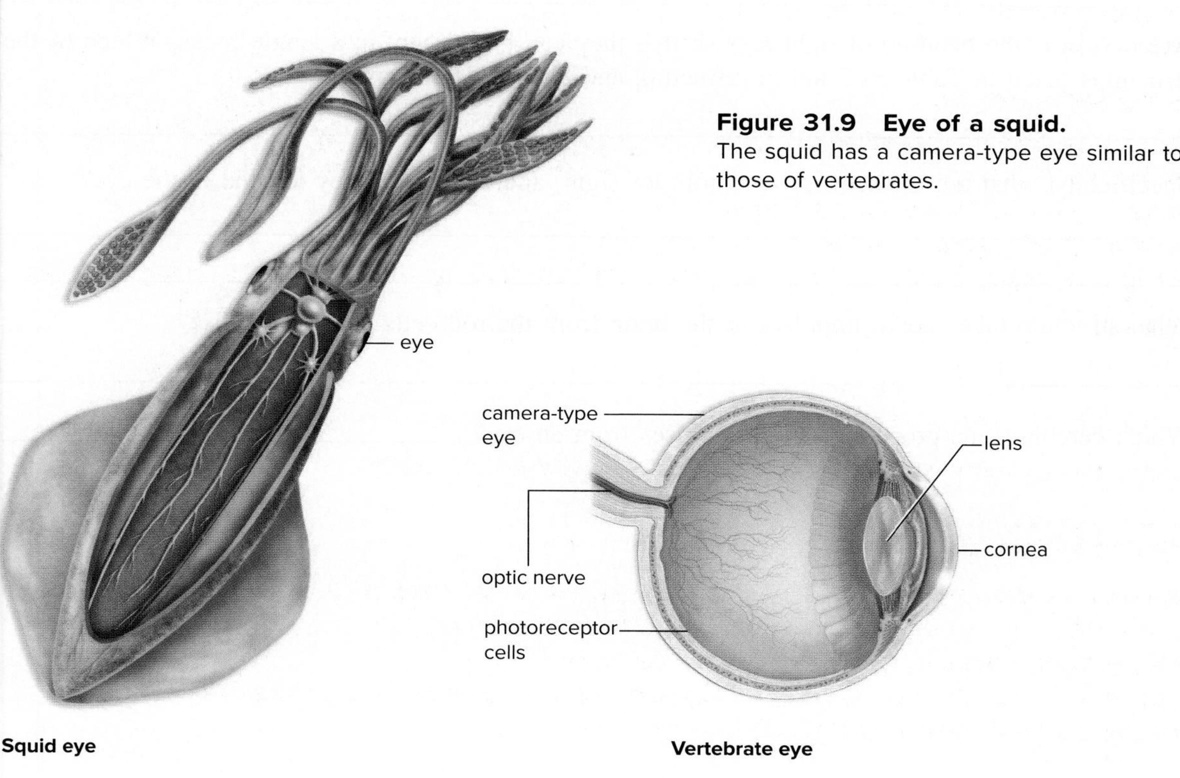

Observation: Invertebrate Eyes

1. Examine the demonstration slide of a compound eye set up under a stereomicroscope or examine a model of a compound eye.
2. Examine the eyes of any other invertebrates on display.

Anatomy of the Human Eye

The human eye is responsible for sight. Light rays enter the eye and strike the **rod cells** and **cone cells,** the photoreceptors for sight. The rods and cones generate nerve impulses that go to the brain via the optic nerve.

Observation: The Human Eye

1. Examine a human eye model, and identify the structures listed and depicted in Table 31.3 and Figure 31.10.
2. Trace the path of light from outside the eye to the retina.

   ____

   ____

   ____

3. During **accommodation,** the lens rounds up to aid in viewing near objects or flattens to aid in viewing distant objects. Which structure holds the lens and is involved in accommodation?

   ____

   ____

4. **Refraction** is the bending of light rays so that they can be brought to a single focus. Which of the structures listed in Table 31.3 aid in refracting and focusing light rays?

   ____

5. Specifically, what are the sensory receptors for sight, and where are they located in the eye?

   ____

   ____

6. What structure takes nerve impulses to the brain from the rod cells and cone cells?

   ____

7. Which cerebral lobe processes nerve impulses from an eye? ____________________

Table 31.3 Parts of the Human Eye

Part	Location	Function
Sclera	Outer layer of eye	Protects and supports eyeball
Cornea	Transparent portion of sclera	Refracts light rays
Choroid	Middle layer of eye	Absorbs stray light rays
Retina	Inner layer of eye	Contains receptors for sight
Rod cells	In retina	Make black-and-white vision possible
Cone cells	Concentrated in fovea centralis	Make color vision possible
Fovea centralis	Special region of retina	Makes acute vision possible
Lens	Interior of eye between cavities	Refracts and focuses light rays
Ciliary body	Extension from choroid	Holds lens in place; functions in accommodation
Iris	More anterior extension of choroid	Regulates light entrance
Pupil	Opening in middle of iris	Admits light
Humors (aqueous and vitreous)	Fluid media in anterior and posterior compartments, respectively, of eye	Transmit and refract light rays; support eyeball
Optic nerve	Extension from posterior of eye	Transmits impulses to occipital lobe of brain

Figure 31.10 Anatomy of the human eye.
The sensory receptors for vision are the rod cells and cone cells present in the retina of the eye.

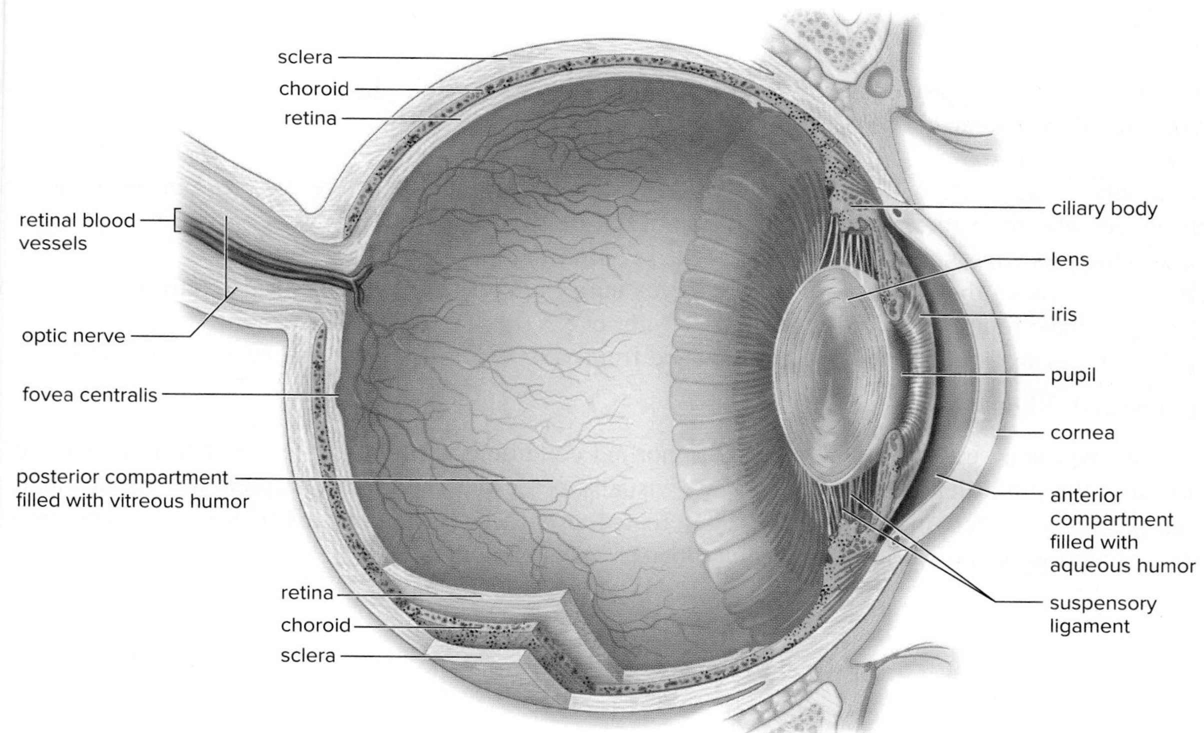

The Blind Spot of the Eye

The **blind spot** occurs where the optic nerve fibers exit the retina. No vision is possible at this location because of the absence of rod cells and cone cells.

Experimental Procedure: Blind Spot of the Eye

This Experimental Procedure requires a laboratory partner. Figure 31.11 shows a small circle and a cross several centimeters apart.

Figure 31.11 Blind spot.
This dark circle (or cross) will disappear at one location because there are no rod cells or cone cells at each eye's blind spot, where vision does not occur.

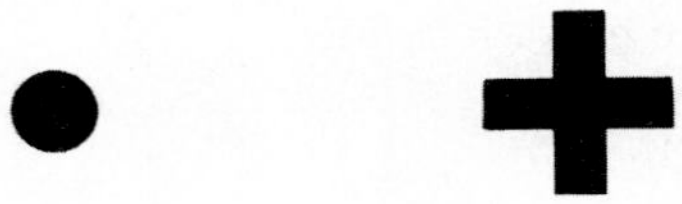

Left Eye

1. Hold Figure 31.11 approximately 30 cm from your eyes. The cross should be directly in front of your left eye. If you wear glasses, keep them on.
2. Close your right eye.
3. Stare only at the cross with your left eye. You should also be able to see the circle in the same field of vision. Slowly move the paper toward you until the circle disappears.
4. Repeat the procedure as many times as needed to find the blind spot.
5. Then slowly move the paper closer to your eyes until the circle reappears. Because only your left eye is open, you have found the blind spot of your left eye.
6. With your partner's help, measure the distance from your eye to the paper when the circle first disappeared. Left eye: ________ cm

Right Eye

1. Hold Figure 31.11 approximately 30 cm from your eyes. The circle should be directly in front of your right eye. If you wear glasses, keep them on.
2. Close your left eye.
3. Stare only at the circle with your right eye. You should also be able to see the cross in the same field of vision. Slowly move the paper toward you until the cross disappears.
4. Repeat the procedure as many times as needed to find the blind spot.
5. Then slowly move the paper closer to your eyes until the cross reappears. Because only your right eye is open, you have found the blind spot of your right eye.
6. With your partner's help, measure the distance from your eye to the paper when the cross first disappeared. Right eye: ________ cm

Why are you unaware of a blind spot under normal conditions? Although the eye detects patterns of light and color, it is the brain that determines what we visually perceive. The brain interprets the visual input based in part on past experiences. In this exercise, you created an artificial situation in which you became aware of how your perception of the world is constrained by the eye's anatomy.

Accommodation of the Eye

When the eye accommodates to see objects at different distances, the shape of the lens changes. The lens shape is controlled by the ciliary muscles attached to it. When you are looking at a distant object, the lens is in a flattened state. When you are looking at a closer object, the lens becomes more rounded. The elasticity of the lens determines how well the eye can accommodate. Lens elasticity decreases with increasing age, a condition called **presbyopia.** Presbyopia is the reason many older people need bifocals to see near objects.

Experimental Procedure: Accommodation of the Eye

This Experimental Procedure requires a laboratory partner. It tests accommodation of either your left or your right eye.

Figure 31.12 Accommodation.
When testing the ability of your eyes to accommodate to see a near object, always keep the pencil in this position.

1. Hold a pencil upright by the eraser and at arm's length in front of whichever of your eyes you are testing (Fig. 31.12).
2. Close the opposite eye.
3. Move the pencil from arm's length toward your eye.
4. Focus on the end of the pencil.
5. Move the pencil toward you until the end is out of focus. Measure the distance (in centimeters) between the pencil and your eye: __________ cm
6. At what distance can your eye no longer accommodate for distance? __________ cm
7. If you wear glasses, repeat this experiment without your glasses, and note the accommodation distance of your eye without glasses: ______________ cm. (Contact lens wearers need not make these determinations, and they should write the words "contact lens" in this blank.)
8. The "younger" lens can easily accommodate for closer distances. The nearest point at which the end of the pencil can be clearly seen is called the **near point.** The more elastic the lens, the "younger" the eye (Table 31.4). How "old" is the eye you tested? __________

Table 31.4 Near Point and Age Correlation

Age (years)	10	20	30	40	50	60
Near point (cm)	9	10	13	18	50	83

31.4 Animal Ears

Ears contain specialized receptors for detecting sound waves in the environment. They also often function as organs of balance.

Anatomy of Invertebrate Ears

Among invertebrates, only certain arthropod groups—crustaceans, spiders, and insects—have receptors for detecting sound waves. The invertebrate ear usually has a simple design: a pair of air pockets enclosed by a tympanum that passes sound waves to sensory neurons.

Observation: An Invertebrate Ear

Examine the preserved grasshopper on display, and with the help of Figure 31.13*a*, locate the tympanum. The tympanum covers an internal air sac that allows the tympanum to vibrate when struck by sound waves. Sensory neurons attached to the tympanum are stimulated directly by the vibration.

Figure 31.13 Evolution of the human ear.
a. A few invertebrates have ears such as that of the grasshopper. **b.** The lateral line of fishes contains hair cells with cilia embedded in a gelatinous cupula. **c.** Hair cells in the human ear have stereocilia embedded in the gelatinous tectorial membrane. When the basal membrane vibrates, bending of the stereocilia generates nerve impulses.

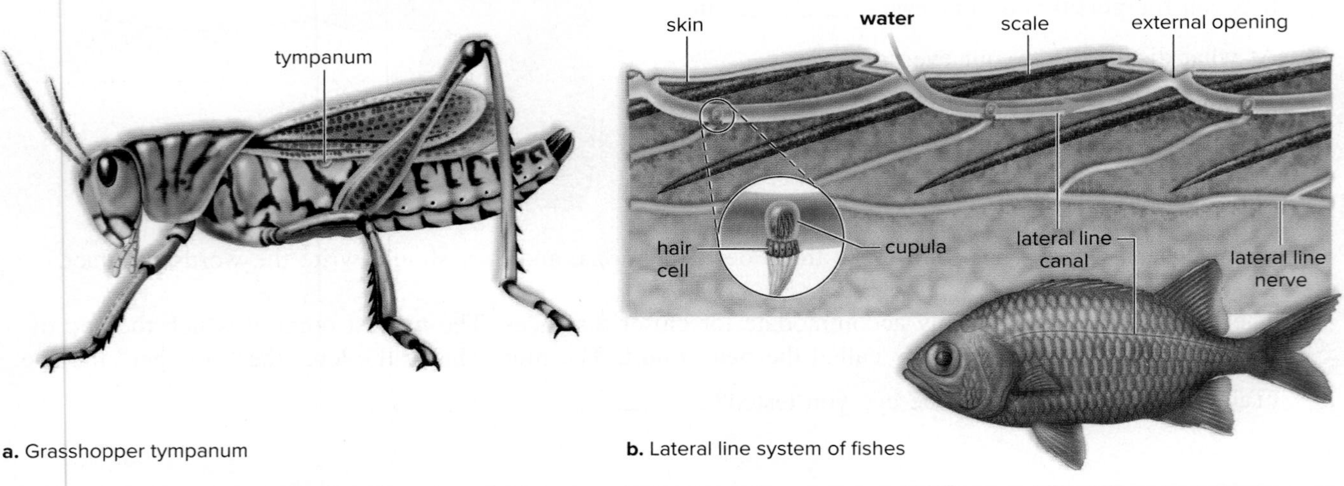

a. Grasshopper tympanum

b. Lateral line system of fishes

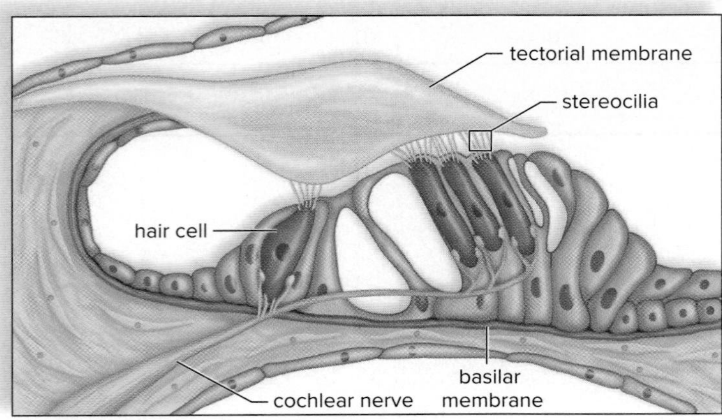

c. Cochlear cross section of the human ear

Anatomy of the Human Ear

The human ear, whose parts are listed and depicted in Table 31.5 and Figure 31.14, serves two functions: hearing and balance.

Observation: The Human Ear

Examine a human ear model, and find the structures depicted in Figure 31.14 and listed in Table 31.5.

Table 31.5 Parts of the Human Ear

Part	Medium	Function	Mechanoreceptor
Outer ear	Air		
Pinna		Collects sound waves	—
Auditory canal		Filters air	—
Middle ear	Air		
Tympanic membrane and ossicles		Amplify sound waves	—
Auditory tube		Equalizes air pressure	—
Inner ear	Fluid		
Semicircular canals		Rotational equilibrium	Stereocilia embedded in cupula
Vestibule (contains utricle and saccule)		Gravitational equilibrium	Stereocilia embedded in otolithic membrane
Cochlea (spiral organ)		Hearing	Stereocilia embedded in tectorial membrane

Figure 31.14 Anatomy of the human ear.
The outer ear extends from the pinna to the tympanic membrane. The middle ear extends from the tympanic membrane to the oval window. The inner ear encompasses the semicircular canals, the vestibule, and the cochlea.

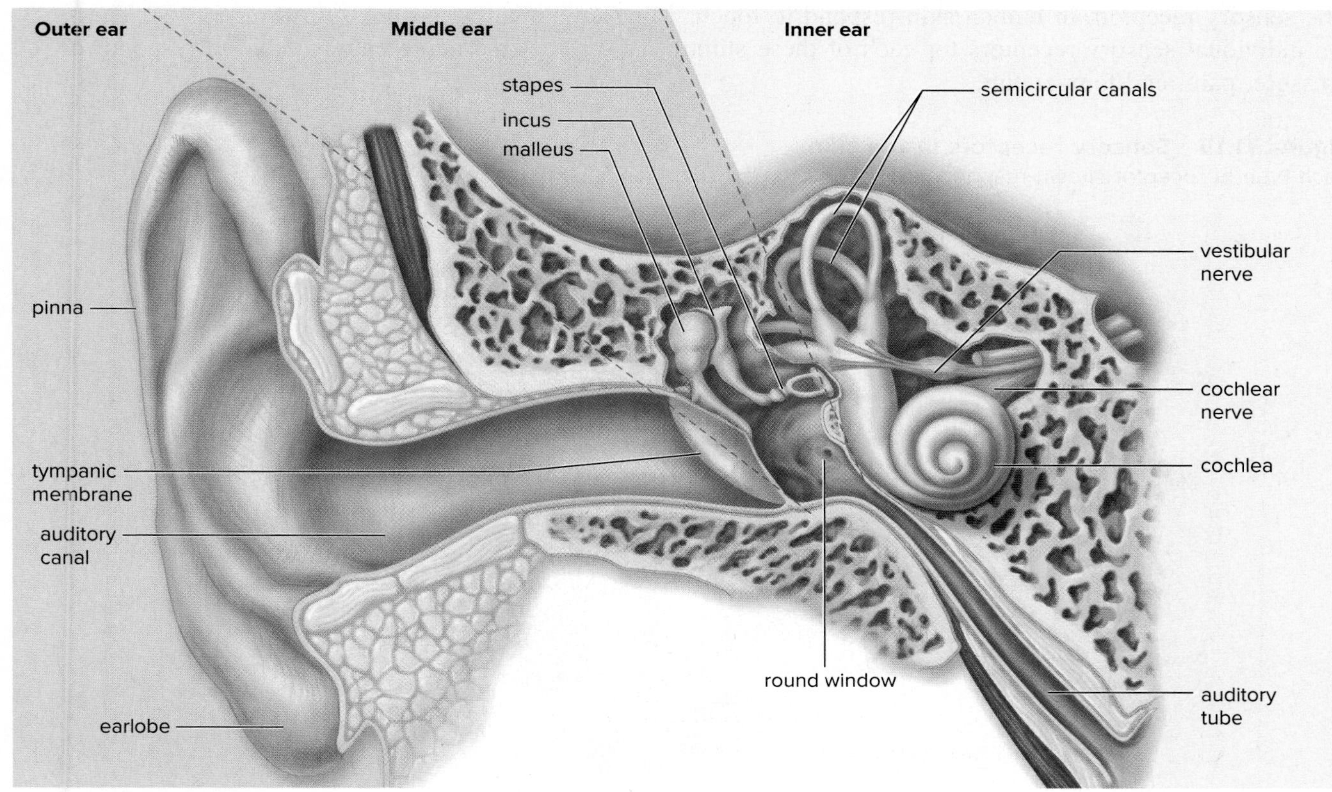

Physiology of the Human Ear

When you hear, sound waves are picked up by the **tympanic membrane** and amplified by the **malleus, incus,** and **stapes.** This creates pressure waves in the canals of the **cochlea** that lead to stimulation of **hair cells,** the receptors for hearing. Hair cells in the utricle and saccule of the vestibule and in semicircular canals are receptors for equilibrium (i.e., balance). Nerve impulses from the cochlea travel by way of the cochlear nerve and the vestibular nerve to the brain and eventually are interpreted by the brain as sound.

Experimental Procedure: Locating Sound

Humans locate the direction of sound according to how well it is detected by either or both ears. A difference in the hearing ability of the two ears can lead to a mistaken judgment about the direction of sound. You and a laboratory partner should perform this Experimental Procedure on each other.

1. Ask the subject to be seated, with eyes closed. Then strike a tuning fork or rap two spoons together at the five locations listed in number 2. Use a random order.
2. Ask the subject to give the exact location of the sound in relation to his or her head. Record the subject's perceptions when the sound is
   a. directly below and behind the head. ______
   b. directly behind the head. ______
   c. directly above the head. ______
   d. directly in front of the face. ______
   e. to the side of the head. ______
3. Is there an apparent difference in hearing between the subject's two ears? ______

31.5 Sensory Receptors in Human Skin

The sensory receptors in human skin respond to touch, pain, temperature, and pressure (Fig. 31.15). There are individual sensory receptors for each of these stimuli, as well as free nerve endings able to respond to pressure, pain, and temperature.

Figure 31.15 Sensory receptors in the skin.
Each type of receptor shown responds primarily to a particular stimulus.

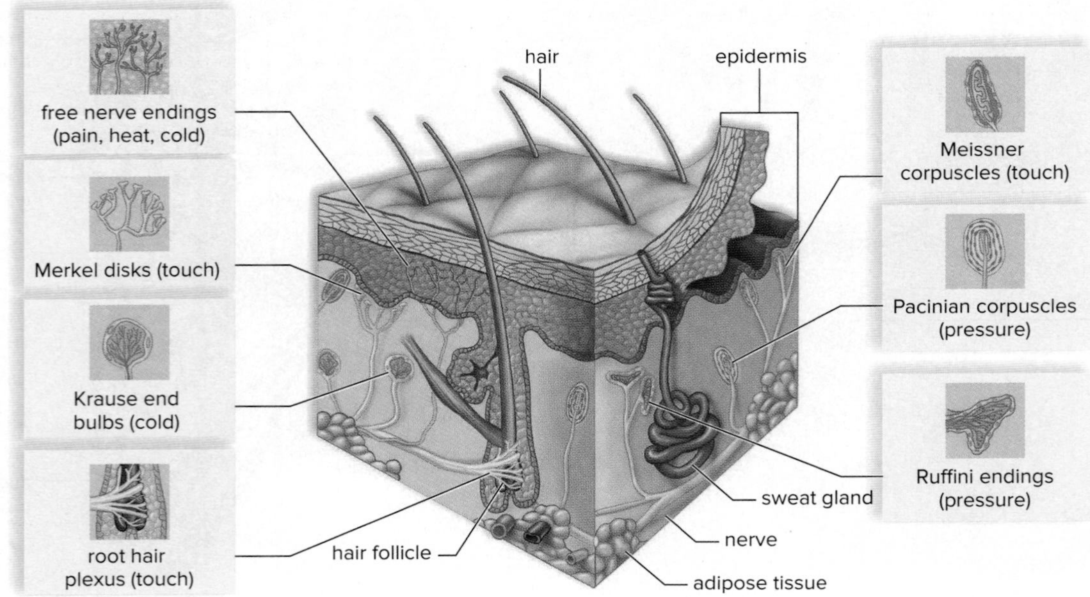

Experimental Procedure: Sensory Receptors in Human Skin

The dermis of the skin contains touch and temperature receptors.

Sense of Touch

You will need a laboratory partner to perform this Experimental Procedure. Enter *your* data, not the data of your partner, in the spaces provided.

1. Ask the subject to be seated, with eyes closed. You are going to test the subject's ability to discriminate between the two points of a hairpin or a pair of scissors at the four locations noted below.
2. Hold the points of the hairpin or scissors on the given skin area, with both points gently touching the subject. Ask the subject whether the experience involves one or two touch sensations.
3. Record the shortest distance between the hairpin or scissor points for a two-point discrimination.

   a. Forearm: __________ mm

   b. Back of the neck: __________ mm

   c. Index finger: __________ mm

   d. Back of the hand: __________ mm
4. Which of these areas apparently contains the greatest density of touch receptors? __________

   Why is this useful? __________
5. Do you have a sense of touch at every point in your skin? __________ Explain. __________

   ____
6. What specific part of the brain processes nerve impulses from touch and pain receptors?

   ____

Sense of Heat and Cold

1. Obtain three 1,000 ml beakers, and fill one with *ice water,* one with *tap water* at room temperature, and one with *warm water* (45°–50°C).
2. Immerse your left hand in the ice-water beaker and your right hand in the warm-water beaker for 30 seconds.
3. Then place both hands in the beaker with room-temperature tap water.
4. Record the sensation in the right and left hands.

   a. Right hand __________

   b. Left hand __________
5. Explain your results. __________

   ____

31.6 Human Chemoreceptors

The taste receptors, called ________________________, located in the mouth, and the smell receptors, called ________________________, located in the nasal cavities, are the chemoreceptors that respond to molecules in the air and water. Nerve impulses from taste receptors go to the ____________________ lobe of the brain, while those from smell receptors go to the ________________________ lobe of the brain.

Experimental Procedure: Sense of Taste and Smell

You will need a laboratory partner to perform the following procedures. It will not be necessary for all tests to be performed on both partners. You should take turns being either the subject or the experimenter. Dispose of used cotton swabs in a hazardous waste container or as directed by your instructor.

Taste and Smell

1. Students work in groups. Each group has one experimenter and several subjects.
2. The experimenter should obtain a LifeSavers® candy from the various flavors available, without letting the subject know what flavor it is.
3. The subject closes both eyes and holds his or her nose.
4. The experimenter gives the LifeSavers® candy to the subject, who places it on his or her tongue.
5. The subject, while still holding his or her nose, guesses the flavor of the candy. The experimenter records the guess in Table 31.6.
6. The subject releases his or her nose and guesses the flavor again. The experimenter records the guess and the actual flavor in Table 31.6.

Table 31.6 Taste and Smell Experiment

Subject	Actual Flavor	Flavor While Holding Nose	Flavor After Releasing Nose
1			
2			
3			

Conclusions: Sense of Taste and Smell

- From your results, how would you say that smell affects the taste of LifeSavers® candy?

  ____

- What do you conclude about the effect of smell on your sense of taste?

  ____

Laboratory Review 31

__________ 1. Is the sciatic nerve part of the central nervous system or the peripheral nervous system?

__________ 2. What part of the brain is divided into right and left hemispheres?

__________ 3. What portion of the brain looks like a tree in cross section?

__________ 4. What is the most posterior portion of the brain stem?

__________ 5. What cerebral lobe is associated with the sense of vision?

__________ 6. What structures protect the spinal cord?

__________ 7. What makes the white matter of the spinal cord appear white?

__________ 8. What type of neuron is found completely within the central nervous system?

__________ 9. What type of neuron is responsible for transmitting nerve impulses from the spinal cord to an effector?

__________ 10. What part of the eye refracts and focuses light rays?

__________ 11. What part of the ear contains the sensory receptors for hearing?

__________ 12. What is the anatomical name for the eardrum that picks up sound waves?

__________ 13. What layer of the skin contains Meissner and Pacinian corpuscles?

__________ 14. What are the names of the chemoreceptors that respond to molecules in air and water?

Thought Questions

15. Explain why people who develop cataracts (cloudy eye lenses) require surgery to implant new lenses.

16. When you stared at the dark cross with your left eye, why did the dark circle "disappear" from view as you moved the illustration toward you?

17. In a drag race, drivers must wait until the green light is illuminated before they can move their vehicle. Explain why a time delay exists, based on the information presented in this exercise.

LABORATORY

32

Musculoskeletal System

Learning Outcomes

32.1 Animal Skeletons

- Compare the features of a hydrostatic skeleton, exoskeleton, and endoskeleton.
- Identify slides and significant features of compact bone, spongy bone, and hyaline cartilage in a human long bone.
- Identify the major bones of the human axial and appendicular skeleton.
- Compare the forelimbs and hindlimbs of vertebrate skeletons.

32.2 Vertebrate Muscles

- Identify slides and significant features of smooth muscle, cardiac muscle, and skeletal muscle.
- Demonstrate that muscles work as antagonistic pairs using examples from frog and human.
- Describe how to demonstrate isometric and isotonic contractions.
- Describe experiments to demonstrate the role of ATP and ions in the contraction of sarcomeres.

Introduction

The term **musculoskeletal** system recognizes that contraction of muscles causes the bones to move. The skeletal system consists of the bones and joints, along with the cartilage and ligaments that occur at the joints. The muscular system contains three types of muscles: smooth, cardiac, and skeletal. The skeletal muscles are most often attached to bones after crossing a joint (Fig. 32.1). In humans, the biceps brachii muscle has two **origins,** and the triceps brachii has three origins on the humerus and scapula. Find the biceps brachii **insertion** on the anterior surface of your elbow after contracting your biceps muscle. Locate the ulna at your posterior elbow. The triceps brachii tendon inserts on the ulna.

Muscles work in antagonistic pairs. For example, when the biceps brachii contracts, the bones of the forearm are pulled upward, while the triceps brachii relaxes; when the triceps brachii contracts, the bones of the forearm are pulled downward, while the biceps brachii relaxes.

Figure 32.1 Muscular action. Muscles, such as these muscles of the arm (which have their origin on the scapula and their insertion on the bones of the forearm), cause bones to move.

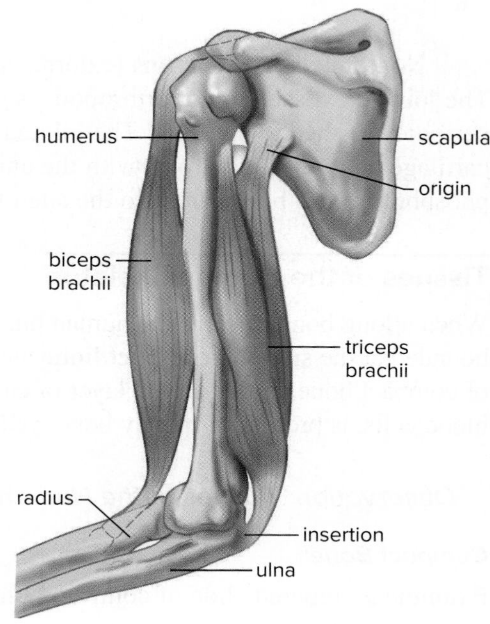

Planning Ahead Use rotation and have some groups start with the frog exercise on page 451 to minimize the number of preserved and dissected frogs needed for this exercise.

32.1 Animal Skeletons

In all animals, skeletons provide a framework and assist movement. To cause body movements, the force of muscular contractions must be specifically directed against other parts of the body. In some animals, such as segmented worms, the pressure of muscular contraction is applied to fluid-filled body compartments (Fig. 32.2); therefore, the animal is said to have a **hydrostatic skeleton.**

Figure 32.2 Earthworm locomotion.
a. Earthworms have a hydrostatic skeleton. Both circular and longitudinal muscles, which work against the fluid-filled coelom, permit them to move forward as illustrated in **b.**

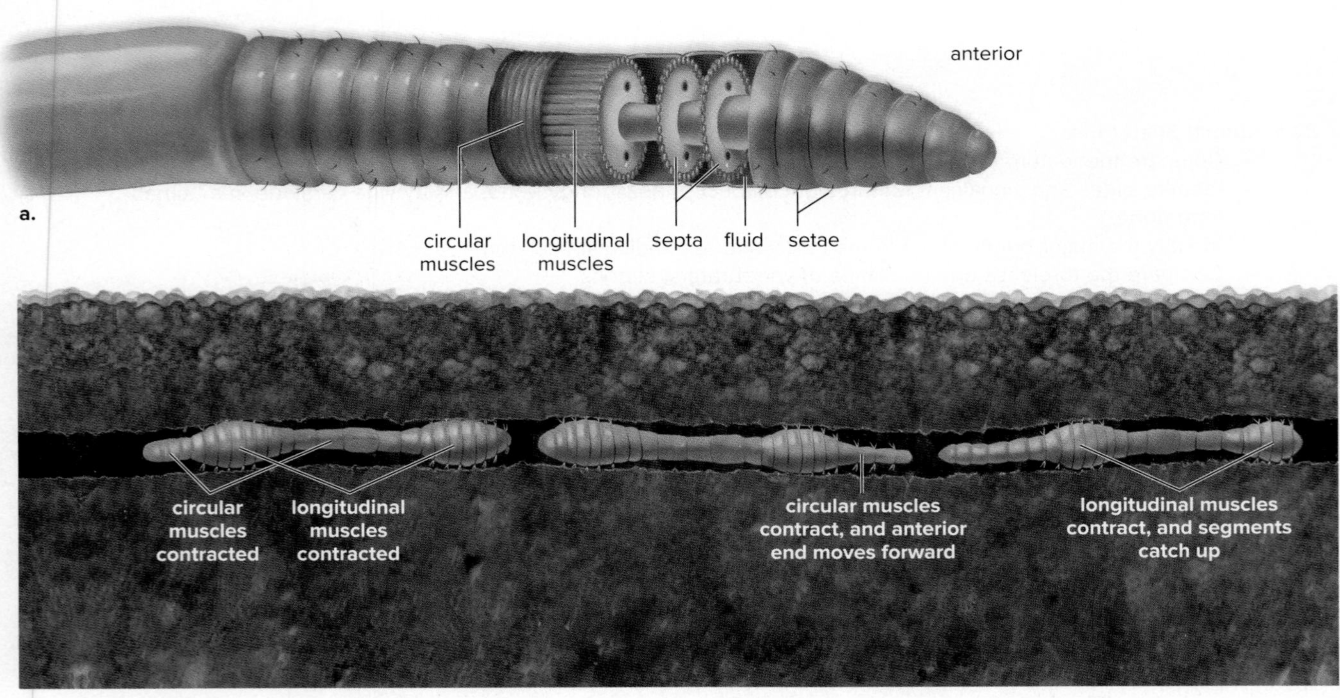

Nonjointed **exoskeletons** (external skeleton) are found among such diverse animals as corals and molluscs. The jointed exoskeleton of arthropods is composed largely of chitin, which, as the animal grows, must be periodically shed by molting. The jointed **endoskeleton** (internal skeleton) of vertebrates is composed of cartilage and bone and grows with the animal. The cartilage and bone tissues are storage areas for calcium and phosphorus. The bones are also the site of blood cell production.

Tissues of the Human Skeleton

When a long bone, such as the human humerus, is split open, as in Figure 32.3*a*, it is not solid but has a cavity bounded at the sides by **compact bone** and at the ends by **spongy bone.** Beyond the spongy bone is a thin shell of compact bone and, finally, a layer of **cartilage.** While **red bone marrow,** a specialized substance that produces blood cells, is present in spongy bone, **yellow marrow,** a fat storage tissue, is present in the medullary cavity.

Observation: Tissues of the Human Skeleton

Compact Bone

Examine a prepared slide of compact bone (Fig. 32.3*b*). Identify:

1. **Osteons** (Haversian system): A series of concentric rings called **lamellae.**
2. **Lacunae:** Cavities within the lamellae that contain **osteocytes** (bone cells).

Figure 32.3 Anatomy of a long bone.
The central shaft of **a.** long bone is composed of **b.** compact bone, but the ends are **c.** spongy bone capped by **d.** cartilage. Spongy bone can contain red bone marrow. (b) ©Ed Reschke/Getty Images; (c) ©Susumu Nishinaga/Science Source; (d) ©Ed Reschke

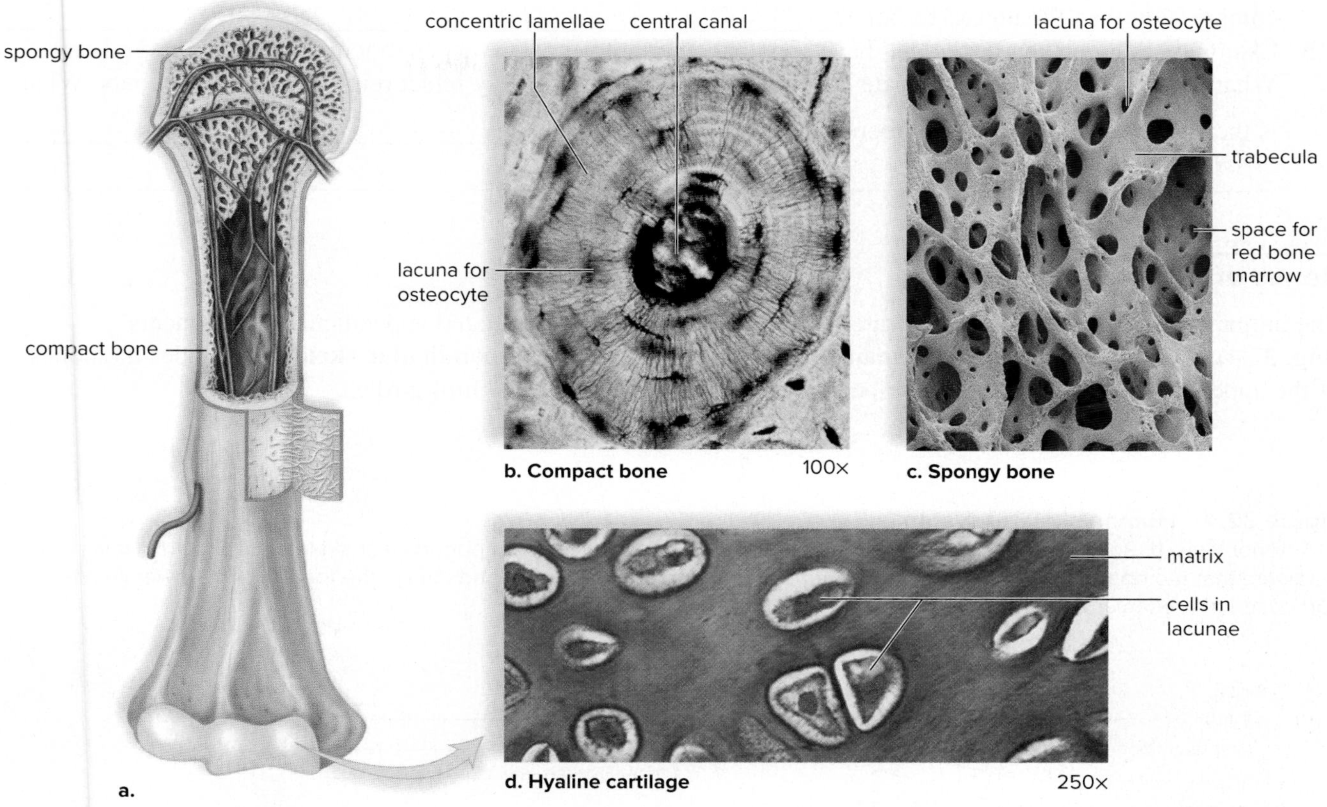

3. **Central canal:** Canal in the center of each osteon.
4. **Canaliculi:** Tiny tubules that allow nutrients to pass between the osteocytes.
5. **Matrix:** The nonliving material maintained by osteocytes. Contains mineral salts (notably calcium salts) and protein.

Spongy Bone

Examine a prepared slide of spongy bone (Fig. 32.3*a* and *c*), and identify the numerous bony bars and plates separated by irregular spaces. These spaces are often filled with red marrow.

Hyaline Cartilage

Examine a prepared slide of hyaline cartilage (Fig. 32.3*d*), and identify:

1. **Lacunae:** Cavities scattered throughout the matrix, which contain chondrocytes (cells that maintain cartilage).
2. **Matrix:** A material that is more flexible because it consists primarily of water and protein.

Experimental Procedure: Compact Bone

Compact bone is the most rigid of the connective tissues. It has an extremely hard matrix of mineral salts, primarily calcium salts, deposited around protein fibers. The minerals give bone rigidity, and the protein fibers provide elasticity and strength, much as steel rods do in reinforced concrete.

1. Examine normal chicken femurs, as well as chicken femurs that have been soaked in 5–10% acetic acid for 24 to 48 hours. Under these conditions, the calcium is removed, but the protein remains.
2. How does the shape of the treated femur compare with that of the untreated femur? ____________________

____

3. Is the treated bone more flexible or less flexible than the untreated bone? ________________________________

4. Is the treated bone able to withstand more direct downward compression or less direct downward compression than the untreated bone? ________________________________

5. Examine other chicken bones that have been baked in an oven at a very high temperature for several days. What remains is calcium phosphate devoid of water and without any intact reinforcing protein fibers. What is the effect of this treatment on bone strength? ________________________________

Bones of the Human Skeleton

The human skeleton, like other vertebrate skeletons, is divided into axial and appendicular components (Fig. 32.4). The **axial skeleton** is the main longitudinal portion. The **appendicular skeleton** includes the bones of the appendages and their supportive pectoral and pelvic (shoulder and hip) girdles.

Figure 32.4 Human skeletal system.
a. Anterior view. **b.** Posterior view. The axial skeleton appears in blue, while the appendicular skeleton is shown in tan. Be aware that the upper limb consists of the humerus in the arm and the radius and ulna in the forearm. The lower limb is composed of the femur in the thigh and the tibia and fibula in the leg.

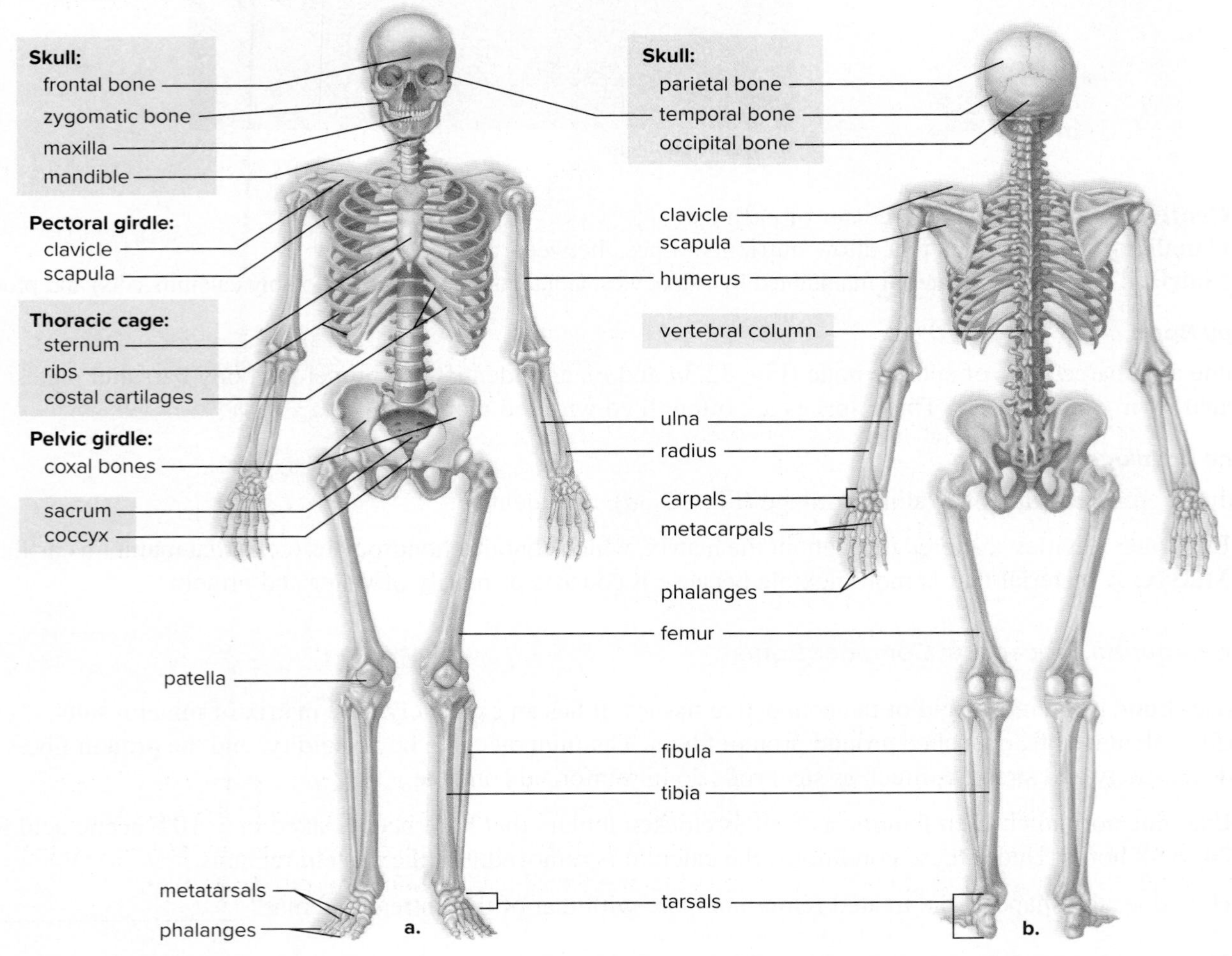

Observation: Axial Skeleton

Examine a human skeleton, and with the help of Figure 32.4, identify the following bones:

1. The **skull** is composed of many small bones fused together. Particularly note the **cranium,** a bone case enclosing the brain and major sense organs; the **facial bones;** and the **mandible** (lower jaw), the only bone of the skull not immovably fused to another.
2. The **vertebral column** provides support and also houses the **spinal cord.** It is composed of many vertebrae separated from one another by intervertebral disks. The vertebral column customarily is divided into five series:
   a. Seven **cervical vertebrae** (forming the neck region).
   b. Twelve **thoracic vertebrae** (with which the ribs articulate).
   c. Five **lumbar vertebrae** (in the abdominal region).
   d. Five fused sacral vertebrae called the **sacrum.**
   e. Several caudal vertebrae (fused in humans to form the **coccyx**).
3. The **ribs** and their associated muscles form a bony case that supports the thoracic cavity wall. The ribs connect dorsally with thoracic vertebrae, and some also are attached by cartilage directly or indirectly to the sternum. Those ribs without ventral attachment are called **floating ribs.**

Observation: Appendicular Skeleton

Examine a human skeleton, and with the help of Figure 32.4, identify the following bones:

1. The **pectoral girdle,** which supports the upper limbs, is composed of the **clavicles** (collarbones) and **scapulae** (shoulder bones).
2. The upper limbs (the arm and the forearm) contain these bones:
   a. **Humerus:** The large long bone of the arm.
   b. **Radius:** The long bone of the forearm, with a pivot joint at the elbow that allows rotational motion.
   c. **Ulna:** The other long bone of the forearm, with a hinge joint at the elbow that allows motion in only one plane.
   d. **Carpals:** A group of small bones forming the wrist.
   e. **Metacarpals:** Slender bones forming the palm.
   f. **Phalanges:** The bones of the fingers.
3. The **pelvic girdle** forms the basal support for the lower limbs, and its lateral half is composed of the **coxal** (hip) **bone.** The female pelvis is much broader and more shallow, and the outlet is larger than that of the male.
4. The lower limbs (the thigh plus the leg) contain these bones:
   a. **Femur:** The long bone of the thigh.
   b. **Tibia:** The larger of the two long bones of the leg.
   c. **Fibula:** The smaller of the two long bones of the leg.
   d. **Tarsals:** A group of small bones forming the ankle.
   e. **Metatarsals:** Slender anterior bones of the foot.
   f. **Phalanges:** The bones of the toes.

Comparison of Vertebrate Skeletons

Vertebrate forelimbs (or hindlimbs) are used for various purposes. Yet they contain the same sets of bones organized in similar ways. The explanation for this unity is that the basic plan originated with a common ancestor, and then it was modified as vertebrates continued along their evolutionary pathway. Structures that are similar because they are inherited from a common ancestor are called **homologous structures.**

Observation: Vertebrate Skeletons

Forelimb Comparisons

1. The bones in the forelimbs of vertebrates are homologous but modifications make them suitable for specific functions (Fig. 32.5). In the first column of Table 32.1, enter the specific function of the forelimb for the animals listed. For example, a frog locomotes by hopping and its forelimb is used for landing. In contrast, a cat walks using only four of its phalanges. The first phalanx is much reduced. A bird uses its forelimb for ____.
2. Keeping function in mind, compare the forelimb bones in the animals listed in Table 32.1 to those of a human. Where *a* is noted, compare the humerus; where *b* is noted, compare the radius and ulna; and where *c* is noted, compare the carpels, metacarpals, and phalanges as a whole. Are the bones fused, longer, stronger, or more delicate than those of a human?
3. Give examples to show that the forelimbs of these animals are modified to suit their means of locomotion.

____

____

Figure 32.5 Vertebrate forelimbs.
The forelimbs of these animals differ according to their specific functions. Homologous structures are color-coded.

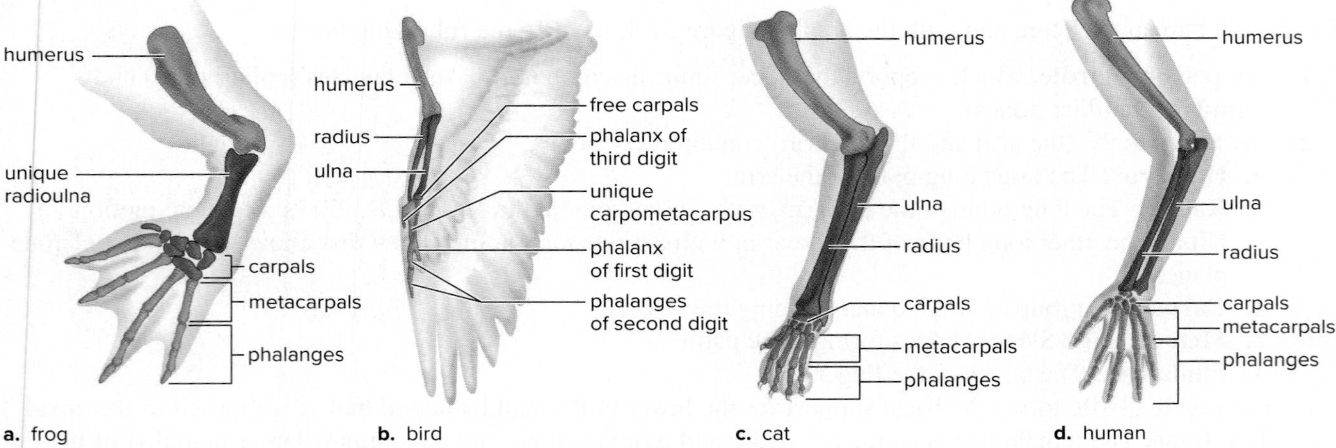

Table 32.1 Function and Comparison of Forelimb Bones to Those of Humans*

Animal	Forelimb Function	Forelimb Bones	Compare to Human Forelimb
Frog		a.	
		b.	
		c.	
Bird		a.	
		b.	
		c.	
Cat		a.	
		b.	
		c.	

* See instructions above in text.

Hindlimbs to Forelimbs Comparison

1. Only the phalanges in the hindlimb of a frog, bird, and cat ever touch the ground (Fig. 32.6), whereas the entire foot of a human touches the ground.
2. *Using Figure 32.5 label the bones of the forelimbs in these animals (Fig. 32.6).*
3. In Table 32.2 compare the structure of the hindlimb bones to that of the forelimb bones in a frog, bird, cat, and human. Explain any difference in the bones of each animal on the basis of the specific locomotor function of the hindlimb. For example, the hindlimbs of a frog allow it to hop while its forelimbs are used for landing. Similarly, the hindlimbs of a cat allow it to leap and the forelimbs allow it to land. However, a cat uses the phalanges of both the forelimbs and hindlimbs for walking.

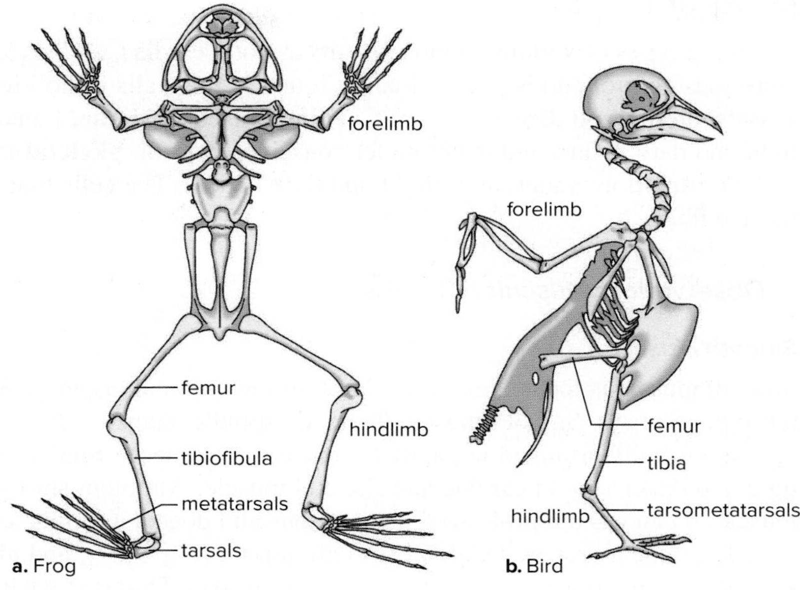

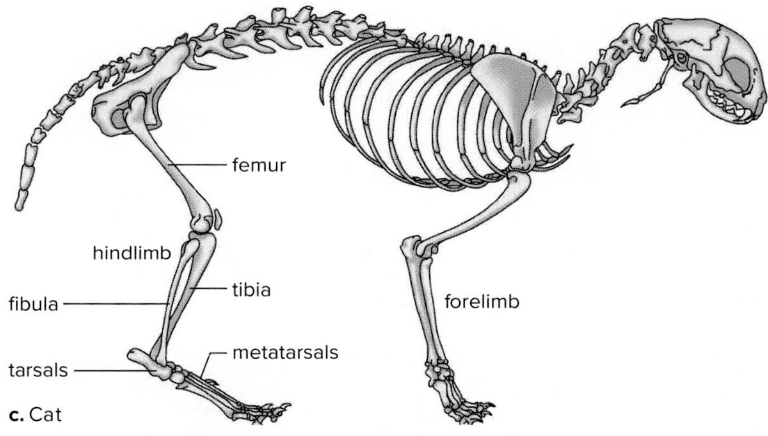

Figure 32.6 Animal skeletons.
The forelimb and hindlimb bones of these animals differ according to the locomotor functions of the limbs.

Table 32.2 Comparison of Hindlimb to Forelimb

Animal	Differences in Hindlimb and Forelimb Bones and Explanation
Frog	
Bird	
Cat	
Human	

32.2 Vertebrate Muscles

Muscle contraction accounts for an animal's ability to respond quickly to outside stimuli. Muscles, along with glands, are called **effectors** because they carry out the orders of the nervous system.

Muscular Tissues

The three types of vertebrate muscles are smooth, cardiac, and skeletal. **Smooth muscular tissue** is not under conscious control and is predominantly found in the walls of hollow organs, such as the bladder, blood vessels, uterus, and digestive tract. **Cardiac muscular tissue,** found in the heart, is striated, contains alternating light and dark bands, and is not under conscious control. **Skeletal muscular tissue** is under conscious control and has **striations** (alternating light and dark bands). The cells that make up muscular tissue are called muscle fibers.

Observation: Muscular Tissues

Smooth Muscle

Smooth muscle is located in the walls of hollow internal organs, and its involuntary contraction moves materials through an organ. Smooth muscle fibers are spindle-shaped cells, each with a single nucleus (uninucleated). The cells are usually arranged in parallel lines, forming sheets. Smooth muscle does not have the striations (bands of light and dark) seen in cardiac and skeletal muscle. Although smooth muscle is slower to contract than skeletal muscle, it can sustain prolonged contractions and does not fatigue easily.

Examine a prepared slide of smooth muscle (Fig. 32.7), and identify the spindle-shaped fibers and the oval-shaped nucleus found in the center of the cytoplasm. Describe what you see:

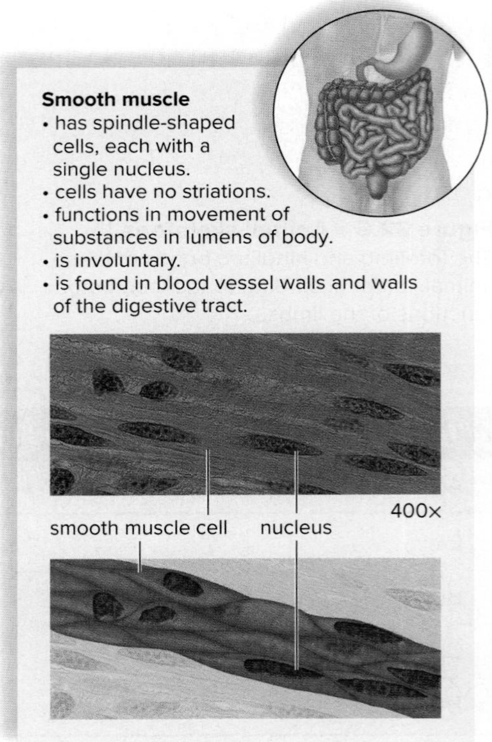

Smooth muscle
• has spindle-shaped cells, each with a single nucleus.
• cells have no striations.
• functions in movement of substances in lumens of body.
• is involuntary.
• is found in blood vessel walls and walls of the digestive tract.
400×
smooth muscle cell
nucleus

Figure 32.7 Muscular tissues.
Smooth muscle is nonstriated and involuntary. This type of muscle is composed of spindle-shaped cells. Icon gives example of smooth muscle location in human body. ©McGraw-Hill Education/Dennis Strete, photographer

Cardiac Muscle

Cardiac muscle forms the heart wall. Its fibers are uninucleated, striated, tubular, and branched, which allows the fibers to interlock at intercalated disks. Intercalated disks permit contractions to spread quickly throughout the heart. Cardiac fibers relax completely between contractions, which prevents fatigue. Contraction of cardiac muscle fibers is rhythmical; it occurs without outside nervous stimulation or control. Thus, cardiac muscle contraction is involuntary.

Examine a prepared slide of cardiac muscle (Fig. 32.8), and identify a single nucleus per cell, the branched fibers joined at intercalated disks (dark lines that pass across the fibers), and striations that show up under high power. Describe what you see:

Cardiac muscle
- has branching, striated cells, each with a single nucleus.
- occurs in the wall of the heart.
- functions in the pumping of blood.
- is involuntary.

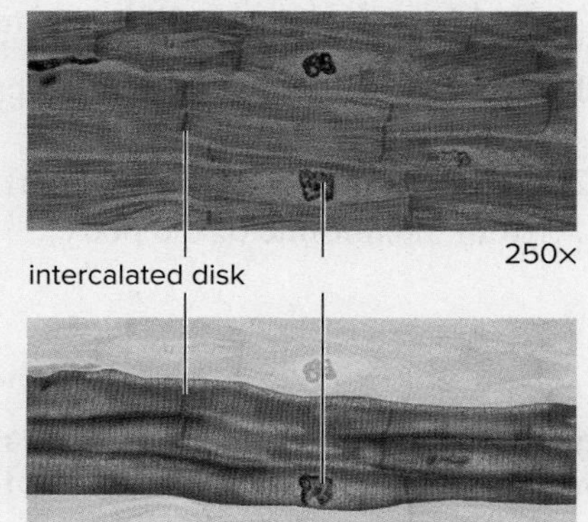

Figure 32.8 Cardiac muscle. Cardiac muscle is striated and involuntary. The branched cells join at intercalated disks. Icon shows where cardiac muscle is located in human body. ©Ed Reschke

Skeletal Muscle

Skeletal muscle fibers are tubular, multinucleated, and striated. They make up the skeletal muscles attached to the skeleton. Skeletal muscle fibers can run the length of a muscle and, therefore, can be quite long. Skeletal muscle is voluntary because its contraction is always stimulated and controlled by the nervous system.

Examine a prepared slide of skeletal muscle (Fig. 32.9), and identify long fibers arranged in a parallel fashion, multinuclei per fiber, and striations. Describe what you see:

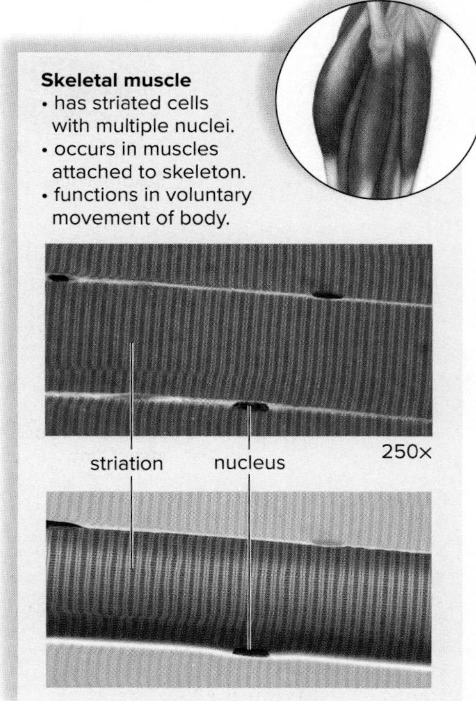

Skeletal muscle
• has striated cells with multiple nuclei.
• occurs in muscles attached to skeleton.
• functions in voluntary movement of body.
striation
nucleus
250×

Figure 32.9 Skeletal muscle. Skeletal muscle is striated and voluntary. The tubular cells contain many nuclei. Icon gives example of skeletal muscle location in human body. ©Ed Reschke

Types of Movement

Skeletal muscles are attached to the skeleton, and when they contract, bones move. **Tendons** most often attach a muscle to the far sides of a joint (the region where two bones meet). One point of attachment is called the **origin** of the muscle; the origin stays stationary when the muscle contracts. The bone attachment that moves is called the **insertion** of the muscle. Because muscles shorten when they contract, they can only pull; they cannot push. Therefore, muscles work in **antagonistic pairs** (Fig. 32.10). That is, contraction of one member of the pair causes a bone to move in one direction, while contraction of the other member of the pair causes the same bone to move in the opposite direction.

In this section, you will be examining two sets of movements:

1. **Flexion** (movement of jointed parts toward each other) and **extension** (movement of jointed parts away from each other).
2. **Adduction** (movement of a part toward the midline of the body) and **abduction** (movement of a part away from the midline of the body).

Figure 32.10 Antagonistic pairs of muscles.
Arrows indicate direction of movement of the lower limb when upper limb muscles contract.

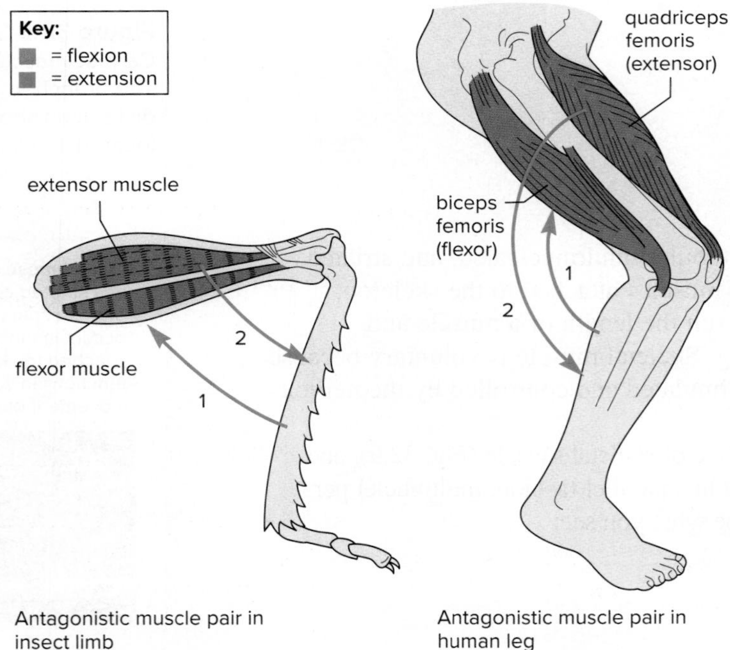

Experimental Procedure: Types of Movement

Frog Muscles

The skin has been removed from the legs of a preserved frog so that you can study the arrangement of muscles (Fig. 32.11). The objective is to discover which of the four movements (flexion, extension, adduction, abduction) are performed by the muscles.

1. Locate the **gastrocnemius,** a large muscle on the back of the shank. Observe the white connective tissue **(fascia)** surrounding the fleshy middle, or **belly,** portion of the muscle. The fascia continues beyond the muscle at either end to form the **tendons,** which attach the muscle to bone.
2. Hold the thigh of the frog rigid with one hand, and grasp the belly of the gastrocnemius with the other. Gently pull the gastrocnemius toward its origin (lower end of the femur, the bone of the thigh). What is the action caused by the gastrocnemius when the thigh is rigid? ______________________________

   ____
3. Next, hold the shank rigid, and again pull the gastrocnemius toward its origin. When the shank is rigid what is the action caused by contraction of the gastrocnemius? ______________________________
4. Locate the frog's **triceps femoris,** a very large ventral muscle of the thigh. It is seen best in the dorsal view. This muscle has two origins on the pelvic girdle and inserts into the femur by way of the broad, white tendon that passes over the anterior surface of the knee. Hold the thigh firmly, and pull on the triceps. Is the triceps femoris a shank extensor or flexor? ______________________________

Figure 32.11 Dorsal view of hindleg muscles in frogs.
The contraction of muscles makes the bones move in particular directions. In this illustration note that the left shows surface muscles and the right shows deep muscles of the frog.

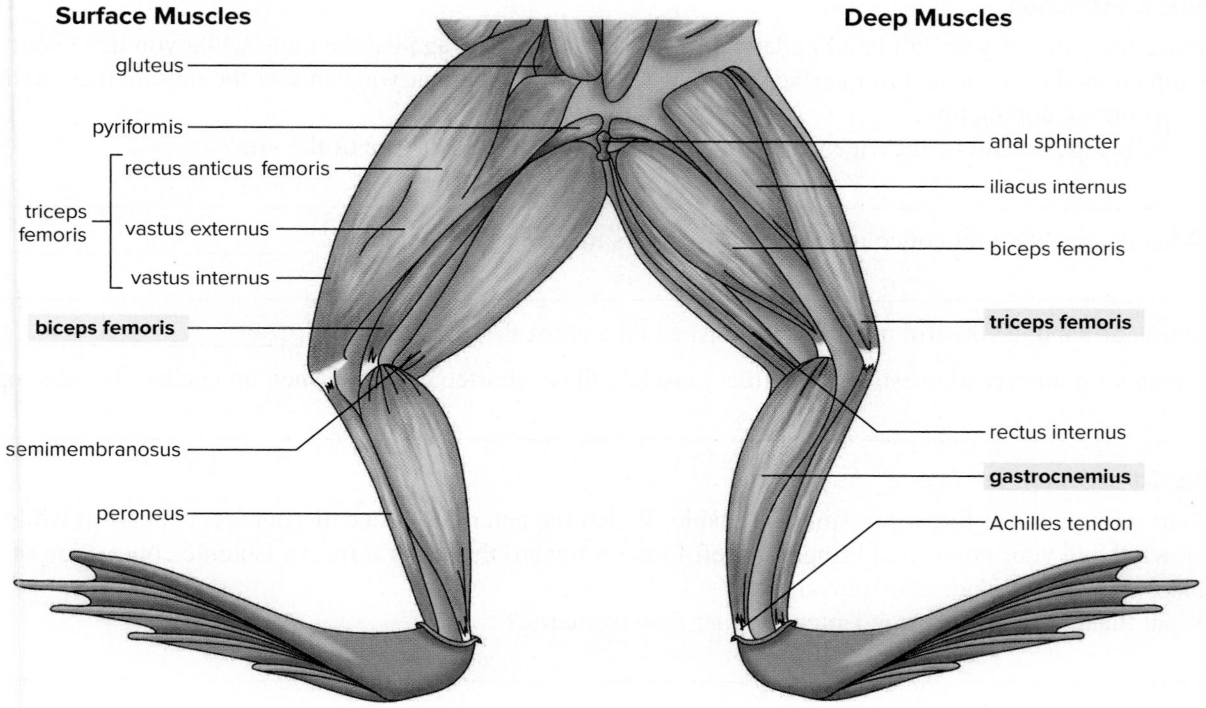

5. Now find your specimen's **biceps femoris,** a narrow muscle seen best in dorsal view just posterior to the triceps. Its origin is on the pelvic girdle. Where does the biceps femoris insert? ______________________

____

6. What action does the biceps femoris cause? ______________________

____

Human Muscles

Locate the following antagonistic pairs (muscles that act in opposition to each other) in Figure 32.12. In each case, state their contrary actions by inserting one of these functions—*flexes, extends, adducts,* or *abducts*—in the following:

1. The biceps brachii ____________________ the forearm.

   The triceps brachii ____________________ the forearm.
2. The sartorius ____________________ the thigh.

   The adductor longus ____________________ the thigh.
3. The quadriceps femoris ____________________ the leg.

   The biceps femoris ____________________ the leg.

Muscular Contractions

Isometric and Isotonic Contractions

During an **isometric contraction,** the length of the muscle does *not* change. During an **isotonic contraction,** the length of the muscle *does* change.

Experimental Procedure: Types of Contraction

Isometric Contraction

1. Place the palm of your left hand underneath a tabletop. Push up against the table while you have your right hand cupped over the anterior surface of your left upper arm so that you can feel the muscle there undergo an isometric contraction.
2. Is the biceps brachii or the triceps brachii located on the ventral surface of the arm?

   ____
3. What change did you notice in the firmness of this muscle as it contracted? ______________________

   ____
4. Did your hand or forearm move as you pushed up against the table? ______________________
5. Given your answer to question 4, did this muscle's fibers shorten as you pushed up against the tabletop?

   ____

Isotonic Contraction

1. Start with your left forearm resting on a table. Watch the anterior surface of your left upper arm while you slowly bend your elbow and bring your left forearm toward the upper arm. An isotonic contraction of the biceps brachii produces this movement.
2. What makes this contraction isotonic rather than isometric?

   ____

Figure 32.12 Human superficial skeletal muscles.
a. Anterior view. **b.** Posterior view. The muscles highlighted here are those noted in the exercise entitled "Human Muscles."

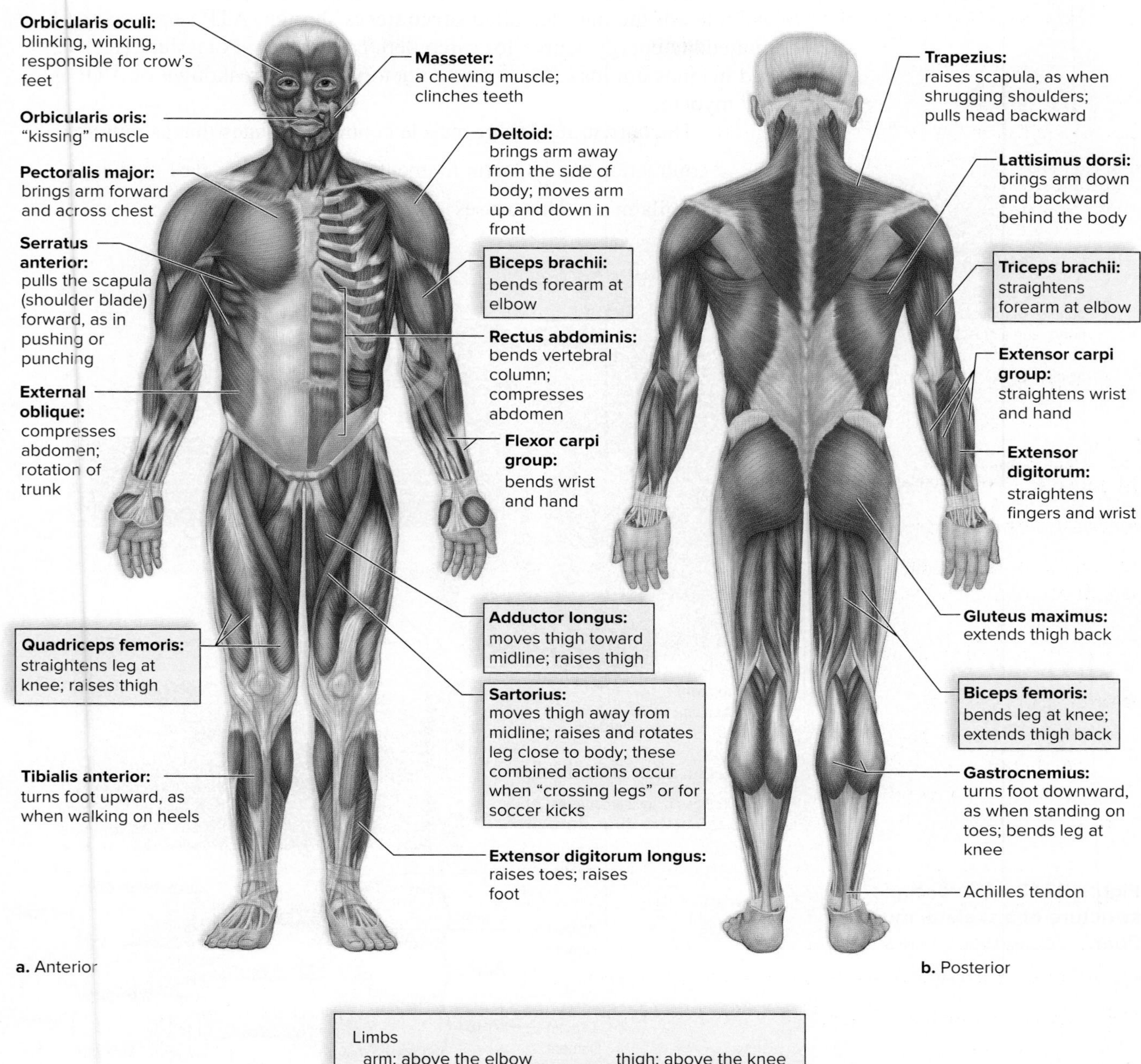

Orbicularis oculi: blinking, winking, responsible for crow's feet
Masseter: a chewing muscle; clinches teeth
Orbicularis oris: "kissing" muscle
Deltoid: brings arm away from the side of body; moves arm up and down in front
Pectoralis major: brings arm forward and across chest
Serratus anterior: pulls the scapula (shoulder blade) forward, as in pushing or punching
Biceps brachii: bends forearm at elbow
Rectus abdominis: bends vertebral column; compresses abdomen
External oblique: compresses abdomen; rotation of trunk
Flexor carpi group: bends wrist and hand
Adductor longus: moves thigh toward midline; raises thigh
Quadriceps femoris: straightens leg at knee; raises thigh
Sartorius: moves thigh away from midline; raises and rotates leg close to body; these combined actions occur when "crossing legs" or for soccer kicks
Tibialis anterior: turns foot upward, as when walking on heels
Extensor digitorum longus: raises toes; raises foot
a. Anterior
Trapezius: raises scapula, as when shrugging shoulders; pulls head backward
Lattisimus dorsi: brings arm down and backward behind the body
Triceps brachii: straightens forearm at elbow
Extensor carpi group: straightens wrist and hand
Extensor digitorum: straightens fingers and wrist
Gluteus maximus: extends thigh back
Biceps femoris: bends leg at knee; extends thigh back
Gastrocnemius: turns foot downward, as when standing on toes; bends leg at knee
Achilles tendon
b. Posterior

Limbs

arm: above the elbow	thigh: above the knee
forearm: below the elbow	leg: below the knee

Contraction of Muscle Fibers

Note in Figure 32.13 that electron microscopy has shown that striations in skeletal muscle are due to the placement of **myosin** and **actin filaments.** During contraction, actin filaments slide past myosin filaments, and units of the muscle called **sarcomeres** shorten. ATP serves as the immediate energy source for sarcomere contraction. Potassium (K^+) and magnesium ions (Mg^{2+}) are cofactors for the breakdown of ATP by myosin.

The current model for muscle contraction states that as muscle contraction occurs, actin filaments ___________ past myosin filaments. This causes ___________ to shorten.

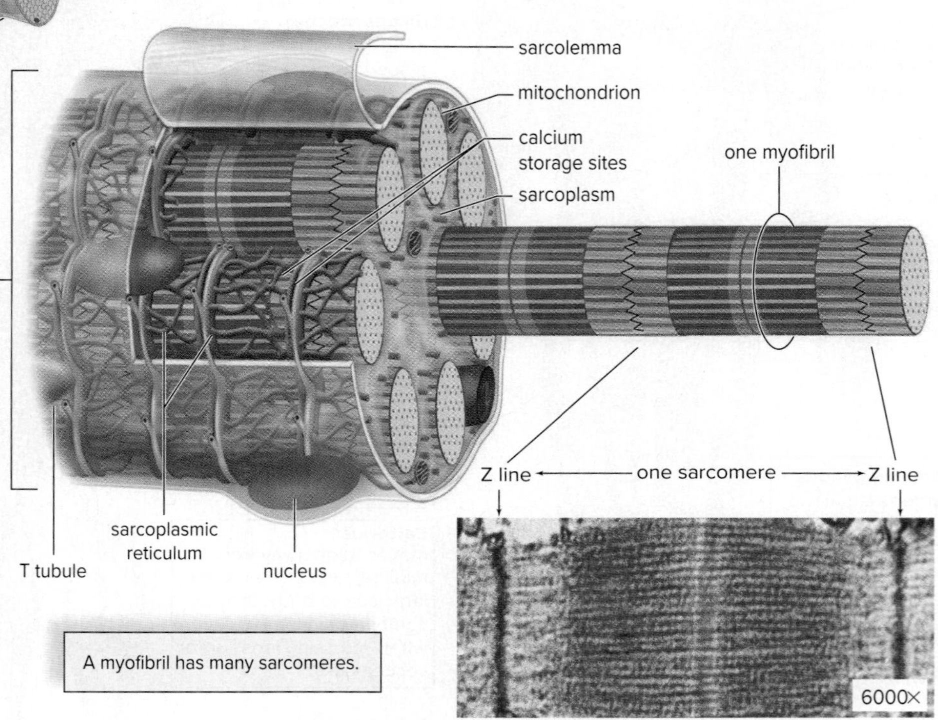

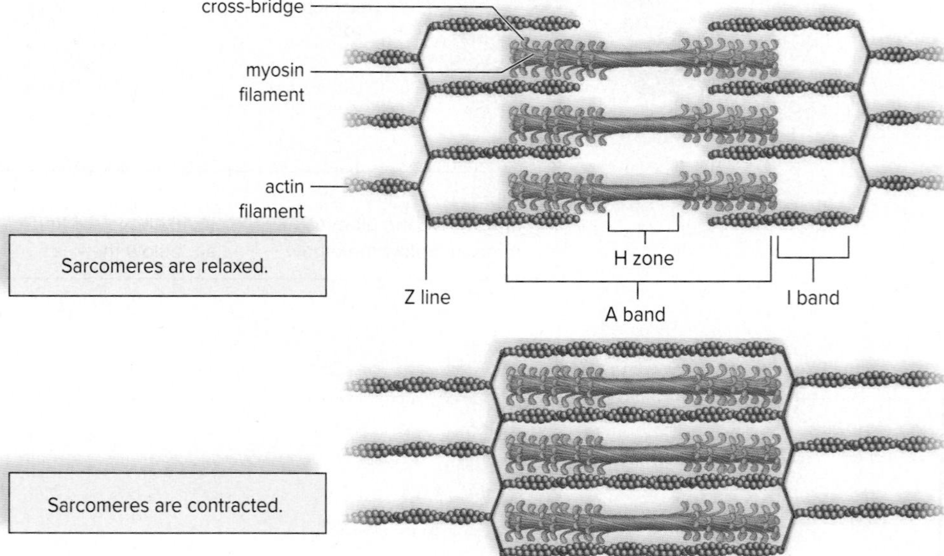

Figure 32.13 Microscopic structure of a skeletal muscle fiber. ©Biology Media/Science Source

Experimental Procedure: Contraction of Muscle Fibers

1. Label two slides 1 and 2. Mount a strand of glycerinated fibers in a drop of glycerol on each slide. Place each slide on a millimeter ruler, and measure the length of the strand. Record these lengths in the first row in Table 32.3.
2. If there is more than a small drop of glycerol on the slides, soak up the excess on a piece of lens paper held at the edge of the glycerol farthest from the fiber strand.
3. To slide 1, add a few drops of salt solution containing potassium (K^+) and magnesium (Mg^{2+}) ions, and measure any change in strand length. Record your results in Table 32.3.
4. To slide 2, add a few drops of ATP solution, and measure any change in strand length. Record your results in Table 32.3.
5. Now add ATP solution to slide 1. Measure any change in strand length, and record your results in Table 32.3.
6. To slide 2, add a few drops of K^+/Mg^{2+} salt solution, and measure any change in strand length. Record your results in Table 32.3.
7. You probably observed muscle contraction (as demonstrated by shortening of the fiber strand) only after both the K^+/Mg^{2+} salt solution and the ATP solution were applied to the muscle. To demonstrate that you understand the requirements for contraction, state the function of each of the substances listed in Table 32.4.

Table 32.3 Glycerinated Muscle Contraction

Solution	*Length (mm)*	
	Slide 1	**Slide 2**
Glycerol alone		
K^+/Mg^{2+} salt solution alone		
ATP alone		
Both salt solution and ATP		

Table 32.4 Summary of Muscle Fiber Contraction

Substance	**Function**
Myosin filament	
Actin filament	
K^+/Mg^{2+} salt solution	
ATP	

Laboratory Review 32

_______________ **1.** Where is the pressure of muscle contraction applied in a hydrostatic skeleton?

_______________ **2.** What are the characteristics of an exoskeleton?

_______________ **3.** What are the characteristics of an endoskeleton?

_______________ **4.** What is produced by the red marrow of the spongy bone?

_______________ **5.** Which mineral provides rigidity to bone?

_______________ **6.** What are osteocytes?

_______________ **7.** Are vertebrae part of the appendicular or axial skeleton?

_______________ **8.** Structures that have a similar origin, but different function are termed _______________.

_______________ **9.** List the three types of muscular tissue.

_______________ **10.** Does contraction of the triceps brachii result in extension or flexion of the forearm?

_______________ **11.** What antagonistic pair of muscles causes adduction and abduction of the thigh?

_______________ **12.** What are the names of the protein filaments that slide across each other during muscle contractions?

_______________ **13.** What molecule supplies energy needed for muscle contractions?

_______________ **14.** What is the name of a muscle contraction where the length of muscle does not change?

Thought Questions

15. Why are the muscular and skeletal systems often considered together?

16. Why do vertebrate forelimbs have a similar underlying structure but different adaptations?

17. Why would a potassium (K^+) deficiency cause muscle weakness and fatigue?

LABORATORY

33 Development

Learning Outcomes

33.1 The First Stages of Development

- Identify the cellular stages of development with reference to slides of early sea star development.
- Identify the tissue stages of development with reference to slides of frog development.

33.2 The Organ Stages of Development

- Associate the germ layers with the development of various organs.
- Describe the significance of the process of induction during development.
- Identify which organs develop first in a vertebrate embryo (e.g., frog, chick, and human).
- Compare the later development of humans to that of chicks.

33.3 Extraembryonic Membranes, the Placenta, and the Umbilical Cord

- Distinguish between and give a function for the extraembryonic membranes, the placenta, and the umbilical cord.

33.4 Human Fetal Development

- Trace the main events of human fetal development.

Introduction

The early development of animals is quite similar, regardless of the species. The fertilized egg, or zygote, undergoes successive divisions by cleavage, forming a mulberry-shaped ball of cells called a morula and then a hollow ball of cells called a blastula. The fluid-filled cavity of the blastula is the blastocoel. Later, some of the surface cells fold inward, or invaginate, eventually forming a double-walled structure. The outer layer is called the ectoderm, and the inner layer is the endoderm. Between these layers, a middle layer, or mesoderm, arises. The embryo is now called a gastrula. In particular, the presence of yolk (nutrient material) influences how the gastrula comes about. All later development can be associated with the three germ layers (ectoderm, endoderm, and mesoderm) that give rise to different tissues and organ systems.

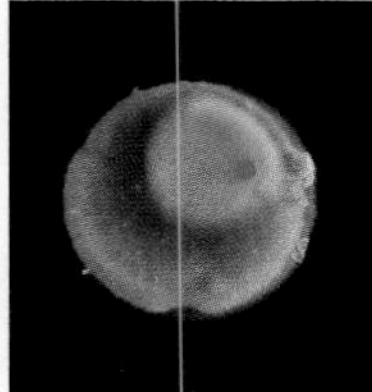
zygote 100×

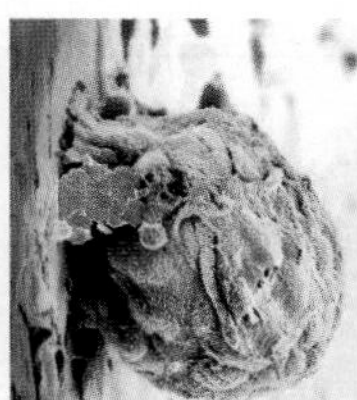
embryo at one week; implants in uterine wall

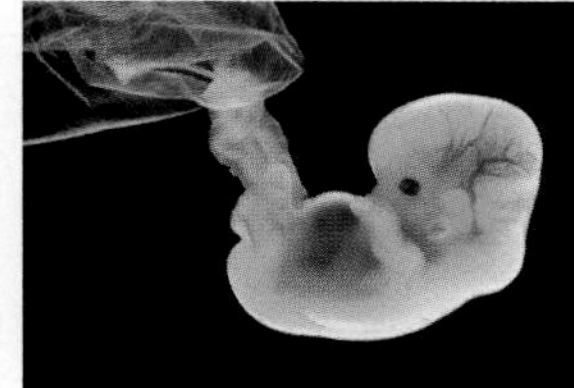
embryo at eight weeks

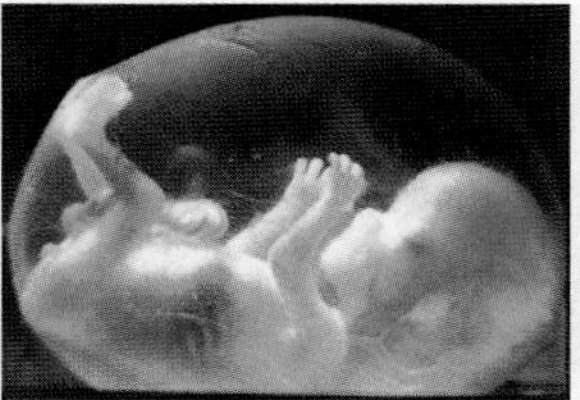
fetus at three months

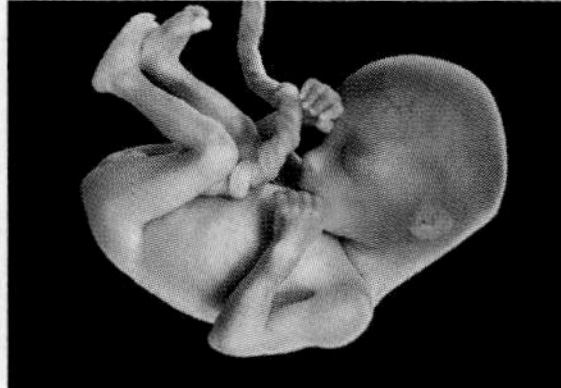
fetus at five months

Development occurs in stages.

(zygote) ©Anatomical Travelogue/Science Source; (embryo, 1 week) ©Bettman/Corbis; (embryo, 8 weeks) ©Neil Harding/Getty Images; (fetus, 3 months) ©Petit Format/Science Source; (fetus, 5 months) ©John Watney/Science Source

33.1 The First Stages of Development

In this section, we will consider the cellular and tissue layer stages of development.

Cellular Stages of Development

The cellular stages of development include the following:

- **Zygote formation:** A single sperm fertilizes an egg and the result is a zygote, the first cell of the new individual.
- **Morula formation:** Zygote divides into a number of smaller cells until there is a cluster of 16–32 cells called a morula.
- **Blastula formation:** The morula becomes a blastula, a hollow ball of cells.

Observation: Cellular Stages of Development in the Sea Star

The cellular stages of development are remarkably similar in all animals. Therefore, we can view slides of sea star development to study the cellular stages of animal development (Fig. 33.1). A sea star is an invertebrate that develops in the ocean and, therefore, will develop easily in the laboratory where it can be observed.

Obtain slides or view a model of sea star development and note the following:

1. **Zygote:** Both plants and animals begin life as a single cell, a zygote. A zygote contains chromosomes from each parent. Explain your answer. ______________________________

   ____

2. **Cleavage:** View slides showing various numbers of cells due to the process of cleavage, cell division without growth until the morula stage. Is the morula about the same size as the zygote? ______________

   Explain your answer. ______________________________

   ____

3. **Blastula:** The cavity of a blastula is called the blastocoel. *Label the blastocoel in Figure 33.1.* The formation of a hollow cavity is important to the next stage of development.

Figure 33.1 Sea star development.
All animals, including sea stars and humans, go through the same cellular stages from cleavage to blastula. (Magnification: zygote is 400×; b–f are 75×)

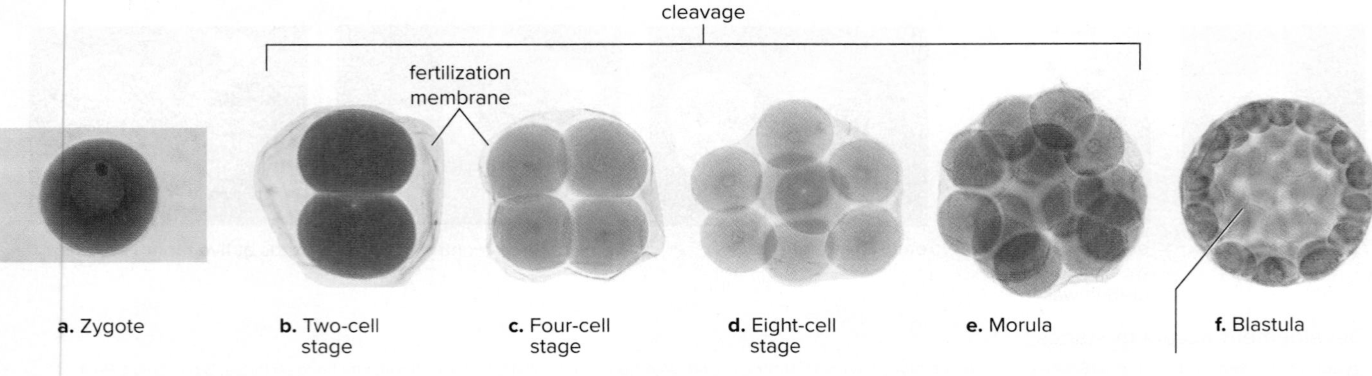

Observation: Cellular Stages of Development in Humans

In Figure 33.2, or in a model of human development, observe the same stages of development already observed in sea star slides. Also, observe that fertilization in humans occurs in an oviduct following ovulation. As the embryo undergoes cleavage, it travels in the oviduct to the uterus.

If the embryo splits at the 2-cell stage, the result is identical twins. (Fraternal twins arise when two separate eggs are fertilized.) How might you account for the development of identical triplets? ______________

____

____

The blastula in humans is called a blastocyst. The blastocyst contains an **inner cell mass** that becomes the embryo, and the outer group of cells (the trophoblast) will become membranes that nourish and protect it. At about day 6, the blastocyst has reached the uterus and implants into the uterine wall, where it will receive nourishment from the mother's bloodstream.

What's the main difference between the cellular stages in a sea star and in a human? ______________

____

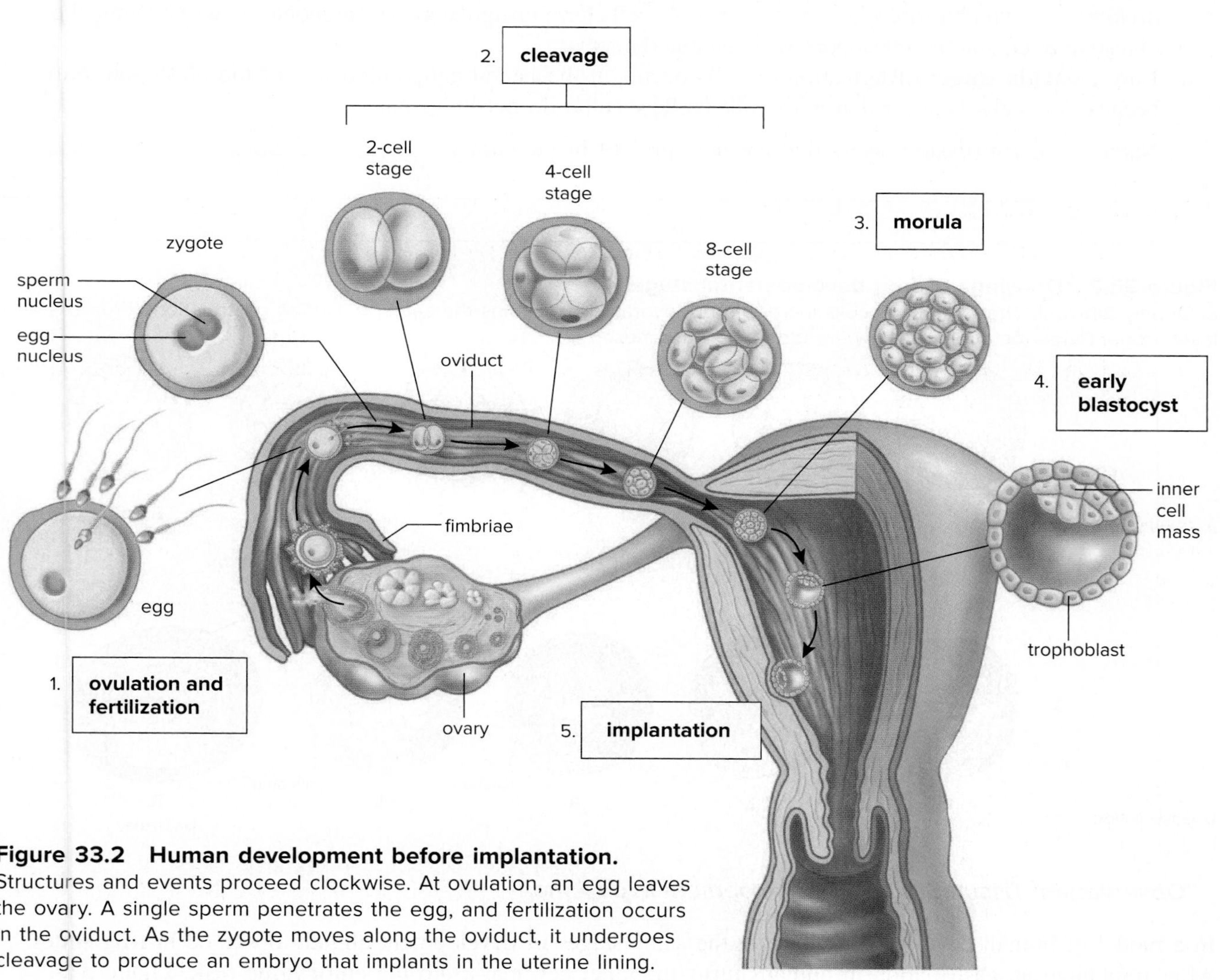

Figure 33.2 Human development before implantation. Structures and events proceed clockwise. At ovulation, an egg leaves the ovary. A single sperm penetrates the egg, and fertilization occurs in the oviduct. As the zygote moves along the oviduct, it undergoes cleavage to produce an embryo that implants in the uterine lining.

Tissue Stages of Development

The tissue stages of development include the following:

- **Early gastrula stage:** This stage begins when certain cells begin to push or invaginate into the blastocoel, creating a double layer of cells. The outer tissue layer is called the **ectoderm,** and the inner tissue layer is called the **endoderm.**
- **Late gastrula stage:** Gastrulation is not complete until there are three layers of cells. The third tissue layer, called **mesoderm,** occurs between the other two layers already mentioned.

Observation: Tissue Stages of Development in Frogs

It is traditional to view gastrulation in a frog. A frog is a vertebrate, and so its development is expected to be closer to that of other vertebrates than is the development of a sea star. In Figure 33.3, note that the yellow (vegetal pole) cells are heavily laden with yolk, and the blue (animal pole) cells are the ones that invaginate into the blastocoel, forming the early gastrula.

1. **Early gastrula stage:** Obtain a cross section of a frog gastrula. Most likely, your slide is the equivalent of Figure 33.3*b,* number 3, in which case you will see two cavities, the old blastocoel and newly forming *archenteron,* which forms once the animal pole cells have invaginated. The archenteron will become the digestive tract, and the blastopore will become the anus.
2. **Late gastrula stage:** Invagination of cells occurs at the lateral and ventral lips of the blastopore only because the cells heavy laden with yolk (yellow cells) do not invaginate.

   Name the germ (tissue) layers that are now present in the embryo. ______________________________

   ____

   ____

Figure 33.3 Drawings of frog developmental stages.
a. During cleavage, the number of cells increases but overall size remains the same. **b.** During gastrulation, three tissue layers form. Blue = ectoderm; yellow = endoderm; red = mesoderm.

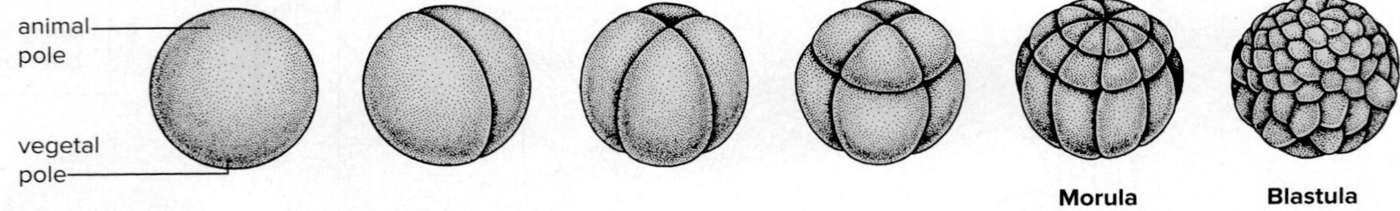

a. Cleavage

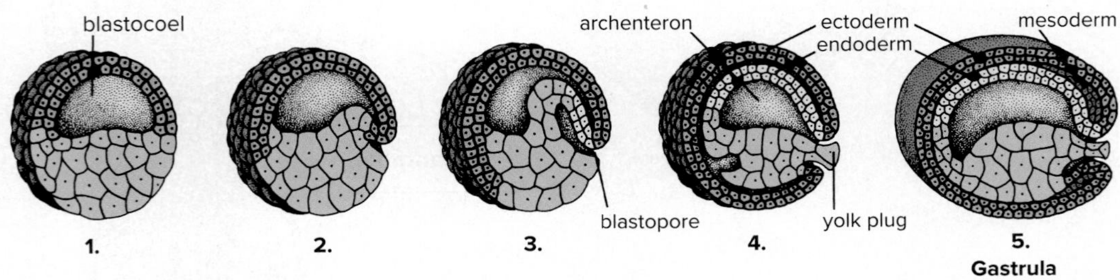

b. Gastrulation

Observation: Tissue Stages of Development in a Human

In a model of human development, observe the same stages of development already observed in frog slides. After implantation, gastrulation in humans turns the inner cell mass into the **embryonic disk.** Figure 33.4 shows the embryonic disk, which has the three layers of cells we have been discussing: the ectoderm, mesoderm, and endoderm. Figure 33.4 also shows the significance of these layers, often called the **germ layers.** The future organs of an individual can be traced back to one of the germ layers.

Figure 33.4 Embryonic disk.
The embryonic disk has three germ layers called ectoderm, mesoderm, and endoderm. Organs and tissues can be traced back to a particular germ layer as indicated in this illustration.

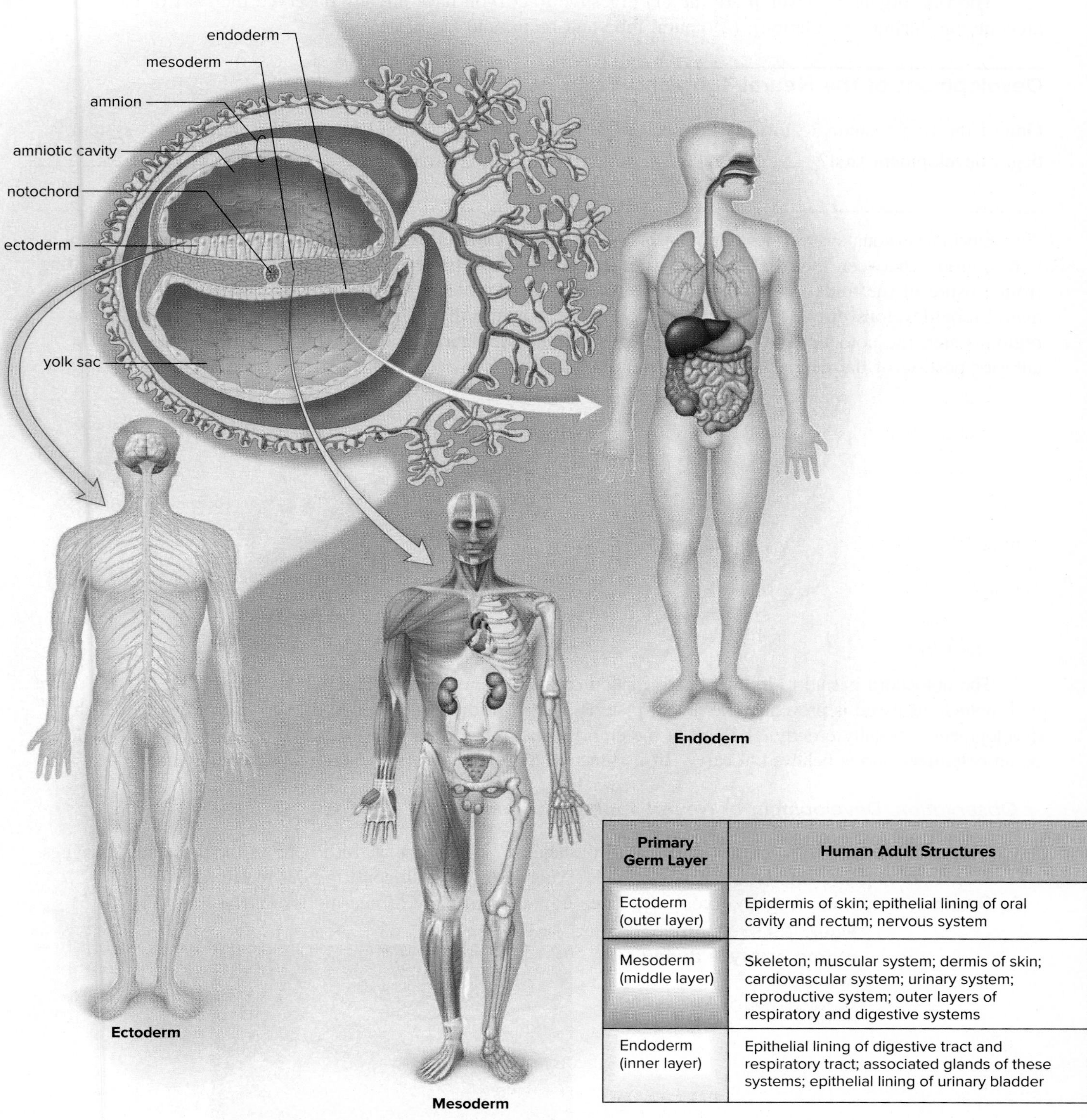

Primary Germ Layer	Human Adult Structures
Ectoderm (outer layer)	Epidermis of skin; epithelial lining of oral cavity and rectum; nervous system
Mesoderm (middle layer)	Skeleton; muscular system; dermis of skin; cardiovascular system; urinary system; reproductive system; outer layers of respiratory and digestive systems
Endoderm (inner layer)	Epithelial lining of digestive tract and respiratory tract; associated glands of these systems; epithelial lining of urinary bladder

33.2 The Organ Stages of Development

As soon as all three embryonic tissue layers (ectoderm, endoderm, and mesoderm) are established, the organ level of development begins. It continues until all organs have formed. Figure 33.4 pertains to the organ stages of development because it shows which germ layers give rise to which organs.

The first organs to develop are the (1) digestive tract (you have already observed the start of the archenteron during gastrulation); (2) neural tube and brain; and (3) heart.

Development of the Neural Tube and Brain

One of the first systems to form is the nervous system. Why might it be beneficial for the nervous system to begin development first? ____

____

During nervous system development in the frog, two folds of ectoderm grow upward as the neural folds with a groove between them. The flat layer of ectoderm between them is the **neural plate.** The tube resulting from closure of the folds is the **neural tube.** An examination of the neurula in cross section shows that the neural tube develops directly above the **notochord,** a structure that arises from mesoderm in the middorsal region. Later, the notochord is replaced by vertebrae and the neural tube is then called the nerve cord. The anterior portion of the neural tube becomes the brain.

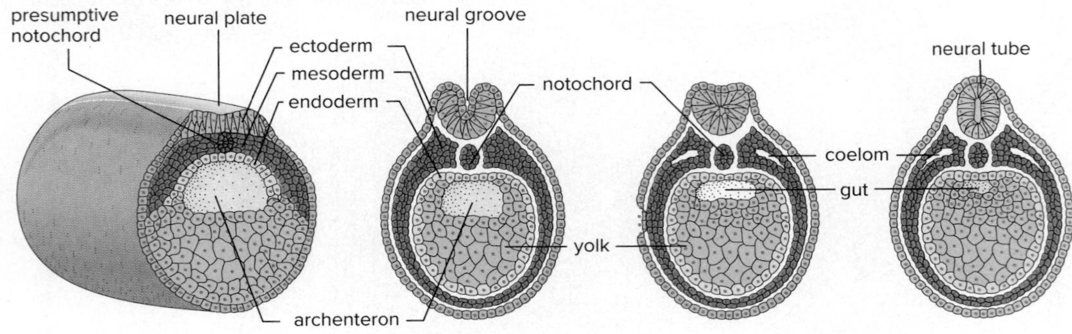

Neurulation

The notochord is said to **induce** the formation of the nervous system. Experiments have shown that if contact with notochord tissue is prevented, no neural plate is formed. **Induction** is believed to be one means by which development is usually orderly. The part of the embryo that induces the formation of an adjacent organ is said to be an **organizer** and is believed to carry out its function by releasing one or more chemical substances.

Observation: Development of Neural Tube and Brain

1. Obtain a slide showing a cross section of a frog neurula stage, and match it to one of the preceding drawings. Which drawing seems to best match your slide? Your instructor will confirm your match for you.
2. Obtain and examine frog embryos for a three-dimensional view of neurulation in the frog (Fig. 33.5).

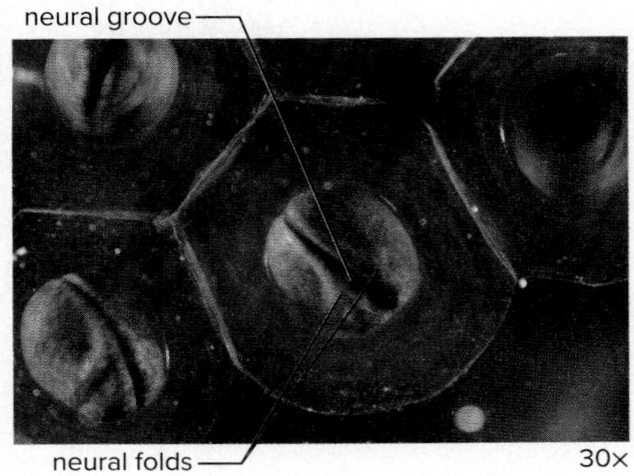

Figure 33.5 Photograph of frog embryos during neurulation.

Development of the Heart

A chick embryo offers an opportunity to view a beating heart in an embryo. Your instructor may show you various stages. In particular you will want to observe the chick embryo from the 48-hour stage up to the 96-hour stage.

Observing Live Chick Embryos

Use the following procedure for selecting and opening the eggs of live chick embryos:

1. Choose an egg of the proper age to remove from the incubator, and put a penciled × on the uppermost side. The embryo is just below the shell.
2. Add warmed chicken Ringer solution to a finger bowl until the bowl is about half full. (Chicken Ringer solution is an isotonic salt solution for chick tissue that maintains the living state.) The chicken Ringer solution should not cover the yolk of the egg.
3. On the edge of the dish, gently crack the egg on the side opposite the ×.
4. With your thumbs placed over the ×, hold the egg in the chicken Ringer solution while you pry it open from below and allow its contents to enter the solution. If you open the egg too slowly or too quickly, the shell may damage the delicate membranes surrounding the embryo.

Observation: Forty-Eight-Hour Chick Embryo

1. Follow the standard procedure for selecting and opening an egg containing a 48-hour chick embryo.
2. The embryo has turned so that the head region is lying on its side. Refer to Figure 33.6, and identify the following:
   a. **Shape of the embryo,** which has started to bend. The head is now almost touching the heart.
   b. **Heart,** contracting and circulating blood. Can you make out a ventricle, an atrium, and the aortic arches in the region below the head? Later, only one aortic arch will remain.
   c. **Vitelline arteries** and **veins,** which extend over the yolk. The vitelline veins carry nutrients from the yolk sac to the embryo.
   d. **Brain** with several distinct regions.
   e. **Eye,** which has a developing lens.
   f. **Margin (edge) of the amnion,** which can be seen above the vitelline arteries (see next section for amnion).
   g. **Somites,** blocks of developing muscle tissue that differentiate from mesoderm, which now number 24 pairs.
   h. **Caudal fold** of the amnion. The embryo will be completely enveloped when the head fold and caudal fold meet the margin of the amnion.

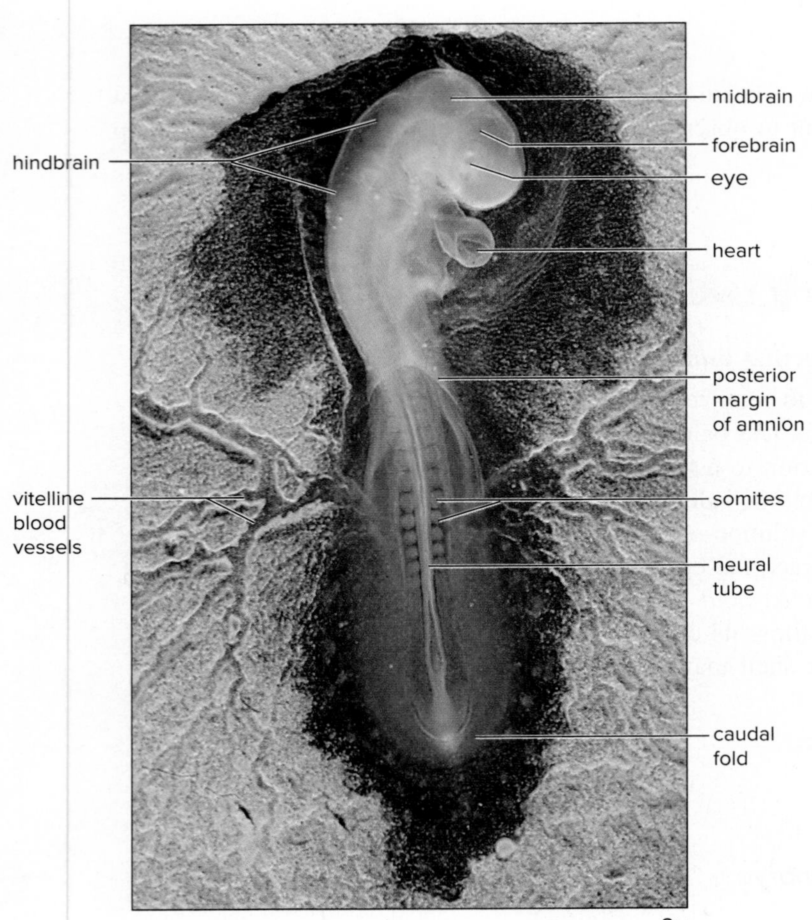

Figure 33.6 Forty-eight-hour chick embryo.
The most prominent organs are labeled.
©Carolina Biological Supply Company/Phototake

Observation: Older Chick Embryos

As a chick embryo continues to grow, various organs differentiate further (Fig. 33.7). The neural tube closes along the entire length of the body and is now called the spinal cord. The allantois, an extraembryonic membrane, is seen as a sac extending from the ventral surface of the hindgut near the tail bud. The digestive system forms specialized regions, and there is both a mouth and an anus. The yolk sac, the extraembryonic membrane that encloses the yolk, is attached to the ventral wall, but when the yolk is used up, the ventral wall closes.

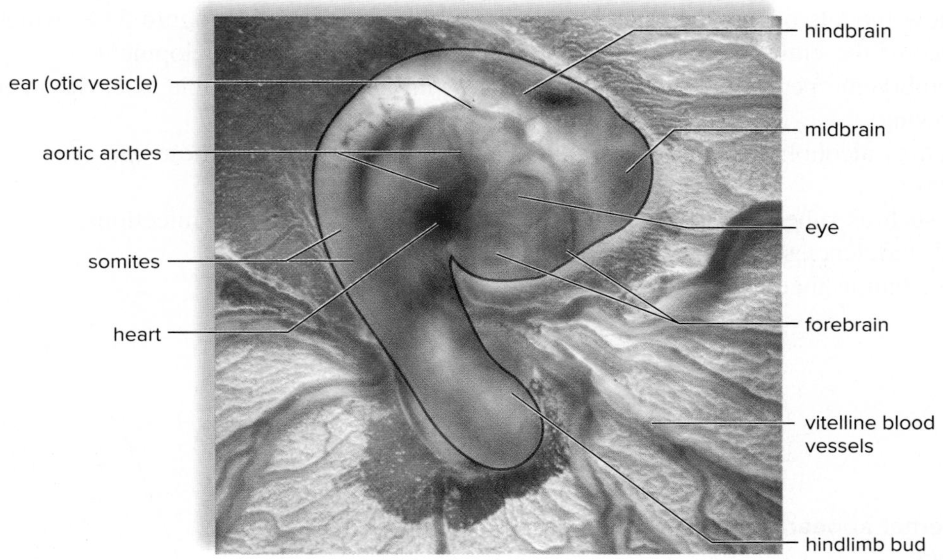

a. Photograph

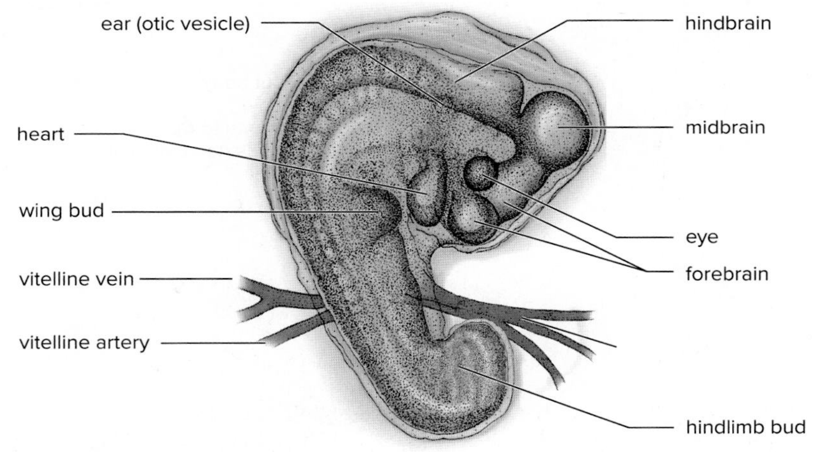

b. Drawing

Key:

Brain region	**Becomes:**
forebrain	cerebrum diencephalon
midbrain	relay station
hindbrain	pons, cerebellum medulla oblongata

Figure 33.7 Ninety-six-hour chick embryo. Brain regions listed in the key can now be seen.

(a) ©Carolina Biological Supply Company/Phototake

Observation: Development of Human Organs

Study models or other study aids available that show the development of the nervous system and the heart in humans and/or show models of human embryos of different ages. Also view Figure 33.8, which depicts the external appearance of the embryo from the fourth to the seventh week of development.

During the embryonic period of development, the growing baby is susceptible to environmental influences, including the following:

- Drugs, such as alcohol; certain prescriptions; and recreational drugs. These can cause birth defects.
- Infections such as rubella, also called German measles, and other viral infections.
- Nutritional deficiencies.
- X-rays or radiation therapy.

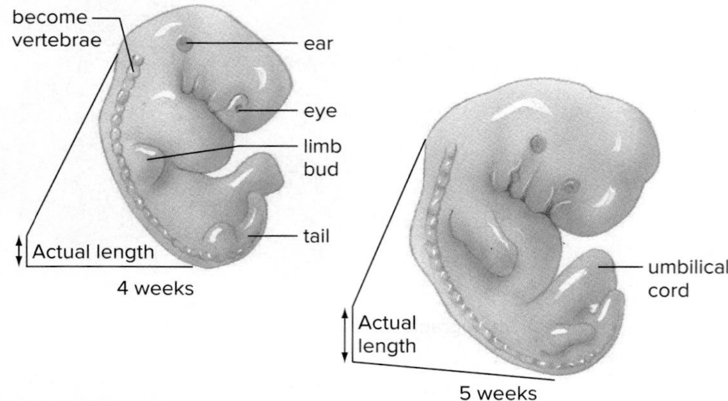

a. Weeks 4 and 5
- Head dominant, but body getting longer.
- Limb buds are visible.
- Eyes and ears begin to form.
- Tissue for vertebrae extend into tail.

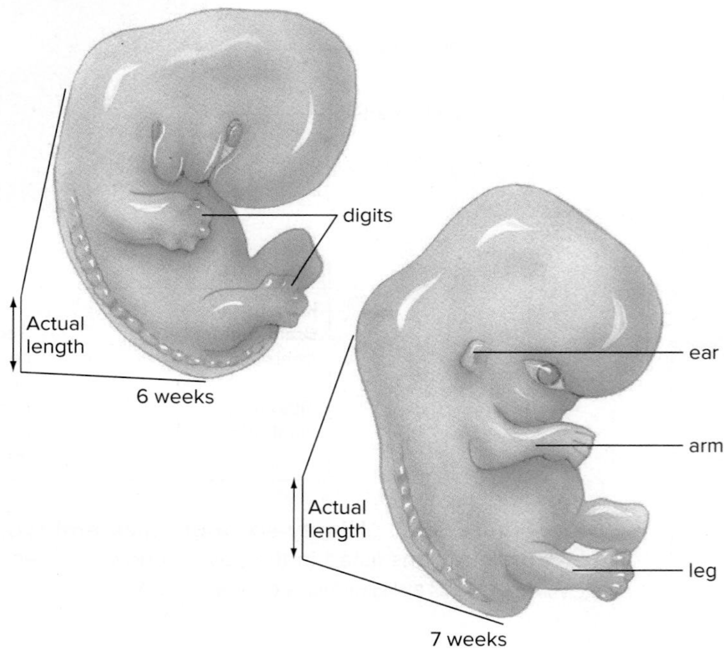

b. Weeks 6 and 7
- Head still dominant, but tail has disappeared.
- Facial features continue to develop.
- Hands and feet have digits.
- All organs are more developed.

Figure 33.8 External appearance of the human embryo.
a. Weeks 4 and 5. **b.** Weeks 6 and 7.

As illustrated in Figure 33.9, the early stages of human development are quite similar to those of the chick. Differences become marked only as development proceeds.

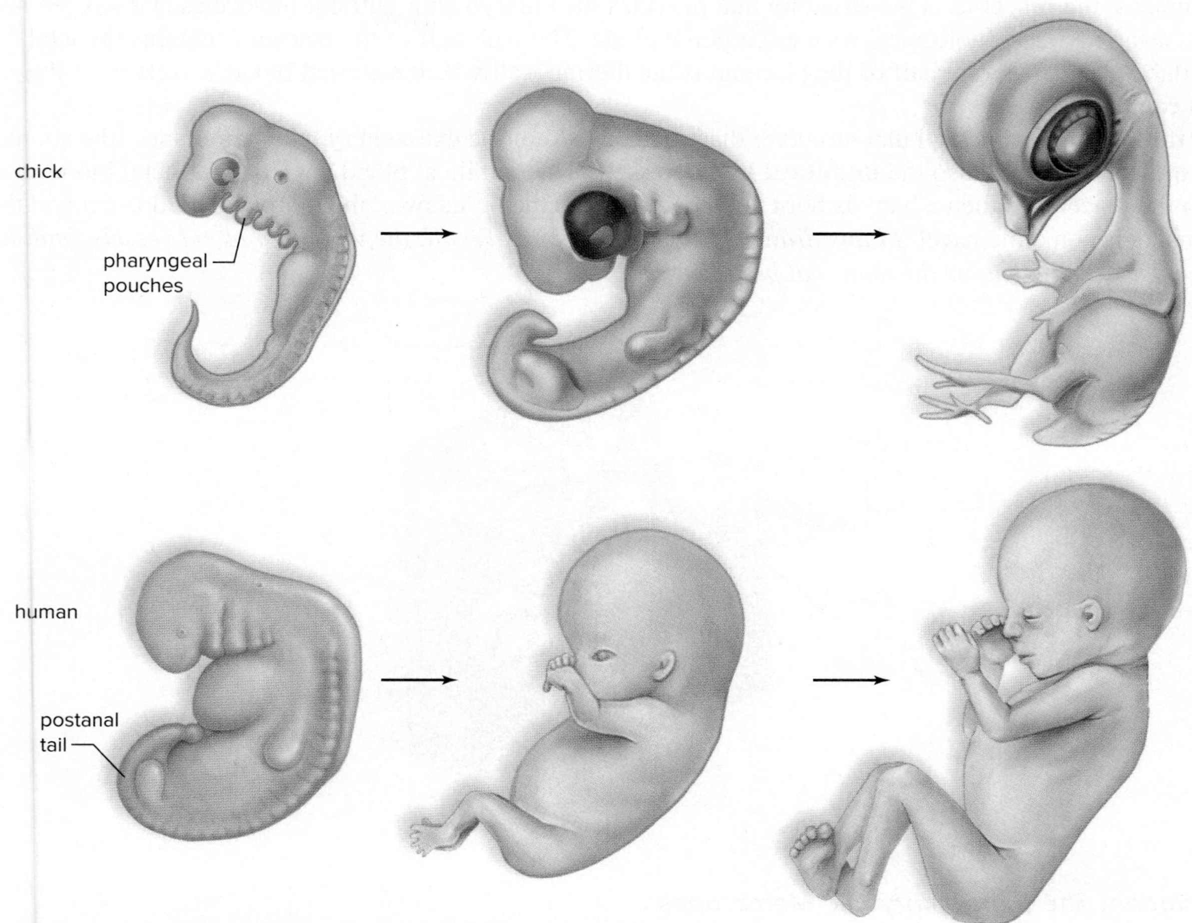

Figure 33.9 Comparison of vertebrate embryos. Successive stages in the development of chick and human. Early stages (*far left*) are similar; differences become apparent as development continues.

33.3 Extraembryonic Membranes, the Placenta, and the Umbilical Cord

- The **extraembryonic membranes** take their name from the observation that they are not part of the embryo proper. They are outside the embryo, and therefore they are "extra."
- In humans, the **placenta** is the structure that provides the embryo with nutrient molecules and oxygen and takes away its waste molecules, such as carbon dioxide. The fetal half of the placenta contains the fetal capillaries. The maternal half of the placenta is the uterine wall where maternal blood vessels meet the fetal capillaries.
- The **umbilical cord** is a tubular structure that contains two of the extraembryonic membranes (the allantois and the yolk sac) and also the **umbilical blood vessels.** The umbilical blood vessels bring fetal blood to and from the placenta. When a baby is born and begins to breathe on its own, the umbilical cord is cut and the remnants become the navel. *In this drawing, label the umbilical cord, the umbilical blood vessels, and the placenta, which contains the maternal blood vessels.*

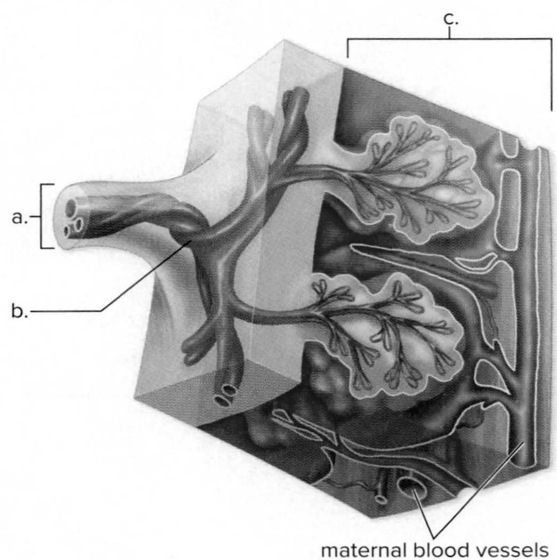

Observation: The Extraembryonic Membranes

In a model, and in Figure 33.10, trace the development of the extraembryonic membranes. Also, note the development of the placenta and the umbilical cord. The extraembryonic membranes are as follows:

- **Chorion:** The chorion is the outermost membrane, and in chicks it lies just below the porous shell, where it functions in gas exchange. In humans, an outer layer of cells surrounding the inner cell mass at the blastocyst stage becomes the chorion. Notice in Figure 33.10 that the treelike **chorionic villi** are a part of the chorion.
- **Amnion:** The amnion forms the amniotic cavity, which envelops the embryo and contains the amniotic fluid that cushions and protects the developing offspring (Fig. 33.11). All animals, whether the sea star, the frog, the chick, or the human, develop in an aqueous environment. Birth of a human is imminent when "the water breaks," and the amniotic fluid is lost.
- **Allantois:** The allantois serves as a storage area for metabolic waste in the chick. In humans, the allantois extends into the umbilical cord. It accumulates the small amount of urine produced by the fetal kidneys and later contributes to urinary bladder formation. Its blood vessels become the umbilical blood vessels.
- **Yolk sac:** The yolk sac is the first embryonic membrane to appear. In the chick, the yolk sac does contain yolk, food for the developing embryo. In humans, the yolk sac contains plentiful blood vessels and is the first site of blood cell formation.

Figure 33.10 Development of extraembryonic membranes in humans.
a. At first, no organs are present in the embryo, only tissues. The amniotic cavity is above the embryonic disk, and the yolk sac is below. The chorionic villi are present. **b, c.** The allantois and yolk sac, two more extraembryonic membranes, are positioned inside the body stalk as it becomes the umbilical cord. **d.** At 35+ days, all membranes are present, and the umbilical cord takes blood vessels between the embryo and the chorion (placenta).

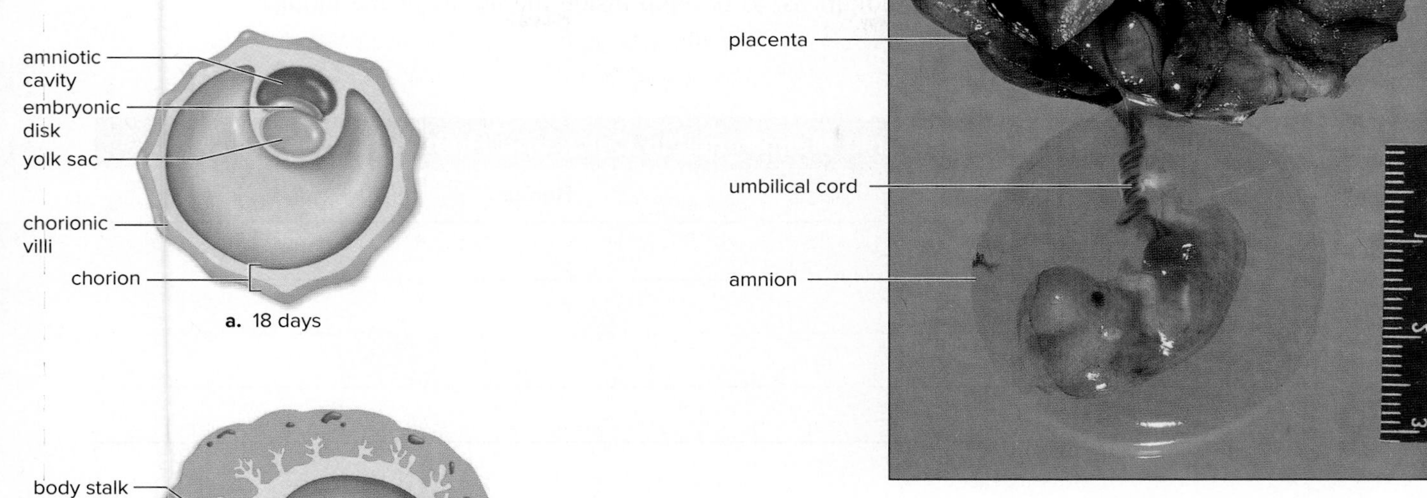

Figure 33.11 Fetus and amnion.
Photograph of a human fetus at 8 weeks. The scale bar is 3 cm.
©Martin Rotker/Phototake

body stalk
allantois
amniotic cavity
embryo
yolk sac
chorionic villi
b. 21 days

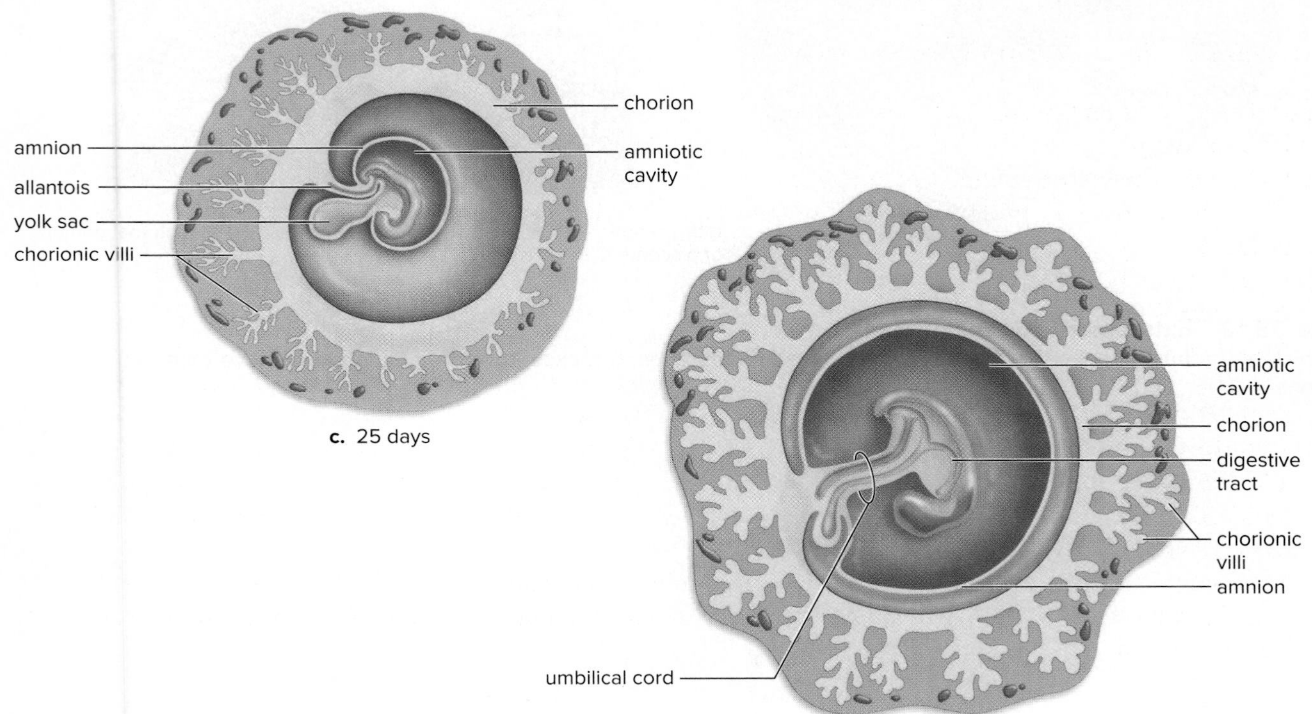

Comparison of Extraembryonic Membranes in Chick and Human

Consult Figure 33.12 and use the information on page 468 to complete Table 33.1, which compares the functions of the extraembryonic membranes in the chick and in the human. Reptiles, which we now know include birds, were the first animals to have extraembryonic membranes. These membranes allowed reptiles to develop on land. They also allow mammals, including humans, to develop inside the uterus of the mother.

Table 33.1 Functions of Extraembryonic Membranes in Chick and Human

Membrane	Chick	Human
Amnion		
Allantois		
Chorion		
Yolk sac		

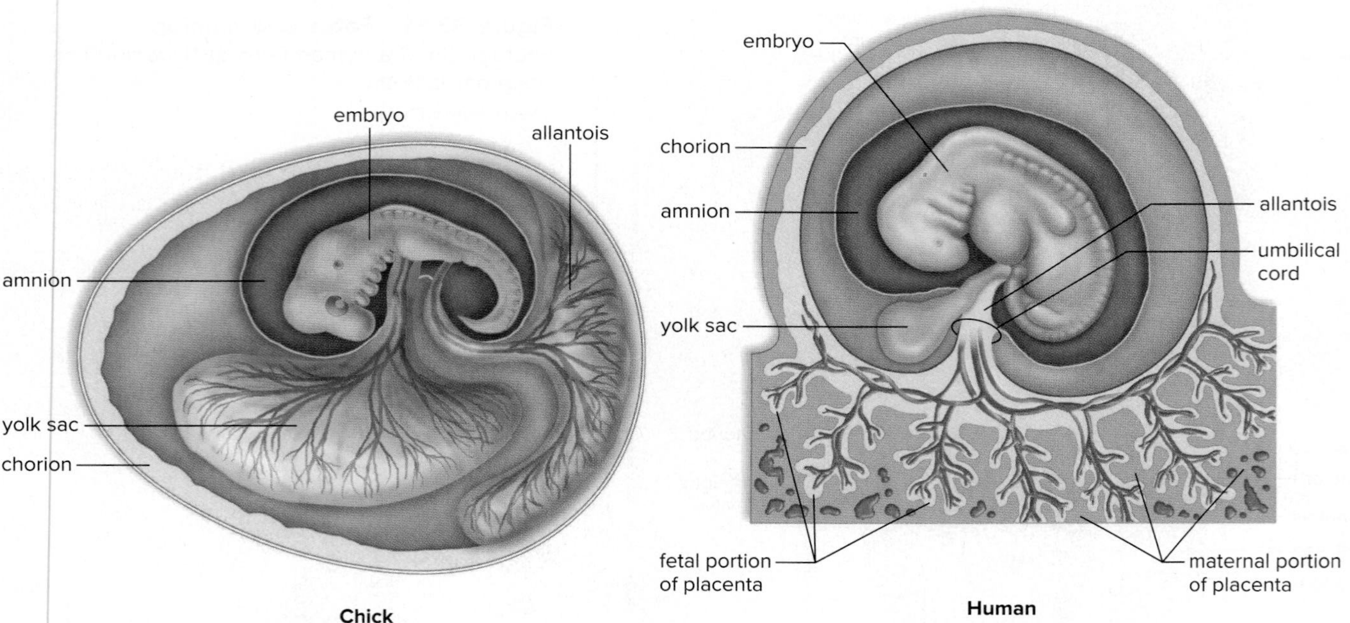

Figure 33.12 Extraembryonic membranes.
The chick and a human have the same extraembryonic membranes, but except for the amnion, they have different functions.

33.4 Human Fetal Development

During fetal development (last seven months), the skeleton becomes ossified (bony), reproductive organs form, arms and legs develop fully, and the fetus enlarges in size and gains weight.

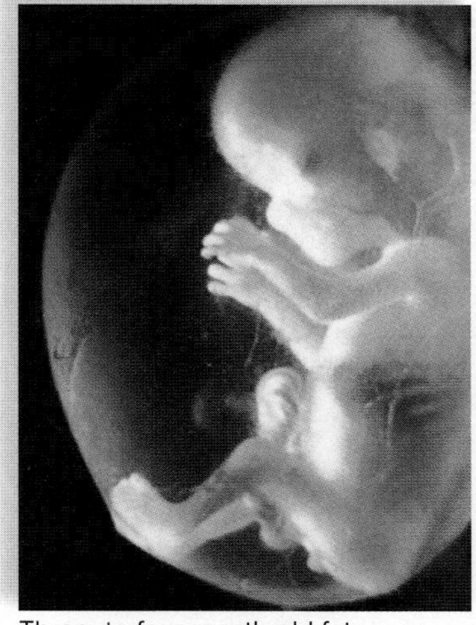

Three- to four-month-old fetus

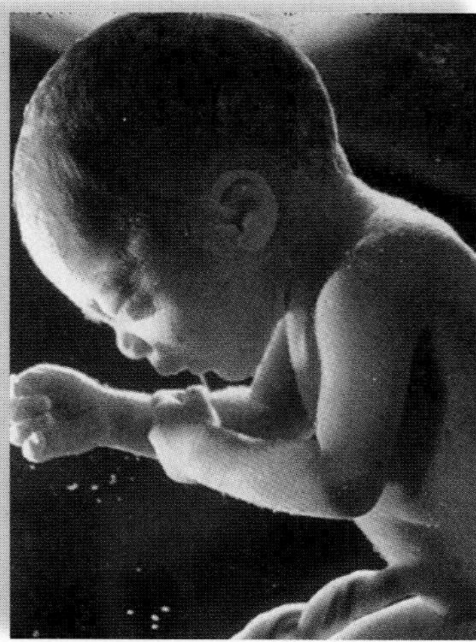

Seven- to eight-month-old fetus

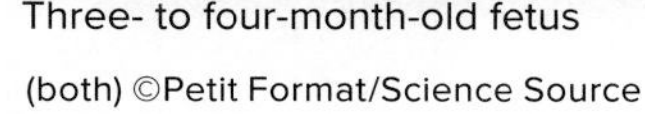

(both) ©Petit Format/Science Source

Observation: Fetal Development

1. Using Table 33.2 and Figure 33.13 to assist you, examine models of fetal development.
2. In Table 33.2, note the following.
   a. **External genitals:** About the third month, it is possible to tell male from female if an ultrasound is done.
   b. **Quickening:** Fetal movement is felt during the fourth or fifth months.
   c. **Vernix caseosa:** Beginning with the fifth month, the skin is covered with a cheesy coating called vernix caseosa.
   d. **Lanugo:** During the sixth and seventh months, the body is covered with fine, downy hair termed lanugo.

Table 33.2 Fetal Development

Month	Events for Mother	Events for Baby
Third month	Uterus is the size of a grapefruit.	Possible to distinguish sex. Fingernails appear.
Fourth month	Fetal movement is felt by those who have been previously pregnant. Heartbeat is heard by stethoscope.	Bony skeleton visible. Hair begins to appear. 150 mm (6 in.), 170 g (6 oz).
Fifth month	Fetal movement is felt by those who have not been previously pregnant. Uterus reaches up to level of umbilicus and pregnancy is obvious.	Protective cheesy coating, called vernix caseosa, begins to be deposited. Heartbeat can be heard.
Sixth month	Doctor can tell where baby's head, back, and limbs are. Breasts have enlarged, nipples and areolae are darkly pigmented, and colostrum is produced.	Body is covered with fine hair called lanugo. Skin is wrinkled and reddish.
Seventh month	Uterus reaches halfway between umbilicus and rib cage.	Testes descend into scrotum. Eyes are open. 300 mm (12 in.), 1,350 g (3 lb).
Eighth month	Weight gain is averaging about a pound a week. Difficulty in standing and walking because center of gravity is thrown forward.	Body hair begins to disappear. Subcutaneous fat begins to be deposited.
Ninth month	Uterus is up to rib cage, causing shortness of breath and heartburn. Sleeping becomes difficult.	Ready for birth. 530 mm (20½ in.), 3,400 g (7½ lb).

Figure 33.13 Human development.
Changes occurring from the fifth week to the eighth month.
(g) ©Chris Downie/Science Source; (h) ©James Stevenson/Science Source

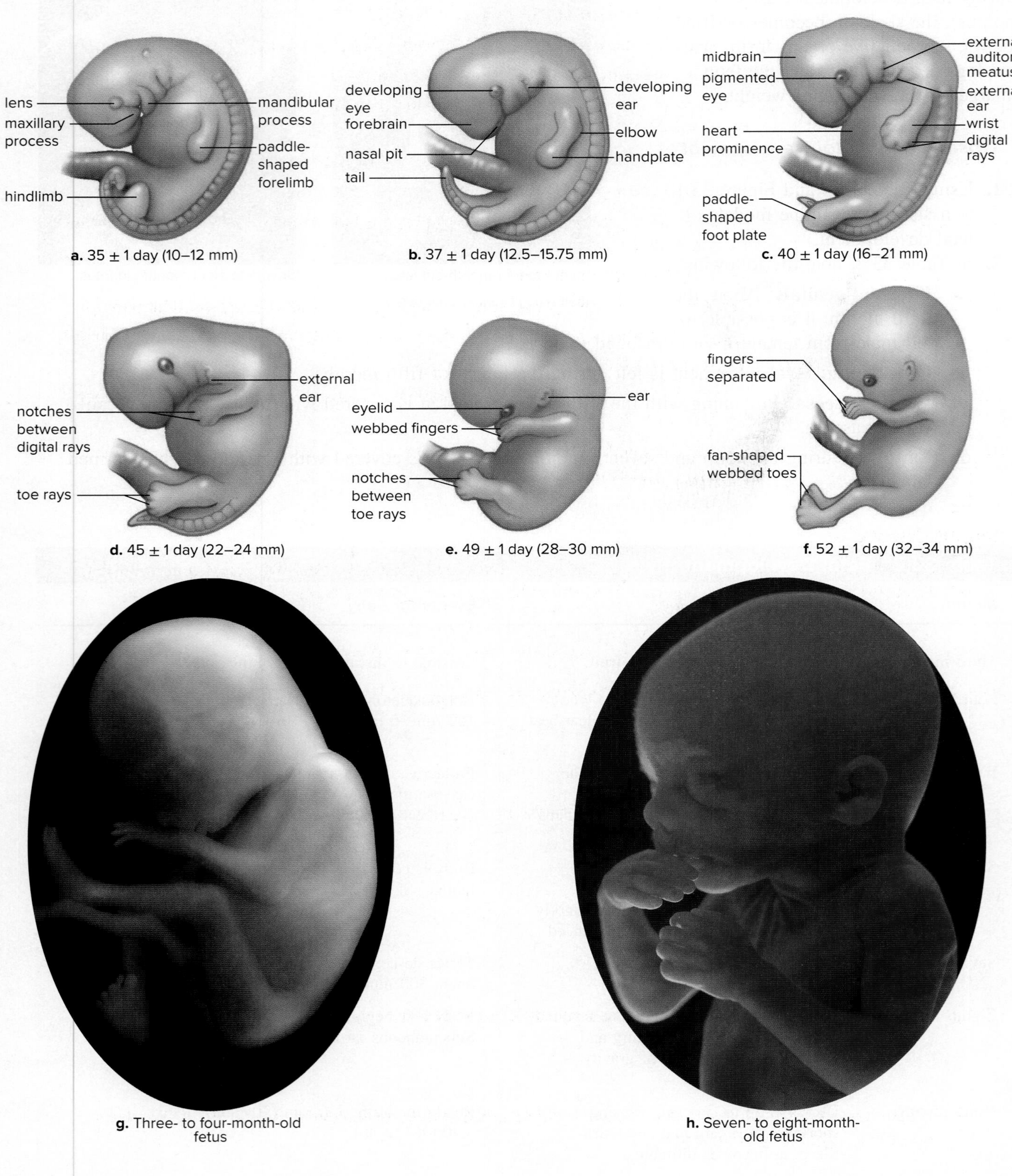

a. 35 ± 1 day (10–12 mm)
b. 37 ± 1 day (12.5–15.75 mm)
c. 40 ± 1 day (16–21 mm)
d. 45 ± 1 day (22–24 mm)
e. 49 ± 1 day (28–30 mm)
f. 52 ± 1 day (32–34 mm)
g. Three- to four-month-old fetus
h. Seven- to eight-month-old fetus

Laboratory Review 33

_______________ **1.** What results from the fertilization of an egg by a single sperm?

_______________ **2.** What is cell division that lacks growth called?

_______________ **3.** What is the solid cluster of 16 to 32 embryonic cells called?

_______________ **4.** What is the hollow cavity of a blastula called?

_______________ **5.** Which part of a human blastocyst becomes the embryo, the inner cell mass or the trophoblast?

_______________ **6.** During what developmental stage do tissues begin to form?

_______________ **7.** What does the archenteron become?

_______________ **8.** What primary germ layer gives rise to the pseudostratified ciliated columnar epithelium that lines the trachea?

_______________ **9.** What replaces the notochord?

_______________ **10.** Which extraembryonic membrane has a similar function in chicks and humans?

_______________ **11.** What extraembryonic membrane is the first site of blood cell formation in humans?

_______________ **12.** Where is metabolic waste stored in a chick?

_______________ **13.** During what month of human development is it possible to distinguish the gender of the fetus?

_______________ **14.** What is the name of the protective cheesy coating that begins to be deposited in the fifth month of human development?

Thought Questions

15. Some invertebrate animals (covered in an upcoming laboratory exercise) have only two primary germ layers, ectoderm and endoderm. What structures are likely to be missing from these animals? Justify your reply.

16. Chorionic villus sampling and amniocentesis are prenatal tests that can be done to check for certain genetic conditions or a chromosomal condition.

a. What part of the human blastocyst gives rise to the chorion and amnion?

b. What do these layers have in common?

17. a. If the notochord is removed, will ectoderm in this location become a neural tube? Explain.

b. If ectoderm above the notochord is removed and replaced with belly ectoderm, will the belly ectoderm become a neural tube? Explain.

LABORATORY

34

Sampling Ecosystems

Learning Outcomes

Introduction

- Define ecology and an ecosystem. Identify the abiotic and biotic components of an ecosystem.

34.1 Terrestrial Ecosystems

- Define and give examples of producers in terrestrial ecosystems.
- Define and give examples of consumers in terrestrial ecosystems.
- Define and give examples of decomposers in terrestrial ecosystems.

34.2 Aquatic Ecosystems

- Define aquatic ecosystem.
- Give examples of producers, consumers, and decomposers in aquatic ecosystems.

Introduction

Ecology is the study of interactions between organisms and their physical environment within an **ecosystem.** The **abiotic** (nonliving) components of an ecosystem include soil, water, light, inorganic nutrients, and weather variables. The **biotic** (living) components can be organized according to the **trophic** (feeding) level in which each organism belongs. This includes producers, consumers, and decomposers.

Producers are autotrophic organisms with the ability to carry on photosynthesis and to make food for themselves (and indirectly for the other populations as well). In terrestrial ecosystems, the predominant producers are green plants, while in freshwater and saltwater ecosystems, the dominant producers are various species of algae.

Consumers are heterotrophic organisms that eat available food. Three types of consumers can be identified, according to their food source:

1. **Herbivores** feed directly on green plants and are termed primary consumers. A caterpillar feeding on a leaf is a herbivore.
2. **Carnivores** feed on other animals and are therefore secondary or tertiary consumers. A blue heron feeding on a fish is a carnivore.
3. **Omnivores** feed on both plants and animals. A human who eats both leafy green vegetables and beef is an omnivore.

Decomposers and **detritivores** are organisms of decomposition, such as bacteria, fungi, and millipedes, that break down **detritus** (nonliving organic matter) to inorganic matter, which can be used again by producers. In this way, the same chemical elements are constantly recycled in an ecosystem.

The trophic structure (feeding relationships) of an ecosystem is represented in the form of a pyramid, such as the one shown in Figure 34.1. **Biomass** is the weight of all the organisms at each trophic level.

Figure 34.1 Ecological pyramid.
Organisms at lower trophic levels are higher in number and have greater biomass than organisms at higher trophic levels. The examples of organisms on the left form an aquatic food chain in which herons eat fish, which eat zooplankton, which eat phytoplankton. The examples of organisms on the right form a terrestrial food chain in which owls eat shrews, which eat beetles, which eat plants.

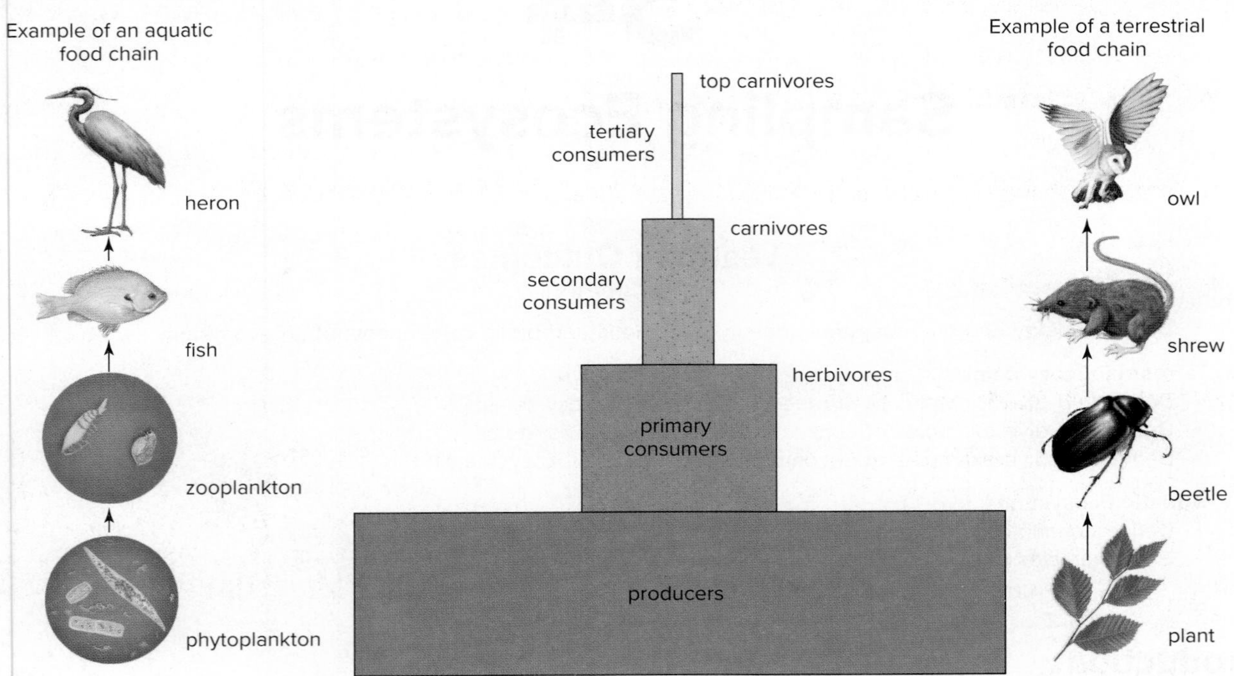

34.1 Terrestrial Ecosystems

Examining an ecosystem in a scientific manner requires concentrating on a representative portion of that ecosystem and recording as much information about it as possible. Representative areas or plots should be selected randomly. For example, random sampling of a terrestrial ecosystem often involves tossing a meterstick gently into the air in the general area to be sampled and then sampling the square meter where the stick lands. A **terrestrial ecosystem** sampling should include samples from the air above and from the various levels of plant materials growing on and in the ground beneath the selected plot.

Study of Terrestrial Sampling Site

The objective of the next Experimental Procedure is to characterize the abiotic and biotic components of a terrestrial ecosystem and to determine how those factors affect the trophic structure of an ecosystem. Although terrestrial ecosystems include deciduous forest, prairie, scrubland, and desert, a weedy field, if dominated by annual and perennial herbaceous plants up to a meter or so in height, is also a good site choice.

1. Gather all necessary equipment, such as metersticks, jars and bags (brown paper and plastic) with labels for collecting specimens, nets, pH paper, thermometers, and other testing equipment, to take with you to the site. Do not forget data-recording materials. Number your collection jars and bags so that you can use these numbers when recording data later (see Table 34.2).
2. When you arrive at the site, take several minutes to observe the general area. Describe what you observe, including weather conditions. ______________________________
____
____

3. Choose two sampling locations at the site. The two locations should differ in significant features (e.g., northern or southern exposure, high- or low-slope position, time since last disturbance, native versus exotic species composition). Formulate hypotheses about differences between the two sampling locations in regard to the following variables:

   a. Air temperature, humidity, and light intensity ______________________________

   ____

   b. Soil temperature, moisture, and pH ______________________________

   ____

   c. Producer biomass and diversity (variety of producers and how many of each type) ______________

   ____

   d. Consumer biomass and diversity ______________________________

   ____

Experimental Procedure: Terrestrial Ecosystems

Your instructor will organize class members into teams, and each team will be assigned specific tasks at each of the two chosen sampling locations. At each sampling location (e.g., shaded versus not shaded), you will randomly choose three 1 m square plots. Within each of these plots, randomly choose one 0.1 m square area (subplot) for all of the following abiotic variables and record your measurements in Table 34.1. For each of the biotic variables, randomly choose separate 0.1 m square subplots to sample and complete Table 34.2. Keep outside the 1 m square plots when sampling to minimize disturbance.

Abiotic Components

1. Measure air temperature, relative humidity, and light intensity at 0.5 m above the surface of the ground. Calculate the average for all three subplots (replicates).
2. Measure soil temperature, soil moisture, and soil pH (and any standing water) at 0.2 m below the surface using a soil corer. Calculate the average for all three replicates.

Biotic Components—Plants

In each subplot:

1. Count the total number of live plants.
2. Harvest each entire living plant and wash any soil from the roots. Place plant material in labeled brown paper bags.

Biotic Components—Animals

In each subplot:

1. Sweep with a net as thoroughly as possible about three to five times to capture different organisms on the vegetation in each plot. Empty the contents of the net into a labeled jar of alcohol for later sorting and identification.
2. Collect leaf litter samples in labeled plastic bags for Berlese (or Tullgren) funnel analysis.

Table 34.1 Abiotic Components of a Terrestrial Ecosystem									
		Location 1				**Location 2**			
	Abiotic Factor								
		a	**b**	**c**	**avg**	**a**	**b**	**c**	**avg**
Air	Temperature								
	Humidity								
	Light intensity								
Soil	Temperature								
	Moisture								
	pH								

Experimental Procedure: Terrestrial Ecosystems

Laboratory Work

1. Examine collected plants and animals using a stereomicroscope, or a compound microscope, if appropriate. Group organisms into different types based on morphological features (morphotypes) and classify them as producers, consumers, or decomposers. Further classify invertebrates into herbivores, detritivores, and carnivores. Complete Table 34.2.
2. Determine the dry biomass (weight) of the plant material, or wet biomass if a drying oven is not available. Dry biomass, although more time consuming to measure, is preferable when comparing biomasses among sites and between trophic levels.
3. Determine the dry biomass of the animal material (or wet weight if plant wet weight was used).
4. Construct graphs comparing the abiotic conditions of each terrestrial sampling location.
5. Construct a pyramid for biomass, morphotype, and the total number of producers, herbivores (including detritivores), and carnivores.
6. Select any three producers and any three consumers from the organisms collected, and explain how each has adapted to its terrestrial environment.

   Producer 1 ______________________________

   Producer 2 ______________________________

   Producer 3 ______________________________

   Consumer 1 ______________________________

   Consumer 2 ______________________________

   Consumer 3 ______________________________
7. Return all organisms and litter samples to their respective collection sites, as explained by your instructor. If any organisms were preserved, ask your instructor what to do with them.

Table 34.2 Biotic Components of a Terrestrial Ecosystem

			Location 1				Location 2			
	Biotic Factor									
			a	**b**	**c**	**avg**	**a**	**b**	**c**	**avg**
Plants	Producers	Total number								
		Number of morphotypes								
		Biomass								
Animals on vegetation	Herbivores	Total number								
		Number of morphotypes								
		Biomass								
	Carnivores	Total number								
		Number of morphotypes								
		Biomass								
Animals in litter	Herbivores	Total number								
		Number of morphotypes								
		Biomass								
	Carnivores	Total number								
		Number of morphotypes								
		Biomass								
	Detritivores	Total number								
		Number of morphotypes								
		Biomass								

Conclusions: Terrestrial Ecosystems

Compare the results of this terrestrial ecosystem analysis with the hypotheses you formulated (see page 477).

- Air temperature, humidity, and light intensity ______________________________

  ____
- Soil temperature, moisture, and pH ______________________________

  ____
- Producer biomass and diversity ______________________________

  ____
- Consumer biomass and diversity ______________________________

  ____

34.2 Aquatic Ecosystems

An **aquatic ecosystem** sampling should include samples from the air above the water column, the column of water itself, and the soil beneath the water column.

Study of Aquatic Sampling Site

The objective of the next Experimental Procedure is to characterize the abiotic and biotic components of an aquatic ecosystem and to determine how those factors affect the trophic structure of an ecosystem. Aquatic ecosystems consist of freshwater ecosystems (e.g., lakes, ponds, rivers, and streams) and marine ecosystems (e.g., oceans). A good site for this study is a large pond, small lake, or reservoir (with a shallow margin having rooted aquatic plants and a deeper zone with water 1–2 m deep).

1. Gather all necessary equipment, such as metersticks, collection jars and bags with labels, nets, pH paper, thermometers, and other testing equipment to take with you to the site. Do not forget data-recording materials. Number your collection jars and bags so that you can use these numbers when recording data in Table 34.3.
2. When you arrive at the site, take several minutes to observe the general area. Describe what you observe, including weather conditions. ____________________

   ____

   ____

3. Choose two sampling locations that differ in significant features (e.g., sheltered by trees versus unsheltered, near stream inflow versus far from stream inflow). Plan to sample conditions near shore (shallow-water zone) and in deeper water away from shore (or only one or the other, if logistics are limiting).

   Formulate hypotheses about differences between the two sampling locations in regard to the following variables:

   a. Temperature, humidity, and light intensity above the surface ____________________

   ____

   b. Temperature, dissolved oxygen, pH, and visibility below the surface ____________________

   ____

   c. Producer biomass and diversity (variety of producers and how many of each type) ____________________

   ____

   d. Consumer biomass and diversity ____________________

   ____

Table 34.3 Abiotic Components of an Aquatic Ecosystem

		Location 1				Location 2			
	Abiotic Factor								
		a	b	c	avg	a	b	c	avg
Air	Temperature								
	Humidity								
	Light intensity								
Water	Temperature								
	Dissolved oxygen								
	pH								
	Visibility								

Experimental Procedure: Aquatic Ecosystem

Your instructor will organize class members into teams, and each team will be assigned specific tasks at each of the two sampling locations. At each sampling location (e.g., shaded versus not shaded), you will randomly choose three 1 m square plots. Within each of these plots, randomly choose one 0.1 m square area (subplot) for all of the following abiotic variables and record your measurements in Table 34.3. For each of the biotic variables, randomly choose separate 0.1 m square subplots to sample and complete Table 34.4. Keep outside the 1 m square plots when sampling to minimize disturbance.

Abiotic Components

1. Measure air temperature, relative humidity, and light intensity at 0.5 m above the surface of the water.
2. Measure water temperature, dissolved oxygen, pH, and visibility at 0.5 m below the surface.

Biotic Components: Plants

In each subplot:

1. Count the total number of live plants in each subplot.
2. Harvest each entire plant and wash any sediment from the roots. Place plant material in labeled brown paper bags.

Biotic Component: Plankton

Lower a plankton net with attached collecting bottle into the water to a depth of 0.5 m. Slowly raise the net vertically two to three times to collect plankton in the water column. Pour the sample into a labeled collecting jar and preserve with Lugol's solution for later sorting and identification.

Biotic Component: Animals

Choose an area of the plot not disturbed by other sampling but near emergent vegetation. Sweep through the water inside the plot three to five times. Be sure to sample around the vegetation and bump the substrate several times to dislodge benthos from the sediment. Empty the contents of the net into a sieve placed over a bucket of water. Examine, wash, and collect any macroinvertebrates and place them in the bucket. Transfer collected organisms into a labeled jar of alcohol for later sorting and identification.

Laboratory Review 34

_______________ **1.** What is being studied when organisms and their physical environment are observed?

_______________ **2.** Are soil, minerals, sunlight, water, and weather variables biotic or abiotic components of an ecosystem?

_______________ **3.** What is the general name for organisms at the first trophic level in an ecosystem?

_______________ **4.** What are the dominant producers in freshwater and saltwater ecosystems?

_______________ **5.** Is a black bear that eats berries, nuts, insects, and fish a herbivore, omnivore, or carnivore?

_______________ **6.** What organisms break down nonliving organic matter into organic matter that producers can use?

_______________ **7.** What term is used for the weight of all the organisms at each trophic level?

_______________ **8.** At what consumer level is a hawk that ate a snake that ate a mouse that ate some grain?

_______________ **9.** What is an example of a terrestrial ecosystem?

_______________ **10.** In a terrestrial ecosystem, what organisms are predominant producers?

_______________ **11.** What is an example of an abiotic component measured above the surface of the ground when surveying a terrestrial ecosytem?

_______________ **12.** What kinds of organisms are most likely to be found in samples of leaf litter?

_______________ **13.** In which type of ecosystem, aquatic or terrestrial, will phytoplankton and zooplankton be found?

_______________ **14.** What is an example of an abiotic component measured below the surface of the water in an aquatic ecosystem?

Thought Questions

15. What is the importance of many bacteria, fungi, and insects to the ability of plants to acquire the elements they need for making organic molecules?

16. Which organisms should have the greatest biomass in a health ecosystem? Justify your reply.

17. a. Will a herbivore ever be a secondary or higher consumer? Explain your reply.

b. Will a carnivore ever be a primary consumer? Explain your reply.

LABORATORY

35

Effects of Pollution on Ecosystems

Learning Outcomes

35.1 Studying the Effects of Pollutants

- Using a hay infusion culture as the experimental material, predict the effect of oxygen deprivation on species composition and diversity of ecosystems.
- Using a hay infusion culture as the experimental material, predict the effect of acid deposition on species composition and diversity of ecosystems.
- Using a hay infusion culture as the experimental material, predict the effect of enrichment on species composition and diversity of ecosytems.
- Using data from a seed germination experiment, predict the effect of acid rain on crop yield.
- Using *Gammarus* as the subject, predict the effect of acid rain on food chains involving animals.

35.2 Studying the Effects of Cultural Eutrophication

- Predict the effect of cultural eutrophication on food chains so that pollution results.

Introduction

This laboratory will consider three causes of aquatic pollution: thermal pollution, acid pollution, and cultural eutrophication. **Thermal pollution** occurs when water temperature rises above normal. As water temperature rises, the amount of oxygen dissolved in water decreases, possibly depriving organisms and their cells of an adequate supply of oxygen. Deforestation, soil erosion, and the burning of fossil fuels contribute to thermal pollution, but the chief cause is use of water from a lake or the ocean as a coolant for the waste heat of a power plant.

When sulfur dioxide and nitrogen oxides enter the atmosphere, usually from the burning of fossil fuels, they are converted to acids, which return to Earth as **acid deposition** (acid rain or snow). Acid deposition kills plants, aquatic invertebrates, and also decomposers, threatening the entire ecosystem.

Cultural eutrophication, or overenrichment, is due to runoff from agricultural fields, wastewater from sewage treatment plants, and even excess detergents. These sources of excess nutrients cause an algal bloom seen as a green scum on a lake (Fig. 35.1). When algae overgrow and die, decomposition robs the lake of oxygen, causing a fish die-off.

Figure 35.1 Cultural eutrophication. Eutrophic lakes tend to have large populations of algae and rooted plants.

35.1 Studying the Effects of Pollutants

We are going to study the effects of pollution by observing its effects on hay infusion organisms, on seed germination, and on an animal called *Gammarus*.

Study of Hay Infusion Culture

A hay infusion culture (hay soaked in water) contains various microscopic organisms, but we will be concentrating on how the pollutants in our study affect the protozoan populations in the culture. We will consider both of these aspects:
species composition: number of different types of microorganisms.
species diversity: composition and the abundance of each type of microorganism.

Experimental Procedure: Effect of Pollutants on a Hay Infusion Culture

During this Experimental Procedure you will examine, by preparing a wet mount, hay infusion cultures that have been treated in the following manner.

1. **Control culture:** This culture simulates the species composition and diversity of an untreated culture. Prepare a wet mount and answer the following questions:
   With the assistance of Figure 35.2 and any guides available in the laboratory, identify as many different types of microorganisms as possible in the hay infusion culture. State whether species composition is high, medium, or low. Record your estimation in the second column of Table 35.1. Do you judge species diversity to be high, medium, or low? Record your estimation in the third column of Table 35.1.
2. **Oxygen-deprived culture:** Thermal pollution causes water to be oxygen deprived; therefore, when we study the effects of low oxygen on a hay infusion culture, we are studying an effect of thermal pollution. Prepare a wet mount of this culture and determine if there is a change in species composition and diversity. Again record the species composition and species diversity as high, medium, or low in Table 35.1.

Figure 35.2 Microorganisms in hay infusion cultures.
Organisms are not to size.

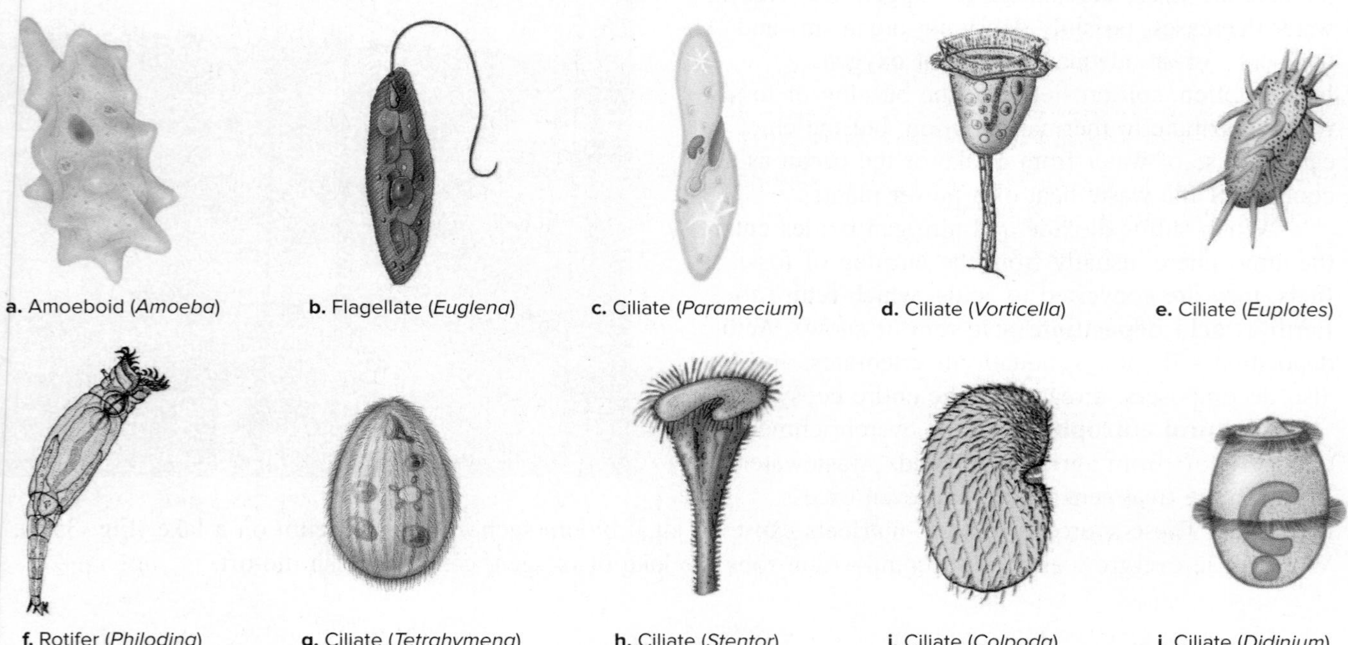

Table 35.1 Effect of Pollution on a Hay Infusion Culture			
Type of Culture	**Species Composition (High, Medium, or Low)**	**Species Diversity (High, Medium, or Low)**	**Explanation**
Control			
Oxygen-deprived			
Acidic			
Enriched			

3. **Acidic culture:** In this culture, the pH has been adjusted to 4 with sulfuric acid (H_2SO_4). This simulates the effect of acid rain on a hay infusion culture. Prepare a wet mount of this culture and determine if there is a change in species composition and diversity. Again record the species composition and species diversity as high, medium, or low in Table 35.1.
4. **Enriched culture:** More organic nutrients have been added to this culture. These nutrients will cause the algae population, which is food for most protozoans, to increase. In the short term, their species composition should increase. Eventually, as the algae die off, decomposition will rob the water of oxygen and the protozoans may start to die off. Prepare a wet mount of this culture and determine if there is a change in species composition and diversity. Again record the species composition and species diversity as high, medium, or low in Table 35.1.

Conclusions

- What could be a physiological reason for the adverse effects of oxygen deprivation on a hay infusion culture? If consistent with your results, enter this explanation in the last column of Table 35.1.
- What could be a physiological reason for the adverse effects of a low pH on a hay infusion culture? If consistent with your results, enter this explanation in the last column of Table 35.1.
- What could be an environmental reason for the adverse effects of an enriched culture? If consistent with your results, enter this explanation in the last column of Table 35.1.

Effect of Acid Rain on Seed Germination

Seeds depend on favorable environmental conditions of temperature, light, and moisture to germinate, grow, and reproduce. Like any other biological process, germination requires enzymatic reactions that can be adversely affected by an unfavorable pH.

Experimental Procedure: Effect of Acid Rain on Seed Germination

In this Experimental Procedure we will test whether there is a negative correlation between acid concentration and germination. In other words, it is hypothesized that as acidity increases, it becomes more likely seeds will ____.

Your instructor has placed 20 sunflower seeds in each of five containers with water of increasing acidity: 0% vinegar (tap water), 1% vinegar, 5% vinegar, 20% vinegar, and 100% vinegar.

1. Test and record the pH of solutions having the vinegar concentrations noted above. Record the pH of each solution in Table 35.2.
2. Count the number of germinated sunflower seeds in each container, and complete Table 35.2.

Table 35.2 Effect of Increasing Acidity on Germination of Sunflower Seeds			
Concentration of Vinegar	pH	Number of Seeds that Germinated	Percent Germination
0%			
1%			
5%			
20%			
100%			

Conclusions: Effect of Increasing Acidity on Germination of Sunflower Seeds

- As you know, each enzyme has an optimum pH. Explain why acid rain is expected to inhibit metabolism, and therefore, seedling development. ____________
- Do the data support or falsify your hypothesis? ____________

Study of *Gammarus*

A small crustacean called *Gammarus* lives in ponds and streams (Fig. 35.3) where it feeds on debris, algae, or anything smaller than itself, such as some of the microorganisms in Figure 35.2. In turn, fish like to feed on *Gammarus*.

Experimental Procedure: Gammarus

- Add 35 ml of spring water to a beaker and record the pH of the water. ____________ pH
- Add four *Gammarus* to the container. Do they all use their legs in swimming? ____________
- Which legs are used in jumping and climbing? ____________
- What do *Gammarus* do when they "bump" into each other? ____________

Control Sample

After observing *Gammarus,* decide what behaviors are most often observed. During a 5-minute time span, total the amount of time spent doing each of these behaviors.

Behaviors	Amount of Time	Total Time
1. ____________	____________	________
2. ____________	____________	________
3. ____________	____________	________

Test Sample

If so directed by your instructor, put a *Gammarus* in a beaker of spring water adjusted to pH 4 by adding vinegar. During a 5-minute time span, total the amount of time spent doing each of these behaviors.

Behaviors	Amount of Time	Total Time
1. ____________	____________	________
2. ____________	____________	________
3. ____________	____________	________

Figure 35.3 ***Gammarus.*** ***Gammarus*** is a type of crustacean classified in a subphylum that also includes shrimp.
(inset) ©William Amos/Photoshot

Conclusion

- Draw a conclusion from this study: ______________________________
____
- Create a food chain that shows who eats whom when the food chain includes algae, protozoans, *Gammarus,* fish, and humans. ______________________________
   a. What would happen to this food chain if the water were oxygen deprived?______________________________
   ____
   Acidic? ______________________________
   Enriched with inorganic nutrients (short term and long term)? ______________________________
   ____

Conclusions: Studying the Effects of Pollutants

- Give an example to show that the hay infusion study pertains to real ecosystems. ______________________________
____
- What are the potential consequences of acid rain on crops that reproduce by seeds? ______________________________
____
On the food chains of the ocean? ______________________________
- How does the addition of nutrients affect species composition and species diversity of an ecosystem over time? ______________________________

35.2 Studying the Effects of Cultural Eutrophication

Chlorella, the green alga used in this study, is considered to be representative of algae in bodies of fresh water. The crustacean *Daphnia* feeds on green algae such as *Chlorella* (Fig. 35.4). First, you will observe how *Daphnia* feeds, and then you will determine the extent to which *Daphnia* could keep the effects of cultural eutrophication from occurring in a hypothetical example. Keep in mind that this case study is an oversimplification of a generally complex problem.

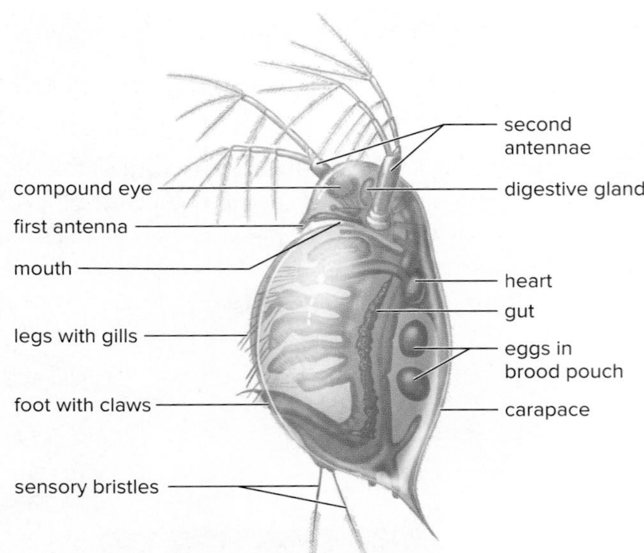

Figure 35.4 Anatomy of *Daphnia*.

Observation: Daphnia *Feeding*

1. Place a small pool of petroleum jelly in the center of a small petri dish.
2. Use a dropper to take a *Daphnia* from the stock culture, place it on its back (covered by water) in the petroleum jelly, and observe it under the stereomicroscope (Fig. 35.4).
3. Note the insectlike carapace and the legs waving rapidly as the *Daphnia* filters the water.
4. Add a drop of *carmine solution,* and observe how the *Daphnia* filters the "food" from the water and passes it through the gut. The gut is more visible if you push the animal onto its side. In this position, you may also observe the heart beating in the region above the gut and just behind the head.
5. Allow the *Daphnia* to filter-feed for up to 30 minutes, and observe the progress of the carmine particles through the gut. Does the carmine travel completely through the gut in 30 minutes? ______________

____

Experimental Procedure: Daphnia *Feeding on* Chlorella

This exercise requires the use of a spectrophotometer. Absorbance will be a measure of the algal population level; the greater the number of algal cells, the greater the absorbance. The higher the absorbance, the greater the amount of light absorbed and *not* passed through the solution.

1. Obtain two spectrophotometer tubes (cuvettes) and a Pasteur pipet.
2. Fill one of the cuvettes with distilled water, and use it to zero the spectrophotometer. Save this tube for step 6.
3. Use the Pasteur pipet to fill the second cuvette with *Chlorella.* Gently aspirate and expel the sample several times (without creating bubbles) to give a uniform dispersion of the algae.
4. Add ten hungry *Daphnia,* and following your instructor's directions, immediately measure the absorbance with the spectrophotometer. If a *Daphnia* swims through the beam of light, a strong deflection should occur; do not use any such higher readings—instead, use the lower figure for the absorbance. Record your reading in the first column of Table 35.3.
5. Remove the cuvette with the *Daphnia* to a safe place in a test tube rack. Allow the *Daphnia* to feed for 30 minutes.
6. Rezero the spectrophotometer with the distilled water cuvette.
7. Measure the absorbance of the experimental cuvette again. Record your data in the second column of Table 35.3, and explain your results in the third column.

Table 35.3 Spectrophotometer Data/*Daphnia* Feeding on *Chlorella*

Absorbance Before Feeding	Absorbance After Feeding	Explanation

Experimental Procedure: Case Study in Cultural Eutrophication

The following problem will test your understanding of the ecological value of a single species—in this case, *Daphnia.* Please realize that this is an oversimplification of a generally complex problem.

1. Assume that developers want to build condominium units on the shores of Silver Lake. Homeowners in the area have asked the regional council to determine how many units can be built without altering the nature of the lake. As a member of the council, you have been given the following information:
   The current population of *Daphnia,* 10 animals/liter, presently filters 24% of the lake per day, meaning that it removes this percentage of the algal population per day. This is sufficient to keep the lake essentially clear. Predation—the eating of the algae—will allow the *Daphnia* population to increase to no more than 50 animals/liter. Therefore, 50 *Daphnia*/liter will be available for feeding on the increased number of algae that would result from building the condominiums.
   Using this information, complete Table 35.4.

Table 35.4 *Daphnia* Filtering

Number of *Daphnia*/Liter	Percent of Lake Filtered
10	24%
50	

2. The sewage system of the condominiums will add nutrients to the lake. Phosphorus output will be 1 kg per day for every 10 condominiums. This will cause a 30% increase in the algal population. Using this information, complete Table 35.5.

Table 35.5 Cultural Eutrophication

Number of Condominiums	Phosphorus Added	Increase in Algal Population
10	1 kg	30%
20		
30		
40		
50		

Conclusion: Cultural Eutrophication

- Assume that phosphorus is the only nutrient that will cause an increase in the algal population and that *Daphnia* is the only type of zooplankton available to feed on the algae. How many condominiums would you allow the developer to build? ______________________________

  ____

- What other possible impacts could condominium construction have on the condition of the lake? __________

  ____

  ____

Laboratory Review 35

_______________ 1. What results when sulfur dioxide and nitrogen oxides, formed by burning fossil fuels, enter the atmosphere?

_______________ 2. What kind of pollution results when water from a lake or the ocean is used to disperse waste heat?

_______________ 3. Does the amount of oxygen dissolved in water increase or decrease when the temperature of the water increases?

_______________ 4. What occurs to a body of water when there is runoff from agricultural fields?

_______________ 5. Which population experiences growth referred to as a "bloom" when excess nutrients enter a body of water?

_______________ 6. After excess nutrients cause algae to overgrow and then die, what causes the fish to die?

_______________ 7. What do we call the number of different microorganisms observed and their relative abundance in the hay infusion culture?

_______________ 8. What chemical was used to simulate the effect of acid rain on a hay infusion culture?

_______________ 9. What does an unfavorable pH adversely affect that inhibits seed germination?

_______________ 10. In the Experimental Procedure on the effect of acid rain on seed germination, if 18 of the sunflower seeds germinated in the 1% vinegar solution, what percent germinated?

_______________ 11. What kind of animal is *Gammarus*?

_______________ 12. What will *Daphnia* do that might prevent cultural eutrophication from occurring?

_______________ 13. Are there more or fewer algal cells present in a water sample as the absorption of light decreases?

_______________ 14. What element from the Silver Lake condominiums' sewage output contributes to the increased algal population?

Thought Questions

15. A "dead zone" forms where the Mississippi River empties into the Gulf of Mexico due to nutrient runoff into the river that is carried to the Gulf. Explain how the dead zone forms and what impact it may have on humans.

16. Pollutants often affect the producers in an ecosystem first and then have far-ranging effects. Use acid rain and a food chain to illustrate your understanding of the impact pollutants have after they enter an ecosystem.

17. Describe how the cultural eutrophication case study illustrates the need to have a balance of population sizes in an ecosystem.

APPENDIX

A

Preparing a Laboratory Report

A laboratory report has the sections noted in the outline that follows. Use this outline and a copy of the Laboratory Report Form on page A–4 to help you write a report assigned by your instructor. In general, do not use the words *we, my, our, your, us,* or *I* in the report. Use scientific measurements and their proper abbreviations. (For example, cm is the proper notation for centimeter and sec is correct for seconds.)

1. **Introduction:** Tell the reader what the experiment was about.
   a. **Background information:** Begin by giving an overview of the topic. Look at the Introduction to the Laboratory (and/or at the introduction to the section for which you are writing the report). Do not copy the information, but use it to get an idea about what background information to include.

For example, suppose you are doing a laboratory report on "Solar Energy" in Laboratory 6 (Photosynthesis). You might give a definition of photosynthesis and explain the composition of white light.

   b. **Purpose:** Think about the steps of the experiment and state what the experiment was about. Tell the independent and dependent variable.

For example, you might state that the purpose of the photosynthesis experiment was to determine the effect of white light versus green light on the photosynthetic rate. The independent variable was the color of light and the dependent variable was the rate of photosynthesis.

   c. **Hypothesis:** Consider the expected results of the experiment in order to state the hypothesis. It's possible that the Introduction to the Laboratory might hint at the expected results. State this in the form of a hypothesis.

For example, you might state: It was hypothesized that white light would be more effective than green light for photosynthesis.

2. **Method:** Tell the reader how you did the experiment.
   a. **Equipment and sample used:** Use any illustrations in the laboratory manual that show the experimental setup to describe the equipment and the sample (subject) used.

For example, for the photosynthesis experiment look at Figure 6.4 and describe what you see. You might state that a 150-watt lamp was the source of white light directed at *Elodea*, an aquatic plant, placed in a test tube filled with a solution of sodium bicarbonate ($NaHCO_3$). A beaker of water placed between the lamp and test tube was a heat absorber.

   b. **Collection of data:** Think about what you did during the experiment such as what you observed or what you measured. Look at any tables you filled out in order to recall how the data were collected and what control(s) were.

For example, for the photosynthesis experiment, you might state that the rate of photosynthesis was determined by the amount of oxygen released and was measured by how far water moved in a side arm placed in a stopper of the test that held *Elodea.* A control was the same experimental setup except the test tube lacked *Elodea.*

3. **Results:** Present the data in a clear manner.
   a. **Graph or table:** If at all possible, show your data in table or graph form. You could reproduce a table you filled in or a graph you drew to show the results of the experiment. Be sure to include the title of the table; do not include any interpretation of the data column in the table.

For example, for the photosynthesis experiment you might reproduce Table 6.3.

Table 6.3 Rate of Photosynthesis (Green Light)

	Data
Gross Photosynthesis (mm/10 min)	
White (from Table 6.2)	*33.5 mm/10 min*
Green	*12.5 mm/10 min*
Rate of Photosynthesis (mm/hr)	
White (from Table 6.2)	*201 mm/hr*
Green	*75 mm/hr*

Or for 5.2 Effect of Temperature on Enzyme Activity you might show this graph as your results.

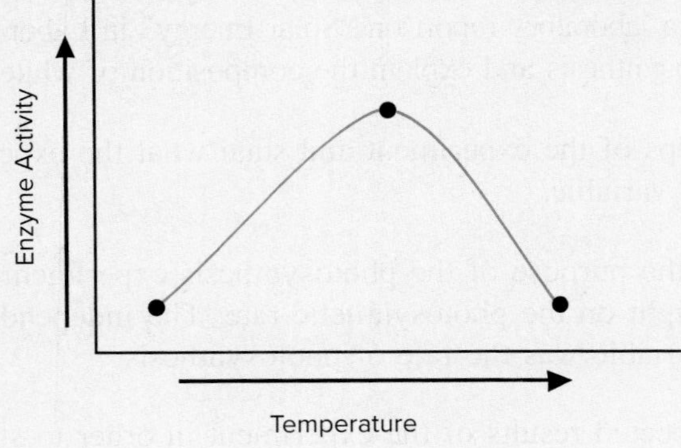

   b. **Description of data:** Examine your data, and decide what they tell you. Then, below any table or graph add a description to help the reader understand what the table or graph is showing. Define any terms in the table that are not readily understandable.

For example, below Table 6.3 you might state that these data indicate that the rate of photosynthesis with white light is faster than with green light. Also, you should define gross photosynthesis. Or below the graph that shows the effect of temperature on enzyme activity, you might state that these data show that the rate of enzymatic activity speeds up until boiling occurs and then it drops off.

4. **Conclusion:** Tell if the data support or do not support the hypothesis.
   a. **Compare the hypothesis with the data:** Do your data agree or disagree with the hypothesis?

For example, for the photosynthesis experiment you might state: These results support the hypothesis that white light is more effective for photosynthesis than green light.

b. Explanation: Explain why you think you obtained these results. Look at any questions you answered while in the laboratory, and use them to help you decide on an appropriate explanation.

For example, the answers to the questions in 6.2 Solar Energy might help you state that white light gives a higher rate of photosynthesis because it contains all the visible light rays. Green light gives a lower rate because green plants such as *Elodea* do not absorb green light.

If your results do not support the hypothesis, explain why you think this occurred.

In this instance you might state that while white light contains all visible light rays and green light is not absorbed by a green plant, the experiment did not support the hypothesis because of failure to use a heat absorber when doing the green light experiment.

Date ______________________

Name ______________________

Laboratory Report for ____

1. Introduction

a. Background information

b. Purpose

c. Hypothesis

2. Methods

a. Equipment and sample used

b. Collection of data

3. Results

a. Graph or table
(Place these on attached sheets.)

b. Description of data

4. Conclusion

a. Compare the hypothesis with the data

b. Explanation

c. Conclusion

APPENDIX

B

Metric System

Unit and Abbreviation	Metric Equivalent	Approximate English-to-Metric Equivalents
Length		
nanometer (nm)	$= 10^{-9}$ m (10^{-3} µm)	
micrometer (µm)	$= 10^{-6}$ m (10^{-3}mm)	
millimeter (mm)	= 0.001 (10^{-3}) m	
centimeter (cm)	= 0.01 (10^{-2}) m	1 inch = 2.54 cm
		1 foot = 30.5 cm
meter (m)	= 100 (10^{2}) cm	1 foot = 0.30 m
	= 1,000 mm	1 yard = 0.91 m
kilometer (km)	= 1,000 (10^{3}) m	1 mi = 1.6 km
Weight (mass)		
nanogram (ng)	$= 10^{-9}$ g	
microgram (µg)	$= 10^{-6}$ g	
milligram (mg)	$= 10^{-3}$ g	
gram (g)	= 1,000 mg	1 ounce = 28.3 g
		1 pound = 454 g
kilogram (kg)	= 1,000 (10^{3}) g	= 0.45 kg
metric ton (t)	= 1,000 kg	1 ton = 0.91 t
Volume		
microliter (µl)	$= 10^{-6}$ liter (10^{-3} ml)	
milliliter (ml)	$= 10^{-3}$ liter	1 tsp = 5 mL
	= 1 cm^3 (cc)	1 fl oz = 30 mL
	= 1,000 mm^3	
liter (l)	= 1,000 ml	1 pint = 0.47 liter
		1 quart = 0.95 liter
		1 gallon = 3.79 liter
kiloliter (kl)	= 1,000 liter	

Common Temperatures

°C	°F	
100	212	Water boils at standard temperature and pressure.
71	160	Flash pasteurization of milk
57	134	Highest recorded temperature in the United States, Death Valley, July 10, 1913
41	105.8	Average body temperature of a marathon runner in hot weather
37	98.6	Human body temperature
13.7	56.66	Human survival is still possible at this temperature.
0	32.0	Water freezes at standard temperature and pressure.

Units of Temperature

°F °C

To convert temperature scales:

$$°C = \frac{(°F - 32)}{1.8}$$

$$°F = 1.8°C + 32$$

Index

Page numbers followed by *f* indicate figures; *t* indicate tables.

B

C

D

E

F

G

H

I

J

K

L

M

N

O

P

Q

R

S

T

U

V

W

X

Y

Z